ORGANISMS AND THEIR EVOLUTION

Agency and Meaning in the Drama of Life

Organisms and Their Evolution

Agency and Meaning in the Drama of Life

Stephen L. Talbott

The Nature Institute

Published by The Nature Institute
Ghent, New York; www.natureinstitute.org
ISBN #978-0-9744906-8-7

Author: Stephen L. Talbott
Book design/Layout: Ann Erwin
Cover Art: Henrike Holdrege. Mixed media

The full text of this book is publicly available as a PDF document at https://bwo.life/bk/pdf/bk.pdf. It is also available, in possibly altered, updated, or corrected form as a web (html) document at https://bwo.life/bk/. This content is freely available, with proper attribution, for noncommercial, personal use (including classroom use). Other forms of availability, if any, will be noted in the Table of Contents at that last web address. The material is part of the Biology Worthy of Life project of The Nature Institute.

The Nature Institute is a non-profit scientific research and educational organization. If you find this book of interest, you may want to subscribe to The Nature Institute's twice-yearly newsletter, *In Context*. You can do so by sending your request to be put on the newsletter's hardcopy or digital mailing list to: info@natureinstitute.org or by going to: https://www.natureinstitute.org/join-our-mailing-list

Table of Contents

Preface i

PART 1: THE LIFE OF ORGANISMS

Chapter 1: The Keys to This Book 1

Theme #1: Narrative
Theme #2: Interiority
Theme #3: Holism
Theme #4: Blindsight
Some definitional hints about key biological terms
Agency
Archetype
Atoms/Molecules
Consciousness
Directive
Integral unity of the organism
Intention/intentional
Material
Material/physical/materialist/materialistic
Meaning
Purpose/purposeful/purposive
Telos-realizing
Where is the evidence? Two concluding notes

Chapter 2: The Organism's Story 19

Introduction
Organisms are agents capable of expressing their own meanings
The end is more constant than the means of attaining it
Every organism is narrating a meaningful life story
Where are we now?

Chapter 3: What Brings Our Genome Alive? 34
Introduction
The genome as you have probably not heard about it
Of dynamism and mystery in the cell nucleus
Does the lawfulness of molecular interactions explain global coherence?
Where are we now?

Chapter 4: The Sensitive, Dynamic Cell 50
Introduction
Does the cell possess its own "senses" and "limbs"?
The sensitive "skin" and organelles of the cell
From information to life
Where are we now?

Chapter 5: Our Bodies Are Formed Streams 61
Introduction
A long way from crystalline order
The unexpected phases of life
Well-structured droplets
How do you regulate flow and phases?
And then there is water — the mediator of flow
Where are we now?

Chapter 6: Context: Dare We Call It Holism? 78
Introduction
Which comes first, the cell or its niche?
"More than the sum of its parts": clarifying a cliché
Every biological context is a complex of embodied ideas
Contextual wholes and living narratives
Where are we now?

Chapter 7: Epigenetics: A Brief Introduction........................ 93
Introduction
An old problem newly recognized
Epigenetics — a useful term?
Unexpected discoveries
Grasping at epigenetic straws — is it really necessary?
Where are we now?

Chapter 8: The Mystery of an Unexpected Coherence 103
Introduction
Flexibility and precision in RNA splicing
The spliceosome
Can DNA coordinate splicing activity?
Shattering the genome
Who's on first — genes or proteins (or neither)?
A sense of the whole
Is an unexpected coherence the problem or the solution?
Where are we now?

Chapter 9: A Mess of Causes 119
Introduction
The seductive appeal of master controllers
More than an innocent abuse of language
Biological clocks: Who is keeping time?
A well-studied worm
Of crosstalk, horror graphs, and collaborative decision-making
The problem of causation is fundamental to biology
Where are we now?

Chapter 10: What Is the Problem of Form? 135
Introduction
The problem of form exists even at the molecular level
Michael Levin: Revolutionary
Whole-part, future-oriented causation
Michael Levin: Counter-revolutionary
In what sense are machines end-directed?
A deep issue, unaddressed
An unquestioned model
Where are we now?

Chapter 11: Why We Cannot Explain the Form of Organisms 153
Introduction
Looking for physical explanations of form
The mastery of genetic switches
Can we trace form to something other than earlier form?
Endless transformations most beautiful
Where are we now?

Chapter 12: Is a Qualitative Biology Possible? 166
Introduction
An animal expressing the character of the tropical forest
The problem of organic form
How to generalize upon a transformational series
The movement is continuous and ideal
The formative movement requires both difference and sameness
An awareness of the movement changes our perception of the leaves
The movement manifests itself through the particulars
The question of causality
Each individual leaf is "coming from" and "passing to"
Whatever specifies the appearance of forms in time has causal significance
A clarification of dynamic form as cause
Threefoldness, the biology of form, and evolution
A starting point: living polarity in the human being
Interpenetration of the three aspects
Threefold organization in mammals
Subtle interweaving
Can evolutionists escape responsibility for explaining these patterns?
Where are we now?

Chapter 13: All Science Must Be Rooted in Experience 201
Introduction
Two distinguishable but indivisible aspects of human experience
Without relations of thought, we have only chaos
We do not see with our eyes alone
How do things around us become what they are?
Do we really want an empirical science?
It is careless thought that deceives us, not our senses
Earth and sun
The atom and beyond
Our "missing" bat sense
Closing thoughts
Where are we now?

TECHNICAL SUPPLEMENT TO PART 1

Chapter 14: How Our Genes Come to Expression 219

Introduction

High expectations: the promise of molecular biology

Complications

Transcription factors and DNA engage in a complex play of form

Genes and proteins interact in tangled causal webs

The cell holds DNA in an intimate and instructive embrace

Getting started is hard to do

Carrying on

Table 14.1: Some factors regulating RNA polymerase elongation

Shaping a significant end

From genetics to epigenetics

MicroRNAs: a large world of tiny regulatory factors

DNA methylation

The nucleosome: a complex marriage of DNA and protein

A tale of tails

A still closer look

Movement and rhythm

A story mostly untold

Concluding thoughts

Where are we now?

Chapter 15: Puzzles of the Microworld 260

Introduction

This is not your familiar Aesop

From here to there — or, down the rabbit hole?

From wine to jelly

Viscous drag

Brownian movement

Chemical energies

Electrical forces

Polymerization

A world hard to get a grip on

What then?

Where are we now?

PART 2: EXTENDING THE ORGANISM'S STORY
Toward an Evolutionary Narrative

Chapter 16: Let's Not Begin with Natural Selection 277

Introduction

What are the "guaranteed results" of natural selection?

Wholeness, unity, type: how not to over-estimate genes

Unanswered questions

The "algorithm" of natural selection is widely treated as if it were an agent

The inadequacy of the theory of natural selection has long been noticed

Misplaced agency

Where are we now?

Chapter 17: Evolution Writ Small 298

Introduction

A "magical" power of self-transformation

Metamorphosis of cells

Organisms manage their own germlines expertly

The unasked question: What can we make of the wisdom of the organism?

Where are we now?

Chapter 18: Teleology and Evolution 312

Introduction: "The life and death of cells"

Are there reasons to reject agency and teleology in evolution?

Individuals versus evolving populations

Do populations lack the necessary causal connections?

Can the presence and nature of agency be explained?

Is the claim of agency anthropomorphic?

What about the differences between development and evolution?

The shortest path to confusion is circular

(1) The problem of the "arrival of the fittest" remains

(2) The explanation assumes what it is supposed to explain

(3) There are no stable mechanisms for selection to work

An aversion to meaning

Where are we now?

Chapter 19: Development Writ Large 336
Introduction
Evolution as a transformation of developmental processes
The heart of the evolving horse
Challenges of the heart
Evolutionists should not forget the directedness of biological activity
A thought experiment
Experiment concluded?
It is, in the primary sense, populations that evolve, not individual organisms
Far from a simple, linear process
A sprawling narrative
Where are we now?

Chapter 20: Inheritance and the Whole Organism.................... 357
Introduction
E. S. Russell picked up Johannsen's problem
The problem of the whole
Is holism incompatible with evolution?
Holism and the independence and integration of parts
Where are we now?

Chapter 21: Inheritance, Genetics, and the Particulate View of Life ... 373
Introduction
DNA as the essential substance of evolution
The aggressive claims of population geneticists
Who defines what counts as an evolutionary process?
Getting to the bottom of things?
How the image of "particles" has distorted the biologist's imagination
Facing up to the gestural reality of DNA
To view from the bottom or view from the top?
Where are we now?

Chapter 22: A Curiously Absolute Demand for Stable Variation 393
Introduction
Richard Dawkins, genes, and the biologist's "ultimate particles"
Development versus evolution
A short critique of Dawkins' view
"Immortal" genetic variation
The embodied organism is not like a cloud
A counter-picture to the gene's-eye view of evolution
Where are we now?

Chapter 23: The Evolution of Consciousness 412
Introduction
There was a primordial unity of inner and outer meaning
What words can teach us about the evolution of consciousness
Europeans and their changing landscape
History of ideas — or an evolution of consciousness
The long historical arc of the evolution of consciousness
An ideal degree of detachment — before and after
From myth to literacy: the coming into focus of the human individual
At the edge of literacy and beyond
The evolution of consciousness, like all evolution, is not a straight line
Where are we now?

PART 3: BEYOND BIOLOGY

Chapter 24: Is the Inanimate World an Interior Reality? 441
Introduction
How the world lends itself to our knowing
We know the world through thinking as well as sensing
Facing up to the interior dimension
The Cartesian diversion — is there a way to bypass it?
Can we recover the unity of the world?
Our cognition does not give us a representation of the world
We cognize the world by participating in its creation
The world as a form of speech
But what about the billions of galaxies?
Where are we now?

INDICES

Index of Names 469

Index of Subjects 474

POSTSCRIPTS

These Postscripts (and perhaps others still to come) are not part of the book itself, but will be made available online as they are written. Links are given below. Future links, if any, will be included in the list given here: https://bwo.life/bk/index.htm#postscripts

Postscript #1: "How Do Biomolecules 'Know' What To Do?" (https://bwo.life/bk/ps/biomolecules.htm).
Postscript #2: "Who Needs Free Will?" (https://bwo.life/bk/ps/free.htm).
Postscript #3: "Still Haunted by Laplace's 'Demon'" (https://bwo.life/bk/ps/laplace.htm).

Preface

I SEE A CROW, perched atop a shagbark hickory tree about fifty meters in front of me. It seems oddly unperturbed on its branchlet as it surfs the tumultuous waves of a stiff wind. The winds, of course, are its own element, just as the twisting currents of a swift stream belong to the trout. The crow, I reflect, is its own sort of master of the wide domain it surveys — a domain whose whole aspect is unfamiliar to me. As I walk, I try to see myself through its unhuman eyes, a small, insignificant figure approaching far below, passing beneath, and then slowly receding into the distance.

I was once taught to see myself this way when in the presence of a bird on high — I, an intruder moving for a few moments through another's native landscape. It was a modest little exercise in becoming detached from one's own point of view. I suppose it's rather easy for us today. We are, after all, heirs of Copernicus, whose one giant leap for mankind sprang from his then-novel capacity to project himself, as an observer, onto the surface of the sun. From that viewpoint he could imagine his own, troubled earth moving serenely through space.

But Copernicus had only to project himself through what was in the process of becoming, for us, "empty space." How much more difficult to insert oneself into the "mind" of a crow! Who is it that looks down at me, and from what strange, inner world does it gaze? What would I *really* be seeing if I could see with crow-vision, so penetrating in its crow-ness, yet so alien to me? I have to admit that there is vastly more of myself projected to the top of that tree than there is of the crow. When the lives of distantly related beings are at issue, isn't getting outside one's own viewpoint all but impossible?

Perplexities of molecular biology

My primary aim in this book is to enable the reader to see organisms — and especially animals, which are my main examples — with new eyes. In place of a systematic survey, I present what might almost be approached as a series of re-visioning exercises whose diverse focal points, so I hope, can merge for the reader into a single, coherent landscape. It will be a landscape viewed, so I also hope, from unexpected angles.

The oddity lies in the fact that I rely rather heavily on topics drawn from molecular biology, a discipline that gives us no real landscape at all — certainly not one based on the kind of direct, sensible experience the founders of the Scientific Revolution craved. The biologist's picture of atoms and particles is synthesized from theoretical constructs and outdated mental pictures that, especially in the physics of the last hundred years, have been thoroughly subverted. So how we should actually picture what I will refer to as the "microworld" is a genuine mystery today.

The problem is that biologists have been content to stick with nineteenth-century images of the solid little "particles" that were debunked in physics long ago. And so they imagine a cell full of little materialized "molecular machines," however tiny, and however ill-matched they may be even to the falsely imagined particles. Where many physicists have acknowledged wide-open questions at the foundation of their discipline, biologists have doubled down on a rather crude, mechanistic materialism.

But the biologists' problem is a problem for this book as well. How can I focus as much as I do on a field of research (molecular biology) that is more or less empty so far as an experience-based (empirical) science is concerned? Am I not just lending further support to a kind of biological fantasy world?

I am inclined to plead guilty to this charge. Of course, I do at times try to warn the reader against misconceptions — for example, in Chapter 15 ("Puzzles of the Microworld") and Chapter 21 ("Inheritance, Genetics, and the Particulate View of Life"). But there are also at least three strong, positive justifications for looking carefully at how biologists appeal to molecular-level research as a bottom-up foundation for understanding organisms. These all have to do with the fact that molecular biology presents to one's imagination a kind of blank slate. Looking at what researchers have projected onto this blank slate can tell us a great deal about the character of biological thought today:

To begin with, we see a seemingly unquenchable thirst for unambiguous (and therefore unbiological) cause-and-effect explanation. These explanations tend to be of an antiquated, billiard-ball sort involving particles that, as physicists have long known, simply aren't there — certainly not in the way they are being imagined within biology. In this way we come to those ubiquitous and hopelessly misconceived "molecular machines" that are supposed to perform the fundamental living work of organisms.

In the second place, because so much of molecular biology is based on non-empirical, unsupportable, and metaphysical (materialist) assumptions, the supposed explanations issuing from molecular biology never add up. When we look at these explanations, we easily recognize the confusion at work in them. (See, for example, Chapter 8, "The Mystery of an Unexpected Coherence," and Chapter 9, "A Mess of Causes.")

Recognizing the confusion can, in the third place, point us in the direction of a more adequate understanding — one that starts with the observable organism rather than a fantastic, non-observable realm littered with metaphysical "projectiles." I gesture toward the grounding principles of such a fuller understanding in Chapter 12 ("Is a Qualitative Biology Possible?") and Chapter 13 ("All Science Must Be Rooted in Experience").

The troubles emerging from biology's deep dive into molecular biology have — rather ironically, in view of initial expectations — dramatically undermined the mechanistic understanding of life. One prominent example is provided by the prevailing image of natural selection as a "tinkerer" working on biological mechanisms over long time spans. Unfortunately, it has become crystal clear that the coordination of scores or hundreds of molecules knowledgeably performing an intricate operation such as RNA splicing or DNA damage repair in the watery medium of the cell's plasms cannot be viewed as the mere activation of a stably existing mechanism that natural selection has somehow been able to tinker with for ages. What needs explanation is not the perfection of theorized ancient mechanisms we cannot see, but rather the *present* wisdom that, as we can see, informs the moment-by-moment activity of those cooperating molecules. These molecules not only find their way, in the fluid milieu and with changing interaction partners, to perform (unforced) an elaborately choreographed operation, but also manage to vary and adapt their activity to the immediate needs of the larger context — for example, by conducting the RNA splicing operation so as to produce the currently needed variant of a protein, rather than, say, the variant produced last time.

Sobering thoughts

I have, throughout the writing of this book, been accompanied by a discomfiting awareness of the difficulty of the task I have set myself. This is presumably due mainly to my own limitations. Seeing things anew — as opposed to collecting more and more data and trying to assemble it into unambiguous demonstrations of truth — is not something I find easy, nor is it something we are generally encouraged to strive for today. The following thoughts, borrowed from others, have, for me, emphasized the great distance from assertive claims of truth to genuine profundity:

♦ The first of these thoughts is an overall conclusion drawn from a study of meaning entitled *Poetic Diction*, written in 1928 by the philologist and student of the evolution of consciousness, Owen Barfield. It expresses a truth also forced upon me directly by many less-than-satisfying efforts at communication. (The phrasing is my own:)

> If a conversation takes place primarily as a logical contest or as a battle of "proofs," rather than as an effort to clarify, shift, and deepen meanings, it is likely to be shallow.

In my run-up to writing this book — and throughout the writing — I have had to suppress my own deeply rooted, almost congenital instincts toward doing intellectual battle. I now know that victory in this particular struggle with myself will never be fully won.

♦ Then there is my vague remembrance of a remark I somehow associated with the late physicist, Georg Maier. It ran more or less like this:

> If you think you have reached a point where you can cleanly explain a profound truth, you do not yet understand it.

After the first appearance of this preface, my colleague, Henrike Holdrege, gave me an actual quotation from Maier, which serves just as well: "the knowing of a phenomenon (appearance) is not at all completed by a successful explanation."

♦ Finally — again from Barfield, and this time as a direct quote wrapped up with a striking metaphor — there is this:

> If you take your view of the world seriously, to air it is tiring. Moreover, in any ordinary conversation you can only do so very superficially, and your own heard superficiality wounds you. The opinions, whether firm or tentative, of a man over fifty who has thought for himself about the nature of man and the universe will have acquired a certain depth and weight that make them ill adapted for point-blank encounter. Submarines rarely engage one another in battle (Barfield 1965, p. 74).

If you want to have a fruitful conversation with someone, the two of you must meet upon some sort of common ground. For if you see things in such fundamentally different ways that every assertion from one side is met by a refusal to accept it on the other side, then there is not much reason to talk. If, on the other hand, the two of you are so close in thought and assumption that you mean the same thing with your words and can work with precisely the same set of facts, then the role of conversation is also limited. All you need to do is to order the facts in such a way as to prove your case to the other person. Nothing really new will arise, because your proof was already implicit in your mutually accepted understanding of things.

But there is a potentially productive middle ground where enough is shared to make conversation possible, and enough is not shared to raise the hope of genuinely new insight. In this case the challenge is to be non-defensive and to hear the other person's words and facts with receptive ears. We can most easily achieve this if we have managed somehow to get outside our own culture's "common sense," much as we today are able to challenge, or even laugh at, the unquestioned wisdom of previous historical eras. Managing to see our culture in such a foreign light, however, can be an almost impossible task. But even a small effort in that direction can be life-changing — like being let out of a prison you hadn't realized you were in.

I do not expect my efforts here to be adequate. But I do hope they may be of some use to those sympathetic readers seeking a new vantage point upon biology — one that, even if at first it presents an unfamiliar and perplexing landscape, at least does not require us to deny the living experience of all creatures, including ourselves.

SOURCES

Barfield, Owen (1965). *Unancestral Voice.* Middletown CT: Wesleyan University Press.

_____ (1973). *Poetic Diction: A Study in Meaning*. Middletown CT: Wesleyan University Press. Originally published in 1928.

Part 1:
THE LIFE OF ORGANISMS

CHAPTER 1

The Keys to This Book

WE BEGIN WITH A VIGNETTE drawn from a single activity of just one from among the millions of species with whom we share the earth. This description is taken from the biologist, novelist, and science philosopher, E. L. Grant Watson, who in turn is compactly summarizing observations by one of the world's great entomologists, who lived during the late nineteenth and early twentieth centuries:

The enigmatic wisdom of the potter wasp

"Among the fascinating stories of animal life told by the French naturalist Henri Fabre is that of the [potter wasp] *Eumenes*. The fertilized female builds a little domed house of sand spicules on some stone or rock foundation. The foundation ring is traced in minute pebbles. On this she builds a series of concentric rings, each diminishing in circumference, so as to enclose a domed space. At the top she leaves a hole. She then begins collecting certain species of small caterpillars. She stings these into a partial paralysis, but does not kill them, for they will be needed as fresh meat for the young she will never see.

Figure 1.1. An unidentified species of *Eumenes.*[1]

"When the wasp has collected either five or ten caterpillars, she prepares to close the dome, reducing the size of the hole. She now goes through a complicated process which would seem to indicate foresight on her part. Yet she has no foresight, only a highly developed instinct. From her ovipositor she excretes a juicy substance, working it with her legs into a narrow, inverted cone. With a thread of the same substance, she stitches the cone to the top of her domed building. Into the inverted cone, she lays an egg. She then seals up the hole, leaving the egg safe within the cone, suspended on a thread. This done, she goes off and builds another dome to repeat the same cycle of events.

"In a short time the egg hatches into a tiny, white grub, so helpless and delicate that if placed among the still-living caterpillars on the floor of the dome, it would inevitably be injured. In its cradle it is safe. When hungry it spins a thin thread of its own, on which it descends and takes a bite of caterpillar. If the wriggling caterpillars appear threatening, it can retreat up the thread, and wait. In this way the grub spends its infancy; but, as it grows stronger, it risks a final descent, and devours, at its leisure, the still living food that mother has so satisfactorily provided.

"From the domes that contain five caterpillars male wasps emerge; from where there are ten caterpillars, the larger female wasps. This raises an interesting question: Does the amount of food determine the sex? The mother wasp, who appears throughout her lifetime as a highly nervous and brilliantly alive creature, has built just the right sort of houses for the offspring she will never see; and has provided just the right amount of food. She is singularly well-adapted for her life; she stings the caterpillars just enough to keep them quiet, but not enough to kill them; she packs each dome with the right amount of food for male or female grub. The suspended cradle protects the tender infant from the rough reactions of the caterpillars while being eaten. Everything is in order, and as the emerging wasp dries her wings in the summer sunshine, she must surely feel that God is in his heaven, and all is well with the world. The caterpillars might harbour different sentiments ..." (Watson 1964, pp. 85–86).

Figure 1.2. Nest of a potter wasp on top of a concrete wall.[2]

AND SO (focusing on the wasp's offspring) we picture in our imaginations a minuscule creature, with the nascent intelligence of an insect newly hatched from its egg, immediately setting out upon a journey by descending an almost invisible, yet reliably strong thread, spun (surely it knows not how or why) by itself — all because it needs a bite of food. And it then quickly retreats back up the thread (itself a remarkable feat) because its existence is threatened by larvae far more massive than itself.

That word "because" — *due to the cause of* — is central to a science always concerned with the causes of things. But the usage here, referring to a creature's *need* and its *effort to preserve its own existence*, is as far removed from the word's preferred scientific employment as the little drama of the potter wasp's performance is from the events of the nonliving world. Inanimate phenomena are not characterized by need, effort, or a drive toward self-preservation.

The seemingly unbridgeable difference between living beings and inanimate phenomena is not something many scientific students of life are fond of. That is why they have invested heavily in an abstract evolutionary drama of nearly miraculous character in order to explain the difference away. As Lila Gatlin, a prominent biochemist, mathematician, and shaper of evolutionary theory in the twentieth century, once acknowledged, "The words 'natural selection' play a role in the vocabulary of the evolutionary biologist similar to the word 'god' in ordinary language" (quoted in Oyama 2000, p. 31). In effect (as we will see in later chapters), the organism's living wisdom has been transferred to an omnipotent "force" of evolution, where it can be kept safely out of sight, obscured behind an elaborate technical and mechanistic terminology.

The story of the potter wasp will strike most people as truly amazing. What is not so often realized is that we can discover more or less the same improbable story in all biological activity once it is investigated deeply enough. Whether we are talking about DNA damage repair, or cell division, or the cell's timely and spatially patterned production of subtle protein variants distinct from anything for which there are unambiguous "instructions" in genes, or the development of any complex organism from a single cell toward maturity, or the annual round trip that migrating arctic terns make from pole to pole, or the endlessly variable and often bizarre reproductive strategies of plants — in all such cases we encounter an intricately organized wisdom that doesn't align well with the restricted explanatory resources available to the contemporary biologist. We do, however, find in the literature many celebrations of the logic of natural selection and how it is supposed to explain everything about life without any problem.

An aim of this book is to recapture the drama of life in the place where it actually occurs — not in the "mechanism" of natural selection, but in organisms themselves — and then to lay bare as clearly as possible the failure of the reigning evolutionary theory to explain the features of that drama. This will be a matter of showing that, in a primary sense, the life of organisms explains evolution, rather than being explained by it.

Meanwhile, all may agree that our wonder at the potter wasp's behavior is perfectly natural. And we can rightly be confident of the further marvels we would encounter if we looked into the wasp's mating and reproductive processes, or inquired how it *perceives* a world and effectively navigates the features of that world. Or how it searches out prey for its young. Or how its body gains and sustains its staggeringly intricate and complex physical form, all the way down to the pattern of its molecular interactions.

We find ourselves woven into a fabric of earthly life that is diverse and luxuriant beyond words, and is nearly incomprehensible in its wondrous displays. But then, too, there is this: the wasp's capacities, like those of countless other creatures, seem in some regards wholly routine, familiar, and even human-like to us. In fact, they so powerfully remind us of our own skills and intelligence that we are continually tempted to commit the sin of projecting our own sort of experience onto other organisms.

On one hand, no scientist would — or should — say, with anything like the human meaning and feeling of the words, "The potter wasp takes great care to make thoughtful provision for its young." On the other hand, we can hardly avoid our scientific responsibility to ask, "How is it that the performance of the potter wasp so forcibly reminds us of what, in our own evolutionary development, has become 'taking great care to make thoughtful provision for our young'? Could the two kinds of behavior arise from wholly disparate roots in the history of life on earth, despite appearances and despite our common evolutionary origin?"

Perhaps the best place to start answering that question is with a resolve not to compromise any side of the truth merely because we are philosophically uncomfortable with its apparent implications. In particular, we ought not to twist our understanding out of shape due to a historically conditioned revulsion against anything like a purposive dimension to life processes. Nor should we be unwilling to acknowledge the ways in which all organisms behave as more or less centered agents in the world. Nor again ought we to respect any presumed rule in biology that says, with blatant self-contradiction, "Some human traits appearing in our evolutionary history are unnatural and cannot be referred to in a properly 'naturalized' science."

Oddly, those who most eagerly remind us that "humans belong to the animal kingdom" often seem the ones most reluctant to embrace the flip side of this truth: all animals have arisen within the same drama of evolving life that, we now know, also happened to be in the business of producing humans. If we want to say that humans share in the nature of all animals, how can we then turn around and ignore the obvious implication that all animals share something of the nature of humans?

Here I would like to summarize, ever so briefly, certain themes suggesting what sort of book you are reading:

Theme #1: Narrative

Meaningful life stories are the primary subject matter of biology

Every organism (in this book I speak primarily of animals) is weaving a life story — or, perhaps better, is actively *participating*, or *caught up*, in a life story, a meaningful narrative. The description of the potter wasp above is one episode in one such story. These stories are future-oriented in a manner roughly analogous to a biographical or historical narrative. That's why so much of biology concerns the *development* of organisms — a word no one would use in quite the same sense for geological strata or clouds. Biological narratives depict the meaningful activity through which organisms progressively express and realize the potential of their own natures.

These narratives feature births and deaths, both at the cellular and the whole-organism levels. Life is partly founded upon continual death. And yet the course of a life has its own unbroken unity and wholeness, whereby the imprint and meaning of the past is borne into the future, as anyone who has inherited an abused pet dog or cat well knows. This imprint of the past may or may not be further worked on and transformed.

An animal's life story is composed of innumerable smaller stories (episodes) woven into the overall narrative. And the episodes may in turn be composed of innumerable gestures (micro-episodes?). Life stories and the meanings constituting them seem to have this pure character: the elements of stories are stories-in-small, just as the elements of meaning are never anything but meanings.

A life story is an end-directed, intentional movement from a beginning to an end. This movement is there for us to observe regardless of whether an organism is self-aware and highly individuated — regardless of whether it subjectively conceives intentions for itself in a human-like fashion. The bare fact of *something like intention* is written all over the potter wasp's behavior.

If our scientific understanding rightly teaches us to avoid the all too natural but wrongheaded idea of a "goal" being "aimed at," this should not scare us away from recognizing the full sense of the wasp's performance in its own evident terms, or prevent us from acknowledging the playing out before our eyes of a set of meanings that we can follow with something like the interest we commonly give to a story. The events hold together and flow from a beginning to an end much like a story. We find ourselves marveling at a remarkably apt wisdom — a wisdom we can hardly understand as anything other than a present, moment-by-moment activity of some form of reasoned, more or less conscious, and generally non-self-aware intention.[3]

We need to distinguish between *possessing* one's intentions in the rather free and conscious way we humans do (when we are fully awake), and being *possessed by* them, which seems to be the potter wasp's case. These are very different conditions, but even the unfree state of *being possessed by* is not a reduction to mere inanimate physical lawfulness. We're still talking about a life story with intentional behaviors, even if the intentions are not fully the organism's own — even, that is, if the organism is not sufficiently individuated to be seen as a potential bearer of its own intentions.

It's worth remembering here how we ourselves can sometimes become aware of meanings and intentions that once lay far below (or above?) our conscious willing and planning, and that therefore possessed us more than the other way around. Whose can we say are those intentions before we become aware of them? It's no easy question, and it presumably has rather different answers depending on whether we are talking about humans or wasps.

Much of what I have just said requires us to acknowledge the organism — and particularly every animal — as a focal *agent*, a being capable of spinning out and inhabiting its own story, and whose causal activity is locally centered and distinct from the more general regularities we observe in the inanimate world. At the same time, every organism is interwoven with that inanimate surround, whose substance and reliable lawfulness it makes into a means for its own existence and self-expression.

Theme #2: Interiority

Every animal's life narrative is an outward expression of interior meaning.

It may be that when humans communicate, there is nothing (apart from certain instances of spoken and written language) more richly and specifically informative than the expressions of the human face. Much of our life is shaped and guided by the facial expressions all around us. What these expressions tell us, however, cannot be reduced to the physical-causal terms of facial musculature, skeleton, and flesh. That which bears the expression is indeed outward, material, and physically lawful. But what is expressed is, we can reasonably say, *interior*. Sadness, pensiveness, elation, doubt, anger, vexation, impatience, uncertainty, satisfaction — these are not physical entities. Or again: while the material embodiment of what is expressed is both real and spatial, what is expressed *through* the outward manifestation is real and meaningful, but not spatial. So the word "interior" is problematic; it suggests a spatial relation, whereas I am using it to suggest something like "not *out there* in a sense-perceptible or spatially locatable sense."

Of course we do, in a sense, "perceive" the interior. We look through and by means of faces as material manifestations in order to see the interior meanings they are expressing. It is much the same with spoken words, whose interior meanings are not revealed so long as we are noticing the words only as sense-perceptible sounds. We are "hearing through" the sounds when we grasp their immaterial and interior meaning.

It is not altogether different when, gazing through a microscope and conducting molecular assays, we "watch" a cell carrying out its *intention* to divide. But the quote marks here are meant to indicate that we're not really directly perceiving intentions or any other interior phenomena with our physical senses. We have to add an interpretive activity — the "seeing" through or "hearing" through — to the immediate physical reports of our eyes and ears. Anyone who thinks this interpretive activity carries us dangerously away from scientifically verifiable reality should cease attending to human language, including the language of scientific description.

As I have already indicated by mentioning the cell, it is not just humans who possess interiors. All living performance expresses one or another form of interiority. In our own case: if I walk through a campground looking for a source of drinkable water, what I am doing can never be captured by what we think of as a purely physical description of the movement of my legs and arms, vocal apparatus, and so on. So, too, with an animal engaged in anything we would call "behavior." The meaning of the behavior — whether a courtship ritual, or burial of food, or tracking of a scent, or flight of a hawk, or the digging of a burrow — can never be described in strictly physical terms. We are always watching an unfolding interior narrative expressed through the outward, physical "face" of events.

Further, as I will try to suggest throughout this book, even our descriptions of cellular and molecular "behavior" refuse to be altogether cleansed of interiority. When we look at cellular goings-on, we can always recognize a *meaning* or an *end*, a *task* or a *function* — what a biological activity is *about* (for example, synthesis of needed proteins, or extraction of usable energy from a substance) — and our biological inquiries are guided by our curiosity about this meaning. We may want to learn, for example, how a particular kind of cell pulls off cell division, or how the mammalian circulatory system meaningfully adjusts to cold weather or high altitude.

A dramatic fact about contemporary biology is that biologists seem to have a horror of interiority, or the non-spatial and non-sense-perceptible. Given that the *life* of animals is through and through an interior business, this horror is not only perplexing, but also devastating for the prospects of a truly biological science.

If there is one central theme of this book, it is that we need to leaven every biological topic, from gene regulation to respiration, blood circulation, and animal behavior, with an understanding of interior meanings. This will lead us to talk about intentions, purposiveness, wisdom, intelligence, agency, needs, and interests, all of which are implicit in nearly all biological description. Yet that description is badly distorted by the undisguised horror of interiority and the attempt to substitute purely physical terms for the interior dimension of life.

Making this point is my primary aim in this book. I am not looking for dramatic new discoveries in biology. I am saying, "Look how all biology is transformed if only we overcome our antipathy toward interiority and acknowledge what is right in front of us." And, in fact, the acknowledgment is already implicit whenever we are doing biology as opposed to physics and chemistry. It is simply not possible to talk about the extraordinarily complex process of DNA repair without taking into account, however subconsciously, the fact that the cell is *attempting as best it can to perform the difficult task of DNA repair* — and must somehow, in some terms, possess the necessary knowledge of health, the practical skill, and the persistent intention required for the task.

Rocks don't have intentions of this sort. How do organisms come to differ from rocks in this matter? And why are we not in a state of wonder about the whole business? This tells us a good deal about how blasé we've become about fundamental issues in biology — and also how we are so intimately aware of the nature of living things that we can't help taking their true nature for granted even when we have been intellectually swallowing absurdities about the non-living character of life. And it may also remind us how many biologists are convinced that natural selection has solved all mysteries such as this difference between rocks and organisms. (See "The shortest path to confusion is circular" in Chapter 18, "Teleology and Evolution.")

The idea of interiority not only overlaps that of narrative (a story is an interior reality), but also that of holism, as we will now see.

Theme #3: Holistic portrayal

The meaningful, narrative character of life demands its own, holistic style of understanding and explanation.

If the organism's life, its biological existence, takes *narrative* form, then our characterization of its life — contrary to conventional notions of explanation — must also take a narrative form. And it could hardly be clearer that the elements of a story, like the elements of an organism's life, can never be considered adequately in isolation from each other. Nothing is absolutely distinct from everything else. The end of a really great novel will be illuminated by its beginning, and the beginning by the end. This interwovenness of interior narratives amounts to a kind of holism, and in this respect a narrative might far more appropriately be compared to a portrait that captures a subject's essential being than to the analysis of a machine into discrete parts and causal relations.

How, in fact, do we come to understand any context of meaning — a dance, a painting, a novel, a human life, the choreography of a developing embryo? Johann Wolfgang von Goethe noted the impossibility of capturing an "inner nature" — say, a person's character — in any kind of direct causal or explanatory way. "But when we draw together his actions, his deeds, a picture of his character will emerge" (Goethe 1995, p. 158). That is certainly how we try to understand each other — and we, too, are organisms.

Our knowledge of the character of wholes is not impotent. If I familiarize myself with the distinctive way of being of a bluejay, I may not be able to predict exactly what it will do or project its flight as a Newtonian trajectory. But my knowledge is nevertheless real. I will, in appropriate circumstances, be able to say, "Yes, that is just like a bluejay" or "No, that is not at all what one would expect of a bluejay in this situation. There is something wrong, or something missing from the picture." With such knowledge I can learn to interact meaningfully with the bird even though I cannot mechanistically predict its behavior. In developing a qualitative portrait, we aim less at exact prediction and control than at understanding and the potentials for working with nature.

The main question about a portrait is how full, how detailed, how multi-faceted a picture we gain. The supposed causes, of course — when properly contextualized and shorn of their strict causal aura — help us to build this picture. There is neither any end to our picture-building, nor an inherent limit to how far we can carry it. And biologists surely *are* carrying it further, even when they think they are fingering clear-cut, explanatory causes.

Moreover, it is clear that we cannot have holism without also applying the remarkable analytical skills that we humans have so fruitfully gained. It is hard even to conceive how one might sketch an organic whole without having a lucid and detailed awareness of its parts. The need is to hold together the complementary movements of thought — the synthetic (holistic) and the analytical.

And, in fact, the meaningful counter-movement to analysis is inescapable — although generally not noticed for what it is. After all, in order to analyze a whole into parts, we must start with an already recognized whole, and then we must recognize each part as possessing a significance of its own — as being a meaningful whole in its own right. This recognition of wholes, however unconscious it tends to be, is fully qualitative, contrary to our usual ideas of science, and it requires a movement of understanding that runs contrary to analysis. I say "fully qualitative" because only qualities can blend or interpenetrate so as to erase the rigid boundaries and mutual "otherness" of things.

The synthetic, or holistic, counter-movement to analysis is implicit in the biologist's frequent citing of the "context-dependence" of biological processes (Chapter 6, "Context: Dare We Call It Holism?"). The problem is that the implication here — the implication that there is a kind of influence or causation running from a collective, complex whole toward its parts — has drawn little reflection and has had little effect on the underlying assumptions of biologists.

"Context" is a word commonly used by geneticists and molecular biologists. But it seems there is little interest in explaining what one actually means by the term.

"Holism," by contrast — and despite its being hardly distinguishable from "contextuality" — has become a kind of "devil word" in biology, a fact ironically coexisting with a refusal to consider the issues implicit in current, context-centered biological language.

In this book "holism" will simply be taken for granted from the beginning. But, unlike "context-dependence" in the existing literature, the meaning of "holism" will be consciously and explicitly drawn out and illustrated as we go along.

Theme #4: Blindsight

A kind of blindsight is evident in much of biology.

Living narratives, as observed, for example, in all animals, are in fact recognized within biology. For example they provide the structure for research projects. These typically have to do (as I mentioned above) with how an organism accomplishes this or that function, or *task*, such as obtaining food, or maintaining bodily temperature at an acceptable level, or, in the case of many cells, achieving cell division. (Rocks and streams do not have tasks.) But something rather like a taboo seems to require biologists to ignore all this in their scientific explanations. They are allowed to discuss only physical "mechanisms" that make *no inherent reference* to — and therefore do not explain — the task-nature of the problems that prompted biological inquiry in the first place.

This might bring to mind the curious and well-known phenomenon called "blindsight." It works like this. Suppose there is a certain life-sized statue on the floor of a museum I am exploring. If, for some reason, I suffer from blindsight about the statue and am asked about it, I might truthfully reply, "What statue? I don't see one." But then, in wandering about the room, I am observed always to walk carefully *around* the statue rather than bump into it. Clearly, in some sense I do see it, even while remaining consciously unaware of (and even denying) what I see.

My suggestion, then, is that something analogous to this phenomenon works powerfully within biology today. Biologists carefully walk *around* the fact of the animal's narrative agency, even while pretending in their explicit theorizing that nothing is there. Every biological question they ask ("How does an organism accomplish this or that?") affirms their *knowledge* of this agency. One result is that much about the true character of animals (and organisms generally) comes through in the biological sciences despite the biologist's explicit denials. Bringing attention to the great mass of obscured truth already "seen," if only blindsightedly, is a lot of what this book is about.

Nevertheless, because of biologists' blindsighted theoretical and philosophical commitments, their science suffers from the deepest possible distortions. They end up with living processes *theoretically* stripped of their life — this despite the fact that they know this life more directly and intimately than they know anything about the non-living world.

What is needed, according to the late Harvard geneticist, Richard Lewontin, is for biologists "to take seriously what we already know to be true" (Lewontin 2000, p. 113).

Some definitional hints about key biological terms

A number of the terms central to this book, while common in normal human discourse, are foreign to conventional biological usage. The strangeness in this, I dare say, is on the part of biology rather than this book. In general, I try to employ the following words in agreement with their routine, non-technical use, and not to tie them down with overly artful precision. I hope that the meanings will become more specific — or more flexible — based on their various contexts of use.

Agency. Humans are agents. We possess agency because we possess an awareness of our world and can *act* in it instead of merely *being shuffled around* along with the furniture of our surroundings. We help to create the situations in which we live, instead of being wholly determined by them. The cells of our bodies clearly can participate in our agency by giving expression to it, as when we move our limbs intentionally. But we would never say of those cells as such that they possess awareness or agency in their own right, as opposed to moving with an agency not fully their own. This is suggestive of the kinds of distinction we must make between ourselves and other organisms, all the way down to single-celled organisms.

I know of no reason *not* to believe that, just as the intention of a human individual can play through trillions of cells, so also, though perhaps in a somewhat different manner, an intention can play through a collection of bacteria in a bacterial film (evident, for example, in "quorum sensing"), as well as through the members of a termite colony, or any species at all so far as its members share a common way of being — and indeed in human society in ways of which we are scarcely conscious.

Agency does not arise from physical interactions among the parts of an organism. Rather, the purposive coordination or organization of such interactions is an expession of agency. The distinction between these two ways of looking at the matter is not often enough appreciated. Nothing about physical laws connects with or supports our understanding of the biological *striving* we so readily observe. This is why the biological literature is awash in references to "emergence," a rather magical term referring to features of life that are thought to "just show up somehow," without specific reference to lawful process.

See also **Intention/intentional** below.

Archetype. The archetypal idea of an organism is its dynamic, adaptive, evolving way of inhabiting, and remaining true to, the character and potentials of its kind. It is simply and obviously not true that the fact of evolution makes a lie of the observable *way of being* (archetype) of any given type of organism.

Atoms/Molecules. You will find comments here and there in this book suggesting something about the unreality of atoms and molecules. The effort is to emphasize that in the submicroscopic realm we are dealing with theoretical constructs that do not have the reality required by a science of the material world — the reality of sense-perceptible experience. The problems arise, as physicists well know, when we endow certain constructs such as *wave* or *particle* with imagery derived from our experience of the material world. Then we are dealing with invented unrealities, and these tend to mock us when we try to make sense of our experiments.

I attempt to show in Chapter 24 that we have little choice but to assume that the reality the world possesses is, according to its own nature, a reality appearing in all the possible forms of *experience*. To make any other assumption is, on its face, to speak ignorantly about *what we do not know from experience*.

Blindsight. See Theme #4 above.

Consciousness. We might say that consciousness is the experience of meaning in an organism's life. Human consciousness can be an experience of meaning of which we are aware. In many other organisms (and in some aspects of ourselves), so far as there is not awareness of the play of meaning, we might speak paradoxically of "subconscious consciousness," or "consciousness of which one is not aware." Or we could switch to *intelligence*, which we can readily imagine as operating without awareness. That is, intelligence can work in us (or an animal), without being *possessed* by us as "our own." Think, for example, of animal instinct or, in humans, the implacable logic of disruptive complexes derived from childhood abusers.

Directive. E. S. Russell, a marine biologist and proponent of "organismal biology" during the first half of the twentieth century, adopted the word "directive," as in the title of his wonderful book, *The Directiveness of Organic Activities* (Russell 1945). He chose this less familiar word in order to encourage in his readers an awareness of the distinctions between human end-directed, or planned, activity and the future-oriented activity of animals such as the nest-building of birds. I will, in part, follow suit, although I will also freely use "directed" or "end-directed" in the conviction that we need to cultivate, not only an awareness of the *differences* between humans and animals, but also of the *connections*.

Biological activity *is* directed in the immediate sense of the word — interiorly and insistently guided in the way the development of a squid or fox or ape is guided from the zygote toward the adult form. This remains true even though the process is not at all consciously directed in the manner of our own willed activity. For that matter, neither is our own human movement from zygote to adult form consciously directed by ourselves. See also ***telos*-realizing** below.

Holism. See Theme #3 above.

Integral unity of the organism. When I use something more or less like this phrase, I intend it as an active concept in Aristotle's sense of "being at work staying itself" (in Joe Sachs' translation of Aristotle's *entelecheia* — Sachs 1998, p. 245). Through this activity, the parts of an organism arise within an integral and differentiating whole; they are not assembled together as pre-existing building blocks in order to make a whole. The integral unity is actively there from the start, and is not at any point imposed from outside. It is a unity because each part reflects — or participates in and remains consistent with — the nature of the whole from which it arises and gains its identity.

Intelligence. See **Consciousness** above.

Intention/intentional. I try to use these words as far as possible in their routine, day-to-day meaning. We recognize intentions by observing the guiding principles and meanings at work in an activity. It needs noting, however, that we humans can intentionally do something not only through careful planning, but also subconsciously ("unconsciously"), as when we notice a traffic irregularity while driving a car and engaging in conversation, despite the fact that we were paying no conscious attention to the road. It is much the same when we ride a bicycle while quite unconscious of any intention to remain upright and balanced on the bike.

We should never ascribe our own, most wide-awake sort of consciousness to other organisms, who seem to function quite well by means of intentions that do not originate reflectively. Also, a great part of human subconscious activity (think of the bicycle-riding example) derives from prior intensely conscious practice.

But we can't say the same of, say, a monarch butterfly's participation in a multi-generational migration from Canada to Mexico. So, just as we shouldn't project our self-aware consciousness upon other organisms, neither should we assume that their subconsciousness is the same as ours.

Nevertheless, in all cases of intentional behavior, I'm not aware of any grounds for taking the intentions to be anything other than a function of mind. The difficult question regarding organisms then becomes "Whose mind?" Whose mind accounts for the archetypal or shared way of being among the members of a species? And I did say *difficult* question. Not all questions currently facing us in biology have obvious answers. But it seems to me important for biologists to notice that, in all organic performance, we *are*, in one way or another, looking at the activity of mind as well as body. This remains true even if the organism is not aware of its intentions as its own. Even in some of our human activity we can sometimes recognize a kind of unconscious, collective "supra-mind," not fully our own, taking hold of us in a crowd, as in a football stadium, or in a highly charged interpersonal confrontation.

Interiority. See Theme #2 above.

Material. Accessible through our physical senses.

Material/physical/materialist/materialistic. I speak broadly of "the material world" as the world we routinely experience, the world we live and move in, the world accessible to our senses. "Material" and "physical" might be taken as rough synonyms, but I preferentially use "material" when speaking about the sensible world as we directly experience it and can know it, and I use "physical" when emphasizing the habits of thought that come to the fore when, as *materialists*, we are thinking falsely and *materialistically* about the nature of the world and trying to conceive it purely in terms of inanimate entities and processes conceived as mindless and having nothing to do with our own interiority. So I might say, on one hand, that an organism adapts to its *material* environment, but, on the other hand, that we are commonly thought to live in a world subject only to *physical* laws. But there is no strict line between these terms, and doubtless no full consistency in my usage.

Problems arise because the idea of the "strictly physical" is incoherent: physical laws are ideal and conceptual, not mindless in the sense of "physical" usually taken for granted. My use of the "physical" is a bit schizoid, since I may use it in the materialistic sense ("inanimate, mind-independent"), especially when trying to represent a materialistic point of view; or I may use it in a more neutral fashion as a synonym for "material" ("sense-perceptible").

Please note that whatever is sense-perceptible — whatever is available as a content of science — has an irreducibly interior character. We possess it only as a content of consciousness. Whatever we perceive and whatever we think, we perceive and think upon the stage of consciousness. (Where else might we become aware of these contents?) And we have good reason for thinking that this appearing as a content of consciousness reflects the material world's inherent and objective nature. For more on this, see Chapter 24, "Is the Inanimate World an Interior Reality?"

Meaning. All coherent descriptive content is meaningful, a fact already implicit in the word "coherent." ("Coherent" in common usage just *means* "hanging together in a meaningful way.") Meaning seems to us problematic only because we have materialist mindsets as a cultural inheritance, and because meaning is so thoroughly inescapable, like a fish's watery environment, that we have a hard time stepping back and seeing it for what it is. The sea of meaning is that from which we are born and in terms of which we continue to live and finally die. We cannot do anything or say anything or pursue any science without the doing, saying, or pursuing being an expression of meaning.

Some people have a very difficult time with any use of the word "meaning" in a scientific context. It's worth setting this difficulty alongside the fact that the things we *know* about the world are generally things we try to communicate in words — which is to say, things we try to express in terms of meanings. Meaninglessness (nonsense) would not yield itself to knowledgeable, scientific articulation.

And so meaning can hardly be questioned. The effort to question or define it — or just point to it — assumes that the person being addressed already possesses a working understanding of meaning, such as the meaning of a pointing finger. Acting out or expressing meanings is pretty much the only thing we do with our lives. The same thing is true of organisms generally, all the way down to one-celled creatures — except that they lack the capacity for conscious reflection upon the meanings at work in their lives. The interesting questions have to do with the different sorts of meaning at work in different kinds of organism.

The fact that we are dealing with the fundamental basis of life when we use the word "meaning" is hardly a reason to avoid it in biology. The (always unsuccessful) effort of avoidance is perhaps the central pathology of contemporary biological thought and practice. In a thousand ways the taboo against any suggestion of meaningfulness makes a fool of scientists and nonsense of their use of scientific language, which is nothing but a highly sophisticated way of expressing the meanings they have discovered in the world (Chapter 23, "The Evolution of Consciousness").

Narrative. See Theme #1 above.

Purpose/purposeful/purposive. We know the routine human meanings of these terms, where "purposeful" and "purposive" are synonymous. As is common in the biological literature, I often use "purposive" to distinguish directive activity in many other organisms from the conscious, self-aware, planned, goal-aimed activity of humans. But I sometimes use "purposeful" when referring to non-human animals, if only to avoid making the human-animal distinction seem absolute or unnatural.

All biological activity is purposeful in a way we have no great difficulty understanding. And we do not require all that subtle an understanding in order to realize that animals in general are not reflecting upon or planning their activity in the self-aware way we sometimes do. Perhaps we can be aided in understanding an animal's purposes by considering the "purposes" of our own cells in carrying out the intentional movements of our bodies. We would not want to say that the cells have purposes of their own in anything like the whole-human sense. But so far as they are capable of being caught up in our purposes and giving perfect expression to them, they themselves clearly have (or have been lent) a kind of purposive character.

***Telos*-realizing.** *Telos* ("end") is often taken to refer to final causation — to the *end* we humans are aiming at when we consciously formulate plans. But, consistent with the Greek term, it may be more useful to take the "end" as a matter of *self-realization*, which is the "being at work remaining oneself" referred to under **Integral unity of the organism** above. Or, we might say, "being oneself ever differently." It's a matter of bringing oneself to ever fuller and ever different expression — taking always a further step in expressing one's own nature. Only in the human case does this involve a creative awareness whereby an action can become intimately *our own*.

Regarding the ideas conveyed by "end," "self-realization," and "holism," we have this incisive comment by the philosopher Ronald Brady: An organism's biological development "does not proceed towards [a] whole, but rather *expresses it*" (Brady 1987). From the very beginning of its life, it is already a whole. It is, however, hard to find words that capture the meaningful coordination of processes in the achievement of a certain result without seeming to imply an external goal. The alert reader will need to make an inner adjustment whenever encountering language that sounds external-goal-directed (unless the language refers to humanly planned activity).

See also under **Directive** above.

Where is the evidence? Two concluding notes

The preceding discussion, especially that of Themes #1 and #2, underscores a truth that is alien to contemporary biology: *We meet in the living world something akin to our own inner being*. However, everything I have hinted at here desperately needs expansion, which is why this book was written.

But while the themes and underlying convictions shaping the character of the book lie far outside mainstream thinking, I offer no new or revolutionary findings in biology or evolutionary theory — and would lack the qualifications for doing so even if that were my inclination. Instead, I merely ask: What would biology and evolutionary theory look like if we overcame our blindsight and reckoned with the stories we actually observe in the life of organisms? Can we allow ourselves to see with restored vision?

And so there will be no occasion for readers to ask, "Where is all the new evidence?" The evidence supporting my contentions here — as I try to show chapter by chapter — amounts to just about *everything* biologists have already recognized as truth, however much they might prefer not to acknowledge the gifts of their own insight. This is why you will not find me straining toward the fringes of biology, but rather citing, with very few exceptions, one fully accredited researcher and theorist after another. The case for a thoroughly disruptive re-thinking of organisms and their evolution has long been staring us in the face.

A second note is not unrelated to the first. Throughout this book I have, to a degree, tuned my vocabulary to the more complex animals with which we are most familiar, although the language could readily be adjusted to reflect the intelligent life processes in bacteria, plants, and other groups. Many will say that this is to ignore what are by far the most abundant creatures on earth. Perhaps so. But I am convinced that, contrary to the usual intuitions, the "higher" organisms are key to understanding the "lower," not the reverse.

This is true in the indisputable sense that the kind of understanding we are looking for emerges only in humans, so that we are the *only* organisms capable of understanding other beings in a scientific manner. But I believe it is also true in the sense that those organisms more fully manifesting the potentials of life do in fact *more fully manifest the potentials of life*.

At the same time, we have no reason to think that the intelligence working through the material limitations of, say, a bacterium is a "lower" or less capable intelligence than that which is at work in ourselves — or that the intelligence at work in the cells of our own bodies is lower than what works in our conscious minds. Actually, our cellular intelligence quite evidently far transcends our conscious capacities. We can say this without doubting that the arrival of a self-aware sort of consciousness is a pivotal development in the evolution of life. It's just that we have no grounds for arrogance regarding our current conscious achievements. These achievements are, in the past, present, and future context of life on earth, doubtless humble indeed!

NOTES

1. Figure 1.1 credit: Rama Warrier (https://commons.wikimedia.org/wiki/File:Potter_wasp_1.jpg), CC BY-SA 4.0.
2. Figure 1.2 credit: Pollinator (https://commons.wikimedia.org/wiki/File:Potter_wasp_nest_6734.jpg), CC BY-SA 3.0.
3. The implanted "wisdom" — a wisdom from the past — that we so easily ascribe to an unconscious machine always has its origin in a prior, conscious, designing activity of a person. And the manifestation of that wisdom in the machine is radically different from its immediate presence in whatever sort of consciousness acts *now* in a living being. Organisms are not designed machines. This truth is underscored time and again in the following chapters. Our construal of organisms on the model of machines — a construal that so much current biological thinking shares with so much intelligent design theory — needs to be overcome.

SOURCES

Brady, Ronald H. (1987). "Form and Cause in Goethe's Morphology," in *Goethe and the Sciences: A Reappraisal*, edited by F. Amrine, F. J. Zucker, and H. Wheeler, pp. 257–300. Dordrecht, Holland: D. Reidel. Available on The Nature Institute's website: https://natureinstitute.org/ronald-h-brady/form-and-cause-in-goethes-morphology

Goethe, Johann Wolfgang von (1995). *Scientific Studies*, edited by Douglas Miller. Vol. 12 in *Goethe: Collected Works*. Princeton NJ: Princeton University Press.

Lewontin, Richard (2000). *The Triple Helix: Gene, Organism, and Environment*. Cambridge MA: Harvard University Press.

Oyama, Susan (2000). *The Ontogeny of Information*, 2nd edition, foreword by Richard C. Lewontin. Durham NC: Duke University Press. First edition published in 1985 by Cambridge University Press.

Russell, E. S. (1945). *The Directiveness of Organic Activities*. Cambridge UK: Cambridge University Press.

Sachs, Joe (1998). *Aristotle's Physics: A Guided Study*. New Brunswick NJ: Rutgers University Press. Originally published in 1995.

Watson, E. L. Grant (1964). *The Mystery of Physical Life.* Hudson NY: Lindisfarne Press. Originally published in 1943.

CHAPTER 2

The Organism's Story

ORGANISMS ARE PURPOSIVE ("teleological") beings. Nothing could be more obvious. The fact of the matter is so indisputable that even those who don't believe it really do believe it. Philosopher of biology Robert Arp speaks for biology as a whole when he writes,

> Thinkers cannot seem to get around [evolutionary biologist Robert] Trivers' claim that "even the humblest creature, say, a virus, appears organized to *do* something; it acts as if it is trying to achieve some purpose," or [political philosopher Larry] Arnhart's observation that … "Reproduction, growth, feeding, healing, courtship, parental care for the young — these and many other activities of organisms are goal-directed."[1]

And yet, despite his acknowledgment that we "cannot get around" this truth, Arp again speaks for almost the entire discipline of biology when he tries, with some delicacy, to get around it: "with respect to organisms, it is useful to think *as if* these entities have traits and processes that function in goal-directed ways."[2] This *as if* is a long-running cliché in biology, designed to warn us that the organism's purposive behavior is somehow deceptive — not quite what it seems. The goal-directedness is, in the conventional terminology, merely *apparent* or *illusory*. Certainly it must not be seen as having any relation at all to human purposive activity — an odd insistence given how eager so many biologists are to make sure we never forget that the human being is "just another animal."

Others have commented on this strange, blindsighted reluctance to acknowledge fully the purposiveness that is there for all to see. The philosopher of science, Karl Popper, said that "The fear of using teleological terms reminds me of the Victorian fear of speaking about sex."[3] Popper may have had in mind a famous remark by his friend and twentieth-century British evolutionary theorist, J. B. S. Haldane, who once quipped that "Teleology is like a mistress to a biologist; he cannot live without her but he's unwilling to be seen with her in public."[4]

We find this same unwilling yet unshakable conviction of purposiveness at the foundations of evolutionary theory. The theory, we are often told, explains the organism's apparent purposes — it "naturalizes" them (explains them away), as those who claim to speak for nature like to say. But at the same time the theory is itself said to be grounded solidly in the fact that organisms, unlike rocks, thunderstorms, and solar systems, *struggle to survive and reproduce*. If they did not spend their entire lives striving toward an end, or *telos*, in this way, natural selection of the fittest organisms (those best qualified to survive and reproduce) could not occur. So it is not at all clear how selection is supposed to explain the origin of such end-directed behavior. (See the discussion of natural selection and teleology in Chapter 18, "Teleology and Evolution.")

A double and conflicted stance toward end-directedness — believing and not believing, acknowledging and explaining away — constitutes, you could almost say, the warp and woof of biology itself. Look for "purpose" in the index of any biological textbook, and you will almost certainly be disappointed. That term, along with others such as "meaning" and "value," is effectively banned. There is something like an allergy or taboo against it.

Yet, in striking self-contradiction, those textbooks are themselves structured according to the purposive activities and meaningful tasks of organisms. Biologists are always working to narrate goal-directed achievements. How is DNA replicated? How do cells divide? How does metabolism supply energy for living activity? How are circadian rhythms established and maintained? How do animals arrive at the evolutionary strategies or games or arms races through which they try to eat and avoid being eaten?

Such questions are endless, and their defining role is reflected on every page of every textbook on development, physiology, behavior, or evolution. A research question is *biological*, as opposed to physical or chemical, only when it is posed in one way or another by the organism's purposive, future-oriented activity.

The puzzle is that, having been aroused by such purposive questions, biologists look for answers rooted in the assumption that organisms have no purposes. The reigning conviction is that explanations of physical and chemical *means* effectively remove any need to deal scientifically with the *ends* that alone could have prompted our search for means in the first place.

My larger argument in this book will be that the biologist's conscious commitment to purely physical and chemical descriptions — which is to say, her conscious refusal of much that she actually knows — has devastating effects upon many fields of biological understanding, and particularly evolutionary theory. It hardly needs emphasizing that *if* organisms really are purposive beings — if the fact of purposive activity is not an illusion — then a biological science so repulsed by the idea of purpose that its practitioners must avert their eyes at the very mention of it … well, it appears that these practitioners must feel threatened at a place they consider foundational. And with some justification, for to admit what they actually know about organisms would be to turn upside down and inside out much of the science to which they have committed their lives.

"Purpose" — an idea that needs careful qualification in different biological contexts — gives us but one of several intimately related avenues of approach to what is distinctive about the life of organisms. In the remainder of this chapter I will briefly sketch a few of these avenues.

Organisms are agents capable of expressing their own meanings

Organisms are agents; they *do* things. The difference between a motionless rock, on one hand, and a motionless cat on the other is that the cat is not merely motionless; it is *resting*, or perhaps preparing to pounce. When it ceases *doing* things, it is no longer alive. Whereas a rock may *be moved* according to universal laws, the cat is *self-moved*; the needs and interests according to which it moves are not the universal laws of its surroundings. In our routine experience we take self-motivated activity to be definitive of living things. If an object moves unexpectedly — without an evident external cause — we immediately begin testing the assumption that it is living.

Figure 2.1. A wildebeest, otherwise known as a gnu.[5]

When an animal responds to a physical stimulus, its response is not in any strict way physically enforced, or directly caused, by the stimulus. Rather, the animal "reads" the meaning of the situation in light of its own concerns, including its needs and interests, and then alters that meaning by responding to it. If the animal *is* physically moved by a stimulus, as when a rolling stone bumps into a leg, we don't consider the movement to be the organism's own act. It is not a *response*, but merely a physically caused *result*.

Figure 2.2. A charging lioness in the Serengeti.[6]

As a useful picture of this, we need only consider how the negligible force producing an image on the retina — say, the image of a charging lion — can set the entire mass of a quarter-ton wildebeest into thundering motion.

The impelling force comes from within, so that the movement seems to originate within the animal itself in a way that we do not see in inanimate objects.

The wildebeest is not forcibly moved by a physical impact, but rather *perceives* something. Further, its perception is at the same time an *interpretation* of its surroundings from its own point of view and in light of its own world of meaning. The "lawfulness" at issue here, such as it is, is far from being universal. It differs radically from one living being to another, so that the retinal image of a charging lion means a very different thing to the wildebeest from what it means to another lion or to a vulture circling overhead. And it produces an altogether different response in these cases.

All this may seem trivially obvious — and so it is. We make sense of biological activity in terms of meanings radically different from the meanings we bring to inanimate events. The truth comes out in a thousand ways, and above all in the choice of language. The words employed for description of animate activity differ dramatically from those applied to inanimate activity.

Think, for example, of a living dog, then of its decomposing corpse. At the moment of death, all the living processes normally studied by the biologist rapidly disintegrate. The corpse remains subject to the same laws of inanimate nature as the live dog. But now, with the cessation of life, we see those laws strictly in their own terms, without reference to life. The dramatic change in our descriptive language as we move between the living and the dead speaks more loudly than any philosophical convictions we may have about life and death.

No biologist who had been studying the *behavior* of the living dog will concern herself with the corpse's "behavior." Nor will she refer to certain physical changes in the corpse as *reflexes*, just as she will never mention the corpse's *responses* to *stimuli*, or the *functions* of its organs, or the processes of *development* being undergone by the decomposing tissues.

Virtually the same collection of molecules exists in the canine cells during the moments immediately before and after death. But after the fateful transition no one will any longer think of genes as being *regulated*, nor will anyone refer to *normal* or *proper* chromosome functioning. No molecules will be said to *guide* other molecules to specific *targets*, and no molecules will be carrying *signals*, which is just as well because there will be no structures *recognizing* signals. *Code, information*, and *communication*, in their biological sense, will have disappeared from the scientist's vocabulary.

The corpse will not produce *errors* in chromosome replication or in any other processes, and neither will it *attempt* error *correction* or the *repair* of damaged parts. More generally, the ideas of *injury* and *healing* will be absent. No structures will inherit features from parent structures in the way that daughter cells *inherit* traits or tendencies from their parent cells, and no one will cite the *plasticity* or *context-dependence* of the corpse's *adaptation* to its environment.

It seems clear that, if they were denied the language of life, biologists (unlike physicists and chemists) would have nothing to say. The science of biology would not exist. And yet the living language is strikingly incompatible with the language of the inanimate. This problem — how can we have a biological science, given current scientific assumptions? — is far from front and center in the training and conversation of biologists. The silence — from all appearances, not even an embarrassed silence — on such a decisive issue on which the very existence of biology as a legitimate science hinges is itself an embarrassment. When investigators do their best to ignore the gap between the living and the dead layers of meaning — for example, when they present their findings *as if* there were nothing to elucidate except physical and chemical interactions — then they are contradicting their own biological descriptions.

It is not that the strictly physical and chemical approaches are inadequate in their own, limited terms. *In such terms* we can be sure that everything being described makes perfect sense, and that the physical picture reveals no mysterious gaps. It's just that, within the arbitrarily imposed limits of physical and chemical description, we will see no living activity. "Physically lawful" describes only those aspects of the animal's body that continue uninterrupted, according to exactly the same laws, when it dies. If we restricted our understanding to *this* characterization, death would not even be a recognizable event.

Of course, in a split-personality, blindsighted sort of way every biologist does recognize death, because she recognizes the distinctive sorts of meaning, including the perceptions, purposes, intentions, and responses, that the once-living dog is no longer expressing. It's just that she typically refuses to let the expressive aspects of the creature's life become uncomfortably explicit, or to influence fundamental theory. Or, when they do affect theory, it must be the organism's physical activity, not its interior life as a perceptive and intentional actor, that enters into scientific consideration. Like the behaviorists of old, we are forbidden to accept the inner, immaterial, and immediately given *reality* of perceptions and intentions, as opposed to various associated physical manifestations.

The end is more constant than the means of attaining it

William McDougall, who lived from 1871 to 1938, was a highly respected (if also rather controversial) British psychologist who, after teaching at Oxford, spent the latter part of his career in the United States. He authored widely used textbooks of psychology and, for several years, occupied William James' chair at Harvard. Then he moved to Duke University where, with J. B. Rhine, he founded the Parapsychology Laboratory. Our present interest, however, is in a 1929 work, where McDougall usefully summarized certain typical features of purposive activity (McDougall 1929, pp. 50–51). He was writing about human behavior, but we can recognize something like these features in all purposive behavior, conscious or otherwise:

- Goal-directed activity tends to be *persistent* and may be repeatedly renewed even after being effectively blocked for a time. If you tie up your hungry dog at some distance from its food bowl, it may cease straining at the leash. But as soon as you grant it freedom, it will again head for the bowl.
- Goal-directed activity is very often *adaptable* to one degree or another. If one strategy fails, the organism may vary it or switch to a different strategy. As many dog owners have discovered after forgetting to give Fido his food, their beloved pet may contrive to enjoy the freshly roasted chicken on the kitchen counter.
- And, as soon as the goal is reached, that particular goal-directed activity *ceases*. Having had its fill, your dog may want to play or else to sleep. But it will not continue its quest for food.

We do not find the same combination of features in the inanimate world. Yet anyone who interacts with animals takes them for granted. Moreover, analogous features are evident even in physiological activity, all the way down to the molecular biology of the cell. In its development "the embryo seems to be resolved to acquire a certain form and structure, and to be capable of overcoming very great obstacles placed in its path." When encountering such an obstacle to its development, the organism "adjusts itself to the changed conditions, and, in virtue of some obscure directive power, sets itself once more upon the road to its goal; which under the altered conditions it achieves only by means of steps that are different, sometimes extremely different, from the normal" (McDougall 1911, pp. 242–43).

When a cell is preparing to divide, it passes through what are known as internal "checkpoints," where the cell responds to the presence or absence of conditions necessary for a successful division. If something is awry, the cell may nevertheless *persist* in the aim of dividing by taking any corrective (*adaptive*) action that happens to be within its power. It then proceeds with its division, and *ceases* the entire, highly coordinated and complex activity once the process is complete. Or, when division is "judged" inadvisable — say, because chromosomes have been irreversibly damaged — the cell may "decide" to self-destruct and offer its resources up for the good of the rest of the organism of which it is a part.

No one will bristle upon hearing that "this cell is preparing to divide." But we would certainly bristle if we heard that "Mars is preparing to make another journey around the sun," or "the nebula has ceased its effort after forming the solar system." A planet moves according to universal laws acting in an unchanging manner. There is no point in its journey when an act is *initiated* or *concluded*, but only the playing out of the immediately preceding forces. There is in this sense nothing new to explain. Biological explanation, by contrast, always involves something new, an element of initiative, a response to circumstances not fully necessitated by the preceding play of physical and chemical processes.

Here's another illustration, drawn from the great English physiologist, Sir Charles Scott Sherrington, writing in 1922. He is talking about what happens when, in some animals, a motor nerve is severed and the portion running from the point of severance to the muscle dies. The living end of the nerve immediately embarks upon a meaningful and unfathomably complex journey:

> The fibre, so to say, tries to grow out to reach to its old far-distant muscle. There are difficulties in its way. A multitude of non-nervous repair cells growing in the wound spin scar tissue across the new fibre's path. Between these alien cells the new nerve-fibre threads a tortuous way, avoiding and never joining any of them. This obstruction it may take many days to traverse. Then it reaches a region where the sheath-cells of the old dead nerve-fibres lie altered beyond ordinary recognition. But the growing fibre recognises them. Tunnelling through endless chains of them, it arrives finally, after weeks or months, at the wasted muscle-fibres which seem to have been its goal, for it connects with them at once. It pierces their covering membranes and re-forms with their substance junctions of characteristic pattern resembling the original that had died weeks or months before. Then its growth ceases, abruptly, as it began, and the wasted muscle recovers and the lost function is restored (quoted in Russell 1945, p. 111).

Here we see again goal-directed persistence over a long period, adaptability in the face of obstacles, and cessation of this particular activity when its end is achieved.

Notice also Sherrington's careful caveat ("so to say") whereby he qualifies the easily anthropomorphized phrase, "*tries* to grow." The care and the qualification are fully justified. But the fact is that such phrasing is pervasive and seemingly unavoidable whenever the researcher would offer informative biological descriptions. This suggests that we owe it to the discipline of biology to explore the nature of our own usage. It pays to know what we are *really* saying, rather than leaving it in a vague and ambiguous cloud of suggestion. Throughout this book we will touch on some of the problems we run into when employing the easily misused language of purposiveness, goals, and intentions.

E. S. Russell, a British marine biologist whose writings during the first half of the twentieth century can sometimes seem more up-to-date regarding the decisive issues of twenty-first century biology than the literature of our own day, summarized the gist of the foregoing discussion with wonderful succinctness: "The end-state is more constant than the method of reaching it" (Russell 1945, p. 110). This suggests that the end-state, understood as somehow implicit in the entire drama leading up to it, plays something like a causal role. It reminds one of the way a complex, well-considered conclusion is implicit in the profound, multivalent play of thought leading up to it, rather than being the mere passive outcome of a deterministic march of "naked" machine logic. (For a fuller treatment of this, see the section on leaf sequences in Chapter 12 ("Is a Qualitative Biology Possible?").

Surely any such causal dimensions involving end-states would have large implications for a science focused on unraveling physical and chemical means while ignoring the ends that express the meaning of the activity.

Every organism is narrating a meaningful life story

The fact of purposive activity; the obvious play of an active agency; the coordination of diverse means toward the realization of interwoven and relatively stable ends; the undeniable evidence that animals perceive a world, interpreting and responding to perceptions according to their own concerns; and the coherence of all this activity in a governing unity that is the unity of a particular way of life — this tells us that every animal is narrating a meaningful life story. This is not something that a rock, say, loosened by ice and tumbling down the steep slope of a mountain ravine, does in anything like the same manner. The pattern of physical events in the organism is raised by its peculiar sort of coherence toward something like a biography whose "logic" unfolds on an entirely different level from the logic of inanimate physical causation. When we tell a living story, the narrative threads convey the meanings of a life — for example, motives, needs, ways of experiencing the world, and intentions — and these are never a matter of mere physical cause and consequence.

So when I speak of the organism's wise and knowing agency, or its purposive striving, I refer, among other things, to *its capacity to weave, out of the resources of its own life, the kind of biological narrative we routinely observe, with its orchestration of physical events in the service of the organism's own meanings.*

We normally feel every birth as a new beginning, full of hope and expectation — a beginning of a sort we do not experience in the genesis of a raindrop or dust devil. Even the first shoot of a bean or squash seed, pushing upward through the soil surface, is the prelude to a narrative promising many vicissitudes — engagements with insects and diseases, complex communal relations with other plants, and confrontations with nurturing or threatening forces of nature. And a death is always the end of one particular story.

E. S. Russell, commenting on descriptions such as that of the chaffinch in Box 2.1, noted the narrative connectedness of the events: "Behaviour is often part

BOX 2.1

The Nesting Cycle of the Chaffinch

From a 1927 description by the British naturalist and ornithologist, Edward Max Nicholson:[7]

"The male must leave the flock, if he has belonged to one, and establish himself in a territory which may at the time be incapable of sustaining him alone, but must later in the season supply a satisfactory food-supply for himself, his mate and family, and for as many birds of other species as overlap his sphere of influence. He must then sing loudly and incessantly for several months, since, however soon he secures a mate, trespassers must be warned off the territory, or, if they ignore his warning, driven out.

"His mate must help with the defence of the territory when she is needed; pairing must be accomplished; a suitable site must be found for the nest; materials must be collected and put together securely enough to hold five bulky young birds; eggs must be laid in the nest and continuously brooded for a fortnight till they hatch, often in very adverse weather; the young are at first so delicate that they have to be brooded and encouraged to sleep a great part of the time, yet they must have their own weight of food in a day, and in proportion as the need of brooding them decreases, their appetites grow, until in the end the parents are feeding four or five helpless birds equal to themselves in size and appetite but incapable of digesting nearly such a wide diet.

"Enemies must be watched for and the nest defended and kept clean. When the young scatter, often before they can fly properly, they need even greater vigilance, but within a few days of the fledging of the first brood a second nest will (in many cases) be ready and the process in full swing over again. All this has to be done in face of great practical difficulties by two creatures, with little strength and not much intelligence, both of whom may have been hatched only the season before."

of a long-range cycle of events, in which one action prepares for and leads on to the next until the end term is reached. Each stage in the chain or cycle is unintelligible to us except in its relation to what has gone before, and, more particularly, to what is yet to come. Such cycles have a temporal unity …" (Russell 1938, pp. 7–8). Present significances are interwoven with and inseparable from the tapestry of past events and their meanings. And future developments, along with whatever new and unpredictable elements they bring, are a continued, improvisational elaboration of the same tapestry of meaning.

In other words, the "end" being approached in many details of an organism's story is not some particular, discrete accomplishment, distinct from the means of getting there, but rather the wholeness and perfection of the entire narrative movement from "here" to "there." Assessing this end is much the same as if we were assessing the meaning of a novel: knowing the ending in isolation would have little significance compared to knowing the larger story of which, so we often feel, it is a necessary and proper part.

We are organisms, but not all organisms are human

Note well, then, that when speaking of the organism's story, we need make no reference to the consciously directed performances of human beings, even though our performances certainly exhibit a narrative character in the sense meant here. When I refer to living activity as "end-directed," I am not suggesting the formulation of a conscious goal that is "aimed at." I mean, rather, something like this:

> The organism's life is a continual *playing forward of meanings within meaningful contexts*. There is a certain directedness to any such play of meaning (as when birds build a nest), but it need not be the directedness of human plan fulfillment.

The directedness of a temporally unfolding play of meaning implies no narrow goal and no conscious planning. But every such play of meaning does have a certain directedness to it. Think of the greatest poems or novels, where nothing is calculated *in order to* reach the conclusion, but the movement is nevertheless from the beginning to the end, not the reverse. This movement expresses the progressive deepening of a meaningful and coherent unity — more like a dance than pursuit of a fixed and predefined goal. And the dance looks ever more improvisational as organisms ascend in the scale of complexity.

I offer here no specific hypotheses to explain the existence of intentional agency and story narration. I only note that the *fact* of the narrative is immediately demonstrable in every organism. There may be huge differences in the nature of the stories that can be told by different kinds of organism, but from the molecular level on up there are always elements of story that we do not find in inanimate things. The narrative of meaningful activity undertaken and accomplished is there to be seen, and is characterized as such, if only inadvertently, in every paragraph of biological description.

Moreover, our recognition of intelligent and intentional activity does not require us to understand its source. Looking at the pages of a book, we have no difficulty distinguishing written marks from deposits of lint and dust, even if we know nothing about the origin of the marks. We can declare a functioning machine to be engaged in a purposive operation, whether or not we have any clue about the engineers who built a mechanistic reflection of their own purposes into it. And if we find live, intelligent performances by organisms, we don't have to know how, or from where, the intelligence gets its foothold before we accept the testimony of our eyes and understanding.

Neither should we expect the stories to be predictable — no more than we expect the ending of a half-read novel to be predictable. We can, however, expect the ending to *make sense*, and even to throw light on everything that went before. The story will *hold together* in a way that unstoried physical events (which have their own sort of meaning) do not.

The storytelling is the being of the organism

If the organism's life is an unfolding story, then we might well take the essence of that life to be the *activity of storytelling* itself — the sort of activity by which the distinctive character of this or that species is sketched and acted out. Organisms, as philosopher Hans Jonas has written, "are individuals whose being is their own doing ... they are committed to keeping up this being by ever renewed acts of it." Their identity is "not the inert one of a permanent substratum, but the self-created one of continuous performance" (Jonas 1968, p. 233).

Or, again, we have the rather different formulation by Paul Weiss, a profound observer of cellular life:

> Life is a dynamic *process*. Logically, the elements of a process can be only elementary *processes*, and not elementary *particles* or any other static units (Weiss 1962, p. 3).

An organism is not, most essentially, its body. After all, its body at one time is never materially identical to its body at a different time. The body reflects, rather, a unique power of activity. It is first of all a result of this activity, while also developing into a further vehicle for it. Organisms, in other words, are *doings* rather than *beings*.

So it is not that an organism's material being determines its doings (as is broadly assumed throughout the biological sciences); rather, its doings are what constitute it as a material being. This means that it is never wholly present to our observation in any outward or material sense. The organism's essential power to act cannot itself be a material product of its activity.

The preeminence of activity in relation to physical substance and structure would, if taken seriously, give us an altogether new science of life. For example, it might have saved us from an entire century of badly misdirected thinking about the causal primacy of DNA and genes. It might also have spared biologists the crude materialism that many physicists long ago gained the freedom to question.

But this is to get ahead of the story. For now, it is enough to mention two questions implicit in the foregoing, while deferring further comment:

Regarding our theory of evolution: If, in reality, every organism's existence is a live, moment-by-moment, improvisational storytelling — a creative and adaptive, irreversible narrative that is always progressing coherently and contextually from challenge to response and adaptation, from initiative to outcome, from nascence to renascence, from immaturity through maturity to regeneration — then an evolutionary theory rooted in notions of random variation and mindlessness is a theory hanging upon a great question mark. "The answer to the question of what status teleology ['end-directedness'] should have in biology" — so the influential biologist and philosopher Francisco Varela came to see at the end of his life — determines "the character of our whole theory of animate nature" (Weber and Varela 2002).

And then there is the question whether the future of individual species, the future of particular ecological settings, the future of life's diversity on earth, and the future of earth itself, all depend on our willingness and ability to attend to the life stories of the beings among whom we live — depend, finally, on our capacity for the awe and reverence that these stories so naturally evoke.

WHERE ARE WE NOW?

An organism's story gives form to its material existence (not the other way around).

We have seen that animals are irreducibly purposive in both their behavior and their physiology, and that the purposive ends they seek are more constant than the means for seeking them. We have also seen that, as living beings rather than merely physical objects, animals are motivated and moved by perceived meanings rather than by impelling physical forces.

Its interpretive activity — activity through which meanings are apprehended and expressed — is what enables an animal to weave the story of its life, as opposed to being moved by a set of physical causes and consequences. A story just *is* an evolving tapestry of interwoven meanings, through which an animal gains its narrative unity and coherence in time. Without such unity, there is no story, and without a story there is no animal.

There are many questions raised by the discussion in this chapter, including these:

- Given that we share common roots with all life, what is the relation between the purposes of organisms in general and our own human purposes?
- Does saying that an organism makes a story of its life imply a form of consciousness? And, if so, what are the different forms consciousness takes in living beings?
- How do our own human purposes relate to the purposiveness in our bodies and cells, through which many of our intentions are carried out?
- In what sense must we consider the world itself as "living," given that it has brought forth and nurtured all living things on earth? Can the world from which we arose somehow be poorer in features and potentials than the creatures it brings forth?

And much more. In many chapters of this book we will tangentially brush up against such questions. And at times — as in Chapter 24 ("Is the Inanimate World an Interior Reality?") — we will face some of them head on. You may also find a few helpful preliminary notes in Chapter 1 ("The Keys to This Book"). But I hope every reader will be left with many open questions, as befits any *living* science.

NOTES

1. Arp 2007. See also Trivers 1985, p. 5. In this same connection, the following comment by Georg Toepfer of Humboldt University in Berlin is significant:

    > Most biological objects do not even exist as definite entities apart from the teleological perspective. This is because biological systems are not given as definite amounts of matter or structures with a certain form. They instead persist as functionally integrated entities while their matter and form changes. The period of existence of an organism is not determined by the conservation of its matter or form, but by the preservation of the cycle of its activities (Toepfer 2012).

    Then there is this from the American philosopher, Susanne Langer:

    > The image of life as motivated activity reflects an aspect of animate nature that has baffled philosophers ever since physics rose to its supreme place among the sciences, because inanimate nature — by far the greatest concern of physics — has no such aspect: the telic phenomenon, the functional relation of needs and satisfactions, ends and their attainment, effort and success or failure. There are no failures among the stars. Rocks have no interests. The oceans roar for nothing. But earthworms eat that they may live, and draw themselves into the earth to escape robins, and seek other worms to mate and procreate. They need not know why they eat, contract, or mate. Their acts are telic without being purposive (Langer 1967, p. 220).

    But, of course, "telic" just means "purposive." What I think she is getting at is the truth that purposive or end-directed activity need not be *consciously* purposive — that is, need not be *planned* in the human sense.

2. Emphasis in original. Hereafter and in all succeeding chapters this can be assumed unless explicitly stated otherwise.
3. Quoted in Niemann 2014, p. 30.
4. Quoted in Mayr 1974. Reports of this remark by Haldane come with many variations. The eminent French biologist, François Jacob, wrote, without attribution: "For a long time, the biologist treated teleology as he would a woman he could not do without, but did not care to be seen with in public" (Jacob 1973, pp. 8–9).
5. Figure 2.1 credit: Chris Eason (https://commons.wikimedia.org/wiki/File:Connochaetes_taurinus_-Kruger_National_Park-8.jpg), CC BY-SA 2.
6. Figure 2.2 credit: Schuyler Shepherd (https://commons.wikimedia.org/wiki/File:Serengeti_Lion_Running_saturated.jpg), CC BY-SA 2.5.
7. Quoted in Russell 1938, pp. 7–8. I have added paragraph breaks. The book by Nicholson is entitled *How Birds Live: A Brief Account of Bird-Life in the Light of Modern Observation*, and was published in London by Williams and Norgate, Ltd., in 1927.

    The engraving of a chaffinch pair and their nest is from a book published in 1866 and titled, *Homes without Hands: Being a Description of the Habitations of Animals, Classed According to Their Principle of Construction*, by John George Wood and others. For more information, see The Internet Archive Book Images.

SOURCES

Arp, Robert (2007). Evolution and Two Popular Proposals for the Definition of Function, *Journal for General Philosophy of Science* vol. 38, pp. 19–30. doi:10.1007/s10838-006-9008-3

Jacob, François (1973). *The Logic of Life: A History of Heredity*, translated by Betty E. Spillmann. New York: Random House. Originally published in French in 1970.

Jonas, Hans (1968). Biological Foundations of Individuality, *International Philosophical Quarterly* vol. 8, no. 2, pp. 231–51.

Langer, Susanne K. (1967). *Mind: An Essay on Human Feeling* vol. 1. Baltimore, MD: Johns Hopkins University Press.

Mayr, Ernst (1974). Teleological and Teleonomic, a New Analysis, in *Methodological and Historical Essays in the Natural and Social Sciences* (vol. 14 of Boston Studies in the Philosophy of Science), edited by Robert S. Cohen and Marx W. Wartofsky. Dordrecht, Holland: Springer Netherlands, pp. 91–117. doi:10.1007/978-94-010-2128-9_6

McDougall, William (1911). *Body and Mind: A History and a Defense of Animism*. New York: Macmillan.

________ (1929). *Modern Materialism and Emergent Evolution*. New York: D. Van Nostrand.

Niemann, Hans-Joachim (2014). *Karl Popper and the Two New Secrets of Life*. Tübingen, Germany: Mohr Siebeck.

Russell, E. S. (1938). *The Behaviour of Animals*, second edition. London: Edward Arnold.

________ (1945). *The Directiveness of Organic Activities*. Cambridge UK: Cambridge University Press.

Toepfer, Georg (2012). Teleology and Its Constitutive Role for Biology as the Science of Organized Systems in Nature, *Studies in History and Philosophy of Biological and Biomedical Sciences* vol. 43, pp. 113–19. doi:10.1016/j.shpsc.2011.05.010

Trivers, Robert (1985). *Social Evolution*. Menlo Park CA: Benjamin-Cummings.

Weber, Andreas and Francisco J. Varela (2002). Life after Kant: Natural Purposes and the Autopoietic Foundations of Biological Individuality, *Phenomenology and the Cognitive Sciences* vol. 1, pp. 97–125. Available at https://link.springer.com/article/10.1023/a:1020368120174

Weiss, Paul (1962). From Cell to Molecule, in *The Molecular Control of Cellular Activity*, edited by John M. Allen, pp. 1–72. The University of Michigan Institute of Science and Technology Series. New York: McGraw-Hill.

CHAPTER 3

What Brings Our Genome Alive?

THROUGHOUT MOST OF the twentieth century, genes were viewed as the "agents" responsible for an organism's development, activity, and evolution. Their agency was said to lie in their ability to "regulate," "organize," "coordinate," and "control" physiological processes, and their changes ("mutations") were the material of evolution. DNA, the bearer of these genes, became the "Book of Life" — the essential maker of organisms and driver of evolution. And this view remains stubbornly entrenched today, despite many changes in our understanding. In 2019 a leading behavioral geneticist could still write a book titled, *Blueprint: How DNA Makes Us Who We Are*.

Nevertheless, the idea that genes are the decisive "first causes" of life — and, more generally, that molecules at the "bottom" ultimately explain everything that happens at larger scales — has come in for a great deal of criticism in recent years. This criticism, as we will see, is fully justified. But the issues can be subtle, as is suggested by an apparent paradox. Philosopher of biology Lenny Moss, who wrote the valuable book, *What Genes Can't Do*, has remarked:

> Where molecular biology once taught us that life is more about the interplay of molecules than we might have previously imagined, molecular biology is now beginning to reveal the extent to which macromolecules [such as DNA], with their surprisingly flexible and adaptive complex behavior, turn out to be more *life-like* than we had previously imagined (Moss 2012).

In a similar vein, I myself have written:

> Having plunged headlong toward the micro and molecular in their drive to reduce the living to the inanimate, biologists now find unapologetic life staring back at them from every chromatogram, every electron micrograph, every gene expression profile. Things do not become simpler, less organic, less animate. The explanatory task at the bottom is essentially the same as what we faced higher up (Talbott 2010).

But if this is really true, what are we to make of Harvard geneticist Richard Lewontin's declaration, itself hardly disputable, that

> DNA is a dead molecule, among the most nonreactive, chemically inert molecules in the living world. That is why it can be recovered in good enough shape to determine its sequence from mummies, from mastodons frozen tens of thousands of years ago, and even, under the right circumstances, from twenty-million-year-old fossil plants.

Many astute observers have echoed Lewontin's remarks, and I have never seen anyone question them, including those who remain enamored of the "Book of Life." So which is it? When we peer at DNA, do we see a dead molecule or a living dynamic? Lewontin himself, in that same passage, pointed toward the answer (we will try to forgive his ill-fitting use of the word "machinery"):

> DNA has no power to reproduce itself. Rather it is produced out of elementary materials by a complex cellular machinery of proteins. While it is often said that DNA produces proteins, in fact proteins (enzymes) produce DNA ... Not only is DNA incapable of making copies of itself, aided or unaided, but it is incapable of "making" anything else (Lewontin 1992).

In other words, the proper functioning of DNA is an achievement of its entire cellular context. If we conceive the molecule in the usual way as a bit of mindless, inherently inert matter, then, according to our own conceptions, we see only dead stuff. But if we conceive the molecule as a system of forces and energies capable of *participating in, and being caught up in*, the creative life of the whole cell and organism, then we can hardly help recognizing — and perhaps even reverencing — the living performance unfolding before our eyes.

Saying this is one thing; making it both meaningful and profound is quite another — and that is one task of the present book.

The genome as you have probably not heard about it

If you arranged the DNA in a human cell linearly, it would extend for nearly two meters. How do you pack all that DNA into a cell nucleus just five or ten millionths of a meter in diameter? According to the usual comparison it's as if you had to cram twenty-four miles (thirty-nine kilometers) of extremely thin thread into a tennis ball. Moreover, this thread is divided into forty-six pieces (individual chromosomes) averaging, in our tennis-ball analogy, over half a mile long. Can it be at all possible not only to fit those chromosomes in the nucleus, but also to keep them from becoming hopelessly entangled?

Obviously it must be possible, however difficult to conceive. The first thing to realize is that chromosomes do not consist of naked DNA. Their actual substance, an intricately woven and ever-changing structure of DNA, RNA, protein, and other molecules, is referred to as *chromatin*. (See Box 3.1 for some basic terminology.) Histone proteins, several of which can bind together in the form of a complex *histone core particle*, are the single most prominent, non-DNA constituents of this chromatin. Every cell contains numerous such core particles — there are some 30 million in a typical human cell — and the DNA double helix, after wrapping a couple of times around one of them, typically extends for a short stretch and then wraps around another one. The core particle with its DNA wrapping is referred to as a *nucleosome* (about which you can read much more

BOX 3.1

Some Standard Terminology

The usual formula has it that DNA makes RNA and RNA makes protein. The DNA double helix forms a kind of spiraling ladder, with pairs of nucleotide bases constituting the rungs of the ladder: a nucleotide base attached to one siderail of the ladder bonds with a base attached to the other siderail. These two bases, commonly referred to as "base pairs" ("letters" of the DNA "text"), are normally complementary, so that, of the four different bases (abbreviated as A, T, C, and G), an A pairs only with a T (and vice versa), just as C and G are paired. Each siderail, with all its attached nucleotide bases, is considered a single *strand* of the double helix. Because the chemical subunits making up the siderails are asymmetrical and oriented oppositely on the two strands, the strands can be said to "point" in opposite directions.

The enzyme that *transcribes* DNA into RNA (thereby *expressing* a gene) must move along the length of the gene in the proper direction, separating the two strands and using one of them, with its sequence of nucleotide bases, as a template for synthesizing a single-stranded RNA *transcript*—a transcript that complements the template DNA strand in much the same way that one DNA strand complements the other. It is by virtue of this complementation that the "code" for a protein is said to be passed from DNA to RNA. Once synthesized, the RNA may pass through the nuclear envelope to the cell's cytoplasm, where it may be *translated* into protein.

It all makes for a neat, if (as told here) greatly simplified, story. For a fuller exploration of technical terms, see the glossary at https://bwo.life/mqual/glossary.htm.

in Chapter 14, "How Our Genes Come to Expression"), and between 75 and 90 percent of our DNA is said to be wrapped up in nucleosomes. This is one way the cell packs its DNA into a surprisingly small volume.

But how is all this material organized so as to serve the infinitely complex requirements of a flatworm, bumblebee, shark, or human? Biologists have spent a good number of years trying to visualize the organization of chromosomes in the cell nucleus, and, unsurprisingly, the picture tends to differ depending on the scale at which you look.[1] Most broadly, the genome appears to be partitioned into two compartments, the "A" compartment, with more "open" (less densely packed) chromatin and more active genes, and the "B" compartment, with more "closed" chromatin and less active genes. Some researchers have pointed to the existence of several subcompartments distinguished by the presence of distinctive features ("marks") on the proteins associated with the DNA.

At a somewhat smaller, megabase scale, there are so-called "topologically associated domains," within which the interactions among loci are more frequent than across such domains. Also at this scale, it is now thought that some chromosome regions form "fractal globules" that are more or less free of knots (Figure 3.1).

And, at a still smaller scale (roughly 200,000 bases), there are loop domains generally associated with active genes (Figure 3.2). For a smaller scale yet — one that is intensely relevant to gene regulation and expression — see the discussion of nucleosomes in Chapter 14.

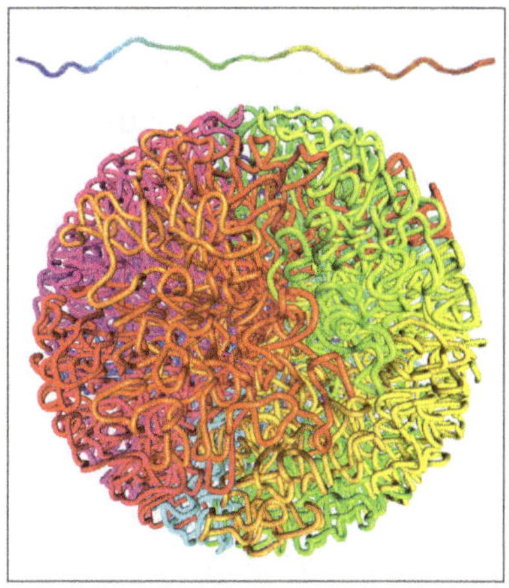

Figure 3.1. A schematic representation of a proposal for how parts of a human chromosome can be organized into an unknotted "fractal globule" in the cell nucleus. The linear chromosome segment at the top of the figure shows, in a miniaturized way, what the unfolded globule might look like.[2]

How all this fits together is, of course, less than fully clear. And things only become more complex when you consider that loci on separate chromosomes often come into complex and intimate relation with each other, in part because of the need to coordinate the expression of genes on different chromosomes. And there are also the chromatin proteins, the modifications of those proteins, and the vast number of associated molecules in the nucleus that influence how genes will be expressed.

The image shown in Figure 3.1 is a geometric idealization. It is designed to show certain principles of the folding of chromosomes at the megabase scale, and is not meant to suggest that any part of any chromosome is organized into a neat sphere.

In reality, the cell nucleus presents us with an almost infinitely complex and dynamic configuration of functionally related regions all the way down to the smallest scale. Different parts of the same chromosome might lie in the "A" and "B" compartments and can move between them. Similarly, loops can form and disappear. And nucleosomes, we will see in Chapter 14, seem almost continually in movement, which is central to gene expression.

There is also continual engagement between the genome and other contents and activities of the nucleus. For example, substantial portions of the "B" compartment reside near, and interact with, the outer envelope of the nucleus, whereas much of the "A" compartment lies more in the interior. During the processes of DNA replication and cell division (mitosis), the entire arrangement, for all its seemingly convoluted complexity, radically transforms into a series of different configurations. (See, for example, Figure 3.3.)

The picture is always dynamic. But it's not so much that chromosomes *move* as that they are *brought into movement*. Particular genes — which is to say, particular parts of chromosomes — can be shifted from one place to another, and the associations thereby formed with other chromosomal regions, whether on the same or different chromosomes, can be decisive for the regulation of gene expression. We can easily wonder how the overall choreography of the cell nucleus and whole cell can be perfectly "calculated" for the management of the 20,000 genes and millions of significant loci in the genome. And the intricately dynamic relationships between different chromosomes give us a glimpse of how misleading an image like that of Figure 3.1 can be, with its geometrically compact, isolated, and static character.

In Figure 3.2 the paired red marks at the point where a loop converges on itself indicate the presence of two copies of a particular protein, one of a number of molecules that play a role in loop formation. (How do they "know" where to place themselves?) Of the two widely separated loci thus brought together, one may be near a gene while the other is near DNA regulatory sequences necessary for that gene to be expressed. Their coming together (or not) is therefore part of how particular genes come to be expressed (or not). And, likewise, the reconfiguration of such loops may be critical for the altered expression of genes as the cellular and organismal context changes.

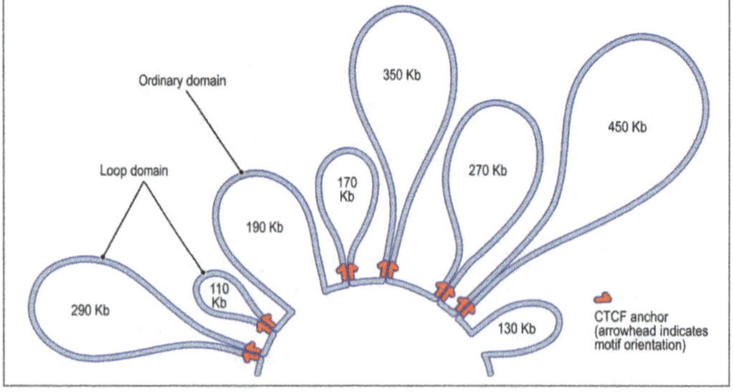

Figure 3.2. A schematic representation of DNA loops. "Kb" stands for "kilobases," or thousands of "letters" of the "genetic code" — in this case, the number of letters strung out along the length of a chromosome loop. For example, the loop at far left is 290,000 letters long.[3] (On terminology, see Box 3.1.)

Note that the two loci where the protein binds a particular loop can be separated on the linear chromosome by hundreds of thousands of genetic "letters." (For comparison, while genes vary greatly in size, they average about 30,000 "letters" in length. And human chromosomes range from about 47 to 247 million "letters.")

So we have seen that there are different ways for genes to be brought into "community," all of which becomes extremely complex, as it surely must, given the diverse uses to which radically distinct cell types must put some of their genes. Investigations into the organization of chromosomes for different functions and at different scales can probably be said to be at an early stage, and the picture will doubtless become still more complex as research proceeds. At present there seem to be no absolute rules of interaction, and the question of clear-cut cause and effect always seems to be in doubt (Chapter 9, "A Mess of Causes"). For example, highly expressed genes are strongly associated with chromosome loops, but they do not absolutely have to be.

Most of the foregoing description has been more or less static. We have so far hardly done more than hint at the true dynamism that enlivens our genetic heritage, but we have perhaps already glimpsed that *gesture* in three-dimensional space is crucial. And the general picture of the genome's dynamic spatial organization has seemed to galvanize molecular biologists. John Rinn, director of the Rinn Lab at Harvard, has said of the nuclear space and its chromosomal drama, "It's genomic origami … It's the shape that you fold [the genome] into that matters" (quoted in Zimmer 2015).

According to a paper from another group of researchers, "A loop that turns a gene on in one cell type might disappear in another. A domain may move from subcompartment to subcompartment as its flavor changes. No two cell types [have their chromosomes] folded alike. Folding drives function."[4] Or, as we might put it, gesturing gives expressive shape to the cell's life. Suhas Rao, the paper's lead author and a researcher at Baylor College of Medicine's Center for Genome Architecture, remarked:

> A loop is the fundamental fold in the cell's toolbox. We found that the formation and dissolution of DNA loops inside the nucleus enables different cells to create an almost endless array of distinct three-dimensional folds and, in so doing, accomplish an extraordinary variety of functions (quoted in Physorg 2014).

Every overall configuration of chromosomes in the nucleus (involving many factors we have not yet considered) represents a unique combination of expressed and repressed genes among our total complement of 20,000 or so genes.[5] Moreover, new features of chromosome spatial and dynamic organization continue to be elucidated on a regular basis, and there appears to be no limit to the variety and scale of these features.

Think about all this dynamic form and movement for a while, and you may find yourself asking, along with me: What possible "mechanism" could ensure the *coherence* of all this movement and gesturing in relation to all the requirements of the trillions of cells in your or my body, or the tissues and organs into which those cells are organized, as we go about our endlessly varying activities under endlessly varying conditions?

Of dynamism and mystery in the cell nucleus

The chromosome, remarked Christophe Lavelle of France's Curie Institute, "is a plastic polymorphic dynamic elastic resilient flexible nucleoprotein complex."[6] There are many activities in which it is caught up, revealing significant form and organization. In order to visualize just one of these activities, consider a long, double-stranded rope whose two strands coil around each other, much like the two strands of a DNA molecule. If you twist a segment of this rope in a manner opposite to its natural spiraling, you will find that the strands tend to separate (that is, loosen, or become less tightly wound). And if you continue to twist, then the rope as a whole will begin to coil upon itself (called "negative supercoiling"). Similarly, if you twist in the same direction as the rope's natural twist, you will tighten the winding of the strands, and if you continue twisting, the rope will again coil upon itself ("positive supercoiling").

The DNA double helix can likewise be loosened by twisting, along with formation of coils, and it can also be tightened and coiled. In fact, it happens that both effects result wherever the enzymes transcribing DNA into RNA are at work. And this twisting in one direction or another in turn either encourages or discourages the expression of nearby genes.[7]

In other words, in addition to the chromosome domains discussed above, there are transient domains established by the twisting (torsional) forces that are communicated more or less freely (and not only by transcribing enzymes) along bounded segments of the chromosome. The loci within such a region share a common torsion, and this can attract a common set of regulatory proteins that read the changes as "suggestions" about activating or repressing nearby genes (Lavelle 2009; Kouzine et al. 2008). The torsion also tends to correlate with the level of compaction of the chromatin fiber, which in turn correlates with many other aspects of gene regulation.

Picture the situation concretely. Every bodily activity or condition presents its own requirements for gene expression. Whether you are running or sleeping, starving or feasting, rousing yourself to action or calming down, suffering a flesh wound or recovering from pneumonia — in all cases the body and many of its different cells have specific, almost incomprehensibly complex and changing requirements for differential expression of thousands of genes. And one thing (among countless others) bearing on this differential expression in all its fine detail is the coiling and uncoiling of chromosomes.

With so much concerted movement going on (including the looping we heard about earlier) how does the cell keep all those "twenty four miles of string in the tennis ball" from getting impossibly tangled? We do at least know some of the players addressing the problem. For example, there are complex protein enzymes called *topoisomerases*, which the cell employs to help manage the spatial organization of chromosomes. Demonstrating a spatial insight and dexterity that might amaze those of us who, even with the benefit of full consciousness, have struggled to sort out tangled masses of thread, these enzymes manage to make just the right local cuts to the strands in order to relieve strain, allow necessary movement of individual genes or regions of the chromosome, and prevent a hopeless mass of knots.

Some topoisomerases cut just one strand of the double helix, allow it to wind or unwind around the other strand, and then reconnect the severed ends. This alters the coiling of the DNA. Other topoisomerases can undo knots by cutting both strands, passing a loop of the chromosome through the gap thus created, and then sealing the gap again.

Imagine trying this with miles of string wrapped around millions of minuscule beads compacted into a few cubic inches of space (tennis ball), with the string all the while looping and squirming like a nest of snakes in order to bring all the right loci together so as to achieve the tasks of the moment. (And how are these tasks "known"?) I don't think anyone would claim to have the faintest idea how this is actually managed in a meaningful, overall, contextual sense, although great and fruitful efforts have been made to analyze the local forces and "mechanisms" at play in isolated interactions.

Does the lawfulness of molecular interactions explain global coherence?

We have scarcely begun to look at the dynamic aspects of the cell nucleus. Not only are chromosomes made to fold, loop, coil, and twist rather like a nest of snakes, but they engage in decisive and changing electrical interactions; they relocate from here to there within the nucleus, partly in order to associate with dynamically assembled collections of molecules important for regulating gene expression; and they are influenced by pushes and pulls from the fibers of the extra-nuclear cytoskeleton (Chapter 4, "The Sensitive, Dynamic Cell").

Or again, DNA is said to "breathe" in rhythmical movements as it tightens and relaxes its embrace of the histone core particles mentioned earlier. And again, it breathes in a different way and in a different sort of rhythm as lengths of the two strands of the double helix alternately separate and reunite. And yet again, there are many profoundly significant structural novelties to which DNA lends itself, beyond the conventional form of the double helix. All this and much more is the cell's way of evoking the genetic performance that it needs — a performance that expresses the cell's own life and that of the organism as a whole.[8]

And so, when researchers refer to the "choreography" of the cell nucleus and the "dance" of chromosomes, as they sometimes do, their language is closer to being literal than many have imagined. If the organism is to survive, chromosomal movements must be well-shaped responses to sensitively discerned needs — all in harmony with innumerable dance partners, and all resulting in every gene being expressed or not according to the meaning of the larger drama. We can hardly help asking: If such a qualitative choreography is how the organism lives and performs at the molecular level, what does this mean for the nature of molecular biological explanation — especially when we are acknowledging an organism's qualitative needs, interests, and purposes?

Yes, the use of terms such as "dance" and "choreography" in molecular biology is rather distinctive. Some might call it eccentric. But this particular eccentricity has for some time now been creeping into the conventional technical literature. We have already heard of "genomic origami." And we have also been told: "The statement, 'genomes exist in space and time in the cell nucleus' is a trivial one, but one that has long been ignored in our studies of gene function" — this according to two leaders of the current work: Job Dekker, head of a bioinformatics lab studying the spatial organization of genomes at the University of Massachusetts Medical School, and Tom Misteli, a research director at the National Cancer Institute. Recent investigations, they say, have taught us that "gene expression is not merely controlled by the information contained in the DNA sequence," but also by the "higher-order organization of chromosomes" and "long-range interactions in the context of nuclear architecture" (Dekker and Misteli 2015).

This last remark may startle some readers into the sudden realization that in all the foregoing there has been scant mention of the famed DNA *sequence* — the supposedly precise logical content of the "coded genetic program" that "makes us who we are." Why is that?

It looks very much as if the chromosome, along with everything else in the cell, is itself a manifestation of life, not a logic or mechanism explaining life. This performance cannot be captured with an abstract code. Gene regulation is defined less by static elements of logic than by the quality and force of the cell's gesturing as it brings its genome into movement. The chromosome becomes an expression of a larger context of living activity. As *Nature* columnist Philip Ball has put it, the clean logic of the DNA code, as it has been commonly formulated, "is so elegant that it risks blinding us to the awesome sophistication of the total process" (Ball 2003).

The fixation upon an abstract, neatly identifiable informational sequence has served well the aim of biologists to find precise, unambiguous, logically clean, and satisfyingly deterministic causal explanations. Nevertheless, what's been happening in rapidly intensifying fashion over the past couple of decades, has been a forced retreat from explanations of this sort. To cite a few key words and phrases from the contemporary literature: everything turns out to be mind-numbingly *complex*, which means, in part, that *context* makes all the difference. We are forced to try to understand how *regulatory networks*, intricate *feedback* loops, and the frequent *difficulty of distinguishing causes from effects* bear upon our biological understanding. Ultimately, we seem to be driven toward *systems biology*, an easily degraded term that many seem to prefer over the embarrassment (and richer meaning) of *holistic biology*.

What is not generally realized, however, is that this retreat *from* simplistic "causal mechanisms" suggests a movement *toward* a kind of explanation biologists have not yet come to terms with. It is, after all, one thing to explain, say, how a topoisomerase enzyme "mechanistically" passes one double-stranded section of DNA through another, and quite a different thing to ask how this activity — which *could* be carried out in countless different patterns — is made to harmonize with everything else going on at the molecular level in order to produce an overall, directed, coherent outcome for the cell as a whole. How might we make sense of the vast coordination of trillions of molecular events in the interest of a larger picture that is subject to continual change, as when a cell initiates the transition leading toward cell division (which changes the *meaning* of everything going on)?

The globular and peculiarly organized aggregation of chromosomes we saw in Figure 3.1 is a long way, for example, from the the chromosomal organization during DNA replication, and likewise from the striking configurations we observe with the mitotic spindle during cell mitosis (Figure 3.3). What is a topoisomerase to do when it is in contact with a particular locus of a DNA molecule — a particular locale among the intricately folded, 6.4 billion nucleotide bases ("letters") of a human cell? How does it connect with the larger drama, so as to play its local role

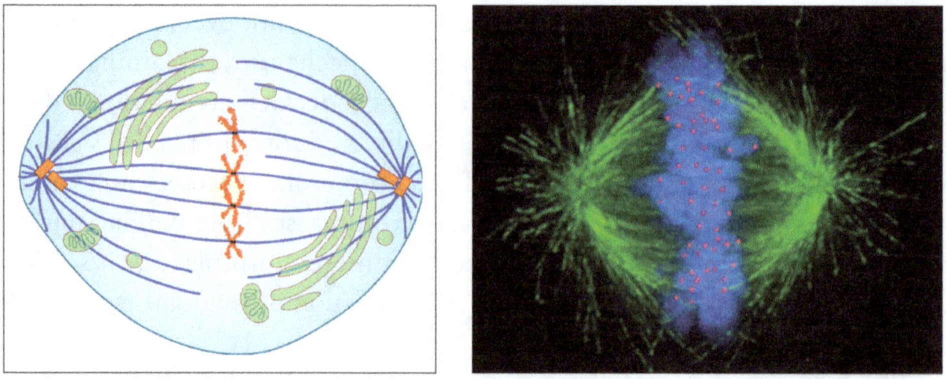

Figure 3.3. (left) A schematic representation of a mitotic spindle in a cell with just four duplicated chromosomes. *(right)* An artificially colored image of the mitotic spindle in a human cell, showing microtubules in green, chromosomes in blue, and kinetochores in red. A kinetochore is a protein structure that temporarily holds a chromosome and its duplicate together while also providing an anchor for a "thread" of the mitotic spindle. In the following phase of mitosis, each chromosome and its duplicate will be pulled apart, destined for different daughter nuclei.[9]

properly? Or is it rather that the larger drama connects immaterially with the individual topoisomerase?[10]

James Wang, the Harvard University molecular biologist who discovered the first topoisomerase, seems to have had some awareness of the problem. Writing about the striking capability of a topisomerase to untie a DNA knot by cutting through the double helix and later putting it back together again — all without disturbing the critical continuity of the original chemical structure — he expresses his wonder:

> When we think a bit more about it, such a feat is absolutely amazing. An enzyme molecule, like a very nearsighted person, can sense only a small region of the much larger DNA to which it is bound, surely not an entire DNA [molecule]. How can the enzyme manage to make the correct moves, such as to untie a knot rather than make the knot even more tangled? How could a nearsighted enzyme sense whether a particular move is desirable or undesirable for the final outcome? (Wang 2009)

Despite his language, Wang presumably knows that a molecule does not *sense* anything at all. And he surely also knows that the topoisomerases always have an adequate *physical* basis for doing what they do in the place where they are. And yet this physically lawful activity (which is what Wang concerns himself with) does not yet get us to an understanding of how the enzymes act in support of radically different *purposes* as a cell proceeds through DNA replication, for example, or gene transcription, or the distinctive phases of cell division.

Wang's reference to whatever is "desirable or undesirable for the final outcome" is what we must ultimately reckon with. That is, the context to which the topoisomerase molecule must conform is, in the end, the activity of the whole organism, with its requirements for specific gene expression in every part of the body. Put simply, the molecule must meaningfully participate in *everything* — organism and environment — without fixed limit.

This points to the need for a kind of explanation biologists in general seem unwilling to acknowledge. For it would, indeed, upset the entire world of conventional biological thought, based as that thought is on local, analyzable, physical cause and effect. "Desirable for the entire context" and, similarly, "undesirable" are not physical categories.[11] Yet here is a perfectly competent physical scientist driven to use such phrases. We should pay attention.

Yes, we have every reason to believe that whatever happens, happens lawfully. But this still leaves us with the question, "How does our understanding of the overall coherence of cellular and organismal processes relate to the lawfulness we unfailingly observe whenever we isolate particular interactions and analyze them in physical and chemical terms?" (Talbott 2024) That lawfulness continues the same throughout all cellular activity of the most diverse sorts, and it does not seem to have any obvious provisions for explaining the unique, ever-varying principles of coordination and coherence governing biological entities ranging from cells to organs to whole organisms to different species within their environments.

WHERE ARE WE NOW?

Yes, the cell's genomic performance is complex!

Where Francis Crick and James Watson (known as the discoverers of the structure of DNA) were looking for a single, univocal code, we now know that a thousand different things are going on. Not only is all the regulatory activity and the resulting, three-dimensional "dance" of our genome exceedingly complex, it also shows us clearly that we are really looking at a whole-cell and whole-organism performance. The genome can do nothing of itself — not even twist itself into coils or "go loopy" — and, in achieving such things, the cell comes at the genome from every possible direction and temporally varies its approach in tune with ever-changing conditions. We will learn more about this complexity in further chapters, especially Chapters 7 ("Epigenetics: A Brief Introduction") and 14 ("How Our Genes Come to Expression"). The question how everything is coordinated in a useful, need-fulfilling, and meaningful way seems continually to encourage biologists to transcend conventional scientific descriptive language, as when they refer to the "three-dimensional dance of chromosomes."

You will have noticed in these first chapters that we seem to be raising a lot of questions! You can count on one thing — the question-raising will never come to an end. This is, in the first place, what all good science should do — raise decisive questions with ever greater clarity. But we can also nourish a hope that is not common in today's science: namely, that by continuing to describe the life of organisms in a revelatory way — acknowledging the narrative and holistic character of beings whose lives manifest from the immaterial "inner" toward the material "outer" — we will find the description itself coming more and more to constitute exactly the sort of *biological* understanding and explanation we can best look for. We will explicitly address this sort of understanding, and how it connects to our ideas of causality, in Chapter 12 ("Is a Qualitative Biology Possible?").

We will also confront — especially in Chapters 13 ("All Science Must Be Rooted in Experience") and 24 ("Is the Inanimate World an Interior Reality?") — how our questions relate to the problem of the thought-infused character of the material world generally. And while just about the whole book raises a question about the relation between isolated and specific living processes, on one hand, and their larger context, on the other, we will try to make this question more pointed in Chapter 6 ("Context: Dare We Call It Holism?") and Chapter 8 ("The Mystery of an Unexpected Coherence").

NOTES

1. Two important efforts to map the spatial arrangement of chromosomes were published in 2009 and 2014: Lieberman-Aiden et al. 2009 and Rao et al. 2014.
2. Figure 3.1 credit: Copyright Miriam Huntley, Rob Scharein, and Erez Lieberman-Aiden, from Lieberman-Aiden et al. 2009. Linear chromosome at top of figure: Ed Yong (https://commons.wikimedia.org/wiki/File:Fractal_globule.jpg), CC BY-SA 3.0.
3. Figure 3.2 credit: from Rao et al. 2014.
4. Rao et al. 2014. The quote comes from the authors' video abstract of their paper in *Cell*.
5. Toward the end of the Human Genome Project in 2000, according to a report in *Nature*, "geneticists were running a sweepstake on how many genes humans have, and wagers ranged from tens of thousands to hundreds of thousands. Almost two decades later, scientists armed with real data still can't agree on the number." Current estimates tend to run between 19,000 and 22,000, but recent criticisms "underscore just how difficult it is to identify new genes, or even to define what a gene is" (Willyard 2018).
6. Lavelle 2009. Nucleoproteins are proteins bound up with DNA or RNA. A nucleoprotein complex would be a complex of DNA or RNA plus protein.

7. To get more specific about it, think of it this way. If, taking a double-stranded rope in hand, you insert a pencil between the strands and force it in one direction along the rope, you will find the strands winding ever more tightly ahead of the pencil's motion and unwinding behind. An RNA polymerase, which must separate the two strands of DNA as it transcribes a gene, can in the right circumstances have an effect rather like the pencil: it will cause negative supercoiling (loosening of the double helix spiral) behind itself, and positive supercoiling ahead. And if, say, negative supercoiling has already occurred in the region being transcribed, the polymerase will find it much easier to separate the two strands and do its work. So in this way the variations in coiling along the length of a chromosome either encourage or discourage the transcription of particular genes.
8. To get a rough sense merely for the number of significant variations in DNA double helix conformation and the kind of effect they can have, here is a statement enumerating such variations and their bearing on a single regulatory feature, namely, the position of certain nucleosomes (referred to as "variant –1 nucleosomes," which themselves play a key role in regulation of gene expression). There is no need to understand the different technical terms in order to get a feel for the complexity of the sculptural details of any particular stretch of DNA, and the kind of role these details can play in relation to gene expression.

    > Variant –1 nucleosomes [that is, nucleosomes at the places on DNA where gene transcription starts] exhibited a preference for sequences with altered features such as propeller twist, opening, electrostatic potential, minor groove width, rise, stagger, helix twist, and shear and roll. Variant –1 nucleosomes that shifted downstream in KDM5B-depleted ES [embryonic stem] cells preferred sequences with increased propeller twist, opening, electrostatic potential, stagger, minor groove width, rise, and buckle, while –1 variant nucleosomes that shifted upstream preferred sequences with decreased propeller twist, opening, electrostatic potential, stagger, minor groove width, rise, and buckle … Combined, these findings suggest that DNA shape predicts sequence preferences of canonical nucleosomes and variant nucleosomes. These results also suggest that histone DNA binding patterns such as bending or electrostatic interactions may be influenced by posttranslational modifications such as H3K4 methylation (Kurup, Campeanu and Kidder 2019).

9. Figure 3.3 credit: left: LadyofHats (https://commons.wikimedia.org/wiki/File:Mitotic_Metaphase.svg), Public Domain via Wikimedia Commons. Right: Afunguy (https://commons.wikimedia.org/wiki/File:Kinetochore.jpg), Public Domain via Wikimedia Commons.
10. The notion of "immaterial" causation is, of course, scarcely allowable in today's science. Or so it is commonly thought. But this seems far from true. Is there not a sense in which every scientist implicitly agrees that ideas possess causal power? What about the *idea* of gravity — the ideal, form-giving aspect of it that we routinely formulate in mathematical thought? Isn't this immaterial idea, or thought, definitively present in all analyses of our movements on earth?

    The idea of gravity is, of course, a long way from the formative ideas we see at work in organisms. But no one has shown us inherent limits upon the kinds of ideas that might be embodied in the phenomena of the material world. In any case, just as we indisputably "see" the mathematics of gravity in earthly motions, we also and with equal persuasiveness "see," for example, the striving for life evident in all organisms. Where physicists prefer to concern themselves with universal laws that apply to objects solely with regard to abstracted quantities such as mass, biologists deal with the behavior arising from within the qualitatively differentiated, more or less individuated "objects" (beings) of their science.

11. We are, of course, told that "desirable" and "undersirable" really refer to whether a trait is or is not conducive to an organism's survival and therefore favored by natural selection. But ask yourself: How does this line of thought make more explicable what we have just heard about the activity of topoisomerases in the cell? To believe that every feature an organism actually possesses must be consistent with natural selection is no ground for saying that the materialistically conceived processes of natural selection can positively account for processes inexplicable in strictly physical terms. See the discussion of natural selection under "The shortest path to confusion is circular" in Chapter 18, ("Teleology and Evolution").

SOURCES

Ball, Philip (2003). Portrait of a Molecule, *Nature* vol. 421 (January 23), pp. 421–22. doi:10.1038/nature01404

Dekker, Job and Tom Misteli (2015). Long-Range Chromatin Interactions, *Cold Spring Harbor Perspectives in Biology* 2015;7:a019356. doi:10.1101/cshperspect.a019356

Kouzine, Fedor, Suzanne Sanford, Zichrini Elisha-Feil et al. (2008). The Functional Response of Upstream DNA to Dynamic Supercoiling *in Vivo, Nature Structural and Molecular Biology* vol. 15, no. 2 (February), pp. 146–54. doi:10.1038/nsmb.1372

Kurup, Jiji T., Ion J. Campeanu and Benjamin L. Kidder (2019). Contribution of H3K4 Demethylase KDM5B to Nucleosome Organization in Embryonic Stem Cells Revealed by Micrococcal Nuclease Sequencing, *Epigenetics & Chromatin* vol. 12, no. 20. doi:10.1186/s13072-019-0266-9

Lavelle, Christophe (2009). Forces and Torques in the Nucleus: Chromatin under Mechanical Constraints, *Biochemistry and Cell Biology* vol. 87, pp. 307–22. doi:10.1139/O08-123

Lewontin, R. C. (1992). The Dream of the Human Genome, *New York Review* (May 28), pp. 31–40. https://www.nybooks.com/articles/1992/05/28/the-dream-of-the-human-genome/

Lieberman-Aiden, Erez, Nynke L. van Berkum, Louise Williams et al. (2009). Comprehensive Mapping of Long-Range Interactions Reveals Folding Principles of the Human Genome, *Science* vol. 326 (October 9), pp. 289–93. doi:10.1126/science.1181369

Moss, Lenny (2003). *What Genes Can't Do*. Cambridge MA: MIT Press.

_______ (2012). Is the Philosophy of Mechanism Philosophy Enough? *Studies in History and Philosophy of Science Part C: Studies in History and Philosophy of Biological and Biomedical Sciences* vol. 43, no. 1 (March), pp. 164–72. doi:10.1016/j.shpsc.2011.05.015

Physorg (2014). Scientists Map the Human Loop-ome, Revealing a New Form of Genetic Regulation (December 11). https://phys.org/news/2014-12-scientists-human-loop-ome-revealing-genetic.html

Rao, Suhas S. P., Miriam H. Huntley, Neva C. Durand et al. (2014). A 3D Map of the Human Genome at Kilobase Resolution Reveals Principles of Chromatin Looping, *Cell* vol. 159, pp. 1665–80 (December 18). doi:10.1016/j.cell.2014.11.021

Talbott, Stephen L. (2010). The Unbearable Wholeness of Beings, *The New Atlantis* no. 29 (fall), pp. 27–51. Available at https://thenewatlantis.com/publications/the-unbearable-wholeness-of-beings. Original version published in *NetFuture* no. 29 (fall), pp. 27–51. Also available at https://bwo.life/mqual/genome_5.htm.

_______ (2024). How Do Biomolecules "Know' What To Do?, *In Context* #51 (spring). Latest version: https://bwo.life/bk/ps/biomolecules.htm

Wang, James C. (2009). *Untangling the Double Helix*. Cold Spring Harbor, NY: Cold Spring Harbor Laboratory Press.

Willyard, Cassandra (2018). New Human Gene Tally Reignites Debate, *Nature* vol. 558 (June 21), pp. 354–55. doi:10.1038/d41586-018-05462-w

Zimmer, Carl (2015). Is Most of Our DNA Garbage? *The New York Times* (March 5). https://www.nytimes.com/2015/03/08/magazine/is-most-of-our-dna-garbage.html

CHAPTER 4

The Sensitive, Dynamic Cell

THROUGHOUT A GOOD PART of the twentieth century, cell biologists battled over the question, "Which exerts greater control over the life of the cell — the cell nucleus or the cytoplasm?" (Sapp 1987) From mid-century onward, however, the badge of imperial authority was, by enthusiastic consensus, awarded to the nucleus, and especially to the genes and DNA within it. "Genes make proteins, and proteins make us" — this has been the governing motto, despite both halves of the statement being false (which will become ever clearer as we proceed).

The question for our own day is, "Why would anyone think that any part of a cell must possess executive *control* over all the other parts?" We have already caught our first glimpse of the performances in the nucleus (see Chapter 3, "What Brings Our Genome Alive?"), and these hardly testify to domination by a single, controlling agent. Now we will broaden our outlook by making a first approach to the rest of the cell — the cytoplasm, along with its organelles and enclosing membrane.

It would be well to remind ourselves before we proceed, however, that, whatever else it may be, an organism is a physical being. Its doings are always in one way or another *physical* doings. This may seem a strange point to need emphasizing at a time when science is wedded to materialism. And yet, for the better part of the past century problems relating to the material coordination of biological activity were largely ignored while biologists stared, transfixed, into the cell nucleus. If they concentrated hard enough, they could begin to hear the siren call of a de-materialized, one-dimensional, informational view of life.[1]

The idea of a genetic *code* and *program* proved compelling, even though the program was never found, and even though the supposedly fixed code became many different codes and these were continually modified by the cell in every phase of its activity. So long as one lay under the spell woven by notions of causally effective information, logic, and code, the complex, causal realities of the material organism tended to disappear from view, or seemed unimportant. An overall, never clearly observed logic was assumed to *govern* all the messy particulars, which did not need to be studied too closely.

Unfortunately for conventional thought, the particulars *did* come into view, however slowly, and however much they were at first ignored. Eventually and inevitably they undermined the much too neat story of a clean, all-determining, informational logic.

Surely, even if genes are not the decisive logical and informational causes usually imagined, they must connect in *some* manner with the features they were thought one-sidedly to explain. But this just as surely means they must connect physically, via movements and transformations of substance testifying to the deeply meaningful, underlying narrative we actually observe in every organism (Chapter 2, "The Organism's Story"). And the picture we were exposed to earlier (in Chapter 3, "What Brings Our Genome Alive?"), detailing some of the significant movements and gesturings of chromosomes, is only the beginning of the story.

Does the cell possess its own "senses" and "limbs"?

Let's continue by taking note of the cytoskeleton (Figure 4.1), which plays a key role in the cell's physical movement. It consists of many exceedingly thin molecular filaments and tubules, visible only under powerful microscopes. Many of these are growing at one end and perhaps shrinking at the other end, or else disassembling altogether even as new filaments are establishing themselves. Through this dynamic activity — this constant growth and dissolution of minuscule fibers — the cell gains its more or less stable shape and organization.

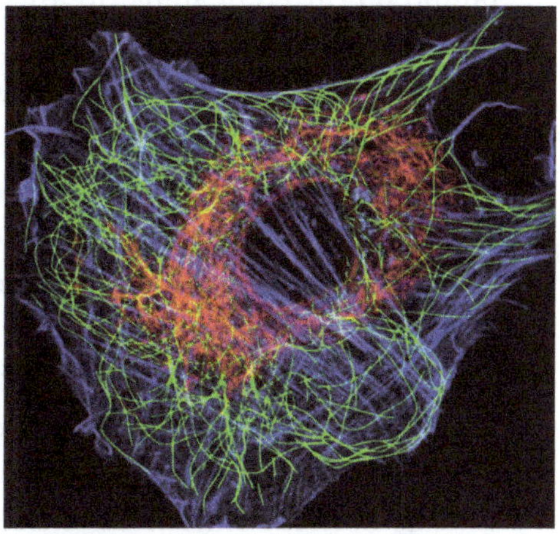

Figure 4.1. A cultured fibroblast cell, specially prepared so as to show features of the cytoskeleton in artificial color: narrow actin filaments (blue); wider microtubules (green); and intermediate filaments (red). The dark and roughly circular (spherical) region near the center is the cell nucleus.[2]

Cellular organelles, to which the cytoskeleton attaches, are positioned and re-positioned as the cytoskeleton somehow "senses" internal needs, while also responding to external stresses such as stretching or compression. Beyond that, the filaments and tubules, by dynamically managing the distribution of forces within the cell as a whole, help to enable and guide its movements so that it can find its proper place among the millions of cells in its nearby environment.

And the cells of our bodies do move. Literal rivers of cells shape the young embryo. So, too, migrating cells in and around a wound cooperate in restoring the damaged architecture. In every tiniest hair follicle niche, as well as throughout our tissues generally, cells move, replace dying neighbors, and reorganize themselves. And even while remaining in one place, cells must continually adapt their form to their immediate environment — certainly a major task in the rapidly growing embryo and fetus. But the stresses and tensions of that environment are in turn the partial result of interconnected cytoskeletal activities in all the cells of the local tissue.

The cytoskeleton not only supports cell migration, but also provides pathways for the orchestrated movement of substances within the cell. A protein molecule is not of much use if it cannot find its way to where it is required. Individual molecules and protein complexes are shifted about along these cytoskeletal pathways, as are the relatively voluminous contents of membrane-enclosed transport structures ("vesicles"). These latter can "bud off" various internal membranes of the cell and then move, along with their cytoplasmic contents, to a particular destination where, having released their contents, they are degraded and recycled.

Such directed movements are essential to the life of the cell. Where an enzyme or signaling molecule goes in a cell is decisive for its function. Some molecules, for example, are outward-bound to, and through, the cell surface on signaling missions to distant reaches of the body. Meanwhile, others are inward-bound on different signaling missions. (Hormones, secreted by cells of a gland at the start of their journey, and then received by cells in various other parts of the body, illustrate both sorts of movement.) Some molecules produced in a cell are destined for a particular locus on the highly differentiated cell membrane, while others are targeted to any of a virtually infinite number of possible stopping places somewhere in the cell's "intricate landscape of tubes, sacs, clumps, strands and capsules that may be involved in everything from intercellular communication to metabolic efficiency."[3]

But the cytoskeleton is not just a cytoskeleton. The filaments and tubules themselves are teeming with associated regulatory molecules. As of 2010, more than 150 proteins capable of binding to just one type of filament — actin — had already been identified. As one researcher has put it: "Despite the connotations of the word 'skeleton,' the cytoskeleton is not a fixed structure whose function can be understood in isolation. Rather, it is a dynamic and adaptive structure whose component polymers and regulatory proteins are in constant flux" (Fletcher and Mullins 2010).

There is scarcely any aspect of cellular functioning in which the cytoskeleton fails to play a role. On the exterior side, it connects with the cell's outer ("plasma") membrane, where it helps to import substances from the environment while also facilitating the adhesion of extracellular molecules and other cells. Through its interaction with the extracellular matrix, it contributes to the mechanical stiffness and coherence of entire tissues. On the interior side, it engages with the nuclear membrane and the specialized filaments underlying that membrane. These filaments are vital regulators of gene expression. In this way the cytoskeleton links various sorts of extracellular signals, both mechanical and biochemical, to the nucleus and its chromosomes, providing a foundation for holistic behavior involving much more than the individual cell.

There are many ways to affect gene expression, and they do not all occur in the cell nucleus. For example, a key part of this expression is the translation of RNA molecules into proteins, which occurs in the cytoplasm. Evidence suggests that "the physical link between cytoskeletal and translational components helps dictate both global and local protein synthesis." But it's not just that the cytoskeleton affects translation. As is all too typical, the causal effects work both ways: "specific translation factors are able to affect the organization of cytoskeletal fibres."[4]

The cytoskeleton plays many other roles, not least by ensuring the proper separation of mitotic chromosomes, the division of a cell into two daughter cells, and the correct allocation of chromosomes to those daughter cells. (See Figure 3.3, where the mitotic spindle, shown in green, consists of cytoskeletal fibers.) It is perhaps unsurprising, then, that some have seen the cytoskeleton, with its nuanced organizational "skills," as the seat of cellular intelligence or the "brain" of the cell. However, we need not invite a misleading anthropomorphism in order to acknowledge the subtle and nuanced organizational activity — the *narratively intelligible* activity (Chapter 2, "The Organism's Story") — realized through the dynamics of cytoskeletal movement.

One thing is certain: neither the cytoskeleton's moment-by-moment dynamics nor the coherent and intelligible aspect of its activity can be ascribed to "instructions" from genes — or even to the physical laws bearing on cytoskeletal proteins. As the matter was summarized by Franklin Harold, an emeritus professor of biochemistry and molecular biology at Colorado State University, "One cannot predict the form or function of these complex [cytoskeletal] ensembles from the characteristics of their component proteins." And yet, Harold went on, "When seen in the context of the parent cell the arrangement of the molecules becomes quite comprehensible." He then raised the obvious question: "How is the cytoskeleton itself so fashioned that its operations accord with the cell's overall 'plan' and generate its particular morphology time after time?"[5]

Harold answered the question merely by expressing confidence that understanding will eventually come. And surely it will. But we can be equally sure that it will not come before we have penetrated more deeply this problem: How does a living context, or whole — in this case, the cell with its "overall plan" — manage to express itself through all its parts?

In an integral, organic whole, we can assume the "viewpoint" of many parts in such a way as to make each one *momentarily* seem to be the coordinating "master" element. This is why the cytoskeleton, just as much as our genes, might appear to explain everything that goes on. With wonderful sensitivity it "feels out" the surfaces of the cell and all its organelles. The balance of forces maintained by the fibers shapes the cell, dynamically positions the organelles, and both guides and helps to power the critical movement of the cell within its environment. As we have seen, the cytoskeleton likewise plays a key role in moving substances to their functional locations within the cell. And it is a decisively important regulator of gene activity.

And yet, this does not make the cytoskeleton a *master* regulator. The truth is simply that, to one degree or another, each part of an organic whole bears that whole within itself — is informed by, and expresses, the whole. The idea of a master regulator arises only when we insist on viewing a specific part in isolation from the whole so as to identify single, local, and unambiguous causal interactions. We then say that this part *makes* certain things happen. The fact that the part is itself made to happen by the very things it supposedly accounts for then tends to be ignored. We lose sight of the fluidity and physical indeterminism of the living context — an indeterminism whose meaning and coherence become visible only when we allow particular physical causes to "disappear" into the unifying *narratives*, or stories, of the organism's life (Chapter 2, "The Organism's Story"). In much the same way, we experience physical sounds and gestures disappearing into the *meaning* of the speech we hear.

The sensitive "skin" and organelles of the cell

Interestingly, the cell membrane ("plasma membrane") is likewise a highly dynamic feature that has been seen as a decisive coordinator of cellular activity, and even as a seat of cellular intelligence. It is here that we see "decisions" continually being made about which substances and signals — from among the endlessly

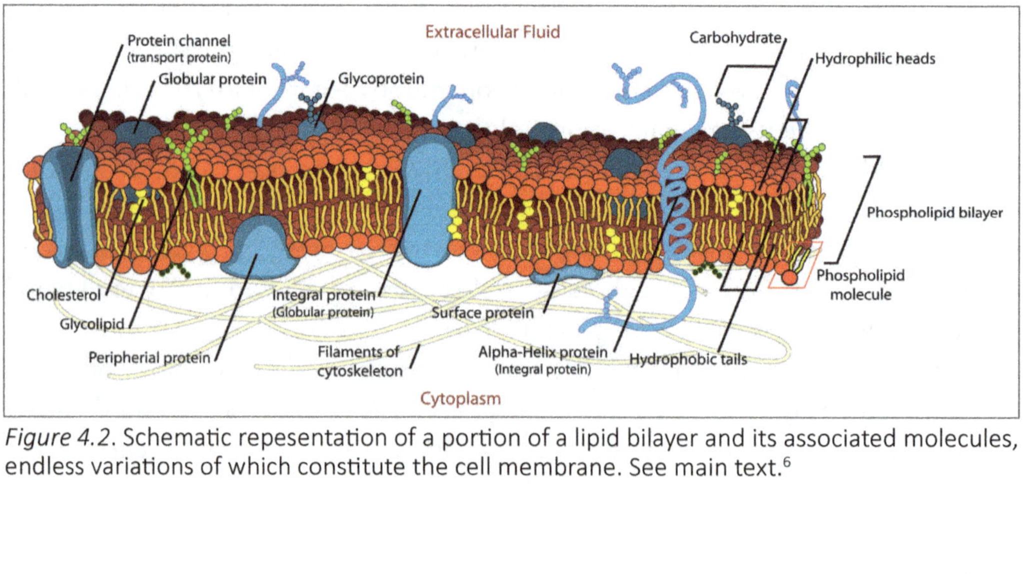

Figure 4.2. Schematic repesentation of a portion of a lipid bilayer and its associated molecules, endless variations of which constitute the cell membrane. See main text.[6]

streaming crowds passing through the neighborhood — are to be admitted into the cell and which ones are "foreign," or else unnecessary at the moment. Here, perhaps more than anywhere else, is where cellular identity is established and "self" is distinguished from "other." This happens partly by means of protein receptors ("sensors") embedded in, or attached to, the lipid matrix of the membrane.

Again, everything flows (which is one reason why any image like the two here is a kind of frozen lie, despite being useful when approached with the right awareness). Molecules continually associate with, and dissociate from, the membrane, even as they undergo various modifications that redirect their functioning. They also migrate within the membrane, forming specialized communities that are in no two locales exactly the same. All the while portions of the membrane, along with cytoplasmic contents, are "pinched off" as more or less spherical vesicles that, once they are fully detached, move elsewhere, either externally to the cell or internally. At the same time, selected vesicles from external sources fuse with the membrane and release their contents into the cell's interior.

Much the same is true of all the interior membranes delimiting the various organelles of the cell (Figure 4.3). These, too, "harbor sensitive surveillance systems to establish, sense, and maintain characteristic physicochemical properties that ultimately define organelle identity. They ... play active roles in cellular signaling, protein sorting, and the formation of vesicular carriers" (Radanović et al. 2018).

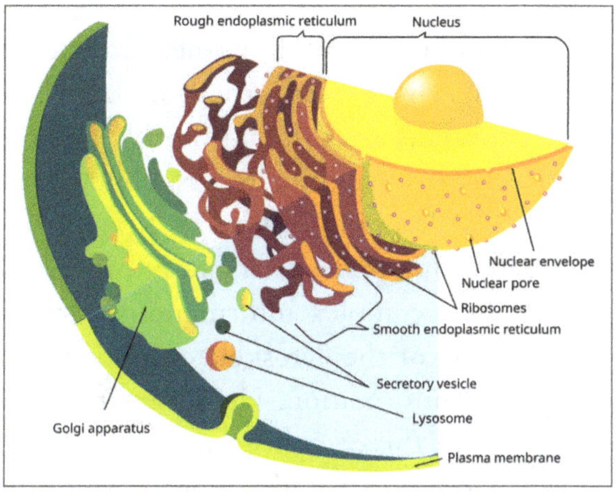

Figure 4.3. Schematic representation of the internal membrane systems of a nucleated cell.[7]

Membranes, then, not only structure the cell into distinctive compartments and organelles, but they also "oversee" the characteristic and essential contents of those compartments and play decisive roles in managing the ceaseless and massive intercommunication among them.

All this finely discriminating activity is going on, as the eminent cell biologist, Paul Weiss, wrote in 1973, while "the cell interior is heaving and churning all the time" (Weiss 1973, p. 40). Everything is watery movement of substances and transformation of organizational structure, and yet the cell's identity and unified character are maintained. The movement expresses the character and constitutes the life of the cell. The intricately choreographed flows and chemical transactions in plasm and membrane are responsive to the ever-unpredictable conditions of the moment, and are the means by which the cell not only stays true to itself, but also remains in harmony with its larger environment.

The dynamics of this material accomplishment are a long way from the clean, informational logic commonly associated with genes. Lenny Moss, a molecular biologist who transformed himself into one of our most insightful philosophers of biology, had this to say about the relation between cellular membranes and genes:

> The membranous system of the cell, the backbone of cellular compartmentalization, is the necessary presupposition of its own renewal and replication. Cellular organization in general and membrane-mediated compartmentalization in particular are constitutive of the biological "meaning" of any newly synthesized protein (and thus gene), which is either properly targeted within the context of cellular compartmentalization or quickly condemned to rapid destruction (or cellular "mischief"). At the level of the empirical materiality of real cells, genes "show up" as indeterminate resources ... If cellular organization is ever lost, neither "all the king's horses and all the king's men" *nor* any amount of DNA could put it back together again.[8]

From information to life

Returning for a moment to our introductory question about the control of the cell by its genes: perhaps we have by now gained a feeling for how the cell and organism as a whole can flexibly and contextually express itself through any one of its parts, including its DNA and chromosomes — a fact we will get much more specific about in Chapter 7 ("Epigenetics: A Brief Introduction") and Chapter 14 ("How Our Genes Come to Expression"). If we think of the genome as an almost infinitely complex informational structure, there is no reason not to think, for example, of the cytoskeleton and membranes of a cell as at least equal bearers of vital information. However, it is also important to recognize the illegitimate aspects of this comparison.

In particular, the concept of information as normally applied to DNA is a quantitative one. It depends on the existence of discrete, iterated elements ("letters" of the "code"), any one of which can take on certain precise values. But everything we know about the "heaving and churning" interior of the cell — including even the coiling and looping of chromosomes we saw in Chapter 3 ("What Brings Our Genome Alive?") — tells us that we are looking at boundless and continuous variations of form and gesture whose depth of meaning is both non-quantifiable and more profoundly expressive than any quantifiable features we can abstract from it.

To ask about the amount of information in various aspects of the cellular performance (including the performance of chromosomes) is rather like asking about the amount of information in Stravinsky's ballet, "The Rite of Spring." It would be one thing to define informational quantities in terms of some more or less arbitrary method of choreographic notation ("code"), and quite another to consider the expressive content of the ballet itself.

So, too, our means for quantifying the informational content of a genomic sequence bears little relation to the material gestures expressing the cell's life. The truth here will become even more vivid when we look (in Chapter 6, "Context: Dare We Call It Holism?") at the context-dependence that biologists freely acknowledge at every turn.

WHERE ARE WE NOW?

The cell is reflected in its parts.

In this chapter (as will happen throughout much of the book) we have had thrown at us the question of the relation between part and whole. Both the cytoskeleton and the collection of cell membranes participate in and seem to represent the whole cell to such a degree that some biologists are inclined to see one or the other — and not the genetic material — as the "controlling" element of the cell. But neither point of view is satisfactory. We are continually forced back to nothing less than the whole itself, not as a mechanistically controlling entity, but rather as the narrator of the ongoing drama that is the organism's life.

We will hear more about the cytoskeleton and membranes in Chapter 5 ("Our Bodies Are Formed Streams"). The main lesson for the moment is that the cell is a material whole in its own right. In a multicellular organism it is, of course, a *relative* whole. But the fact is — as we will take specific note of in Chapter 6 ("Context: Dare We Call It Holism?") — even every *organism* is a relative whole: it is not only reflected in its parts, but it is itself caught up in, and is a reflection of, its encompassing community and environment.

By noting the complementary manner in which the cytoskeleton and membranes work together to express a cell's living character, we can get a feeling for the integral unity of a biological whole. Perhaps (although I do not discuss this here) the linear, ray-like character of the cytoskeletal filaments and the more globular, enclosing character of the membranes tell us something about the fundamental polarity out of which the living unity of the cell arises.[9]

NOTES

1. "In 1989, the gene was in the process of being disembodied, increasingly informaticized and formalized; it was very much swimming upstream to suggest that genes had physical embodiment and dynamic behaviours" (Landecker 2015).
2. Figure 4.1 credit: courtesy of Harald Herrmann, University of Heidelberg, Germany.
3. Kwok 2011. Here is a further description (from Plankar et al. 2012) of the various roles of the cytoskeleton:

   > The cytoskeleton, in addition to its classical structural-mechanical role, integrates many signalling pathways, influences the gene expression, coordinates membrane receptors and ionic flows, and localizes many cytosolic enzymes and signalling molecules, while at the same time it represents an immense, electrically active catalytic surface for metabolic interactions. Together with cell adhesion molecules and the extracellular matrix, it forms a tensionally integrated system throughout the tissues and organs, which is able to coordinate gene expression via mechano-transduction. Given the strong relationship between mechanical and electromagnetic excitations in the microtubules (piezoelectricity) and their well-established organising potential, a weakened EM field may thus influence both cell and tissue aspects of carcinogenesis.

4. Kim and Coulombe 2010. The use of words such as "dictate" to suggest unambiguous, one-directional causation is extremely common in all the literature of molecular biology. And almost as common is the immediate contradiction of this language, as we see here. For more on this, see Chapter 9 ("A Mess of Causes"). There is also this from two biologists in McGill University's Department of Physiology, pointing to the two-way interaction between the cytoskeleton and the ubiquitous signaling activity through which the cell's diverse activities take place:

   > Filamentous actin, microtubules, and intermediate filaments regulate cell shape, motility, transport, and interactions with the environment. These activities rely on signaling events that control cytoskeleton properties. Recent studies uncovered mechanisms that go far beyond this one-directional flow of information. Thus, the three branches of the cytoskeleton impinge on signaling pathways to determine their activities (Moujaber and Stochaj 2019).

5. Harold 2001, p. 125. Harold makes his question more emphatic with a little elaboration:

   > How, for instance, do [the cell's] famously fluid membranes hold their shape? How does the endomembrane system as a whole acquire its spatial orientation and location, while the cell of which it is a part grows, divides and moves around? … In a nutshell, the cytoskeleton is responsible for the *mechanical intergration of cellular space*; unpacked, this phrase covers a host of actions and interactions, mediated by a large and growing ensemble of proteins. … [Moreover, the cytoskeleton itself] is subject to frequent remodelling. Mitosis, for instance, entails the dissolution of much of the cytoskeleton; its components are redeployed in the service of cell division, and subsequently reconstituted in their former order. Everything is in flux, but in a regulated purposeful manner (pp. 123–24, emphasis added).

6. Figure 4.2 credit: LadyofHats Mariana Ruiz (https://commons.wikimedia.org/wiki/File:Cell_membrane_detailed_diagram_en.svg), Public Domain via Wikimedia Commons.
7. Figure 4.3 credit: LadyofHats Mariana Ruiz (https://en.wikipedia.org/wiki/File:Endomembrane_system_diagram_en_(edit).svg), Public Domain via Wikimedia Commons.
8. Moss 2003, p. 95. Pages 76–98 in Moss' book provide an excellent overview of the dynamics associated with cellular membranes. There is also this from Harayama and Riezman 2018: "We are beginning to understand why even small changes in lipid structures and in composition can have profound effects on crucial biological functions":

    > Although our knowledge of lipid metabolism and function has improved, we have so far revealed only the tip of the iceberg. We have only a limited understanding of the biological consequences of slight structural differences in lipids, but the known cases suggest that small structural changes will be very important. Many of these cases were unpredictable when the research started, suggesting that exciting new findings lie ahead.

9. One thing these opposing qualitative characters remind me of is a rather bold saying by Samuel Taylor Coleridge at the beginning of the famous Chapter XIII of *Biographia Literaria*:

    > Grant me a nature having two contrary forces, the one of which tends to expand infinitely, while the other strives to apprehend or *find* itself in this infinity, and I will cause the world of intelligences with the whole system of their representations to rise up before you (Coleridge 1906).

    And this in turn might remind us of a remark by Jakob Boehme (whose work was important to Coleridge):

    > Nothing without contrariety can become manifest to itself; for it has nothing to resist it, it goes continually of itself outwards, and returns not again into itself (quoted in McFarland 1981, pp. 323–24).

    And again from Boehme: if a thing has only one will and "finds not a contrary will, which gives occasion to its exercising motion, it stands still" (ibid.). We need only think of the importance of gravity and the resulting friction of our feet upon the ground, to realize that we walk and move forward by "pushing off" against the force of gravity. We could ourselves accomplish no movement forward if we were floating in space with nothing to resist us.

    This lack of contrariety sounds rather like a fanciful picture of a cell with growing and unrestrained cytoskeletal fibers, but no enclosing membrane, or like the inertness of a cell with enclosing membrane but no dynamic cytoskeleton.

SOURCES

Coleridge, Samuel Taylor (1906). *Biographia Literaria*. London: J. M. Dent & Sons. Originally published in 1817. Available at https://archive.org/details/cu31924102776196/

Fletcher, Daniel A. and R. Dyche Mullins (2010). Cell Mechanics and the Cytoskeleton, *Nature* vol. 463 (January 28), pp. 485–92. doi:10.1038/nature08908

Harayama, Takeshi and Howard Riezman (2018). Understanding the Diversity of Membrane Lipid Composition, *Nature Reviews Molecular Cell Biology* vol. 19, pp. 281–96. doi:10.1038/nrm.2017.138

Harold, Franklin M. (2001). *The Way of the Cell: Molecules, Organisms and the Order of Life*. Oxford: Oxford University Press.

Kim, Seyun and Pierre A. Coulombe (2010). Emerging Role for the Cytoskeleton as an Organizer and Regulator of Translation, *Nature Reviews Molecular Cell Biology* vol. 11 (January), pp. 75–81. doi:10.1038/nrm2818

Kwok, Roberta (2011). The New Cell Anatomy, *Nature* vol. 480 (December 1), pp. 26–28. doi:10.1038/480026a

Landecker, Hannah (2015). Commentary: The Information of Conformation, *International Journal of Epidemiology*, pp. 1–2. doi:10.1093/ije/dyv022

McFarland, Thomas (1981). *Romanticism and the Forms of Ruin: Wordsworth, Coleridge, and Modalities of Fragmentation*. Princeton NJ: Princeton University Press.

Moss, Lenny (2003). *What Genes Can't Do*. Cambridge MA: MIT Press.

Moujaber, O. and U. Stochaj (2019). The Cytoskeleton as Regulator of Cell Signaling Pathways, *Trends in Biochemical Sciences*. doi:10.1016/j.tibs.2019.11.003

Plankar, M., E. Del Giudice, A. Tedeschi, and I. Jerman (2012). The Role of Coherence in a Systems View of Cancer Development, *Theoretical Biology Forum* vol. 105, no. 2 (January 1), pp. 15–46.

Radanović, Toni, John Reinhard, Stephanie Ballweg et al. (2018). An Emerging Group of Membrane Property Sensors Controls the Physical State of Organellar Membranes to Maintain Their Identity, *Bioessays*. doi:10.1002/bies.201700250

Sapp, Jan (1987). *Beyond the Gene: Cytoplasmic Inheritance and the Struggle for Authority in Genetics*. Oxford: Oxford University Press.

Weiss, Paul (1973). *The Science of Life: The Living System — A System for Living*. Mount Kisco, NY: Futura Publishing.

CHAPTER 5

Our Bodies Are Formed Streams

> The method of nature: who could ever analyse it? That rushing stream will not stop to be observed. We can never surprise nature in a corner; never find the end of a thread; never tell where to set the first stone. The bird hastens to lay her egg: the egg hastens to be a bird … [The world's] smoothness is the smoothness of the pitch of the cataract. Its permanence is a perpetual inchoation. Every natural fact is an emanation, and that from which it emanates is an emanation also, and from every emanation is a new emanation. If anything could stand still, it would be crushed and dissipated by the torrent it resisted …
>
> – Ralph Waldo Emerson, "The Method of Nature"
> (in Emerson 1908, p. 43)

IN THIS MATERIALIST ERA, we like our reality *hard* and our truths *weighty* and *rock solid*. We may accept that there are states of matter less substantial than rocks, but in our imaginations we turn even fluids and gases into collections of tiny *particles*. Similarly, in our reconstructions of physiological processes, *material structures* come first, and only then can movement, flow, and meaningful activity somehow occur.

How, after all, can there be movement without *things* to do the moving? (It's easy to forget that energy, fields, and forces are not things!) Ask someone to describe the circulatory system, and you will very likely hear a great deal about the heart, arteries, veins, capillaries, red blood cells, and all the rest, but little or nothing about the endless subtleties of circulatory *movement*. And yet, embryological development shows that

> the body does not behave like a plumber, first connecting the water pipes in a house and then turning the water on … the first blood-like liquid … simply trickles through gaps in the tissues … Preferred channels develop only very gradually as blood cells are deposited along the edges and eventually merge into the beginnings of vessel walls (Schad 2002, p. 80).

The situation loosely reminds one of college campuses when new lawn is laid down. Landscapers typically wait to see where human traffic creates clear pathways through the grass before "solidifying" the paths with concrete.

Moreover, "when blood vessels first start to form, the heart does not yet exist ... early blood flow stimulates the development of the heart" (Schad 2002, pp. 82–83). Again, form arises from movement. Thus, the spiraling fibers of the heart muscle that help to direct the blood in its flow are themselves a congealed image of the swirling vortex of blood within. This kind of mutuality holds even for the heart's basic structural divisions:

> Before the heart has developed walls (septa) separating the four chambers from each other, the blood already flows in two distinct "currents" through the heart. The blood flowing through the right and left sides of the heart do not mix, but stream and loop by each other, just as two currents in a body of water. In the "still water zone" between the two currents, the septum dividing the two chambers forms. Thus the movement of the blood gives the parameters for the inner differentiation of the heart, just as the looping heart redirects the flow of blood[1] (Holdrege 2002, p. 12).

There is no escaping the fact that we begin our lives in a thoroughly fluid and plastic condition. Only with time do relatively solid and enduring structures precipitate out as tentatively formed "islands" within the streaming rivers of cells that shape the life of the early embryo. Movement gives rise to structures, structures do not give rise to movement. As adults, we are still about seventy percent water.

One might think quite differently based on the scientific rhetoric to which we are daily exposed. This could easily lead us to believe that the real essence and solid foundation of our lives was from the beginning rigidly established inside those very first cells. There we find DNA macromolecules that, in a ceaseless flood of images, are presented to us as crystalline forms in the shape of a spiraling ladder — a ladder whose countless rungs constitute the fateful stairway of our lives. So, too, with the proteins and protein complexes of our bodies: we have been told for decades that they fold precisely into wondrously efficient *molecular machines* whose all-important functions are predestined by the DNA sequence.

The trouble is, biological researches of the last few decades have not merely hinted at an altogether different story; they have (albeit sometimes to deaf ears) been trumpeting it aloud as a theme with a thousand variations. Even the supposedly "solid" structures and molecular complexes in our cells — including the ones we have imagined as strict determinants of our lives — are caught up in functionally significant movement that the structures themselves can hardly have originated. (See Chapter 3, "What Brings Our Genome Alive?" and Chapter 4, "The Sensitive, Dynamic Cell.")

Nowhere are we looking either at a static sculpture or at controlling molecules responsible for the sculpting. In an article in *Nature* following the completion of the Human Genome Project, Helen Pearson (2003) interviewed many geneticists in order to assemble the emerging picture of DNA. One research group, she reported, has shown that the molecule is made "to gyrate like a demonic dancer." Others point out how chromosomes "form fleeting liaisons with proteins, jiggle around impatiently and shoot out exploratory arms." Phrases such as "endless acrobatics," "subcellular waltz," and DNA that "twirls in time and space" are strewn through the article. "The word 'static' is disappearing from our vocabulary," remarks cell biologist and geneticist Tom Misteli, a Distinguished Investigator at the National Cancer Institute in Bethesda, Maryland.

Everywhere we look, shifting form and movement show themselves to be the "substance" of biological activity. The physiological narratives of our lives play out in gestural dramas that explain the origin and significance of structures rather than being explained by those structures.

Hannah Landecker, a professor of both genetics and sociology at UCLA, having looked at the impact of recent, highly sophisticated cellular imaging techniques on our understanding, has written: "The depicted cell seems a kind of endlessly dynamic molecular sea, where even those 'structures' elaborated by a century of biochemical analysis are constantly being broken down and resynthesized." And she adds: "It is not so much that the structures begin to move, but movements — for example in the assembly and self-organization of the cytoskeleton — begin to constitute structure" (Landecker 2012). See Figure 5.1.

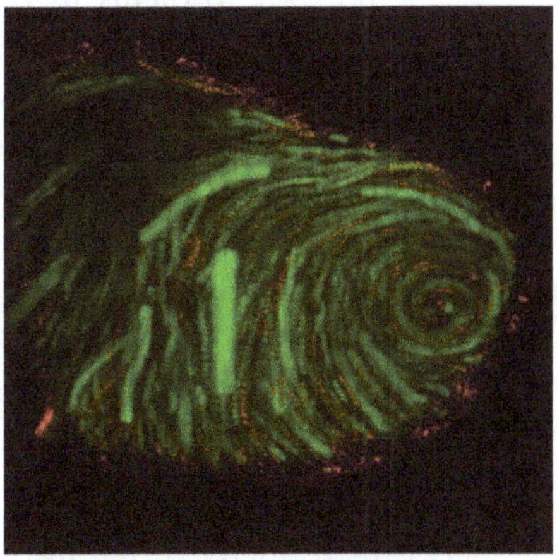

Figure 5.1. Multiple, superimposed images from a movie, showing movements in a fruit fly oocyte (a developing egg). Yolk granules are stained green, and tiny red fluorescent polystyrene beads have been injected into the egg to show the dynamism of flow in the egg body over time.[2]

And a team of biochemists from Duke and Stanford Universities point out how inadequate is our knowledge of the action of biomolecules when all we have is a frozen structure of the sort commonly reported in the literature. "In reality," they say, "all macromolecules dynamically alternate between conformational states [that is, between three-dimensional folded shapes] to carry out their biological functions":

> Decades ago, it was realized that the structures of biomolecules are better described as "screaming and kicking," constantly undergoing motions on timescales spanning twelve orders of magnitude, from picoseconds [trillionths of a second] to seconds (Ganser et al. 2019).

Why, after all, should we ever have expected our physiology to be less a matter of *gesturings* than is our life as a whole?

A long way from crystalline order

According to the old story of the machine-organism, a protein-coding DNA sequence, or gene, is not only mirrored in an exact messenger RNA (mRNA) sequence, but the mRNA in turn is translated into an exact amino acid sequence in the resulting protein, which finally folds into a fixed shape predestined by that sequence. It was a picture of perfect, lawful, lockstep necessity, leading from DNA through mRNA to a final, functional protein.

"There is a sense," wrote Richard Dawkins, "in which the three-dimensional coiled shape of a protein is determined by the one-dimensional sequence of code symbols in the DNA." Further, "the whole translation, from strictly sequential DNA read-only memory to precisely invariant three-dimensional protein shape, is a remarkable feat of digital information technology" (Dawkins 2006, p. 171).

And these proteins in turn were thought to carry out their functions by neatly engaging with each other in a machine-like manner, snapping into place like perfectly matched puzzle pieces or inserting into each other like keys in locks.

We now know, and already knew when Dawkins published those words, that everything about this narrative was wrong — and not only the parts about DNA and RNA. Among proteins (those "workhorses of the cell") every individual molecule lives in transformational movement — as a dynamic ensemble of rapidly "morphing," or interconverting, conformations — and therefore does not have a "precisely invariant three-dimensional shape."

But there is much more that wholly escaped Dawkins' computerized imagination. Quite apart from the fact that each protein molecule rapidly shifts between distinctly different, folded structures, we now know that *intrinsically disordered proteins* — proteins that, in whole or in part, have no particular, inherent structure at all — are crucial for much of a cell's functioning. Researchers refer to "fluid-like" and "surface-molten" proteins (Grant et al. 2010; Zhou et al. 1999). This is why biophysicist Konstantin Turoverov and his Russian and American colleagues tell us that "the model of the organization of living matter is changing to one described by highly dynamic biological soft matter." For decades, they note, protein interactions were "considered to be rigid, where, for a given protein, a unique 3D structure defined a unique biological activity." However,

> it is now realized that many protein functions rely on *the lack of specific structure*. This recognition has changed the classical consideration of a functioning protein from a quasi-rigid entity with a unique 3D structure resembling an aperiodic crystal into a softened conformational ensemble representation, with intrinsic disorder affecting different parts of a protein to different degrees[3] (Turoverov et al. 2019, emphasis added).

Clearly, the finally achieved protein need not be anything like the predetermined, inflexible mechanism with a single, well-defined structure imagined by Dawkins. Proteins can be true shape-shifters, responding and adapting to an ever-varying context — so much so that (as the noted experimental cell biologist, Stephen Rothman, has written) the "same" proteins with the same amino acid sequences can, in different environments, "be viewed as totally different molecules" with distinct physical and chemical properties (Rothman 2002, p. 265).

Many intrinsically unstructured proteins are involved in regulatory processes, and often serve as Proteus-like hub elements at the center of large protein interaction networks (Gsponer and Babu 2009). They also play a decisive role in molecular-level communication within and between cells, where their flexibility allows them to modulate or even reverse the typical significance of a signal,[4] in effect transforming *do this* into *don't do this* or *do that* (Hilser 2013).

But the troubling question arises: if unstructured proteins, or unstructured regions in proteins, are not "pre-fitted" for particular interactions — if, in their "molten" state, they have boundless possibilities for interacting with other molecules and even for reversing the effects of those other molecules — how do these proteins "know" what to do at any one place and time? (Talbott 2024) Or, as one pair of researchers put it, "How is the logic of molecular specificity encoded in the promiscuous interactions of intrinsically disordered proteins?" (Zhu and Brangwynne 2015) In a following section ("The unexpected phases of life") we will look at one of the most recent and dramatic developments in cellular physiology, which has seemed to many biologists to offer an approach to this problem.

But first we should note the continuing mechanistic bias in the negative descriptors, "disordered" and "unstructured," which I have grudgingly adopted from the conventional literature. Contrary to this usage, the loose, shifting structure of a protein need be no more disordered than the graceful, swirling currents of a river or the movements of a ballet dancer. Given the many living processes these proteins harmoniously support and participate in (including, in fact, the movements of the ballet dancer), it would be strange to assume that their performance is anything *less* than graceful, artistic, purposive, and meaningful.

Fluid, "living" molecules do not lend themselves to the analogy with mechanisms, which may explain why the mistaken idea of precisely articulated, folded parts was so persistent, and why the recognition of unstructured proteins was so late in coming. Indeed, this recognition has only recently been dawning upon the biological community as a whole, a fact that led to this lament as late as 2008 at a conference on "bioinformatics and bioengineering" at Harvard Medical School:

> Experimentalists have been providing evidence over many decades that some proteins lack fixed structure or are disordered (or unfolded) under physiological conditions. In addition, experimentalists are also showing that, for many proteins, their functions depend on the unstructured rather than structured state; such results are in marked contrast to the greater than hundred year old views such as the lock and key hypothesis. Despite extensive data on many important examples, including disease-associated proteins, the importance of disorder for protein function has been largely ignored. Indeed, to our knowledge, current biochemistry books don't present even one acknowledged example of a disorder-dependent function, even though some reports of disorder-dependent functions are more than fifty years old (Dunker et al. 2008).

The unexpected phases of life

It has become increasingly clear in recent years that, quite apart from its cytoskeleton and membrane-bound organelles (Chapter 4, "The Sensitive, Dynamic Cell"), the fluid cytoplasm in each cell is elaborately and "invisibly" organized. Various macromolecular complexes and other molecules, in more or less defined mixes, congregate in specific locations and sustain a collective identity, despite being unbounded by any sort of membrane. Here we're looking at significant structure, or organization, without even a pretense of mechanically rigid form. How do cells manage that?

The problem was framed this way by Anthony Hyman from the Max Planck Institute of Molecular Cell Biology and Genetics in Dresden, Germany, and Clifford Brangwynne from the Department of Chemical and Biological Engineering at Princeton University:

> Non-membrane-bound macromolecular assemblies found throughout the cytoplasm and nucleoplasm … consist of large numbers of interacting macromolecular complexes and act as reaction centers or storage compartments … We have little idea how these compartments are organized. What are the rules that ensure that defined sets of proteins cluster in the same place in the cytoplasm?

Even more puzzling, a "compartment" can maintain its functional (purposive) identity despite the rapid exchange of its contents with the surrounding cytoplasm. "Fast turnover rates of complexes in compartments can be found throughout the cell. How do these remain as coherent structures when their components completely turn over so quickly?" (Hyman and Brangwynne 2011).

Well-structured droplets

Part of the picture that has recently come into focus has to do with the phases of matter and the transitions between these phases. (Think, for example, of the solid, liquid, and gaseous phases of water, or of solutions and gels — matter in different states.) For example, it's possible for well-defined droplets of one kind of liquid to occur within a different liquid, like oil droplets in water.

We now know that molecular complexes containing both RNA and protein often gather together to form distinctive RNA-protein liquids that separate out as droplets within the larger cytoplasmic medium. Like liquids in general, these droplets tend toward a round shape, can coalesce or divide, can wet surfaces such as membranes, and can flow. The concentration of particular molecules may be much greater in the droplets than in the surrounding fluid, conferring specific and efficient functions upon the assemblies.

Enzymes and reactants can rapidly diffuse within the liquid droplet, while also moving with relative ease across the boundary between droplet and surrounding medium. Yet this boundary can remain distinct until phase-changing environmental conditions occur — conditions that might involve slight changes in temperature, pH, salt concentration, electrical charge, molecular densities, the addition of small chemical groups to proteins, degradation of proteins, the activity of gene transcription, or still other factors.

In this way, a very subtle change — originating, say, from an extracellular influence — can yield a dramatic transformation of cytoplasmic organization, just as a slight change in the temperature or salinity of water can shift an ice-forming condition to an ice-melting one, or vice versa.

Moreover, these phase-separated droplets can be highly organized internally: "multiple distinct liquid phases can coexist and give rise to richly structured droplet architectures determined by the relative liquid surface tensions" (Shin and Brangwynne 2017). Also, some droplets may become gel-like,[5] while others may form more or less solid granules. Many such droplets may pass through stages, from more liquid to more solid, before dispersing. They form in response to particular needs, perform their work, and then pass away. Others are more or less permanent. Phase separation has been called "a fundamental mechanism for organizing intracellular space" (Shin and Brangwynne 2017) — one where "function derives not from the structures of individual proteins, but instead,

BOX 5.1

On Shape-Shifting Blobs

Here are a few comments from an article in *Nature* titled "The Shape-Shifting Blobs That Rule Biology" (Dolgin 2022):

"For years, if you asked a scientist how they pictured the inner workings of a cell, they might have spoken of a highly organized factory, with different departments each performing specialized tasks in delineated assembly lines.

Ask now, and they might be more inclined to compare the cell to a chaotic open-plan office, with hot-desking zones where different types of cellular matter gather to complete a task and then scatter to other regions.

Everywhere scientists look in cells, throngs of proteins and RNA seem to be sticking together, coalescing into pearl-like droplets distinct from their surrounding environment. These dynamic compartments allow cells to perform essential functions, ranging from gene control and DNA repair to waste disposal and stress responses. They are often fleeting, and are unhindered by an enclosing membrane—unlike many other cellular components, such as mitochondria, which are membrane-bound. When a droplet is no longer needed, it vanishes."

"One particular scaffolding protein seems to be the epicentre of stress-granule assembly. When the cell encounters adversity, this protein, called G3BP1, changes shape, prompting nearby RNA molecules to link up with it and promote clustering."

"A catch-all name for these compartments: biomolecular condensates. The name left open how these assemblages of proteins and nucleic acids took shape or became undone. 'It was deliberately supposed to be mechanism-free' [explained one biologist] ... Further experiments and theory showed that a huge number of forces work together to create condensates."

"'There isn't a cellular process that's been studied that is not now known to involve condensates' [biologist Rick] Young says. 'It involves damn near everything.'"

from dynamic material properties of entire [protein aggregates] acting in unison through phase changes" (Halfmann 2016).

We also know now that weak, transient interactions among intrinsically unstructured proteins and RNAs can result in crucial, flexible "scaffolds" that help to assemble these phase-separated aggregates, drawing in a set of functionally related molecules. "Weak," "transient," and "flexible" in my description here might be taken as indicators of the living, responsive, and non-machine-like character of the activity.

When *things happen* in the cell, phase transitions often play decisive roles, as a University of Colorado group discovered when looking at phase transitions in a roundworm. According to the researchers, these transitions "are controlled with surprising precision in early development, leading to starkly different supramolecular states" with altered organization and dynamics. "Reversible interactions among thousands of [these phase-separated] complexes," the authors found, account for "large-scale organization of gene expression pathways in the cytoplasm" (Hubstenberger et al. 2013).

How do you regulate flow and phases?

All this is, if you think about it, an amazing departure from the kind of picture once burned into the minds of biologists such as Richard Dawkins, from whom we heard some errant words above. Once there were dreams of compelling digital instructions in DNA; of machine-like interactions between molecules; of deterministic formation and functioning of proteins; of the cell as a collection of distinct, well-defined structures; and of cellular processes with fully predictable outcomes. But this dream has faded in the clear daylight of an entirely different reality where, among many other things, we watch a subtle and almost incomprehensible play of material changes of state.

These state changes can be affected by infinitely varying factors, such as the momentary interaction between a few molecules of a particular sort, the "minor" modification of a molecule, the increasing concentration of molecules in a particular location, or the slight temperature change of a degree or two — the kind of change that, in the larger world of nature, can freeze the surface of a lake where, a few days previously, fish routinely breached the surface to feed on insects.

> Ice cools a drink, water carves a canyon, steam powers a locomotive ...
> But ice brings down power lines, water floods towns, steam scalds skin.
> The context for these states matters, and there can be consequences if
> the appropriate state is perturbed or dysregulated. Now more than ever,
> we understand that physical states dictate biological function, and ...

> recent papers have highlighted, at the subcellular and tissue levels, the importance of understanding those states and the conditions in which they occur (Szewczak 2019).

We heard it asked earlier how intrinsically unstructured proteins "know" what to do at any one place and time. The old model assumed, rather puzzlingly, that random encounters between freely diffusing molecules accounted for many of the biological interactions we observe. But numerous researchers are now embracing the emerging picture of biological phase transitions as offering a very different understanding. Peter Tompa, a structural biologist from Vrije Universiteit Brussel in Belgium, sees certain phase transitions as directing "the movement of regulatory proteins in and out of organized subcellular domains" — part of the systematic maintenance of order in the cell[7] (Tompa 2013).

This is all well and good, but does it tell us (as is often implied) what "controls" and "directs" molecular engagements in relation to the distinct needs of the cell at different locations and times? If the organization of phase-separated aggregates is what coordinates the activity of proteins, then we shouldn't have to ask, as researchers are now asking, "Why do some proteins localize to only the nucleolus, while others can be found in both the nucleolus and Cajal bodies?" (Zhu and Brangwynne 2015) (Cajal bodies, like the nucleolus, are non-membrane-bound organelles found in the cell nucleus.) And, even if that question had a ready answer, the more fundamental issue would remain: if we assume that phase-separated droplets lead to properly coordinated protein interactions, then what explains the well-timed and intricately organized formation, structuring, and dissolution of the condensates?

This illustrates how (to get ahead of ourselves just a little bit) all attempts to answer questions of regulation in strictly physical terms never do really answer them. Rather, they lead only to an elucidation of previous physical states that again raise the same broad questions. There is no way to step outside the endlessly regressing physical explanations except by truly stepping outside them — except, that is, by turning to a different sort of explanation possessing a certain "finalistic" aspect. This is where we attend to the play of intentions and end-directed activities that are implicit in the stories we find ourselves looking at.

After all, questions about biological regulation are questions about the *significant patterning* of living events, and these just *are* questions about a story — about the relation of continually adjusted means to the needs, strivings, and qualities of a particular life. It is no surprise, then, that our answers must be gained in the way we come to understand a story — for example, in the way we make sense of a journey rather than in the way we grasp the physical mechanics of walking.

And then there is water – the mediator of flow

I have long thought that some day water will be seen as the single most fundamental, "information-rich" physical constituent of life, and that revelations in this regard will outweigh in significance even those concerning the structure of the double helix. Not many biologists today would countenance such a suggestion, and I am not going to mount a serious defense of it here, if only for lack of ability. Time will decide the matter soon enough. But I was particularly pleased to find that the widely read and respected *Nature* columnist, Philip Ball, once entitled a piece, "Water as a Biomolecule." In it he wrote:

> Water is not simply "life's solvent," but rather an active matrix that engages and interacts with biomolecules in complex, subtle and essential ways ... Water needs to be regarded as a protean, fuzzily delineated biomolecule in its own right (Ball 2008a; see also Ball 2008b).

In another paper, Ball (2011) summarized some work bearing on the role of water in biological contexts. The main topic had to do with the relation between water, the binding cavity of an enzyme, and the substrate molecule to which the enzyme binds. It turns out, according to the authors of a study Ball cites, that "the shape of the water in the binding cavity may be as important as the shape of the cavity." Ball goes on to remark:

> Although all this makes for a far more complicated picture of biomolecular binding than the classic geometrical "lock and key" model, it is still predicated on a static or quasi-equilibrium picture. That, too, is incomplete.

Then he cites another paper on enzyme-substrate binding. There it is revealed that, before the binding is complete, water movement near the enzyme is retarded. "Crudely put, it is as if the water 'thickens' towards a more glassy form, which in turn calms the fluctuations of the substrate so that it can become locked securely in place. It is not yet clear what causes this solvent slowdown as a precursor to binding; indeed, the whole question of cause and effect is complicated by the close coupling of protein and water motion and will be tricky to disentangle. In any event, molecular recognition here is much more than a case of complementarity between receptor and substrate — it also crucially involves the solvent."

All this suggests to Ball that "changes in protein and solvent dynamics are not mere epiphenomena, but have a vital role in substrate binding and recognition."

Structural biologists Mark Gerstein and Michael Levitt (the latter a 2013 Nobel laureate in chemistry) wrote a 1998 article in *Scientific American* entitled

"Simulating Water and the Molecules of Life." In it they mentioned how early efforts to develop a computer simulation of a DNA molecule failed; the molecule (in the simulation) almost immediately broke up. But when they included water molecules in the simulation, it proved successful. "Subsequent simulations of DNA in water have revealed that water molecules are able to interact with nearly every part of DNA's double helix, including the base pairs that constitute the genetic code" (Gerstein and Levitt 1998).

Early attempts to simulate protein molecules rather than DNA produced an analogous difficulty, with the same, water-dependent resolution. Gerstein and Levitt concluded their article with this remark:

> When scientists publish models of biological molecules in journals, they usually draw their models in bright colors and place them against a plain, black background. We now know that the background in which these molecules exist — water — is just as important as they are.

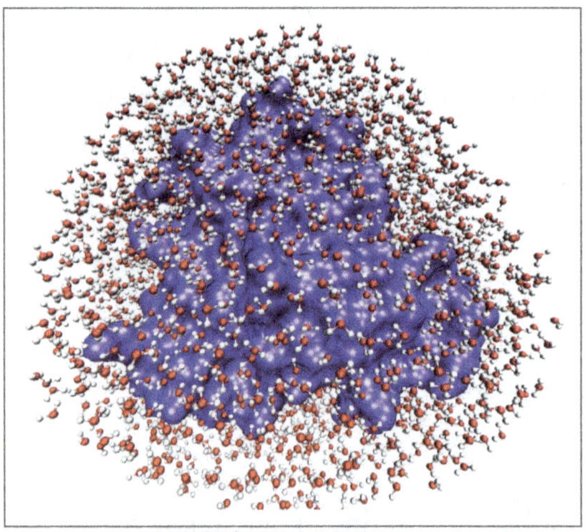

Figure 5.3. A representation of the hydration shell of myoglobin, where blue is the myoglobin protein and the small, red-and-white figures stand for water molecules.[8]

That was twenty-five years ago. Today the background remains to be filled in, even if we are now seeing signs of change. Philip Ball (who cites that Gerstein/Levitt remark, and who reproduces two images like the one in Figure 5.3) has more recently noted "an interesting sociological question," namely, "why certain communities in science decide that particular aspects of a problem are worth devoting a great deal of attention to while others become minority concerns, if not in fact regarded as somewhat suspect and disreputable." He adds:

> Why should we place so much emphasis, for example, on determining crystal structures of proteins and relatively little on a deep understanding of the [water-related] forces … that hold that structure together and that enable it to change and flex so that the molecule can do its job? (Ball 2013)

Certain peculiar historical episodes have contributed to the disreputability of water as a "molecule of life." (Too many researchers have thought they glimpsed something about water that went beyond current principles of understanding, so that work of this sort came to be seen as mystically tainted or "on the fringe.") But surely part of the answer to Ball's question has to do with the longstanding distortion of biology due to the emphasis upon code and mechanism. It is much easier to imagine the step-by-step execution of a computer-like code or the clean insertion of a key into a lock than it is to come to terms with fluid transformations — that is, with what is actually life-like.

The high era of molecular biology that followed upon discovery of "the" structure of the double helix was indeed the Age of Simplicity. We can be thankful that the feverish enchantment of fixed code and crystal is now giving way to an increasing recognition of movement, flow, dynamically flexible interaction, and the continual transfiguration of form — prime narrative elements in the organism's story.

WHERE ARE WE NOW?

Organisms are activities, not things.

Many observers have sensed, whether vividly or dimly, that the modern fixation upon things rather than activities — on what has already become rather than the process of becoming — severely distorts our sense of reality. But it is hard for us today to step fully out of this distortion. And nowhere is that distortion more destructive than in the science of life.

Perhaps for that very reason the distortion is also more visible in the science of life. And thanks to new imaging technologies, the visibility is now quite literal. At the cellular level, novel techniques are enabling us to see not only frozen, crystallized structures, but living movement. DNA, RNA, and proteins are being reconceived as "biological soft matter," subject to continually changing form so that molecular performances become more like improvised dances than automatic lock-and-key mechanical interactions. "Disordered" or "unstructured" sequences in proteins are now seen as decisive for coordinated activities throughout the cell, from gene regulation to signaling across membranes.

Still more dramatically, molecular biologists have in recent years become almost transfixed by the novel importance of phase transitions — for example, the forming and dissolving of distinctive, membraneless droplets within the fluid cell, whereby specialized and localized functional capacities are maintained despite the rapid passage of molecules in and out of the droplets.

And perhaps most important of all is the nascent recognition — which still hasn't taken widespread hold in biology — that the amazing functional plasticity of water may be key to just about everything that goes on in a cell.

All this points us to the question of coherence: how are the virtually infinite "degrees of freedom," so evident in the free flows of the cell, disciplined and subordinated to the larger purposes of the cell, whether they be gene expression or intercellular communication or metabolism or cell division. In the next chapter ("Context: Dare We Call It Holism?") and in Chapter 8 ("The Mystery of an Unexpected Coherence") we will try to get some clearer views of this larger, meaningful picture.

NOTES

1. The twentieth-century American philosopher, Susanne Langer, clearly grasped the essence of the matter in her own discussion of the heart's development and functioning. The heart, she said,

   > begins to form early in embryonic life, apparently serving no purpose until the incipient vascular system is ready to act with it. In the earliest phases, however, a characteristic function of periodic contraction, the so-called "pulse," appears in many evolving tissues, some of which will cease to exhibit it later, while others will join the cardiac development, so their rhythms will become entrained by larger ones and finally by the [entire] circulatory pulse.

   This preliminary beating, which comes early in the heart's formation, "illustrates a basic characteristic of organic function, namely, that its integated activities are often detectable before their special mechanisms have even begun to appear." This is a powerful reminder that, in an organism's development, the part "descends from," or is differentiated within, its larger context, which is ultimately the whole organism. Speaking further of the heart's development, Langer wrote:

   > Nothing could demonstrate more aptly the primacy of acts in biological existence, and their gradual concentration in those regions of an organism where they can expand, dominate and integrate most fully. This order of development, from differentiating function to specialized location (tissue determination) and finally specialized form (cell determination), has been noted many times by embryologists. [American zoologist] Charles Manning Child remarked, fifty years ago, that "differences in reaction or in capacity to react very commonly exist in different parts even before visible differentiation occurs, or in cases where it never occurs."

   Langer reinforces these remarks by citing the embryologist and author of *Form and Causality in Early Development*, Albert M. Dalcq, to the effect that, to begin with, the unity of the nervous system "is not so much spatial as functional … The nervous system does not really originate from a unique and continuous layer of cells." And the American developmental biologist, Clifford Grobstein, whose life spanned much of the twentieth century, concluded from his experimental studies of development in young embryos that "when nervous tissue 'self-differentiates' … the cells themselves have not yet acquired fixity of type as nerve cells. … some stabilization at the tissue level seems to precede stabilization at the cell level" (Langer 1967, pp. 200, 401–02).

For a more recent discussion of the heart, see the impressive evidences and analysis in Branko Furst's technical treatise on *The Heart and Circulation: An Integrative Model* (Furst 2020).

2. Figure 5.1 credit: Courtesy of Margot Quinlan (copyright).
3. A terminological issue: Turoverov and colleagues speak more specifically of "highly dynamic biological soft matter positioned at the edge of chaos." The abstract and perhaps rather tiresome notion of "the edge of chaos" is better captured in this context by a picture of life-like processes — powerfully organized, but in a dynamic manner that continually adapts to circumstances from a purposive, and therefore not *physically* predictable, center of agency. The predictability, such as it is, lies in the reasonable expectation of coherence in the interweaving meanings we observe. (See Chapters 2 and 8.)
4. Biologists often speak of communication in terms of *signals* and *signaling*, where *signal* can hardly be distinguished in any absolute way from *cause*. However, "signals" tend to be spoken of where there are repeated, more or less stereotypical sequences ("pathways") of molecular interaction between different cells, leading to more or less consistent consequences. This happens, for example, when a gland secretes a hormone ("signal") that subsequently has effects in other parts of the body.

   Wikipedia offered this definition of "cell signaling" in August, 2019: "Cell signaling is part of any communication process that governs basic activities of cells and coordinates multiple-cell actions. The ability of cells to perceive and correctly respond to their microenvironment is the basis of development, tissue repair, and immunity, as well as normal tissue homeostasis." This easy acknowledgment of "communication," "coordination," "governance," "perception," and "correct response" — all within a science that, on the surface, refuses the normal and unavoidably immaterial meaning of these terms — illustrates the biologist's blindsight described in Chapter 1.
5. A sol-gel transition occurs when a solution (in which one substance is dissolved in another) passes into a gel state. The latter consists of a solid molecular lattice that is expanded throughout its volume by a fluid — water, in the case of a hydrogel. The fluid may constitute over 99% of the volume of the gel, yet the solid lattice prevents the gel from flowing like a liquid.
6. Figure 5.2 credit: from Gerum et al. 2013, CC BY-SA 3.0.
7. Here is one of innumerable examples of the role of phase separation in physiological processes: "Cells under stress must adjust their physiology, metabolism, and architecture to adapt to the new conditions. Most importantly, they must down-regulate general gene expression, but at the same time induce synthesis of stress-protective factors, such as molecular chaperones … [We] propose that the solubility of important translation factors is specifically affected by changes in physical–chemical parameters such [as] temperature or pH and modulated by intrinsically disordered prion-like domains. These stress-triggered changes in protein solubility induce phase separation into aggregates that regulate the activity of the translation factors and promote cellular fitness" (Franzmann and Alberti 2019).
8. Figure 5.3 credit: From Frauenfelder et al. 2009.

SOURCES

Ball, Philip (2008a). Water as a Biomolecule, *ChemPhysChem* vol. 9, pp. 2677–85. doi:10.1002/cphc.200800515

________ (2008b). Water as an Active Constituent in Cell Biology, *Chemical Reviews* vol. 108, no. 1, pp. 74–108. doi:10.1021/cr068037a

________ (2011). More Than a Bystander, *Nature* vol. 478 (October 27), pp. 467–68. doi:10.1038/478467a

________ (2013). Concluding Remarks: *Cum Grano Salis, Faraday Discussions* vol. 160, pp. 405–14. doi:10.1039/c2fd20126g

Dawkins, Richard (2006). *The Blind Watchmaker*, third edition. New York: W. W. Norton. First edition published in 1986.

Dolgin, Elie (2022). The Shape-Shifting Blobs That Rule Biology, *Nature* vol. 611 (November 3), pp. 24–27. doi:10.1038/d41586-022-03477-y

Dunker, A. Keith, Christopher J. Oldfield, Jingwei Meng et al. (2008). The Unfoldomics Decade: An Update on Intrinsically Disordered Proteins, *BMC Genomics* vol. 9 (Suppl 2):S1. doi:10.1186/1471-2164-9-S2-S1

Emerson, R. W. (1908). *The Conduct of Life: Nature and Other Essays*. New York: E. P. Dutton. Available at https://archive.org/details/conductoflifenat00emeriala

Franzmann, Titus M. and Simon Alberti (2019). Protein Phase Separation as a Stress Survival Strategy, *Cold Spring Harbor Perspectives in Biology* vol. 11, no. 6 (June). doi:10.1101/cshperspect.a034058

Frauenfelder, Hans, Guo Chen, Joel Berendzen et al. (2009). A Unified Model of Protein Dynamics, *PNAS* vol. 106, no. 13 (March 31), pp. 5129–34. doi:10.1073/pnas.0900336106

Furst, Branko (2020). *The Heart and Circulation: An Integrative Model*, 2nd edition. New York: Springer.

Ganser, Laura, Megan L. Kelly, Daniel Herschlag and Hashim M. Al-Hashimi (2019). The Roles of Structural Dynamics in the Cellular Functions of RNAs, *Nature Reviews Molecular Cell Biology* (Aug). doi:10.1038/s41580-019-0136-0

Gerstein, Mark and Michael Levitt (1998). Simulating Water and the Molecules of Life, *Scientific American* vol. 279, no. 5 (November), pp. 100–05. doi:10.1038/scientificamerican1198-100

Gerum, R. C., B. Fabry, C. Metzner et al. (2013). The Origin of Traveling Waves in an Emperor Penguin Huddle, *New Journal of Physics* vol. 15 (December). https://iopscience.iop.org/article/10.1088/1367-2630/15/12/125022

Grant, Barry J., Alemayehu A. Gorfe and J. Andrew McCammon (2010). Large Conformational Changes in Proteins: Signaling and Other Functions, *Current Opinion in Structural Biology* vol. 20, pp. 142–47. doi:10.1016/j.sbi.2009.12.004

Gsponer, Jörg and M. Madan Babu (2009). The Rules of Disorder Or Why Disorder Rules, *Progress in Biophysics and Molecular Biology* 99, no. 2–3 (February-May), pp. 94–103. doi:10.1016/j.pbiomolbio.2009.03.001

Halfmann, Randal (2016). A Glass Menagerie of Low Complexity Sequences, *Current Opinion in Structural Biology* vol. 38, pp. 18–25. doi:10.1016/j.sbi.2016.05.002

Harold, Franklin M. (2001). *The Way of the Cell: Molecules, Organisms and the Order of Life*. Oxford: Oxford University Press.

Hilser, Vincent J. (2013). Structured Biology: Signalling from Disordered Proteins, *Nature* vol. 498 (June 20), pp. 308–10. doi:10.1038/498308a

Holdrege, Craig, editor (2002). *The Dynamic Heart and Circulation*. The chapters originally written in German were translated by Katherine Creeger. Fair Oaks CA: AWSNA.

Hubstenberger, Arnaud, Scott L. Noble, Cristiana Cameron and Thomas C. Evans (2013). Translation Repressors, an RNA Helicase, and Developmental Cues Control RNP Phase Transitions during Early Development, *Developmental Cell* vol. 27 (October 28), pp. 161–73. doi:10.1016/j.devcel.2013.09.024

Hyman, Anthony A. and Clifford P. Brangwynne (2011). Beyond Stereospecificity: Liquids and Mesoscale Organization of Cytoplasm, *Developmental Cell* vol. 21 (July 19), pp. 14–16. doi:10.1016/j.devcel.2011.06.013

Landecker, Hannah (2012). The Life of Movement: From Microcinematography to Live-Cell Imaging, *Journal of Visual Culture* vol. 11, pp. 378–99. doi:10.1177/1470412912455622

Pearson, Helen (2003). Beyond the Double Helix, *Nature* vol. 421 (January 23), pp. 310–12. 10.1038/421310a

Rothman, Stephen (2002). *Lessons from the Living Cell: The Limits of Reductionism*. New York: McGraw-Hill.

Schad, Wolfgang (2002). A Dynamic Morphology of the Cardiovascular System, in Holdrege (2002), pp. 77–97.

Shin, Yongdae and Clifford P. Brangwynne (2017). Liquid Phase Condensation in Cell Physiology and Disease, *Science* vol. 357, no. 6357 (September 22). doi:10.1126/science.aaf4382

Szewczak, Lara (2019). Just Solid or Liquid Enough, *Cell* vol. 178, no. 4 (August 8), pp. 763–64. doi:10.1016/j.cell.2019.07.027

Talbott, Stephen L. (2024). How Do Biomolecules "Know" What To Do? https://bwo.life/bk/ps/biomolecules.htm

Tompa, Peter (2013). Hydrogel Formation by Multivalent IDPs: A Reincarnation of the Microtrabecular Lattice?, *Intrinsically Disordered Proteins* 1:e24068 (January–March). doi:10.4161/idp.24068

Turoverov, Konstantin K., Irina M. Kuznetsova, Alexander V. Fonin et al. (2019). Stochasticity of Biological Soft Matter: Emerging Concepts in Intrinsically Disordered Proteins and Biological Phase Separation, *Trends in Biochemical Sciences* vol. 44, no. 8 (August 1), pp. 716–28. doi:10.1016/j.tibs.2019.03.005

Zhou, Yaoqi, Dennis Vitkup and Martin Karplus (1999). Native Proteins Are Surface-Molten Solids: Application of the Lindemann Criterion for the Solid versus Liquid State, *Journal of Molecular Biology* vol. 285, pp. 1371–75. doi:10.1006/jmbi.1998.2374

Zhu, Lian and Clifford P. Brangwynne (2015). Nuclear Bodies: The Emerging Biophysics of Nucleoplasmic Phases, *Current Opinion in Cell Biology* vol. 34, pp. 23–30. doi:10.1016/j.ceb.2015.04.003

CHAPTER 6

Context: Dare We Call It Holism?

THE CENTRALITY OF LIVING WHOLES within biology seems beyond argument. These have not been "put together" or built by an external agency. They are never the *results* of a physical activity that starts with non-wholes. Biology gives us nothing but beings that have never existed except as wholes possessing the formative powers that enable them to pass through further stages of physical development.

The one-celled zygote is already a functioning whole. It does not gain further cells through the addition of "building blocks" assembled by an engineer or designer, but rather through an internal power of reorganization and subdivision in which the entire organism participates. All the parts are orchestrated in a unified performance that yields (through division of existing cells) new cells, and particular kinds of cells, just where they are needed. The orchestrating power of the whole can hardly be determined by the particular parts it in this way brings into being and orchestrates.

Where the physicist may prefer unambiguous, isolated, and well-defined "point" causes, the biologist never has such causes to theorize about. A biological whole is never absolute, and never perfectly definable as distinct from its environment. Further, its actions are always multivalent, and they interpenetrate one another, like the meanings of events in a story.

The wonderfully insightful, twentieth-century botanist, Agnes Arber (Arber 1985, p. 59), captured well the polar tension between organic wholeness, on one hand, and contextual embeddedness, on the other:

> The biological explanation of a phenomenon is the discovery of its own intrinsic place in a nexus of relations, extending indefinitely in all directions. To explain it is to see it simultaneously in its full individuality (as a whole in itself), and in its subordinate position (as one element in a larger whole).

Every ecological setting, every organism within that setting, every organ within the organism, and every cell within the organ is a whole providing a context for its own interrelated parts, and at the same time is itself contextually embedded within larger wholes. "Context," "whole," and "part" can never be rigid, absolute terms in biology. They are bound up with interweaving spheres of activity.

We need to gain some practice in thinking, not with the single, distinct point-causes of the physicist (or at least the classically minded physicist), but rather with the actual narrative qualities of biological activity. The perplexing issues surrounding attempts at holistic thought may thereby lend themselves more easily to our efforts at understanding.

Which comes first, the cell or its niche?

Every cell in an organism lives in a sufficiently distinctive way — is enough of a whole in its own right — to pose the question of identity. What makes this particular cell of my skin a keratinocyte rather than a melanocyte? Does identity imply constancy of cellular character? To what degree does a cell's environment — the larger community of cells — shape its identity and character?

Current researches are showing us how the fate of any given cell in our bodies is bound up with that of nearby cells in the same local environment, or "niche."[1] Consider, for example, the basal stem cells of the mammalian airway. (*Basal cells* constitute an internal layer of the epithelium immediately above the basement membrane, and they help to anchor the epithelium to the tissue below. *Stem cells* are relatively undifferentiated cells capable, at need, of dividing and differentiating into more specialized cell types.)

One research group found that when airway basal stem cells were in demand as a result of injury, there was a "surprising increase in the proliferation of committed secretory cells." It turned out that many of these latter, fully differentiated (specialized) cells, were, so to speak, reversing their specialization and becoming basal stem cells. The "de-differentiated" cells "were morphologically indistinguishable from stem cells and they functioned as well as [normal stem cells] in repairing epithelial injury ... This capacity of committed cells to de-differentiate into stem cells may have a more general role in the regeneration of many tissues" (Tata et al. 2013).

Further, direct contact with a single basal stem cell was enough to prevent secretory cells from de-differentiating and becoming stem cells. Clearly, then, the identity of these fully matured secretory cells is not rigidly fixed, and at the same time their transformation potential is delicately sensitive to context. We can hardly separate the question of a cell's identity from that of the niche's identity, or from the changing needs of the moment.

This point is driven home by a second study concerning mouse hair follicles.[3] The researchers explored how a cell's location within various compartments of the niche affects its fate. For example, stem cells in the bulge (see Figure 6.1) tend to stay quiescent — that is, they remain in a resting state without cell division — whereas those in the hair germ are continually differentiating into more specialized

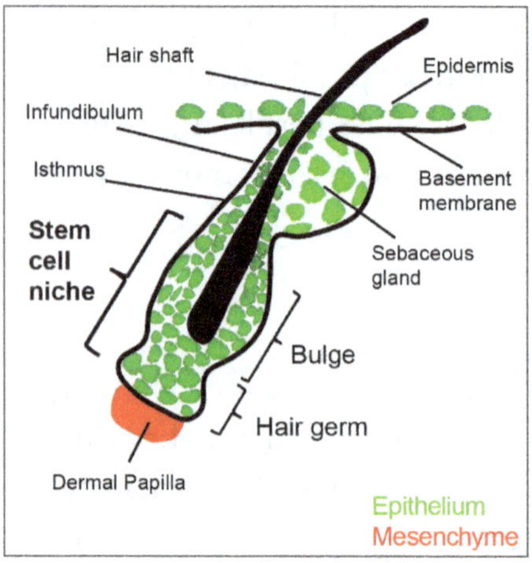

Figure 6.1. Mouse hair follicle.[2]

cell types. And even within the bulge, stem cells in the upper half remain much more consistently quiescent, whereas those in the lower half are more proliferative.

Dramatically, the authors also show that "hair follicle stem cells are dispensable for regeneration, and that epithelial cells, which do not normally participate in hair growth, re-populate the lost stem-cell compartment and sustain hair regeneration" — provided, however, that "the overall integrity of the niche is maintained." When the stem cell population in the bulge or hair germ is destroyed by laser ablation, distant epithelial cells flow toward the damaged compartment and go through a transformation of identity enabling them to replace the lost cells. As the authors summarize it, "The overall structure and function of the tissue is maintained because cells are capable of adopting new fates as dictated by their new niche microenvironment."

Clearly, the different elements of the hair follicle niche are not rigidly fixed entities. Rather, their changing forms and relationships are choreographed by the larger environment. So the goings-on in the hair follicle niche illustrate very well how the context helps to "decide" what sorts of elements it will have, how they will be formed and transformed, how they will flow toward the places where they are needed, and how they will come into mutual relationship. Nothing could be further from the common picture of an organism being constructed, bottom-up, from an available collection of well-defined building blocks capable of determining outcomes. It appears, rather, that the desirable outcome determines the "building blocks."

And we need to remember that the humble hair follicle represents just one of millions of distinct niches within a mouse or human being. Liver, kidney, heart — every organ embraces countless micro-environments, none of which is exactly like any other. In every one of those micro-environments a unique, evolving collection of cells is caught up in the wholeness of its governing context. And the same demand for flexible coordination, but now at a higher level, is repeated as all those niches are assimilated to the unity of a single organ, and again in the way the organs are brought into harmony within the functioning of the whole organism.

Embryos in general exhibit this power of flexible coordination to an extraordinary degree. Harvard biologist Richard Lewontin once described how you can excise the developing limb bud from an amphibian embryo, shake the cells loose from each other, allow them to reaggregate into a random lump, and then replace the lump in the embryo. A normal leg develops. Somehow the form of the limb as a whole is the ruling factor, redefining the parts according to the larger pattern. Lewontin went on to remark:

> Unlike a machine whose totality is created by the juxtaposition of bits and pieces with different functions and properties, the bits and pieces of a developing organism seem to come into existence as a consequence of their spatial position at critical moments in the embryo's development.

A developing organism, Lewontin adds, "is like a language whose elements ... take unique meaning from their context" (Lewontin 1983).

"More than the sum of its parts": clarifying a cliché

The foregoing illustrates how a biological context defines its parts — lends them their meaning — not the other way around. However hard it may be for us to conceive, there seems to be something fundamentally *causal* about a context. It *governs* its parts, bringing them into existence, transforming them, and coordinating their activity. But there is another, closely related way to look at the matter. For this we can turn to the preeminent cell biologist, Paul Weiss, whose work extended from the 1920s into the 1970s, when he was awarded the National Medal of Science by President Jimmy Carter.

As a life-long observer of cells and tissues, Weiss pointed out something obvious, simple, and yet revolutionary for today's biology. When we examine the form and physiology of an organism, we see how "certain definite rules of order apply to the dynamics of the *whole* system... reflected [for example] in the orderliness of the overall architectural design, which cannot be explained in terms of any underlying orderliness of the constituents" (Weiss 1971, p. 286).

That is, despite the countless processes going on in the "heaving and churning" interior of the cell (Weiss 1973, p. 40), and despite the fact that each process might be expected to "go its own way" according to the myriad factors impinging on it from all directions, the actual result is quite different. Rather than becoming progressively disordered in their mutual relations (as indeed happens after death, when the whole dissolves into separate fragments), the processes come together in a larger unity. The behavior of the whole "*is infinitely less variant from moment to moment than are the momentary activities of its parts*":

> Small molecules go in and out, macromolecules break down and are replaced, particles lose and gain macromolecular constituents, divide and merge, and all parts move at one time or another, unpredictably,

> so that it is safe to state that at no time in the history of a given cell, much less in comparable stages of different cells, will precisely the same constellation of parts ever recur ... Although the individual members of the molecular and particulate population have a large number of degrees of freedom of behavior in random directions, the population as a whole is a system which restrains those degrees of freedom in such a manner that their joint behavior converges upon a nonrandom resultant, keeping the state of the population as a whole relatively invariant (Weiss 1962, p. 6).

Tuning in to this basic picture — if we could really take it seriously — might change just about everything in biology. It is therefore worth hearing the voice of one other competent authority who emphatically echoed Weiss' remarks, despite writing from a very different specialist's angle. In a 1985 paper Guenter Albrecht-Buehler, a biophysicist at Northwestern University, wrote this:

> During the course of uncountable simultaneous biochemical and biophysical events, information is copied, received, transmitted, and stored as hormones, mRNAs, neurotransmitters, pre- and postsynaptic potentials, post-translational modifications, covalent links between macromolecules. Different kinds of cyclic events keep time and synchronize other cellular events, eventually generating mitotic cycles and circadian rhythms. Materials such as proteins and RNAs or much larger structures such as viruses or organelles that carry the information around are transported, produced, exchanged, recycled, modified, inhibited, or stored.

And he concluded: "All these events occur in a microscopically small world of violent and random thermal fluctuations. Yet, cytoplasm can keep its complex actions accurate in the midst of drowning thermal noise all around and within" (Albrecht-Buehler 1985).

We might say that a given type of cell (or tissue, or organ, or organism) insists upon maintaining its own recognizable identity with "unreasonable" tenacity, given the relative freedom, in purely physical terms, of its molecular constituents as they make their way through a watery medium.

The same principle holds when we look at the "erratic" placement of individual cells within a larger tissue that exhibits beautiful order. In Weiss' summary:

> While the state and pattern of the whole can be unequivocally defined as known, the detailed states and pathways of the components not only are so erratic as to defy definition, but, even if a Laplacean spirit[4] could trace them, would prove to be so unique and nonrecurrent that they would be devoid of scientific interest.[5]

"This," Weiss remarks, "is exactly the opposite of a machine," where the pattern of the product "is simply the terminal end of a chain of rigorously predefined sequential operations of parts. In a [biological] system, the structure of the whole coordinates the play of the parts; in the machine, the operation of the parts determines the outcome" (Weiss 1973, p. 41). So it happens, in Weiss' words, that "molecules can contribute to 'the control of cellular activity' only insofar as 'cellular control' prevails over their individual activities" (Weiss 1962, p. 62). In other words, it doesn't make much sense to speak at all of molecules as *controllers* of cellular activity.

Everything here might well remind us of what we heard from E. S. Russell in Chapter 2: in living activity, the end is more constant than the means. Or: the *purposive* end is more constant than the *physical* means. And it can hardly be disputed: the imposition of order upon the cell that we have been describing seems impossible to understand without an element that looks much more like intention than like the physical lawfulness through which that intention realizes itself. (As I have tried to make clear since the first chapter of this book, and hope to make clearer still in what follows, I am not necessarily speaking of human-like awareness, concious planning, and intention.)

So, anyway, it turns out, with a touch of irony, that *less* change is what shows the whole cell, or any organic whole, to be *more* than the sum of its parts. It's as if there were an active, coordinating agency subsuming all the part-processes and disciplining their separate variabilities so that they remain informed by, and caught up in, the greater unity. The coordination, the ordering, the continual overcoming of otherwise disordering impacts from the environment so as to retain for the whole a particular character or organized way of being, expressively unique and different from other contexts — this is the "more" of every organism and every context that cannot be had from the mere summing of discrete, causal parts.

So the center holds, and this ordering center — this whole that is more than the sum of its parts — cannot itself be just one or some of those parts it is holding together. When the organism dies, the parts are all still there, but the whole is not.

Every biological context is a complex of embodied ideas

Curiously, "holism" has almost become a dirty word in biology. It commonly connotes loose thinking, vagueness, obscurantism, and perhaps even an unfortunate tendency toward mysticism. I say "curiously" because the fact is that biologists speak incessantly about holism. You might almost think they were, in recent years, becoming fanatical about it. It's just that they prefer to honor holism under the more acceptable (and obscure) slogan, "context matters." This latter idea occurs like a mantra in the contemporary technical literature, so that it would be hard to find any physiological or behavioral process that is not routinely (and rightly) said to be "context-dependent" or "context-specific."[6]

Strangely, despite the almost universal employment of the pregnant term "context" in molecular biology, it is rarely if ever defined, and the unwelcome meanings implicit in the use of the term seem never to become explicit. And yet the word can hardly mean anything at all if it is not a close synonym for "larger whole." The frequent appeal to context as a decisive determining factor, then, looks rather like an under-the-table invocation of the unmentionable concept of holism. It allows biologists to import the seemingly inescapable idea of the causal whole into their descriptions and theorizing, while outwardly pursuing a style of explanation that pretends to disdain holism in favor of purely physical analysis into parts — the parts whose sum supposedly gives us the whole.

Let's not forget: when we say that what happens in a cell is "context-dependent," we are talking about a watery expanse populated by untold billions of molecules in unsurveyable variety. The need is for just the right combinations of molecules to do just the right things "in the moment" — and to do them *in light of the overall state of the entire cell within its particular tissue*. Is this cell just now committing itself to cellular division? Then what these molecules here and those molecules over there must do is now being radically redefined. Their new "assignments" depend not only on their location in the cell, but also on their necessary functional participation in lengthy, complex, temporal sequences of interaction that require the choreographing of countless other molecules as well.

Something is always going on contextually, and all the molecular interactions, taken together, must reflect whatever that something happens to be — must reflect the meaning of the encompassing narrative.

Appeals to context are necessary because transient, local, physical causes are unable to explain the purposive and narrative *significance* of whatever is going on. A broader, orchestrated performance is always playing out — a performance to which local processes are made to conform. This seems to imply that a kind of cause is being directed from the context, or whole, toward the parts. (See Chapters 9, "A Mess of Causes," and 10, "What Is the Problem of Form?" — and especially

BOX 6.1

Some Call It Holism

The yearning for a means to recognize and understand the whole organism seems to have surfaced with more or less intensity throughout all modern history — but perhaps at no time more insistently and wisely than during the first half of the twentieth century. The great neurosurgeon, Kurt Goldstein, wrote in his masterpiece, *The Organism*, first published in 1934:

> By virtue of [an] isolating, dismembering procedure one can readily abstract and single out from living phenomena those phenomena on the physico-chemical "plane." But the attempt to reintegrate the elements thus abstracted, to reorganize these split-off segments into the reality of living nature, is doomed to fail. ... it is not possible to comprehend the whole on the basis of the parts (Goldstein 1995, p. 378).

Likewise, the leading embryologist, F. R. Lillie, wrote in 1906:

> Cells are subordinate to the organism, which produces them, and makes them large or small, of a slow or rapid rate of division, causes them to divide, now in this direction, now in that, and in all respects so disposes them that the latent being comes to full expression ... The organism is primary, not secondary; it is an individual, not by virtue of the cooperation of countless lesser individualities, but an individual that produces these lesser individualities on which its full expression depends (quoted in Russell 1930, pp. 243–244).

The eminent physiologist, John Scott Haldane:

> When we endeavour to treat physiological phenomena as separable events we only reach unintelligible chaos to which there is no end. When we seek to understand them as manifestations of life regarded as a whole we find that we can make them intelligible and predictable (Haldane 1931, p. 69).

The mathematician and student of form, D'Arcy Thompson:

> We tend, as we analyze a thing into its parts or into its properties, to magnify these, to exaggerate their apparent independence, and to hide from ourselves (at least for a time) the essential integrity and individuality of the composite whole (Thompson 1917, p. 712).

And, finally, the neurophysiologist Sir Charles Sherrington:

> The living creature is fundamentally a unity. In trying to make the "how" of an animal existence intelligible to our imperfect knowledge, we have, for purposes of study, to separate its whole into part-aspects and part-mechanisms, but that separation is artificial. It is as a whole, a single entity, that the animal, or for that matter the plant, is finally and essentially to be envisaged (quoted in Russell 1930, p. 166).

There are countless other examples of the same sort. And today we see a resurgent defense of holism in some quarters. But I think it is fair to say that the proponents of holism have yet to make their point in a way that faces up to all the issues and effectively communicates these to the larger scientific community. And two of the key issues, which will figure in later chapters, are (1) the necessity to speak of a type of causation other than physical causation, and (2) the role of idea, or thought, in the phenomena of life.

the section, "The problem of organic form," in Chapter 12, "Is a Qualitative Biology Possible?") The parts, being *caught up* in the form and activity of the larger context, receive from it their shifting identities and meanings.

The word *context* refers etymologically to that which has been *woven together*, and is commonly applied today to language, thought, and intention, and, in general, to the connections that weave through the elements of meaningful conceptual expression, making a coherent whole out of them. But, whether we have recognized it or not, this meaning remains precisely the same when the word is used in biology. Narrative connections of language, thought, and intention are what make a biological context the characteristic whole that it is, lending it its form and meaning.

All this may be easier to grasp by looking at our own experience.

The activity on and around a football field during a game differs as a context from the activity on and around a baseball field. The difference could neither exist nor be articulated if it weren't for the distinct *ideas* and *intentions* establishing the two contexts, including all the rules of the games, the organization of leagues into teams, and the competitive framework. Those ideas and intentions enable us to predict the kinds of activity we will observe. And the predictions are possible because, one way or another, the activities occur with implicit reference to the ideas. The physical facilities, the equipment, and the players — all the activities, viewed physically — do not *produce* the ideas; rather, the activities are *governed* by the ideas. Even the very same physical checkerboard differs as a context, depending on whether the governing ideas are those of checkers or chess.

When a player runs from first base toward second and the catcher throws the ball to the second baseman, we understand the narrative sense of it because we understand the relevant ideas and intentions. Similarly, when we say that the fate of two nearly identical cells will diverge radically depending on their locations in the hair follicle niche, or that a particular chromosomal modification is "cell-type-specific" — neuron or muscle cell? — we have in mind the distinct *character* of the different contexts, their unique ways of being, and what they *need* in order to proceed through their different stories within the still larger context of the organism as a whole.

And so — as we already saw in Chapter 2 ("The Organism's Story") — whenever we speak of beings rather than things, we necessarily turn to a language of directed intention (*respond*, *develop*, *adapt*, *regulate*, and so on); a normative and aesthetically colored language (everything relating to health and disease, order and disorder, rhythm and dysrhythmia, harmony and disharmony, error and error correction); and a language of wholeness (*context, coordination*, *integration*, *organization*).

Not surprisingly, then, the biologist finds herself directly invoking the language of meaning in terms such as *message*, *information*, *communication*, and *signal*. But, again, she usually tries to do so in a mathematized, *de-meaned* manner intended to conceal the inwardness of the organism. Yet her recourse to the ubiquitous idea of context is a dead giveaway: if the word does not signify an ideational, aesthetic, and directive coherence, it refers to nothing living at all. Things just "being there" without an integral unity expressed in relational ideas — things without a describable role in a story that matters — do not make a living context.[7]

It is impossible to fit the idea of context-dependence into the biologist's usual style of causal explanation. Where do we locate a causal arrow pointing from the contextual whole to a particular part? The arrow needs to originate *everywhere*. The fact is, we cannot locate it in space at all. We recognize it, not as a link in a linear causal chain, but rather in the way we recognize a possible implication of a broad understanding. When biologists speak of context-dependence, they are speaking of a play of ideas. It is one of the ways they import the organism's interior into their thinking "under the radar."

Contextual wholes and living narratives

We saw in "The Organism's Story" that every animal's life has a narrative quality through which the meanings of its life are expressed. It pursues its own needs and interests; it interprets, responds to, and re-shapes its environment; and it exhibits a certain biological end-directedness or purposiveness in its activity, from the molecular level on up. It is always "up to something" or "going somewhere."

So long as we are content to look only at the sum of physical causes, we miss this narrative. The same would be true if we read a novel while attending only to the physical causes underlying all the events. If we then said that the unbroken connection of these causes proves the absence of the narrative — the absence of the meaningful living activity and the ideas that the story is actually *about* — then we would merely be asserting our fierce determination to see nothing but physical causes.

At this point the reader, perhaps growing impatient, has every right to ask: "If the organism's life is a narrative in the sense you speak of, then who is the narrator? And how does that narrator achieve its 'governing' or 'coordinating' role — if, indeed, you are not suggesting some sort of vital force?"

These are essential questions, and in this chapter I have tried to take the first steps toward an answer. We have seen (in Chapter 2, "The Organism's Story") that biologists in general, despite the "fierce determination" just referred to — and however unconscious they are of the meanings of their words — compulsively refer to organic *contexts* as if they somehow possessed governing powers, and as if organisms really *are* engaged in purposive, or end-directed, activity. So it appears that the questions are not only mine. They are endemic to biology. Whose are the ends or purposes we find ourselves recognizing in the life of an animal?

On my part, I have tried to begin an answer with two basic recognitions. One is that every context, so far as it is a unitary and integral whole, just *is* regarded, for practical explanatory purposes, as the narrator of the story being enacted within its own domain. It informs its parts and disciplines their activity in harmony with the needs and character of the whole. If there seems to be something *causal* about this disciplinary power (and there certainly does), it differs in some respects from the causation we are familiar with in the inanimate world.

One difference is that this biological causation can never be understood in absolute and unambiguous terms. As when reading a novel, we find that unfolding events make ever clearer sense out of an organism's life. But the interwoven and qualitative *reasons* for things never have the simple character of a mathematical law. Moreover (as we heard from Agnes Arber at the outset of this chapter), every causal context is embedded in still larger contexts, without which it cannot be fully understood. Biological causes always interpenetrate one another.

The second key recognition is that a living context, or whole, exists only by virtue of a certain "inwardness" — an inwardness associated with ideas and intentions, and therefore with volition, cognition, and intelligence. This inwardness may not sit well with the prevailing materialism in today's biology, but we will find in Chapter 12 (the section on "The problem of organic form") that it gives us the key for understanding how a context might be said to *causally govern* its constituent elements.

But even before further explanation, the tantalizing fact remains: every activity with a narrative character shares at least one thing with a human-written novel: it has its own immaterial causal basis, or meaning, which alone enables us to understand and explain what is going on. This remains true even if the meanings at issue are radically different between, say, a giraffe and a politician. An implicit belief in the meaning of organic activity is what leads biologists to classify their research projects according to the significant performances of organisms — from gene expression and cell division to pursuing prey and protecting the young.

Calling these activities "functions" may be more comforting to some (because less alive and more machine-like), but what we here denote by a function typically just *is* a meaningful performance — a form of contextualized self-expression or self-realization. We would never in the same way say of a volcano (and we ought not to say of a machine) that it is *expressing* or *realizing* its own nature by how it *responds* to its surroundings.

Meanings require, and are given by, contexts. It seems way past time in biology not only to implicitly recognize the contextual meanings of biological activity, but also to say a conscious "yes" to them and thereby free ourselves to consider the difference between understanding the animate and inanimate worlds.

WHERE ARE WE NOW?

Seeing the organism in context

In Chapter 2 ("The Organism's Story") the organism's life was described as story-like in the sense that it is a progressive unfolding in time of interwoven meanings expressing the needs, interests, and intentions representing a particular way of being. These meanings cease to exist, as we noted using a dog as an example, at the moment of death.

From there we went on to Chapter 3 ("What Brings Our Genome Alive?"), where we learned about the all-but-incomprehensible skill and wisdom with which a cell manages its own genome, and to Chapter 4 ("The Sensitive, Dynamic Cell"), where this same sort of skill and wisdom finds its expression in connection with the cytoskeleton and cellular membranes.

Then, in Chapter 5 ("Our Bodies Are Formed Streams"), we learned how all this happens in the plastic context of the cell, where organizing powers are reflected in the "dance" of chromosomes, in "molten" regions of proteins, in continual phase transitions in the cellular plasm, and, more generally, in a free movement of molecules within the fluid cell. And now we have seen (with the help of Paul Weiss) how this relatively disordered movement contrasts with the overall order of the cell, which is imposed as if "from above."

All this has led to our present concern with the *governing* role of contexts with respect to their parts — "governing" in the sense of imposing order and meaning upon the collective parts and, in fact, bringing those parts into existence as integral and well-coordinated participants in the meaning of the whole. We heard a fine example of this priority of context over parts in the description of the changing identity of cells within the hair follicle niche.

We hardly need to be told that, in the pages of their professional journals today, biologists constantly mention (although as if merely "in passing") the context-dependence of everything that happens in the organism. The unfortunate fact is that they really do pass such phrases by without making much of an effort to characterize what they mean by "context." So the charge of mysticism, if applied at all, should probably be leveled at the ubiquitous but unclarified use of the term "context" in today's biology.

Nevertheless, the term *can* be used with a perfectly clear awareness of its profound meaning — even if that meaning doesn't sit well with conventional thought today. We will have plenty of occasion in the chapters ahead for further reference to both "contextuality" and "holism," which are very nearly perfect synonyms. In particular, we will deal specifically with the importance of *organizing ideas* for our understanding of biological contexts in Chapter 21 ("Inheritance, Genetics, and the Particulate View of Life"), as well as in the section, "The problem of organic form," in Chapter 12 ("Is a Qualitative Biology Possible?").

NOTES

1. See, for example, the brief article summaries in Tan 2013.
2. Figure 6.1 credit: From Rompolas, Mesa and Greco 2013, with permission.
3. Rompolas et al. 2013. For an updated and more detailed report on this same research, see Xin et al. 2018.
4. Pierre-Simon, Marquis de Laplace (1749–1827) was a mathematician and physicist. The reference is to what is often referred to as "Laplace's demon," although Laplace himself did not use the word "demon":

   > We ought then to regard the present state of the universe as the effect of its anterior state and the cause of the one which is to follow. Given for one instant an intelligence which could comprehend all the forces by which nature is animated and the respective situation of the beings who compose it — an intelligence sufficiently vast to submit these data to analysis — it would embrace in the same formula the movements of the greatest bodies of the universe and those of the lightest atom; for it, nothing would be uncertain and the future, as the past, would be present to its eyes … The curve described by a single molecule in air or vapour is regulated in a manner just as certain as the planetary orbits; the only difference between them is that which comes from our ignorance (Laplace 1951, p. 4).

5. Here's a way to picture the situation concretely. Imagine there were bricks with a built-in mechanism for placing themselves in position on a wall under construction. Imagine further that this placement was somewhat random, so that individual bricks stuck out this way here, and that way there, and tended in general to be precariously balanced, sometimes even falling off. If this degree of randomness held true as course after course of bricks were laid down, and if it applied to the courses themselves as well, we know very well that the variability successively added to variability would eventually bring the entire, unsightly structure to the point of collapse.

   In the real-life case, the mason continually makes fine adjustments to compensate for the low-level variability that remains even under his steady hands. In this way, he reliably guides the structure toward its finished form. But he acts upon the wall as an outside force, and we see no outside force acting upon the organism so as to coordinate and continually "discipline" the degrees of freedom evident in its physiological activity. The only powers we can observe are those intrinsic to the organism itself. Our observation cannot be doubted, and we will have no hope of understanding if we lose sight of this observation due to its troubling nature.

6. Doing a quick hyperbole-check on myself, I find that a google scholar search on

   "context dependent" gene cell

   yields over 491,000 results. (I included the terms "gene" and "cell" in the search in order to ensure that the results were mostly from biological texts, with an emphasis on molecular biology.) It would be a worthwhile exercise for the reader to execute that search at scholar.google.com and then browse down through some of the article titles and excerpts.
7. It is worth mentioning here that inanimate phenomena are surely contextual as well, even if they lack those focal, organic centers of need and interest that can *respond* to context. The physicist Georg Maier once pointed this out by mentioning that the warm air in a closed room collects near the ceiling, while the air in the earth's atmosphere generally becomes cooler with elevation. The two contexts make a big difference.

   Of course, the ideas evident in inanimate phenomena are very different from the ones we find in the phenomena of life. But ideas do remain ideas: we may prefer to formulate the regularities in physical events as high, mathematical abstractions, but they are ideas nonetheless. The seemingly willful blindness to this obvious fact of the ideational character of physical reality may be the central pathology of contemporary science.

SOURCES

Albrecht-Buehler, Guenter (1985). Is Cytoplasm Intelligent Too?, Chapter 1 in *Cell and Muscle Motility*, edited by Jerry W. Shay, pp. 1–21. Boston: Springer.

Arber, Agnes (1985). *The Mind and the Eye: A Study of the Biologist's Standpoint*, with an introduction by P. R. Bell. Cambridge: Cambridge University Press. Originally published in 1954.

Goldstein, Kurt (1995). *The Organism: A Holistic Approach to Biology Derived from Pathological Data in Man*, with a foreword by Oliver Sacks. New York: Zone Books. Originally published in German in 1934, and in English in 1939.

Haldane, J. S. (1931). *The Philosophical Basis of Biology*. Donnellan Lectures, University of Dublin, 1930. Garden City NY: Doubleday, Doran & Company.

de Laplace, Pierre Simon, Marquis (1951). *A Philosophical Essay on Probabilities*, translated from the sixth French edition by Frederick Wilson Truscott and Frederick Lincoln Emory with an introductory note by E. T. Bell. New York: Dover. Originally published in 1814.

Lewontin, Richard C. (1983). The Corpse in the Elevator, *New York Review of Books* vol. 29, nos. 21–22 (January 20), pp. 34–37. https://www.nybooks.com/articles/1983/01/20/the-corpse-in-the-elevator/

Rompolas, Panteleimon, Kailin R. Mesa and Valentina Greco (2013). Spatial Organization within a Niche as a Determinant of Stem-Cell Fate, *Nature* vol. 502 (October 24), pp. 513–18. doi:10.1038/nature12602

Russell, E. S. (1930). *The Interpretation of Development and Heredity: A Study in Biological Method*. Reprinted in 1972. Freeport NY: Books for Libraries Press.

Tan, Jiaying (2013). Stemness in the Niche, *Cell* vol. 155 (December 19), pp. 1439–40. doi:10.1016/j.cell.2013.12.003

Tata, Purushothama Rao, Hongmei Mou, Ana Pardo-Saganta et al. (2013). Dedifferentiation of Committed Epithelial Cells into Stem Cells *in Vivo, Nature* vol. 503 (November 14), pp. 218–23. doi:10.1038/nature12777

Thompson, D'Arcy Wentworth (1917). *On Growth and Form*. Cambridge: Cambridge University Press. Available at https://www.gutenberg.org/ebooks/55264

Weiss, Paul (1962). From Cell to Molecule, in *The Molecular Control of Cellular Activity*, edited by John M. Allen, pp. 1–72. The University of Michigan Institute of Science and Technology Series. New York: McGraw-Hill.

_______ (1971). *Within the Gates of Science and Beyond: Science in Its Cultural Commitments*. New York: Hafner.

_______ (1973). *The Science of Life: The Living System — A System for Living*. Mount Kisco, NY: Futura Publishing.

Xin, Tianchi, David Gonzalez, Panteleimon Rompolas and Valentina Greco (2018). Flexible Fate Determination Ensures Robust Differentiation in the Hair Follicle, *Nature Cell Biology* vol. 20 (December 1), pp. 1361–69. doi.org/10.1038/s41556-018-0232-y

CHAPTER 7

Epigenetics: A Brief Introduction

YOU AND I HARBOR TRILLIONS of "sub-creatures" in our bodies. I am not referring to the microorganisms in our guts, but rather the cells we consider our own — the constituents of our muscles and brains, our livers and bones, our lenses and retinas. Each of these cells, embedded in its supportive environment, sustains a dauntingly complex and unique way of life. If (which is impossible) we had first discovered such cells floating singly in a pool of water and had observed them through a microscope, we would have judged them to be distantly related organisms. Phenotypically (that is, in visible form and function) one cell type in the human body can differ from another as much as an amoeba differs from a paramecium.

And yet, all the cells in the human body have descended from a single cell (zygote) with a single genome.[1] And just as hundreds of different cell types have arisen from that one zygote, so, too, have the multicellular, intricately organized entities we know as lung, heart, eye, kidney, and pancreas, along with all our other organs. Supremely interdependent as these are, each is nevertheless a functioning organic world of altogether distinctive character.

For the past century these facts of development have been thought to present a (largely ignored) problem for the gene-centered view of life. The developmental biologist Frank Lillie, who had directed the prestigious Marine Biological Laboratory at Woods Hole, Massachusetts, and would go on to become president of the National Academy of Sciences, remarked in 1927 on the contrast between "genes which remain the same throughout the life history" of an organism, and a developmental process that "never stands still from germ to old age." In his view, "those who desire to make genetics the basis of physiology of development will have to explain how an unchanging complex can direct the course of an ordered developmental stream" (Lillie 1927, pp. 367–68).

This ordered developmental stream, of course, includes generation of the hundreds of different cell types in our bodies. It is hard to understand how a single genomic "blueprint" — or any other way of construing a fixed genetic sequence — could by itself provide the definitive causal basis for these hundreds of radically distinct ways of living. If the blueprint is compatible with all of them, do we have compelling grounds for thinking that it fundamentally *determines* any one type of cell, or organ, let alone all of them together — or, for that matter, any one type of animal? One might reasonably expect that other, contextual factors direct the developmental process toward particular outcomes of such different sorts.

A more balanced understanding arises when we watch how every cell displays its character through its life as a whole. That character, in all its qualitative richness, somehow seems decisive. In the case of each cell type, DNA is *caught up in* a seamless and integral way of being. When we grasp this integral nature, we quickly realize that the idea of DNA as the crucial causal determinant of the whole is an impossible one. As a specific kind of liver cell passes through its developmental lineage, it must sustain its entire organization in a coherent and well-directed manner from one cell generation to the next — including, for example, the cytoskeletal and cell membrane organization described in Chapter 4 ("The Sensitive, Dynamic Cell"). It must also bring about and orchestrate the elaborate performances of its chromosomes we saw in Chapter 3 ("What Brings Our Genome Alive?") — performances that are unique to each type of cell and that chromosomes themselves have no way to set in motion.

Every individual part, including DNA, is shaped by, and gives expression to, the character of a larger whole. Only when we recognize that genes participate in a living whole can we find an answer to Lillie's challenge "to explain how an unchanging complex can direct the course of an ordered developmental stream." The answer — so we will find — is that there is no unchanging complex. Genes, like all parts of a cell or organism, gain their identities and meanings only within the context of innumerable, interpenetrating, living narratives expressing diverse physiological characters.

An old problem newly recognized

Passing from Lillie's day to our own, we find a peculiarly late-arriving acknowledgment of old problems. Here is where we encounter that rather mysterious and too often abused keyword of contemporary molecular biology: *epigenetics* (along with its companion, *epigenome*). The discipline of epigenetics drives today's effort to come to terms with the relationship between genes and the organisms that put those genes to such diverse uses.

But today the question has gained additional dimensions. The Human Genome Project and its successors surprised many by revealing an unexpectedly low number of human genes relative to many other organisms — roughly the same number, for example, as in the simple, one-millimeter-long, transparent roundworm, *Caenorhabditis elegans*. Many began to ask: If genes really do account for the organism in all its complexity, how can it be that a primitive worm boasts as many genes as we do? "As far as protein-coding genes are concerned," wrote

Ulrich Technau, a developmental biologist from the University of Vienna, "the repertoire of a sea anemone ... is almost as complex as that of a human" (Technau 2008, p. 1184).

A further revelation only compounded the difficulty: our own genome was found to have a great deal in common with that of many animals. It was reported (rather crudely) that we share about 98.5% of our genome with chimpanzees. A good deal of verbal hand-wringing and chest-beating ensued. How could we hold our heads up with high-browed, post-simian dignity when, as the *New Scientist* reported in 2003, "chimps are human"? If the DNA of the two species is more or less the same, and if, as nearly everyone seemed to believe, DNA is destiny, what remained to make us special? Such was the fretting on the human side, anyway. To be truthful, the chimps didn't seem much interested.

All this news conspired to bring epigenetics to the fore. In 2010 the editors of the journal *Nature* wrote:

> By 2004, large-scale genome projects were already indicating that genome sequences, within and across species, were too similar to be able to explain the diversity of life. It was instead clear that epigenetics ... could explain much about how these similar genetic codes are expressed uniquely in different cells, in different environmental conditions and at different times (*Nature* editors 2010).

And in 2015 a contributor to the same journal described a huge, epigenome-centered project, sponsored by the US National Institutes of Health, which was "likely to provide a leap forward in pinning down one of the central mysteries of biology: how do cells with the same genetic instructions take on wildly different identities?" (Ledford 2015) Lillie's old question had finally come center stage. But had the meaning of the question really been recognized? And what, after all, *is* this mysterious thing called *epigenetics*?

Epigenetics – a useful term?

Etymologically, the word *epigenetics* suggests something like "on top of genetics" or "added to genetics." In common technical use, the word refers today to "heritable changes in gene function that are not due to changes in DNA sequence" — where the *DNA sequence* is a succession of nucleotide bases constituting the "letters" of the so-called genetic code, and *heritable* applies not only to what can pass from parent organisms to their progeny, but also what passes from any given cell of an existing organism to its daughter cells. In other words, *epigenetic* refers to that which is not rock-bottom genetics — not genetics proper (which conventionally centers on the DNA sequence alone) — while yet somehow bearing on *functional* genetics, both within organisms and between generations.

The common usage, however, remains deceptively gene-centered. This is shown by the prevailing notion that epigenetics has to do only with secondary "annotations" of the primary "genetic program." For example, researchers, having discovered certain chemical transformations of both DNA and the overall substance of chromosomes, typically refer to these transformations as innocent-sounding and transient "marks" on an otherwise fundamental and essentially unchanging entity.

It is hard to avoid the suspicion that biologists refer to the chemical transformations as mere marks only because they have concluded up front that whatever cells do with their genome cannot be considered genuinely transformative and creative — cannot redefine what a gene is. They prefer to keep the genome a kind of static, "eternal" essence (see Chapter 22, "A Curiously Absolute Demand for Stable Variation") that, unlike every other part of an organism, need not continually *become* what it is or else cease to live.

Rather than think of epigenetics as the application of incidental marks, we could conceive it more realistically as encompassing all the ways DNA is caught up in the activity of its larger context and brought into service of the whole. I say "more realistically" because there is, in fact — as two molecular biologists have phrased it in the journal *Nature* — "an avalanche of biochemical evidence revealing a complex and versatile array of molecular mechanisms that regulate gene expression without changing DNA sequences" (Cervantes and Sassone-Corsi 2019).

In other words, what genes mean to the organism is not merely a matter of the DNA sequence or a "genetic code." It is more a question of the many different ways an organism can employ its genes.

So the word *epigenetics* may usefully remind us that what is "on top of" DNA is nothing less than the functioning organism as a whole. But a word that threatens to encompass just about everything begins to lose its value as a special term. And this in turn suggests that we could just as well retire the word "epigenetics" and get on with describing how organisms carry out their organically integrated lives — express their own character — in part by "reconceiving" their genes in terms of that character.

Unexpected discoveries

In the mammalian genome, chromosomes normally come in pairs, one inherited from the mother and the other from the father. Any given gene occurs twice, with separate versions (*alleles*) located on the two chromosomes of a pair.

These two alleles may or may not be identical. For example, there are mice that, in their natural ("wildtype") state are dark-colored — a color that is partly dependent on a gene known as *Kit*. The mice normally have two identical copies of this gene. When, however, one of the *Kit* alleles is mutated in the laboratory a certain way, the mouse shows white feet and a white tail tip.

That result was perfectly natural (if you call such artificial gene manipulations "natural"). But it is also where the story becomes interesting. Scientists at the University of Nice-Sophia Antipolis in France took some of the mutant, white-spotted mice and bred them together (Rassoulzadegan et al. 2006). In the normal course of things, some of the offspring were again wildtype animals — neither of their *Kit* alleles was mutant.

However, to the researchers' surprise, these "normal," wildtype offspring maintained, to a variable extent, the same white spots characteristic of the mutants. It was an apparent violation of Mendel's laws of inheritance: while the genes themselves were passed between generations properly, their effects did not follow the "rules." A trait was displayed despite the absence of the gene previously associated with it. Apparently something in addition to the genes themselves — something "epigenetic" — figured in the inheritance of the mice offspring, producing the distinctive coloration.

Another group of researchers, led by Michael Skinner at the University of Washington, looked at the effects of the fungicide vinclozolin on laboratory rats (Anway et al. 2006; Crews et al. 2007). Banned in Scandinavia and Europe at the time, but allowed on some crops in the U.S., vinclozolin is known as an endocrine-disrupting chemical. If pregnant female rats are exposed to it while their embryos are undergoing sexual organ differentiation, the male offspring develop serious problems as adults — death of sperm-generating cells, lowered sperm count and motility and, later, immune abnormalities and various diseases including cancer. The remarkable thing is that the effects were found to be transmitted over four generations without weakening. That is, acquired characteristics — deficiencies in embryos brought on by fungicide exposure — were inherited by offspring who were not subject to the same exposure.

Puzzling results such as these put the question, "Are genes equivalent to destiny?" in a new light. In 2007 a team of researchers at Duke University reported that exposure of pregnant mice to bisphenol A (a chemical that was then used in many common plastics such as baby bottles and dental composites) "is associated [in the offspring] with higher body weight, increased breast and prostate cancer, and altered reproductive function." The exposure also shifted the coat color of the mice toward yellow — a change again found to be transmitted across generations despite its not being linked to a gene mutation. Moreover, the changes brought on by the chemical were negated when the researchers supplemented the maternal diet with folic acid, a B vitamin (Dolinoy et al. 2007).

And so an epigenome that responds to the environment can respond to healthy as well as unhealthy influences. As another early illustration of this: researchers at McGill University in Montreal looked at the consequences of two kinds of maternal behavior in rats. Some mother rats patiently lick and groom their newborns, while others generally neglect their pups. The difference turns out to be reflected in the lives of the offspring: those who are licked grow up (by the usual measures) to be relatively confident and content, whereas the neglected ones show depression-like symptoms and tend to be fearful when placed in new situations.

This difference is correlated with different levels of activity in particular genes in the hippocampus of the rats' brains. Not that the gene sequences are themselves mutated in the usual sense. Rather, the researchers found that various epigenetic modifications in the hippocampus alter the way in which the genes are employed (Weaver et al. 2004). Other investigations have pointed toward similar changes in the brains of human suicide victims who were abused as children (Poulter et al. 2009).

What has perhaps excited the general public most is this application of epigenetic studies to human beings. Take, for example, the frequently cited Dutch Hunger Winter during the winter and spring of 1944–45. The much-studied effects of this famine were found to extend, not only to the children of women who were pregnant during the months of hunger, but also to their grandchildren.

Such findings seemed to suggest that our environments and our responses to those environments can play a major, heritable role in shaping our lives. This encouraged in many the hopeful thought, "Maybe we are not really just gene-driven machines" — a thought that surely is true enough, but also rather strange. I will try to explain.

Grasping at epigenetic straws – is it really necessary?

Those early discoveries in epigenetics — especially when treated more expansively and brought more up-to-date (Chapter 14, "How Our Genes Come to Expression") — are truly profound and far-reaching in their implications. But they are profound only in the way everything about the character of organic life we have been discussing in the preceding chapters is profound.

Genes as self-sufficient or definitive First Causes simply don't exist. They never did have a reasonable place in our conceptualization of living beings — something that early twentieth-century critics of gene theory clearly saw (Russell 1930). Every organic process, including every genetic process, is an expression of the life of the whole cell and whole organism. In other words, the only genetics we have is epigenetics.

All this is to say that the crucial thing, if we want to transcend the notion of organisms (or ourselves) as gene-driven machines, is to rise above the entire, spirit-killing picture of mechanistic, gene-programmed life processes. We need to recognize this picture for the fantasy it really is.

Anyone who doubts the scale of the challenge in this need only look at what began happening quickly after the discovery of "epigenetic" effects. No sooner had certain gene-regulatory "marks" been found on key elements of the chromosome than some began to suggest that they constituted just another "code" — *an epigenetic code* (Strahl and Allis 2000). An epigenetic "program" was said to contain "instructions" for "control of gene expression." And so an editorial entitled "Time for Epigenetics" in the *International Journal of Biochemistry & Cell Biology* told us that

> The genome and epigenome together *determine* the phenotype and hence, the function and characteristics of a cell at any given point in development and during differentiation. At the core of gene regulation are elaborate *molecular programs* that alter the packaging of DNA into chromatin, thereby regulating DNA accessibility to transcription complexes and providing cues to the activation or repression of *gene regulatory programs* (Altucci and Stunnenberg 2009; emphasis added).

In other words, the attempt is to assimilate epigenetics to the existing understanding of genetic "programs" and "instructions." The programs and the instructions simply become a little larger and more complex, but the same basic understanding of ourselves as collections of molecular automatons remains.

Or, again, we hear that the epigenome involves a "re-wiring of transcription factor circuits," (Tsankov et al. 2015) as if there were some fixed and standard genetic wiring scheme waiting to be rewired. But — as if biology as a discipline were somehow "of two minds" about such things — the authors of this paper healthily refer to the rewiring as "context-dependent" and "dynamic." So the terminology appears to be impossibly conflicted. If in fact the governing context is always to some degree fluid, dynamic, and shifting, where do we ever see anything remotely analogous to wires constraining all the relevant molecules to go where they need to go, and to do so in the right time, in the right quantities, and with the right molecular partners?

The picture of a wired cell may sound conveniently causal, but it makes no sense. Biologists are sooner or later going to have to decide which half of their descriptive language they are going to side with — the "wired" half or the "dynamic and context-dependent" half. Meanwhile, those of us trying to decipher what "epigenetics" really means can usefully remind ourselves that the deeper issue has to do with the overall terms of the description ultimately decided upon, not with particular "epigenetic" insights that are too eagerly assimilated to traditional, machine-based understanding.

Nothing is *merely genetic*. Every so-called genetic activity is an expression of its entire context, and therefore is altogether epigenetic. Genetics cannot be abstracted from the rest of the organism. So we can safely say, "All genetics is epigenetics."

WHERE ARE WE NOW?

Bringing back the organism

As we move along, we have been seeing more and more how the "molecule's eye view," whatever it may tell us about the physics and chemistry of molecules, is hardly definitive of biological meaning, for which a wider perspective is required. In the end, the meaning of things depends on what the cell or organism is *doing* in its coordination of countless diverse but interwoven processes. An organism just *is* its unified doings, its consistent way of living in its world.

It is perhaps in the field of genetics that biologists have most stoutly resisted this recognition of integral wholeness and significant context. Genes, conceived as First Causes, must exist in exalted isolation. But because of the intensity of research focused on genetics, it is also in this field that the illusions of strictly physical and chemical explanation of the organism are being most strikingly dispelled — even if geneticists are proving slowest at accepting the fact.

The brief introduction to epigenetics in this chapter will be greatly expanded in Chapter 14, "How Our Genes Come to Expression." There I try at least to suggest the endless web of pathways through which the cell brings about its almost infinitely complex patterns of gene expression.

Then we will deepen this picture by bringing the gene into connection with heredity and evolution in Chapter 20 ("Inheritance and the Whole Organism") and Chapter 21 ("Inheritance, Genetics, and the Particulate View of Life"). We will learn how it is that genes rendered the organism invisible to the evolutionist's sight — and how false that substitution of genes for organisms has proven.

Finally, we heard Frank Lillie saying above, "those who desire to make genetics the basis of physiology of development will have to explain how an unchanging complex can direct the course of an ordered developmental stream."

We will get an entirely different view upon this statement when we discuss some work by the philosopher, Ronald Brady, in Chapter 12 ("Is a Qualitative Biology Possible?"). There we will encounter the perhaps initially disorienting truth that the "ordered developmental stream" is not what actually needs explaining, whether by genes or anything else. Order, after all, is itself the kind of explanatory understanding the scientist is always looking for. To recognize the *order* of the developmental stream is to recognize an organism's explanatory principles. So the ideal formative movement of development might be regarded as itself the real "first cause" of the organism's features, including its genetic features. The main thing in the way of our accepting this truth is our habit of taking material things as the explanation of movement.

NOTES

1. While this has often been taken to mean that the genome in all our cells is the same, we now know that this is far from the truth. Many people, in fact, possess some cells derived from entirely different bodies. For example, an embryo or fetus may assimilate cells from its mother, and there can be an exchange of cells between fraternal twins in the womb, even if they are oppositely gendered. Also, many gene mutations occur in cells during development and afterward, so probably no two cells in our body have *exactly* the same genome.

   More important is the fact that, as shown by the radically different cell types in any one person's body, the *functional* genome differs radically from cell to cell.

SOURCES

Altucci, Lucia and Hendrik G. Stunnenberg (2009). Time for Epigenetics, *International Journal of Biochemistry & Cell Biology* vol. 41, no. 1, pp. 2–3. doi:10.1016/j.biocel.2008.09.010

Anway, Matthew D., Charles Leathers and Michael K. Skinner (2006). Endocrine Disruptor Vinclozolin Induced Epigenetic Transgenerational Adult-Onset Disease, *Endocrinology* vol. 147, no. 12 (December), pp. 5515–23. doi:10.1210/en.2006-0640

Crews, David, Andrea C. Gore, Timothy S. Hsu and Michael K. Skinner (2007). Transgenerational Epigenetic Imprints on Mate Preference, *PNAS* vol. 104, no. 14, pp. 5942–46. doi:10.1073/pnas.0610410104

Cervantes, Marlene and Paolo Sassone-Corsi (2019). Modification of Histone Proteins by Serotonin in the Nucleus, *Nature* vol. 567 (March 13), pp. 464–65. doi:10.1038/d41586-019-00532-z

Dolinoy, Dana C., Dale Huang and Randy L. Jirtle (2007). Maternal Nutrient Supplementation Counteracts Bisphenol A-induced DNA Hypomethylation in Early Development, *PNAS* vol. 104, no. 32 (August 7), pp. 13056–61. doi:10.1073/pnas.0703739104

Ledford, Heidi (2015). The Genome Unwrapped, *Nature* vol. 528 (December 3), pp. S12–13. doi:10.1038/528S12a

Lillie, Frank R. (1927). The Gene and the Ontogenetic Process, *Science* vol. 46, no. 1712 (October 21), pp. 361–68. https://www.jstor.org/stable/1651140

Nature editors (2010). Time for the Epigenome, *Nature* vol. 463, no. 7281 (February 4), p. 587. doi:10.1038/463587a

Poulter, M., L. Du, I. Weaver et al. (2009). GABAA Receptor Promoter Hypermethylation in Suicide Brain: Implications for the Involvement of Epigenetic Processes, *Biological Psychiatry* vol. 64, no. 8, pp. 645–52. doi:10.1016/j.biopsych.2008.05.028

Rassoulzadegan, Minoo, Valérie Grandjean, Pierre Gounon et al. (2006). RNA-mediated Non-Mendelian Inheritance of an Epigenetic Change in the Mouse, *Nature* vol. 441 (May 25), pp. 469–74. doi:10.1038/nature04674

Russell, E. S. (1930). *The Interpretation of Development and Heredity: A Study in Biological Method*. Reprinted in 1972. Freeport NY: Books for Libraries Press.

Strahl, B. D. and C. D. Allis (2000). The Language of Covalent Histone Modifications, *Nature* vol. 403, no. 6765 (Jan 6), pp. 41–45. doi:10.1038/47412

Technau, Ulrich (2008). Small Regulatory RNAs Pitch In, *Nature* vol. 455 (October 30), pp. 1184–85. doi:10.1038/4551184a

Tsankov, Alexander M., Hongcang Gu, Veronika Akopian et al. (2015). *Nature* vol. 518 (February 19), pp. 344–49. doi:10.1038/nature14233

Weaver, Ian C. G., Nadia Cervoni, Frances A. Champagne et al. (2004). Epigenetic Programming by Maternal Behavior, *Nature Neuroscience* vol. 7, no. 8 (August), pp. 847–54. doi:10.1038/nn1276

CHAPTER 8

The Mystery of an Unexpected Coherence

WE HEARD IN CHAPTER 2 ("The Organism's Story") that living activity has a certain future-oriented (purposive or directive) character that is missed by causal explanations of the usual physical and chemical sort. The end is always more constant than the continually adjusted means. This is true whether we are talking about a dividing cell, the achievement of adult form through development, or the strategy for taking a prey animal for food (or avoiding being taken).

An animal's end-directed activity may, of course, be very far from what we humans know as conscious aiming at a goal. But all such activity, human or animal, displays certain common features distinguishing it from inanimate proceedings: it tends to be *persistent*, so that it is resumed again and again after being blocked; it likewise tends to be *adaptable* — strategies are changed in the face of altered circumstances; and the entire activity *ceases* once the end is achieved.

This flexible directedness — this interwoven play of diverse ends and means within an overall living unity — is what gives the organism's life its peculiar sort of multi-threaded, narrative coherence. Life becomes a story. Events occur, not merely from physical necessity, but because they hold significance for an organism whose life is an unfolding pattern of significances. We are always looking at the *moment-by-moment expression of a present wisdom* — not the automatic playing forward of a pre-existent mechanism.

The idea of a thoughtful wisdom, like the related idea of a governing context (Chapter 6), is a mystery for all attempts at purely physical explanation. This is why even the explicit acknowledgment of an organism's *striving for life* — central as it may be for evolutionary theory — is discouraged whenever biologists are describing organisms themselves. It sounds too much as if one were invoking inner, or soul, qualities rather than material causes — acknowledging a *being* rather than a thing. And it is true that our physical laws as such, however combined, nowhere touch the idea of *striving*.

Biologists much prefer to identify discrete, definitive causes. The cell nucleus with its genome has long been viewed as the seat of such causation. But, as we saw in our discussion of DNA (Chapter 3, "What Brings Our Genome Alive?") and epigenetics (Chapter 7, "Epigenetics: A Brief Introduction"), the single-minded pursuit of genetic causes has forcibly redirected our attention to epigenetics, where we have discovered that genes are circumscribed and given their meaning by the directive life of the entire cell and organism.

In what follows below we will consider this directive coherence in a more detailed way by taking up one of the many activities of the cell that are often considered under the heading of "epigenetics." Then we will look at a startling phenomenon that, already on its face, renders absurd the idea of central genetic control. In both cases we will be focused on molecular-level activity, which is precisely where we have been most strictly taught to expect the absence of any coherence other than that of "blind mechanism."

Flexibility and precision in RNA splicing

The discovery of *RNA splicing* in the late 1970s was one of the transforming moments in the history of molecular biology.[1] To put it in informal terms: the cleanly autocratic mastery of DNA gave way to massive presumption by various scruffy elements of the cellular "rabble." The idea had originally been that a molecule of messenger RNA (mRNA) was produced as a direct image of the "instructions" in a protein-coding gene and was then exported from the cell nucleus to the cytoplasm. There it yielded passively to *translation*, a process whereby a protein was supposedly produced according to the exact specifications of the "genetic code" previously copied from DNA into the mRNA.

Our growing knowledge of RNA splicing has, together with many other developments in molecular biology, exploded just about every aspect of this picture. We now know that, via an elaborately orchestrated improvisational drama, many so-called *epigenetic* elements in the cell (Chapters 7, "Epigenetics: A Brief Introduction," and 14, "How Our Genes Come to Expression") converge to decide what use will be made of any particular gene.

In particular, the cell has innumerable ways to obtain and sculpt its proteins. RNA splicing is just one of these — a massive reconfiguration process whereby a cell decides which portions of an initially produced (*precursor*) RNA to cast aside for other uses, and which ones to "splice" together into a mature mRNA. As we have come to expect by now, these choices are strongly context-dependent, with different protein variants being produced in different kinds of cell or tissue, or under different cellular conditions.

This splicing involves much more than a minor stitch or two. The large human dystrophin gene (whose malfunction is related to some forms of muscular dystrophy) is said to require 16 hours for its transcription from DNA into RNA. Of this time, 15 hours and 54 minutes is required for transcripton of the non-protein-coding RNA sequences that will have to be spliced out of the RNA in order to obtain a mature messenger RNA. That may be a somewhat extreme case, but it remains true that the sequences to be discarded are "commonly orders of magnitude longer" than the remaining portions fit for the synthesis of protein (Papasaikas and Valcárcel 2016).

But perhaps the most dramatic transformation involves the sequence remaining after removal of the non-protein-related ("noncoding") content. The splicing activity can often select from among the parts of this sequence in differing ways, thereby determining which protein-coding portions of the precursor molecule will be included in the mature mRNA. The protein eventually resulting will vary depending on these *alternative splicing* decisions. (The variably spliced versions of a protein are referred to as *isoforms*.)

The mRNAs generated from over 90 percent of mammalian genes are thought to be alternatively spliced, contributing greatly to physiological complexity (Gehring and Roignant 2021). According to one paper, "As cells differentiate and respond to stimuli in the human body, over one million different proteins are likely to be produced from less than 25,000 genes" (de Almeida and Carmo-Fonseca 2012).

Further, "even relatively modest changes in alternative splicing can have dramatic consequences, including altered cellular responses, cell death, and uncontrolled proliferation that can lead to disease" (Luco and Misteli 2011). The title of one technical paper makes the point vividly: "Cell Death or Survival Promoted by Alternative Isoforms of [the protein] ErbB4" (Sundvall et al. 2010).

You have doubtless heard many times how a mutation or engineered alteration of such-and-such a gene "causes" this or that result. How often, by contrast, do you hear that a slight change in the way your cells orchestrate the sculpting of this or that protein can make the difference between life and death?

The spliceosome

The central player in the sculptural drama of splicing is known as the *spliceosome*, which is not so much a rigidly fixed thing or structure as it is a complex performance. The performers include a few critically important small RNAs and over 150 proteins.[2] Together — although in several, separate, coordinated groups that must continually reconfigure themselves during the process — they excise the protein-unrelated pieces of the RNA and then stitch together a selection of the ones remaining. Misjudging any of the potentially many places to cut the mRNA — shifting the point of severance by a single "letter," or nucleotide base, out of (in many cases) thousands — could possibly render the resulting mRNA useless for producing protein, if not downright harmful.

We heard a little bit in Chapter 3 ("What Brings Our Genome Alive?") about the puzzle of topoisomerases. In a way that is difficult to fathom, these molecules make cuts in the DNA double helix in order to release knots and "untangle" the seemingly indecipherable spatial complexity of chromosomes (46 in the human case) that are tightly packed into the cell nucleus. But the challenge for the spliceosome as it does its work seems no less daunting. And the fact that there is indeed coherently describable *work* to do already takes us beyond normal physical explanation to the idea of an unfolding meaning.

The key, chemically active part of the spliceosome complex "is short lived and reconstructed from individual pieces for each splicing event" (Papasaikas and Valcárcel 2016). This is the part that actually cuts and stitches together the RNA once the end-points for the next excision are chosen. Moreover, few of the scores of proteins required for the activity stay together throughout the intricate work on a single RNA. "At all transitions in the splicing process, the spliceosome's underlying RNA-protein interaction network is compositionally and conformationally remodeled and at each step there is a massive exchange of [spliceosomal] proteins" (Wahl and Lührmann 2015).

But there is more. In multicellular organisms the mRNA being remodeled possesses particular sequences that are supposed to act as signposts for "attracting" the elements of the spliceosome to the correct sites for cutting and stitching. But these signposts are often ambiguous or contradictory, and provide only more or less vague hints.[3] This is despite the extraordinary complexity of the task facing the spliceosome, and the large number of segments that commonly require removal.

"It has been proposed," write two researchers, "that thousands of different sequences" can function as a certain kind of directive for the spliceosome, but these sequences are highly variable, having only a few loci in common. Further, many sequences that look rather like splice sites are ignored by the spliceosome, while other sequences, despite lying at a distance from the splice sites, nevertheless contextually influence site recognition. So it appears that "hundreds of regulatory motifs may need to be integrated" (and understood) in order for the spliceosome to accomplish its surgery in harmony with current cellular needs (Papasaikas and Valcárcel 2016).

Using the thing-oriented (rather than process-oriented) language available to us, it is difficult not to speak of the spliceosome as a fixed structure, and equally difficult to avoid suggesting that it has a specific and well-defined task. What we see, however, is a remarkable plasticity. This is illustrated, for example, by the fact that "nearly all 'activators' of splicing can, in some cases, function as repressors, and nearly all 'repressors' have been shown to function as activators … it is clear that context affects function" (Nilsen and Graveley 2010).

This context-sensitivity extends to the very definitions of the various tasks, which can look utterly different, and require wholly different approaches and capabilities on the part of the spliceosome, depending on the situation. Is the task to skip the next protein-coding segment of the RNA? Is it to make sure that a choice is made between two such segments — to retain only one and remove only one? Is it to choose an alternative location for the beginning or end of a particular segment? Is it, in at least some cases, to make the radical choice of preserving a non-protein-coding segment in the final mRNA?

Each of these operations demands a different sort of coordination among the many molecules involved, and the ways of approaching the work can vary, one might almost say, "wildly." "Mechanisms of alternative splicing are highly variable, and new examples are constantly being found."[4] So there is not just one "spliceosome machine" (as some would like to call it), and not just one task. The numerous molecules participating (or capable of participating, but "electing" not to) in the various splicing operations face the challenge of working together in an unimaginably sophisticated manner that somehow reflects the wider context and the needs of the cell.

Who will disagree with the researchers who write, in what might even be an understatement: "Working in a highly orchestrated manner, [the many parts of the spliceosome] perform incredible feats of molecular gymnastics with each round of splicing" (Chen and Moore 2014)?

And further: everything could go backward

The entire problem is perhaps most vividly framed when we consider one further fact about RNA splicing. Not only is the spliceosome "a remarkably dynamic and flexible molecular machine; its transitions are so malleable that the whole reaction can eventually be reversed to generate precursor mRNA from spliced products" (Papasaikas and Valcárcel 2016). More particularly:

> Rather than being the one-way pathway typically drawn in textbooks, almost every step in the spliceosome cycle is readily reversible … [For example, regarding the first and second chemical steps in splicing], not only can the spliceosome catalyze both chemical steps in forward and reverse, it can even convert spliced products … back into unspliced precursor mRNA! (Chen and Moore 2014)

That is, the splicing choreography can take an already spliced RNA along with the sections previously removed from it, and reinsert those sections into the RNA.

The reversibility and flexibility underlying the finely gauged, discriminating, and "perceptive"[5] activity of RNA splicing are hard to overestimate. Plasticity is layered upon plasticity, and complexity upon complexity. For example, many of the individual protein "surgical assistants" coming together in continually different ways in the spliceosome are themselves subject to modifications that are often decisive for how they will function within their current context. And these modifications, too, are dynamic and reversible.

They are also mutually entangled, with one kind of modification in one protein likely affecting, or being affected by, diverse modifications in other proteins. The untraceable lines of cause and effect blur into — and become subordinate to — the overall storyline.[6]

Can DNA coordinate splicing activity?

Despite the fact that a specific splicing process could, with perfect *physical* propriety, go in an infinite number of different directions, it produces, from among all the present possibilities, the particular result that fits the ever-changing cellular context at the present moment. Splicing must, in some extremely significant sense, be guided by this context. If it were somehow being "dictated" to by a specific element or group of elements in the cell, those elements would have to have incorporated within themselves an effective sense for the current state of the entire cell. But then, why not just recognize that a biological whole, in one way or another, informs *all* its parts?

It is worth noticing the great distance between, on one hand, what RNA splicing shows us and, on the other hand, the idea of DNA as a decisive cause of the cell's life (or even merely DNA as a strict determinant of protein synthesis). The notion of a decisive physical cause immediately comes up against questions such as the following:

- Does DNA single-handedly "dictate" that the splicing operation on a particular RNA *this time* should differ in such-and-such a way from how it was done *last time*?

- Does DNA (or, for that matter, any other cellular feature) have any possibility of determining the specific and crucial, well-timed chemical modifications or changes in form of just one of the proteins involved in the splicing activity, let alone the mutually interacting modifications that must occur in a great number of them as the splicing "surgery" proceeds?

- Does DNA enforce the way these proteins (and other molecules) come together in distinct configurations at one point in the process, or dissociate at other points, or come together in a new configuration at yet another point — all in the temporal order required for the success of the overall procedure?

- In sum, are there computer-like lines of communication through which coordinating *instructions* can be conveyed from DNA to the individual protein and RNA molecules?

And what we have said about DNA and splicing can also be said about DNA and just about any of the innumerable other molecular processes of the cell, from metabolism, to energy management, to establishment and management of the diverse structural features of the cell, to gene expression, to cell division and much more. Further, all the complexities of each of these spheres of activity must be harmonized with those of the other spheres so as to yield the overall integral unity of cell, organ, and organism — this in the face of the fact that many molecular players are common to the different processes.

Shattering the genome

Our second case is a long way from RNA splicing — and also, it might seem at first, from the human being.

A dose of ionizing radiation equal to 10 grays (a measure of absorbed radiation) is lethal to the human body. Most bacteria cannot survive 200 grays. But then there is the bacterium known as *Deinococcus radiodurans*: it can endure over 17,000 grays and get along just fine. Never mind that its genome is thoroughly shattered by the assault.

Here's what happens. Ionizing radiation can damage DNA in various ways, perhaps worst of all by causing double-strand breaks. These are breaks across both strands of the double helix. The familiar bacterium, *E. coli*, not at all untypically, dies when it suffers about four double-strand breaks per each of its four-to-eight circular DNA molecules. *Deinococcus radiodurans*, by contrast, can survive over a thousand double-strand breaks. This means that it continues life after its genome is broken into many hundreds of fragments. It does so by proceeding to put its genome back together again when living conditions improve — a daunting task, to say the least.

Figure 8.1. A tetrad (group of four) *Deinococcus radiodurans*.[7]

Deinococcus radiodurans is one of a small class of single-celled organisms with extreme radiation tolerance. Actually, it tolerates various other extreme conditions as well — some of which, such as dessication, likewise reduce its DNA to genomic shards. It can, for example, survive in a waterless desert for years until moistened again — which could happen, for example, when winds lift it in a cloud of dust from the Sahara, high into the atmosphere (where it is exposed to damaging ultraviolet radiation 100 to 1000 times that on earth's surface), and across the Atlantic ocean to the South American jungles. *D. radiodurans* can be found on Antarctic ice, on dry frozen marble, and in the farthest depths of the sea.

Who's on first — genes or proteins (or neither)?

Biologists have been intrigued by this peculiar survivor (along with some of its kin) for several decades, and of late they have clarified its story considerably. A central feature of that story is striking, because it points toward a truth about organisms in general, not merely those with extreme survival capabilities. The key finding is this: damage to DNA is not, in the most direct sense, what proves lethal about radiation. The primary issue, instead, is damage to proteins. As long as its proteins remain functional, a cell can reassemble even a badly fractured genome; but with damaged proteins, a cell is done for, with or without an intact genome.

D. radiodurans employs a number of strategies for preserving its rather commonplace "proteome," or total inventory of proteins. These strategies include (1) preventing the oxidative damage that results from radiation, a goal it achieves in good part by means of an especially rich supply of antioxidants; (2) eliminating, before they can cause mischief, any proteins that do get damaged, while recycling their constituents; (3) scavenging amino acids and peptides (protein constituents) from the local environment, a capability that, together with the recycling, supports (4) newly synthesizing any proteins that need replenishing.

The proteome thus preserved is then able to go about the task of reconstructing a shattered genome — a task whose complexity at the molecular level is stunning. (Many a bright but befuddled graduate student has twisted his imagination into knots while trying to picture the various textbook processes of DNA damage repair in human cells.) Nevertheless, the task is accomplished in the cells of all organisms. What distinguishes *D. radiodurans* is its ability to carry out this task to an exceptional degree by maintaining its store of proteins intact under extreme duress.

In sum, according to Anita Krisko and Miroslav Radman, researchers at the Mediterranean Institute for Life Sciences who have been studying *D. radiodurans*, "biological responses to genomic insults depend primarily on the integrity of the proteome ... This conclusion is the consequence of the fact that dedicated proteins repair DNA, and not vice versa." Moreover, "this paradigm is fundamental in its obviousness (no living cell can function correctly with an oxidized proteome) and, if it is true, must be universal, that is, hold also for human cells."

All this says something powerful about the longstanding genocentric (gene-centered) bias of biologists. Krisko and Radman delicately hint at the issue when they write in their paper:

> The science of molecular biology was dominated by the notion of information, its storage, transmission, and evolution as encrypted in the nucleotide sequence of nucleic acids [that is, DNA and RNA sequences]. But the biological information is relevant to life only to the extent of its translation into useful biological functions performed, directly or indirectly, by proteins (Krisko and Radman 2013).

This truth, as they also point out, applies to our understanding of cancer and its treatment, which have long been focused on DNA abnormalities. But instead, "an effective cancer therapy by tumor cell killing should target the proteome, or both the proteome and genome, rather than the genome alone." Which is almost to say: it should reckon with the coherent living character of the organism as a whole.

A sense of the whole

It was always a strange thing when biologists, attempting to penetrate the thickly matted tapestry of cellular activity at one or another point and disentangle the threads for analysis, decided that one type of element — the gene or DNA sequence — was the place where all the activity logically begins and from where it is controlled. There is in fact no starting place and no part acting as controller. Any attempt to think in such terms immediately crashes against the facts of cellular behavior. *Deinococcus radiodurans* no more shows proteins to be singularly "controlling" elements than it does DNA.

The work on *D. radiodurans* can remind us that the activity of an organism always reflects something like an immanent "sense of the whole." Surely the protein molecules in this bacterium do not "know" what their "goal" should be in dealing with all those disordered snippets of DNA. But if the overall living context (Chapter 6, "Context: Dare We Call It Holism?") remains sufficiently intact, then the mysterious power of self-realization that we have been gently stalking in these several chapters — the power sustaining the coherent storyline of a life — continues to assert itself. The narrative, whatever its unexpected twists and turns, remains unbroken. If parts can be more fully constituted from their shattered fragments, it is because a functioning whole, with its innate intelligence, was already there.

The information we conceive as *statically* encoded in DNA is a kind of bland abstraction from the living intelligence at work in cellular *processes*. When we occupy ourselves one-sidedly with genocentric information, it is (to employ a rough analogy) as if we elevated a notebook containing selected words, phrases, definitions, and grammatical hints to a pinnacle high above *Moby Dick* or *Faust* or *War and Peace*, worshipping the former as "information" while ignoring the informed and meaningful *activity* through which inert words and phrases are woven into soul-stirring tales.

A phrase-book or dictionary can be an essential resource, but it is the organism (*Deinococcus radiodurans* in the case we have been considering) that uses the dictionary to weave its own story — and even reconstructs the dictionary when the pages fall into a disorganized heap on the floor.

Is an unexpected coherence the problem or the solution?

The problem of what it actually *means* to say, "Molecules accomplish the work of splicing and DNA reconstruction" presents us with one of those vast blanks in scientific understanding that are easily papered over today with informational generalities and convenient pictures of tiny machines busily, and in a "mechanistically" respectable fashion, carrying on the work of a cellular factory.

We already heard about the essential problem from cell biologist Paul Weiss (Chapter 6, "Context: Dare We Call It Holism?"), who spoke about the many degrees of freedom possessed by the cell's constituents in their watery medium, and about how these degrees of freedom are so remarkably constrained and disciplined toward the expression of biological order at higher levels of observation. The University of Massachusetts geneticist, Job Dekker, was apparently nodding toward the same problem when he asked: "How do cells ensure that genes only respond to the right regulatory elements while ignoring the hundreds of thousands of others?" (Dekker et al 2013)

It's a good and obvious question. An editor of *Science* amplified it this way: "If you think air traffic controllers have a tough job guiding planes into major airports or across a crowded continental airspace, consider the challenge facing a human cell trying to position its proteins." A given cell, he noted, may make more than 10,000 different proteins under any particular set of conditions, and it typically contains more than a billion individual protein molecules at any one time. "Somehow, a cell must get all its proteins to their correct destinations — and equally important, keep these molecules out of the wrong places" (Travis 2011).

And once more: after a study showed that 70 percent of mRNAs in a cell are specifically localized, Robert Singer of Albert Einstein College of Medicine in New York City called it a "staggeringly large number." He went on: "It's almost as if every mRNA coming out of the nucleus knows where it's going" (quoted in Travis 2011).

Dekker, after posing the problem of a nucleus crowded with diverse regulatory factors bearing on gene expression in different ways, immediately went on to offer what he thought was at least part of the solution to the problem:

> Recent work has revealed a surprisingly simple strategy for matching genes to only some regulatory elements, which involves the spatial organization and folding of chromosomes inside the nucleus.

Certainly this folding, which we encountered in Chapter 3 ("What Brings Our Genome Alive?"), is an important aspect of the cell's performance. But this doesn't resolve, in a mechanistic fashion, the problem Dekker started with. To explain the achievement of crucial regulatory connections in the nucleus by citing chromosomal foldings that bring genetic loci and regulatory molecules together in just the right way is merely to push the problem back one step. We still have to ask the same sort of question with which we began: How are the foldings achieved with such evident wisdom?

It would help if we could get clear about the fact that there are two profound, and profoundly different, descriptive challenges posed by a cell's impressively coherent activities. One has to do with the underlying physical and chemical processes. The other concerns the coordination of those processes as an expression of the organism's needs and interests, intentions and meanings — its entire qualitative way of being. Severe confusions arise when we say that science must concern itself only with the first challenge, while assuming that the second one, if it can even legitimately be referred to, is automatically taken care of by our answer to the first.

Biologists, in their own fashion, do notice the second question. They notice it, as I have repeatedly mentioned, in their putting of questions to themselves ("How does the cell do X?"), where the question generally refers to a meaningful accomplishment. They notice it in their acknowledgment that organisms *behave* and undertake *tasks*, something solar systems and lake-bottom sediments never do. And they notice it when they grant that every organism acts as *if* it were a purposive being, even if they immediately feel compelled to explain away this purposiveness by appealing to natural selection (Chapters 2, "The Organism's Story," and 18, "Teleology and Evolution"). What is not so often noticed is the fact that an organism's purposive way of being and its pursuit of its own interests require a distinctive manner of understanding that *cannot be assimilated to our understanding of inanimate objects*.

Is the entire matter really so vexing? The mystery of the unexpected coherence that molecular biologists confront, for example, in RNA splicing and DNA damage repair is, from a perfectly reasonable point of view, neither a mystery nor unexpected. The problem arises only at the moment when we refuse to accept life as a foundational fact of the universe and unreasonably demand that an organism's living performances be explained in an inanimate manner. Then, and only then, do we find it difficult to make sense of things.

But, fortunately, researchers never can wholly resist the urge to make good sense of things — which is, of course, the necessity of all science. They seek an understanding of whatever issue they are working on by looking for the *coherence* and *meaning* of events. This is necessary in order to provide at least some minimal context for their physical analyses. And it is so natural that it easily occurs without any conscious effort. What then happens, and what so badly distorts the practice of biology, is that this recognized coherence and meaning must be squeezed out of the resulting explanation, which is allowed to proceed solely in terms of physics and chemistry understood meaninglessly in purely materialistic terms. The result is rarely pretty.

Listen to how Dekker concludes his reflections about the puzzle of genes and the "hundreds of thousands" of regulatory elements they may or may not interact with: "Future studies will no doubt unveil how [certain chromosome domains] are established and how they insulate genes from the wrong crowd."

There you see the uncomfortable conflation of two different explanatory challenges: those of physics and chemistry on one hand, and those of living activity on the other. In appealing to future studies, Dekker speaks as though he were unaware of the gap between the idea of physical lawfulness allowed in those studies, on one hand, and that of the "wrongness" of a molecular crowd, on the other. Part of that gap consists of the fact that the lawfulness of events does not explain how those events are meaningfully coordinated, as when genes are insulated from the wrong crowd.

Efforts at reductionism — efforts to reduce biological meaning to the terms of physical lawfulness — never make any progress. Yes, we have dramatically extended our tracing of physical lawfulness in the cell. But, for all the flood of physical data today, the needs, interests, tasks, intentions, and meanings of the organism never become less necessary for structuring our understanding.

What actually tends to happen, however, is not particularly helpful. Once the clarification of physically lawful processes reaches a certain point, the biologist's deeply ingrained habit of ignoring all questions of meaning leads to the conviction that nothing remains to be understood. And this occurs despite the continuing use, "right under the biologist's nose," of a vocabulary of life and meaning well designed to bridge (and conceal) the gap between lawfulness and adequate understanding. (See, for example, the discussion in Chapter 2 of the different vocabularies applied to living and dead dogs.)

Paul Weiss (Chapter 6, "Context: Dare We Call It Holism?") in addressing the larger coherence of the "heaving and churning" cell, did not merely stare, transfixed, at the problem of order within "chaos." He tried to formulate its essence as clearly as possible, often resorting to statements such as this: "The resultant behavior of the population [of cellular constituents] as a whole is infinitely less variant from moment to moment than are the momentary activities of its parts." And so "the system *as a whole* preserves its character" (Weiss 1962, p. 6). And again: When we examine the form and physiology of an organism, we see how "certain definite rules of order apply to the dynamics of the *whole* system ... reflected [for example] in the orderliness of the overall architectural design, which cannot be explained in terms of any underlying orderliness of the constituents" (Weiss 1971, p. 286).

What was the *constraining* power through which all those molecules, possessing all those degrees of freedom at their own level, yielded to a consistent order at a higher level — a *physically* unexpected coherence? This was the question Weiss' life-long observation of living cells continually brought him up against. But he was too honest to frame an answer in terms of the science of his day. His virtue lay in nevertheless not shrinking from the problem. He spent a long career investigating and describing the physically lawful performances of cells, but he did not pretend that, in doing this, he was *explaining* the order he observed.

I suspect that, with continuing observation and faithful description, the "problem" of order and wisdom (thought-fullness) in cells will more and more fade into nothingness. It is indeed only the effort at reductionism that creates the problem. Cease that effort, and all we have left is the routine scientific task of accurate conceptualization and description. Physicists, after arriving at concepts of law, force, field, and all the rest, do not often complain, "Those are not material things; how can we possibly deal with them?" They simply continue investigating, describing, and thinking until an overall, coherent picture — a harmonious conceptual interweaving — is achieved. That is what *making sense* of the world means.

It would be strange if the initially surprising discovery of living and coherent order in the cell persisted as a problem; another name for the discovery of order is, after all, "science." I suppose that the unexpectedness of at least some forms of order has been part of the scientist's experience all along. But when we live with it long enough, the unexpected becomes expected. In the end, it simply further strengthens our inalienable sense that we live in a world of coherent meaning.

But this happy ending will not be fully realized in biology until we acknowledge that there are many different ways phenomena can add up to a coherent picture in this cosmos of ours. A sloth is not a lion (Holdrege 2021), ice is not water vapor, and an animal is not a rock. Reductively forcing one sort of coherence into the mold of another by violence is never the answer.

WHERE ARE WE NOW?

The organism's coherence need not be mysterious.

We have arrived at a simple truth: the biologist's sense of threatening mystery (or "mysticism") when confronted with the intentional, purposive, and meaningfully expressive aspects of an organism's life typically arises from the unshakable conviction that there needs to be an essentially inanimate explanation of animate beings. As an *insistence*, this is mere dogma. The requirement of science is that we open-mindedly describe every aspect of every phenomenon in its own terms. It does not require a lot of reflection to see, for example, that organic processes of development and self-realization do not have strictly physical descriptions. Inanimate objects do not persistently and directively engage in efforts to develop and realize themselves.

But this does not mean we are headed toward some kind of mystical conception of the organism. As we will see increasingly in coming chapters, the different aspects of the organism (including the more-than-physical — ideal or archetypal — aspects) require only what all science requires: description in terms that are faithful to the phenomena themselves. To describe the marvelous living coherence of molecular processes in an organism's cells is no more mystical than to describe the very different but just as marvelous coherence of the laws of physics. It merely requires a willingness to embrace what we see, rather than recoil from it.

What I have said in this chapter will raise the question for many readers, "Is merely describing what we see in its own apparent terms an adequate foundation for science?" The question will be appproached in Chapter 11 ("Why We Cannot Explain the Form of Organisms") and addressed more fully in Chapter 12 ("Is a Qualitative Biology Possible?"). An even more fundamental question has to do with the role of thought both in our descriptions of the world and in the world itself. Is the refusal to accept thinking and thought as natural aspects of the world the deepest root of the biologist's unwillingness to take organisms at face value? I will take this up in Chapters 13 ("All Science Must Be Rooted in Experience") and 24 ("Is the Inanimate World an Interior Reality?").

NOTES

1. "RNA splicing was initially discovered in the 1970s, overturning years of thought in the field of gene expression" (Clancy 2008).
2. Estimates of the number of proteins participating in the spliceosome vary widely. Some have said there are more than 300, and others "only" 80 — a good indication of a fluidity of structure that is hard to nail down.
3. The "vagueness" here may be a function of the researcher's habit of looking for a precise digital code. Yet, without such a code the cell seems to manage the "life or death" business of splicing with great reliability. This is presumably due to the fact that the relevant business of the governing wisdom is the reading of the contextual meaning of the situation, not the mechanical interpretation of an exact digital code. A comparison might be the confirmed party-goer who navigates the exceedingly subtle, complex, and expressive landscape of a cocktail party without giving it a second thought — and without reading off from a "cheat sheet" a set of rigidly encoded, step-by-step instructions ensuring social harmony.
4. Wikipedia article, "Alternative Splicing," accessed May 11, 2019.
5. Obviously, I am not referring to our own conscious perceptive capacities. But neither am I referring to something less effective in its own way than our power of perception. Whatever brings the biologically coherent and needful results out of the currently inconceivable, creative "chaos" of the cellular plasm is far beyond our efforts to follow, let alone to reproduce. We have to think of a capacity higher than anything we consciously possess, even if — as the psychosomatic unity of the organism suggests — our consciousness is somehow contiguous with this higher capacity.
6. There are many other aspects of RNA splicing not considered here — for example, the role played by certain metal ions in the shift between different spliceosomal protein conformations (and therefore between different protein functioning). Such ions are a long way from the macromolecules in which biologists normally invest their sense of cellular information, and yet their well-informed role is crucial to cellular activity.
7. Figure 8.1 credit: BQUB24-Diraheta (https://commons.wikimedia.org/wiki/File:Deinococcus_radiodurans_dibuix.jpg), CC BY-SA 4.0.

SOURCES

Chen, Weijun and Melissa J. Moore (2014). The Spliceosome: Disorder and Dynamics Defined, *Current Opinion in Structural Biology* vol. 24, pp. 141–49. doi:10.1016/j.sbi.2014.01.009

Clancy, Suzanne (2008). RNA Splicing: Introns, Exons and Spliceosome, *Nature Education* vol. 1, no. 1. https://nature.com/scitable/topicpage/rna-splicing-introns-exons-and-spliceosome-12375

de Almeida, Sérgio F. and Maria Carmo-Fonseca (2012). Design Principles of Interconnections between Chromatin and pre-mRNA Splicing, *Trends in Biochemical Sciences* vol. 37, no 6 (June), pp. 248–53. doi:10.1016/j.tibs.2012.02.002

Dekker, Job, Joanna Wysocka, Iain Mattaj et al. (2013). Nuclear Biology: What's Been Most Surprising?, *Cell* vol. 152 (March 14), pp. 1207–08. doi.org/10.1016/j.cell.2013.02.041

Gehring, Niels H. and Jean-Yves Roignant (2021). Anything but Ordinary – Emerging Splicing Mechanisms in Eukaryotic Gene Regulation, *Trends in Genetics* vol. 37, no. 4 (April), pp. 355–72. doi:10.1016/j.tig.2020.10.008

Holdrege, Craig (2021). *Seeing the Animal Whole — And Why It Matters*. Great Barrington MA: Lindisfarne.

Krisko, Anita and Miroslav Radman (2013a). Biology of Extreme Radiation Resistance: The Way of *Deinococcus radiodurans*, *Cold Spring Harbor Perspectives in Biology*. doi:10.1101/cshperspect.a012765

Luco, Reini F. and Tom Misteli (2011). More Than a Splicing Code: Integrating the Role of RNA, Chromatin and Non-coding RNA in Alternative Splicing Regulation, *Current Opinion in Genetics and Development* vol. 21, pp. 366–72. doi:10.1016/j.gde.2011.03.004

Nilsen, Timothy W. and Brenton R. Graveley (2010). Expansion of the Eukaryotic Proteome by Alternative Splicing, *Nature* vol. 463 (January 28), pp. 457–63. doi:10.1038/nature08909

Papasaikas, Panagiotis and Juan Valcárcel (2016). The Spliceosome: The Ultimate RNA Chaperone and Sculptor, *Trends in Biochemical Sciences* vol. 41, no. 1 (January), pp. 33–45. doi:10.1016/j.tibs.2015.11.003

Sundvall, Maria, Ville Veikkolainen, Kari Kurppa et al. (2010). Cell Death or Survival Promoted by Alternative Isoforms of ErbB4, *Molecular Biology of the Cell* vol. 21, no. 23. doi:10.1091/mbc.e10-04-0332

Travis, John (2011). How Does the Cell Position Its Proteins? *Science* vol. 334 (November 25), pp. 1048–49. doi:10.1126/science.334.6059.1048

Wahl, Markus C. and Reinhard Lührmann (2015). SnapShot: Spliceosome Dynamics I, *Cell*, vol. 161, no. 6, (June 4), p. 1474.e1. doi:10.1016/j.cell.2015.05.050

Weiss, Paul (1962). From Cell to Molecule, in *The Molecular Control of Cellular Activity*, edited by John M. Allen, pp. 1–72. The University of Michigan Institute of Science and Technology Series. New York: McGraw-Hill.

Weiss, Paul A. (1971). *Within the Gates of Science and Beyond: Science in Its Cultural Commitments*. New York: Hafner.

CHAPTER 9

A Mess of Causes

THE DIFFICULTIES IN TALKING about causes in biology have been recognized for over two centuries.[1] It's just that the issues were largely set aside in the era of molecular biology due to the expectation that our rapidly growing powers of minute analysis would bring full causal understanding. Biology would soon be rid of its troublesome language of life in favor of well-behaved molecular mechanisms. And yet today, after several decades of stunning progress in molecular research, the struggle to fit our understanding of living activity into the comfortable garb of familiar causal-physical explanation looks more hopeless than ever.

On one hand, most biologists seem unaware that there is a problem here — or, at least, they are unwilling to betray their awareness in professional circles. On the other hand (as we will see in this chapter), their scientific descriptions could hardly signal more dramatically the failure of the usual causal explanations. We seem to be looking here at another illustration of blindsight.

In Chapter 7 we considered epigenetics, which is commonly taken to be about the way epigenetic "marks" on chromosomes alter gene expression. But no sooner did epigenetics gain biologists' attention than researchers began puzzling over the question, "Do epigenetic marks alter gene expression, or do changes in gene expression alter the marks?" (see Box 9.1) And the question is still with us. According to Luca Magnani, a cancer researcher at Imperial College London,

> It's an absolutely legitimate question and we need to address it. The answer is either going to kill the field [of epigenetics], or make it very important (quoted in Ledford 2015).

"Either kill the field or make it very important." The comment expresses absolute confidence that we can discover unambiguous causation, which will in turn settle the matter: either epigenetic changes *cause* gene activity (in which case they are very important), or they are *mere effects* of that activity, with little causal significance of their own. It must be one way or the other. The general idea is that, if something is to contribute to scientific understanding, it must be the indisputable cause of an indisputable effect. And yet, as we will now see, this stubborn insistence on causal clarity continually prods biological researchers (we will focus on molecular biology) to offer embarrassingly incoherent explanations.

The seductive appeal of master controllers

Consider the following remarks about a protein known rather blandly as "p53." The remarks issue from a perfectly reputable source who is clearly aware of the subtleties and interwoven intricacies of coordinated, molecular-level activity in the cell. And yet this expert is lured by the mirage of unambiguous causation into offering a wondrously self-annihilating description:

> The tumor suppressor p53 is a master sensor of stress that controls many biological functions, including [embryo] implantation, cell-fate decisions, metabolism, and aging … Like a complex barcode, the ability of p53 to function as a central hub that integrates defined stress signals into decisive cellular responses, in a time- and cell-type dependent manner, is facilitated by the extraordinary complexity of its regulation. Key components of this barcode are the autoregulation loops, which positively or negatively regulate p53's activities.

To start with, then, we have a *master sensor* (p53) that *controls* various fundamental cellular processes, and yet is itself wholly dependent on the signals it receives and is subject to "extraordinarily complex" *regulation* by certain autoregulation loops. While all these loops regulate p53 (some positively and some negatively), one of them, designated "p53/mdm2,"

> is the master autoregulation loop, and it dictates the fate of an organism by controlling the expression level and activity of p53. It is therefore not surprising that this autoregulation loop is itself subject to different types of regulation, which can be divided into two subgroups … (Lu 2010).

So the *master controlling* sensor is itself subject to a *master controlling* process (one of several regulatory loops) that *dictates* the fate of the organism. But this master loop, it happens, is in turn *regulated* in various manners (as the author goes on to say in the rest of the article) by a whole series of "multi-layered" processes, including some that are themselves "subject to direct regulation by mdm2" — that is, they are regulated by an element of the regulatory loop they are supposed to be regulating.

It is hard to believe that the confusion here is unavoidable. By now every biologist knows how regulatory processes extend outward without limit, connecting in one way or another with virtually every aspect of the cell. But this only underscores the undisciplined terminological confusion continuing to corrupt molecular biological description today. When key regulators are in turn regulated, and controllers have their fates underwritten or redirected by other players, where within the web of mutual interaction can we single out a *master* controller capable of *dictating* cellular fates? And if we can't, what are reputable scientists doing when they claim to have identified such a controller, or, rather, various such controllers?

More than an innocent abuse of language

Here is a comment from another paper on p53:

> Following DNA damage, the transcription factor p53 determines whether cells undergo apoptosis [self-induced cell death] or cell cycle arrest and DNA repair. To enable different cellular outcomes, p53 is regulated through its temporal expression dynamics and post-translational modification, and by interactions with chromatin, chromatin regulators and transcription factors.[2]

Here again we have the same terminological confusion, with p53 *determining* cellular outcomes, while it is itself regulated by many pervasive cellular processes. But the authors conclude their paper with these remarkably sensible statements:

> The large number of p53 regulatory mechanisms and their cooperation in triggering specific expression programmes remain open areas for investigation. Systematic measurements in multiple conditions together with models integrating the multiple layers of regulation on p53 activity will be required to decipher the complexity of p53 function.

Why not leave the matter there, with this admirable spirit of openness to the research results as given, together with an acknowledgment of almost unsurveyable complexity? Why are so many researchers driven to paste on top of this picture a contradictory assertion of open-and-shut causal determination?

And I do mean *driven*. How else to explain a comment that could serve as a fitting postscript to our discussion of RNA splicing in Chapter 8. Brenton Graveley, a geneticist at the University of Connecticut Stem Cell Institute, reported in 2011 on the discovery of a splice variant of the protein known as FOX-P1 — a variant that has a role in the generation of stem cells. After usefully elucidating some of what goes on, he offers this as his conclusion:

> What controls the [FOX] splicing switch? What splicing factors are responsible for flipping this switch, and how are their expression and activities regulated? Answering these questions is like hunting down the "chicken-or-the-egg" paradox, but they will ultimately uncover the master regulator of stem cell pluripotency (Graveley 2011).

So in the very act of acknowledging the fundamental "chicken-or-egg" paradox of all biological causation, he reflexively reverts to a kind of creedal affirmation of a still-hidden, but eventually-to-be-found Master Regulator.

If all those who use the language of biologically omnipotent control are really trying to describe something like "important influencers," then that's perfectly fine. But influence is not about mechanism and control; the *things* at issue just don't have controlling powers. Nor, despite Graveley's suggestion, is it about a simple flipping of yes-or-no switches. What we see, rather, is a continual mutual adaptation, interaction, and coordination explicable only in terms of the functional ideas through which we grasp the contextualized *meaning* of what is going on (Chapter 6, "Context: Dare We Call It Holism?").

What we see, that is, once we start following out all the interactions at a molecular level, is not some mechanism *dictating* the fate or *controlling* an activity of the organism. Rather (as I have been emphasizing throughout the preceding chapters), we observe an organism-wide, narrative coherence — a functional, end-directed, story-like coherence that we cannot elucidate in terms of strictly physical interactions that make no reference to the *meaning* of events. Only so far as they are caught up in and sensitive to this functional story do the individual molecular players find their proper roles.

The misrepresentation of this organic and rational coherence in favor of supposed controlling mechanisms is not an innocent inattention to language; it is a fundamental misrepresentation of reality at the central point where we are challenged to understand the character of living things.

Biological clocks: Who is keeping time?

Pick any topic in biology and you will encounter an egregious failure to "tie down" biological causes. Clockwork mechanisms are nowhere to be found — a fact that becomes particularly poignant in the investigation of "biological clocks" such as the circadian (daily) rhythms that figure so prominently in human and other forms of life. Biologists, of course, set out to identify the "master clock mechanism" that was presumed to "control" these rhythms, and, yes, they found a rhythmic feedback loop involving genes and transcription factors in a certain area of the brain that seemed the perfect candidate. It quickly came to be viewed as the decisive governor of circadian rhythms in the body:

> In mammals, the anatomical structure in the brain that governs circadian rhythms is a small area consisting of approximately 15,000 neurons localized in the anterior hypothalamus, called the suprachiasmatic nucleus (SCN). This "central pacemaker" in the SCN

> receives signals from the environment and, in turn, coordinates the oscillating activity of peripheral clocks, which are located in almost all tissues (Berger and Sassone-Corsi 2016).

And yet (as this statement already indicates), ongoing research has revealed distinct "clocks" in different mammalian organs and tissues, and indeed in every cell. These "clocks," it turns out, are not merely on the receiving end of a central, governing coordination, but rather are themselves participants in that coordination, and also, it now seems, are interwoven with just about all aspects of the organism's physiology — metabolism, reproduction, cell growth and differentiation, immune responses, central nervous system functions ...

In each of these areas the quest for causes and master controllers leads to the usual perplexity about who's doing what to whom. For example: "Although metabolism is thought to be primarily downstream of the cellular clock, numerous studies provide evidence that metabolic cycles can operate independently from or even influence circadian rhythms" (Kumar and Takahashi 2010). At the molecular level, one research team remarks that the enzymatic function of a certain clock protein "may be controlled by changing cell energy levels, or conversely, could regulate them" (Doi et al. 2006). In general: "It seems that connections between the circadian clock and most (if not all) physiological processes are bidirectional" (Yang 2010).

What we're gaining from all this research is a wonderful portrait of the organism as a rhythmic being. Investigators have not found controlling mechanisms that single-handedly establish or govern the circadian rhythms of the organism, but rather are discovering how those rhythms come to expression at every level and in every precinct of the organism — perhaps more centrally here and more peripherally there, but altogether in a single, organism-wide harmony that is also linked to environmental rhythms. There is no sensible way, as a scientist, to speak of particular mechanisms that *explain* this harmony. Instead, every isolated "mechanism" is found to be a *reflection* of the harmony, and we thereby gain further, detailed understanding of how the whole organism functions as a being in time.

Is any of this a surprise? Should we expect, say, that a "master regulator" of digestion exists? Would it be the stomach? The small intestine? The large intestine? The pancreas? The liver and gall bladder? The metabolism taking place in every cell? The brain that sends various coordinating nervous signals to different organs? The mouth that initiates everything? We would certainly look more to the stomach than, say, to the heart, but the fact remains that the organism as a whole is the closest thing we have to a "master regulator." What we see in the separate, "mechanistic" clocks and regulators of circadian rhythms is simply the functioning of those rhythms in the most recognizable or most focal places. But they merely put on more obvious display the rhythmic functioning of the entire body.

A well-studied worm

Or, we can choose a different example. If there was any place where biologists expected a causal explanation of the organism to emerge clearly, it was in the study of *Caenorhabditis elegans*, a one-millimeter-long, transparent roundworm whose private molecular and cellular affairs may have been more exhaustively exposed than those of any other organism. The adult hermaphrodite has exactly 959 cells, each precisely identified as to origin and type: for example, 302 cells belong to the nervous system. The developmental fate of every somatic cell, from egg to adult, had already been mapped out by 1980. But this mapping and the associated molecular studies did not produce the expected explanations.

Sydney Brenner — who received a 2002 Nobel prize for his work on *C. elegans* — acknowledged that development "is not a neat, sequential process ... It's everything going on at the same time." Even regarding the carefully mapped cell lineages of this "simple" roundworm, "there is hardly a shorter way of giving a rule for what goes on than just describing what there is." In other words, the only "rule" for the development of this worm is the entire developmental description of it.

When critics suggested he had not really come to an understanding of the worm, but had "only" described it, Brenner wisely responded, "I'm not sure that there necessarily is anything more to understand than what it is." British science writer Roger Lewin quoted this remark by Brenner in an article titled, "Why Is Development So Illogical?" with the subtitle, "The more biologists learn about development, the less it appears that organisms are assembled by neat, sequential processes; we should not be surprised" (Lewin 1984). Actually, it's not even true that organisms are *assembled* from pre-existing parts. They grow from within through processes of self-transformation, not mechanical assembly.

The difficulties of linear, causal explanation encountered by the *C. elegans* researchers were not accidental. You can't explain an organism of meaning, and you don't need to. You need only allow it, like any meaningful text, to speak ever more vividly and profoundly, in ever greater detail, so as to yield up its unique and unrepeatable story.

The separate processes do not make tidy explanations because they are not really separate and are not doing just one thing. They are harmonizing with everything else going on in the organism. We gain understanding when we learn to recognize this harmony in every aspect of the organism. Various analyses can play a crucial role in bringing clarity to our understanding, but the full picture takes shape only when the analytical threads are woven back into the larger fabric of meaning.

Of crosstalk, horror graphs and collaborative decision-making

Molecular biologists speak about *signals* arriving at *receptors* at the cell surface. The signals bear *messages*, which are then transferred (as it often happens) to a series of further *messengers* internal to the cell, who may, among other possibilities, finally convey the message to the cell nucleus. There the message may be *interpreted* to require the increased or decreased *expression* of a gene *coded* for a particular protein. The most noticeable players in the signaling are protein molecules.

The terminology so naturally resorted to here vividly invokes language, meaning, and communication — something we saw exemplified in Chapter 2. But, of course, due to the biological blindsight described in Chapter 1, this usage is typically treated as "mere metaphor." Moreover, the entire signaling process has been understood in a digital or computer-like fashion. But the truth turns out to be that we are watching something much more like an ever-shifting dialogue among innumerable shades of meaning than like a set of definitive, staccato, digital pulses.

It remains true that signaling pathways are vital means of communication within and between cells. But the communication is much more fully dimensioned — much more richly meaningful and less narrowly deterministic — than it was once thought to be.

In the conventional machine model of the organism, signaling pathways were straightforward, with a clear-cut input at the start of the pathway leading to an equally clear-cut output at the end. Not so today, as a team of molecular biologists at the Free University of Brussels found out when they looked at how these pathways interact or "crosstalk" with each other. Tabulating the cross-signalings between just four such pathways yielded what they called a "horror graph," and quickly it began to look as though "everything does everything to everything" (Dumont et al. 2001).

Perhaps a horror graph is what a flow of contextualized meaning looks like when we are expecting one-dimensional, reduced, carefully coded, mathematically analyzable information. By contrast, some researchers now imagine (if rather fancifully) the "collaborative" reality of signaling pathways "as a table around which decision-makers debate a question and respond collectively to information put to them" (Levy et al. 2010).

Even considering a single membrane receptor bound by a hormonal or other signal, you can find yourself looking, conservatively, at some two billion possible states, depending on how that receptor is modified by its interactions with other molecules. Despite previous belief, there is no simple binary rule distinguishing deactivated receptors from those activated by some combination of signals in a particular context. "The activated receptor looks less like a machine and more like a … probability cloud of an almost infinite number of possible states, each of which may differ in its biological activity" (Mayer et al. 2009).

Our problem lies in adequately imagining the reality. When a single protein can combine with several hundred different modifier molecules, leading to practically infinite combinatorial possibilities, and when that protein itself is an infinitesimal point in the vast, turbulent molecular sea of continual exchange that is the cell, and when the cell is one instance of maybe several trillion cells of some 250 different major types in the human body — in muscle and bone, liver and brain, blood and artery — well, it would be understandable if some researchers preferred not to stare too long at this picture.

Nevertheless, we should keep in mind that the "collaborative" process mentioned above involves not just one table with "negotiators" gathered around it, but countless tables with countless participants, and with messages flying back and forth in countless patterns as countless "decisions" are made in a manner somehow subordinated to the unity and multidimensioned interests of the organism as a whole.

In other words, not only are the elements of an individual signaling pathway extremely flexible and adaptive; the individual pathway itself, once thought of as discrete and well-defined, doesn't really exist — certainly not as a separate "mechanism." Researchers now speak of the "multi-functionality" or "functional pleiotropism" of signaling nodes, pointing out that signaling networks have "ways of passing physiologically relevant stimulus information through shared channels" (Behar and Hoffmann 2010).

Whenever we imagine a biological process aimed at achieving some particular result, we need to keep in mind that every element in that process is likely playing a role in an indeterminate number of other significant, and seemingly goal-directed, activities. The mystery in all this does not lie primarily in isolated "mechanisms" of interaction. The question, rather, is why things don't fall completely apart — a question we had occasion to ask in Chapter 6, in connection with the idea that the whole is more than the sum of its parts.

In sum: messages do not fly back and forth as metaphors or disembodied abstractions. They move as dynamically sculpted currents of force and energy. Their meanings are mimed or gestured — neither translated into, nor reduced to, a kind of expressionless Morse code, nor impelled along precisely incised channels like computer instructions. And what holds them together amid the ceaseless flows and crosstalk and molecular transformation is the unity of meaning that is the whole organism. This unity is there for us to observe directly, and we all can recognize it, whether with blindsight or otherwise.

Box 9.1 illustrates the problems we've been discussing, with specific reference to a single aspect of cellular function: the molecular biology of gene expression.

BOX 9.1

Cause — or Effect?

Ambiguities related to gene expression

In trying to articulate the idea of a machine-organism, molecular biologists are forever chasing "causal mechanisms," forever lamenting the difficulty of teasing apart cause and effect, and forever failing to see that the sought-for mechanisms don't exist in any stable and reliable sense.

"Together, these results further emphasize the role for RNA polymerase in shaping the chromatin landscape of the genome and point toward the difficulty in disentangling cause and effect in the relationship between chromatin and transcription" (Weiner et al. 2010).

"Epigenetic modifications in Alzheimer's disease: cause or effect?" — title of a paper. The conclusion: "Further studies are necessary" (Piaceri 2014).

Concerning "the pan-placental downregulation of H3K9ac [an epigenetic modification of chromatin] in gestational diabetes mellitus": "Whether this is cause or effect of the metabolic disorder needs to be investigated further" (Hepp et al. 2018).

"A long-standing question is whether [cell] replication timing dictates the structure of chromatin or vice versa. Mounting evidence supports a model in which replication timing is both cause and consequence of chromatin structure by providing a means to inherit chromatin states that, in turn, regulate replication timing in the subsequent cell cycle" (Gilbert 2002).

"While several studies using next-generation sequencing have revealed genome-wide associations between epigenetic modifications and transcriptional states, a direct causal relationship at specific genomic loci has not been fully demonstrated ..." (Fukushima et al. 2019).

continued next page

"Despite the difficulties in proving cause and effect, these examples convincingly illustrate how chromatin crosstalk can functionally increase the adaptive plasticity of the cell exposed to the changing microenvironment" (Göndör and Ohlsson 2009).

"A related unresolved question is whether chromatin loops are the cause or the effect of transcriptional regulation" (Deng and Blobel 2010).

"The enthusiasm for establishing whether epigenetic mechanisms link the environment with disease development must be tempered by the knowledge that the epigenome is dynamic and has as much potential to respond to disease as respond to the environment. Therefore it is very difficult to disentangle cause from consequence when studying epigenetic variation and disease" (Relton 2012).

"Despite abundant evidence that most kinds of tumor cells carry so-called epigenetic changes, scientists haven't yet worked out exactly whether such glitches are a cause or a consequence of disease" (Kaiser 2010).

"The clarification of the cause-and-effect relationship of nuclear organization and the function of the genome represents one of the most important future challenges. Further experiments are needed to determine whether the spatial organization of the nucleus is a consequence of genome organization, chromatin modifications, and DNA-based processes, or whether nuclear architecture is an important determinant of the function of the genome" (Schneider and Grosschedl 2007).

"Although there is widespread agreement that genome form [such as folding and topological domains] and function [gene expression] are intimately connected, their causal relationship remains controversial" (Stadhouders et al. 2019).

"The spatial organization of the genome into compartments and topologically associated domains can have an important role in the regulation of gene expression. But could gene expression conversely regulate genome organization? ... Recent evidence suggests a dynamic, reciprocal interplay between fine-scale genome organization and transcription, in which each is able to modulate or reinforce the activity of the other" (Steensel and Furlong 2019).

"Transcription itself alters loops and consequently requires their continual reformation. Together, [this and other chromatin] properties suggest extensive feedback between chromatin structure and gene activity, rather than a simple cause-and-effect relationship" (Misteli and Finn 2021).

The problem of causation is fundamental to biology

The powerful compulsion to identify decisive causes, even at the expense of painfully self-contradictory language, strongly suggests that a one-sided and unrealizable ideal of biological explanation is at work. Under its influence we aim to discover a physical lawfulness reflecting, above all, our experience with machines — a lawfulness of precise, unambiguous *control*, where one thing can be said, without unwelcome qualification, to make another thing happen.

Think of a machine. Having conceived what we want it to do, we design it to be a closed system whose intended functioning is more or less immune to contextual interference. And we try to do much the same in many scientific demonstrations. For example, we can create a vacuum in a chamber, and then release a leaf from the top of the chamber. It falls like a stone.

Of course, leaves in nature often travel upward. But the experiment in the chamber enables us to observe the singular and lawful play of gravity, without any disturbing "interference" from the resistance or movement of air. We can then — and only then — say that gravity appears to *make* the leaf fall, just as the simple laws governing the gears and springs of a mechanical watch *make* the watch perform as a reliable keeper of time.

But when the biologist tries to see an animal in the same mechanistic light, as a closed system without interfering factors, the attempt fails miserably. This is because, for the animal, *contextual interference is the whole point*. As the meaning of its activity shifts from moment to moment, so, too, does the contextual significance of all the details of its life.

For example, when a deer is grazing in a meadow, its glimpse of a vaguely canine form in the distance changes the meaning of everything from the flowers and grass the deer was eating, to its own internal digestive processes, to the expression of its genes. This happens, not because the distant form is exerting some strange physical force upon the deer, but because that form becomes part of a now suddenly shifted pattern of meaning.[3]

Or (to focus on the cellular level): when a cell enters into mitosis, just about every detail of its physiology and chemistry takes on an altered meaning in light of the changing narrative context. Everything is now heading toward a different outcome. Molecules that had been participating in one set of interactions (and could easily still do so in purely physical terms) now enter into very different intermolecular relations. Similarly with a cell experiencing heat shock, oxygen deprivation or other stress, a cell coming into contact with new neighbors, or a cell proceeding along a path of embryonic differentiation.

Certainly we can still identify unambiguous causes in the organism. It is always possible to narrow the conditions of our experiments so severely that a consistent "causal arrow" for a particular interaction emerges *under those conditions*. But the whole point of life's adaptability is to seek (or help create) *altered conditions* according to present needs and interests. This is why there can be no fixed syntax, no mechanical constancy of relations among the parts. The organism is forever abandoning the coordinating principles of its old context in favor of a new and ever-changing meaning. Its story is always evolving.

I titled this section, "The Problem of Causation is Fundamental to Biology." The problem I had in mind was that of getting clear about the nature of causation in biology. It differs from the problem of causation in the physical sciences. Organisms manifest a fluid, integral, harmonizing sort of causation that is more like a play of the multi-dimensional *reasons* for things than a set of one-dimensional mechanical interactions. It is more like the rich interplay of meaning in an unfolding poem than a rigid syntax or logic.

And yet, despite all this, biologists seem fixated on the "fundamental issue" of distinguishing clear-cut cause from clear-cut effect in the usual physical sense. That's why we hear statements like this:

> Despite intensive studies of genome organization in the past decade, a fundamental issue remains regarding genomic interactions and genome organization as a cause or a consequence of gene expression. This problem is also pertinent to RNAs, which may have regulatory functions in transcription rather than being simply products of transcription (Li and Fu 2019).

Unfortunately, there is little if any effort to elucidate just what hangs upon this "fundamental issue" — or what might be the implications of the fact that the issue appears irresolvable so long as we insist upon unambiguous physical causation as the basis for biological understanding.

WHERE ARE WE NOW?

We need a biology beyond definitive causes.

If the preoccupation with controlling causes reflects, as I have now suggested, an unrealizable ideal of biological explanation, then it also reflects a more or less false understanding of biological reality. I have, in the preceding chapters, been trying to point toward some primary aspects of a more adequate understanding — one that needn't bring us into conflict with what we know. Here is a brief retrospective:

♦ It has turned out, as we saw in "What Brings Our Genome Alive?" and "Epigenetics: A Brief Introduction," that genes — those supposed *prime causes* of the organism's life — are in fact the focus of almost incomprehensible powers of coordination working from the whole of the cell and organism into the cell nucleus. And the principle of coordination was equally evident in "The Sensitive, Dynamic Cell," where we looked at the membranes and cytoskeleton of cells.

♦ We have seen, courtesy of the work of the twentieth-century cell biologist, Paul Weiss, that molecules interacting according to physical law in the fluid medium of the cell possess countless "degrees of freedom" that must be curtailed, or disciplined, by the cell as a whole. Similarly, vast numbers of cells must be "held together" according to the functional needs of particular organs. And so, too, the disparate organs and organ systems are harmoniously subordinated to the needs and interests of the organism as a whole. (See especially the chapter, "Context: Dare We Call It Holism?")

♦ We have also seen (in that same chapter) that biologists incessantly appeal to the "context-dependence" of biological activity. The appeal amounts to a tacit recognition of a kind of causation that works "downward" from the integral unity of a larger whole, into the parts. This causal unity is inseparable from the *ideas* that define a context and hold its elements meaningfully together, thereby posing "The Mystery of an Unexpected Coherence" (Chapter 8).

♦ Again, in the present chapter, we have been alerted to the confusion of causes that makes it impossible to explain organisms in the usual causal terms. That is, it is impossible to explain them *biologically* in this way, as opposed to merely elucidating their physics and chemistry. The life-like coordination of physical interactions involves what I referred to above as the "multi-dimensional *reasons* for things rather than a set of one-dimensional mechanical interactions. It is more like the rich interplay of meaning in an unfolding poem or novel than a rigid syntax or logic."

♦ And, finally, it is hard not to notice that all these themes come together in what we can usefully think of as the organism's story. That is, every organism spins, or cooperates in spinning, the narrative of its own life. Future-directedness, purposiveness, context-sensitivity, the meaningful interweaving of ideas, the subordination of isolated events and physical causes to the needs, interests, and purposes of active agents — these features that we have noted in organisms are also the features of stories (Chapter 2).

In the next two chapters, dealing with problems of form, we will see how the form of organisms offers us an avenue toward biological understanding that can be a useful corrective to the usual preoccupation with cause and effect. Then, in Chapter 12 ("Is a Qualitative Biology Possible?"), we will work at reconceiving biological causation as a matter of form and idea.

NOTES

1. In his 1790 work, *Kritik der Urteilskraft* (subsequently published in English as *Critique of Judgment*), the philosopher Immanuel Kant wrote of the organism that "every part not only exists *by means of* the other parts, but is thought as existing *for the sake of* the others and the whole … also [the] parts are all organs reciprocally *producing* each other" (Kant 1790, Div. 1, para 65).

   In speaking of purely physical causation, we certainly would not say that parts exist *for the sake of* each other. But Kant's treatment of these issues was central to a great deal of biological discussion during the following decades — and still surfaces frequently today, at least among philosophers of biology. But the technically oriented training of biologists themselves no longer encourages a familiarity with decisive issues at the foundation of their own discipline.
2. The quotation is from a Table of Contents description in *Nature Reviews Molecular Cell Biology* for Hafner et al. 2019.
3. I make this same point with the wildebeest and lion in the chapter on "The Organism's Story."

SOURCES

Behar, Marcelo and Alexander Hoffmann (2010). Understanding the Temporal Codes of Intra-Cellular Signals, *Current Opinion in Genetics and Development* vol. 20 (December), pp. 684–93. doi:10.1016/j.gde.2010.09.007

Berger, Shelley L. and Paolo Sassone-Corsi (2016). Metabolic Signaling to Chromatin, *Cold Spring Harbor Perspectives in Biology* vol. 8, no. 11. doi:10.1101/cshperspect.a019463

Deng, Wulan and Gerd A. Blobel (2010). Do Chromatin Loops Provide Epigenetic Gene Expression States?, *Current Opinion in Genetics and Development* vol. 20, pp. 548–54. doi:10.1016/j.gde.2010.06.007

Doi, Masao, Jun Hirayama and Paolo Sassone-Corsi (2006). Circadian Regulator CLOCK Is a Histone Acetyltransferase, *Cell* vol. 125 (May 5), pp. 497–508. doi:10.1016/j.cell.2006.03.033

Dumont, Jacques E., Fréderic Pécasse and Carine Maenhaut (2001). Crosstalk and Specificity in Signalling: Are We Crosstalking Ourselves into General Confusion?, *Cellular Signalling* vol. 13, pp. 457–63. doi:10.1016/S0898-6568(01)00168-1

Fukushima, Hiroto S., Hiroyuki Takeda and Ryohei Nakamura (2019). Targeted in Vivo Epigenome Editing of H3K27me3, *Epigenetics and Chromatin* vol. 12, no. 17. doi:10.1186/s13072-019-0263-z

Gilbert, David M. (2002). Replication Timing and Transcriptional Control: Beyond Cause and Effect, *Current Opinion in Cell Biology* vol. 14, no. 3 (June 1), pp. 377–83. doi:10.1016/S0955-0674(02)00326-5

Göndör, Anita and Rolf Ohlsson (2009). Chromosome Crosstalk in Three Dimensions, *Nature* vol. 461 (September 10), pp. 212–17. doi:10.1038/nature08453

Graveley, Brenton R. (2011). Splicing Up Pluripotency, *Cell* vol. 147 (September 30), pp. 22–24. doi:10.1016/j.cell.2011.09.004

Hafner, Antonina, Martha L. Bulyk, Ashwini Jambhekar and Galit Lahav (2019). The Multiple Mechanisms That Regulate p53 Activity and Cell Fate, *Nature Reviews Molecular Cell Biology* vol. 20, no. 4 (April). doi:10.1038/s41580-019-0110-x

Hepp, Paula, Stefan Hutter, Julia Knabl et al. (2018). Histone H3 Lysine 9 Acetylation is Downregulated in GDM Placentas and Calcitriol Supplementation Enhanced This Effect, *International Journal of Molecular Sciences* vol. 19, no. 4061 (December 14). doi:10.3390/ijms19124061

Kaiser, Jocelyn (2010). Genes Link Epigenetics and Cancer, *Science* vol. 330, no. 6004 (October 29), p. 577. doi:10.1126/science.330.6004.577

Kant, Immanuel (2000). *Critique of Judgment*, translated by J. H. Bernard. Amherst NY: Prometheus Books. Originally published in German in 1790. Available at https://www.gutenberg.org/ebooks/48433

Kumar, Vivek and Joseph S. Takahashi (2010). PARP around the Clock, *Cell* vol. 142 (September 17), pp. 841–43. doi:10.1016/j.cell.2010.08.037

Ledford, Heidi (2015). The Genome Unwrapped, *Nature* vol. 528 (December 3), pp. S12–S13. doi:10.1038/528S12a

Levy, Emmanuel D., Christian R. Landry and Stephen W. Michnick (2010). Signaling through Cooperation, *Science* vol. 328 (May 21), pp. 983–84. doi:10.1126/science.1190993

Lewin, Roger (1984). Why Is Development So Illogical? *Science* vol. 224 (June 22), pp. 1327–29. doi:10.1126/science.6374894

Li, Xiao and Xiang-Dong Fu (2019). Chromatin-Associated RNAs as Facilitators of Functional Genomic Interactions, *Nature Reviews Genetics* vol. 20 (September), pp. 503–19. doi:10.1038/s41576-019-0135-1

Lu, Xin (2010). Tied Up in Loops: Positive and Negative Autoregulation of p53, *Cold Spring Harbor Perspectives in Biology* (December 9). doi:10.1101/cshperspect.a000984

Mayer, Bruce J., Michael L. Blinov and Leslie M. Loew (2009). Molecular Machines or Pleiomorphic Ensembles: Signaling Complexes Revisited, *Journal of Biology* vol. 8, no. 9, article 81. doi:10.1186/jbiol185

Misteli, Tom and Elizabeth H. Finn (2021). Chromatin Architecture Is a Flexible Foundation for Gene Expression, *Nature Genetics* vol. 53, no. 4 (April), pp. 426–27. doi:10.1038/s41588-021-00813-2

Piaceri, Irene, Beatrice Raspanti, Andrea Tedde et al. (2015). Epigenetic Modifications in Alzheimer's Disease: Cause or Effect?, *Journal of Alzheimer's Disease*, vol. 43, no. 4, pp. 1169–73. doi:10.3233/JAD-141452

Relton, Caroline (2012). Epidemiology and Epigenetics — A Marriage Made in Heaven?, *Oxford University Press Blog* (March 28). https://blog.oup.com/2012/03/epidemiology-epigenetics-journal/

Schneider, Robert and Rudolf Grosschedl (2007). Dynamics and Interplay of Nuclear Architecture, Genome Organization, and Gene Expression, *Genes and Development* vol. 21, pp. 3027–43. doi:10.1101/gad.1604607

Stadhouders, Ralph, Guillaume J. Filion and Thomas Graf (2019). Transcription Factors and 3D Genome Conformation in Cell-Fate Decisions, *Nature* vol. 569 (May 16), pp. 345–54. doi.org/10.1038/s41586-019-1182-7

Steensel, Bas van and Eileen E. M. Furlong (2019). The Role of Transcription in Shaping the Spatial Organization of the Genome, *Nature Reviews Molecular Cell Biology* vol. 20, pp. 327–37. doi:10.1038/s41580-019-0114-6

Weiner, Assaf, Amanda Hughes, Moran Yassour et al. (2010). High-Resolution Nucleosome Mapping Reveals Transcription-Dependent Promoter Packaging, *Genome Research* vol. 20, pp. 90–100. doi:10.1101/gr.098509.109

Yang, Xiaoyong (2010). A Wheel of Time: The Circadian Clock, Nuclear Receptors, and Physiology, *Genes and Development* vol. 24, pp. 741–47. doi:10.1101/gad.1920710

CHAPTER 10

What Is the Problem of Form?

IT IS WELL KNOWN that amphibians such as frogs and salamanders have a remarkable ability to regenerate severed limbs. What may not be so commonly realized is that, if you graft the tail bud of a salamander onto the flank of a frog tadpole at the place where a limb would normally form — and also near the time when metamorphosis of the tadpole into a frog will occur — the grafted organ first grows into a salamander-like tail, and then, in some cases, more or less completely transforms into a limb, albeit a dysfunctional one. Among other changes, the tip of the tail turns into a set of fingers (Farinella-Ferruzza 1956).

The experiment can remind us how biologists commonly try to learn about life by severely disrupting it. But the current point is that, in this particular experiment, the transformation of the tail into an approximate limb cannot be the result of local causes, since the local environment of the fingers-to-be is a tail, not a limb. The power of transformation is, in a puzzling manner, holistic. The part is caught up within the whole and moves toward its new identity based, not merely on local determinants, but also on the form and character of a whole that is *not yet physically all there*.

This may remind us of the rather different experiment we heard Harvard biologist Richard Lewontin describing in Chapter 6 ("Context: Dare We Call It Holism?"): if a researcher excises the developing limb bud from an amphibian embryo, shakes the cells apart, allows them to reaggregate into a random lump, and then replaces the lump in the embryo, a normal leg develops. This shows that the currently unrealized form of the limb as a whole is the ruling factor, redefining the parts according to the larger, developing pattern (Lewontin 1983).

But how can this be? How can spatial position within a not yet fully realized form physically determine the future and proper sculpting of that form, and do so even when parts are surgically jumbled?

In one way or another, the problem is universal. A key feature of holistic, end-directed, living processes is that the end plays a role in shaping the means. (See many of the preceding chapters, and especially Chapter 2, "The Organism's Story," Chapter 6, "Context: Dare We Call It Holism?," and Chapter 8, "The Mystery of an Unexpected Coherence.") Tadpoles with faces engineered to be highly abnormal "nevertheless largely become normal [adult] frogs: the craniofacial organs move in abnormal paths until a proper frog face morphology is achieved" (Levin 2020).

In other words, the means are modified, even becoming entirely unprecedented if necessary, in order to achieve a characteristic result.

We find the same principle when we look at cascades of gene expression, such as the sequential expression of the various genes that have been said to "determine" left-right asymmetry of the vertebrate body. The normal expectation would be that if one blocks or changes the expression of earlier genes in the sequence, the disorder should accumulate and be magnified, perhaps explosively, in downstream gene expression, since proper cues for the later steps are missing. But

> Surprisingly, this is not actually what occurs: each subsequent step has fewer errors than the previous step, suggesting that the classic linear pathway picture is importantly incomplete. Embryos recognize transcriptional deviations from the correct pattern and repair them over time ... The existence of corrective pathways in embryogenesis and regeneration raises profound questions about the nearly ubiquitous stories our textbooks and "models" tell about the molecular explanations for specific events (Levin 2020).

All this may remind us of E. S. Russell's remark that in biology "the end-state is more constant than the method of reaching it" (Chapter 2, "The Organism's Story"). We also see here the principle that cell biologist Paul Weiss enunciated so clearly at mid-twentieth century, when he pointed out that the whole "*is infinitely less variant from moment to moment than are the momentary activities of its parts.*" At the lowest level of biological activity, molecules in the watery medium of a cell have degrees of freedom (possibilities of movement and interaction) that would spell utter chaos at higher levels if it were not for the fact that the lower-level activity is "disciplined" from above.

Weiss' point is that, whatever the level we analyze, from macromolecular complexes, to organelles, to cells, to tissues, to individual organs, to the organism as a whole, we find the same principle: we cannot reconstruct the pattern at any level of activity by *starting from parts and their interactions*. There are always organizing principles that must be seen working from the whole into the parts. (See the discussion of Weiss in Chapter 6, "Context: Dare We Call It Holism?")

One further example. During development, the lens of an amphibian eye derives from the outer layer of cells in the developing head, at the point where an outgrowth of the brain comes into contact with the epidermal cells. But if an already developed lens is removed from one of these animals, something truly remarkable happens: a new lens forms from the upper edge of the iris, a structure that has nothing to do with lens formation in normal development. The procedure runs like this (Gilbert 1994, p. 40):

1. Cells from the upper part of the iris — cells that have already reached an endpoint of differentiation — begin multiplying;

2. these multiplying cells then proceed to dedifferentiate — that is, to lose their specialized character, including the pigmentation that gives the iris its color;

3. the newly multiplied, iris-derived cells migrate so as to form a globe of dedifferentiated tissue in the proper location for a lens; and finally,

4. they start producing the differentiated products of lens cells, including crystallin proteins, and are thereby transformed into transparent lens cells — all in the nuanced spatial pattern required for the formation of a proper lens.

And so, lacking the usual resources and the usual context for formation of a lens, the animal follows an altogether novel path toward the restoration of normal form and function.

One sometimes gets the feeling that a single, well-documented example of developmental plasticity of this sort, if taken seriously enough and contemplated deeply enough, could transform all of biology and deliver biologists from the worn-out fantasy of the mechanistic organism. But it doesn't happen. As some have observed, paradigms of explanation, once established, can be very difficult to overturn merely by citing evidence contradicting them.

In any case, it is impossible to believe that these complex and intricately coordinated responses to the loss of the lens were somehow already *physically* determined or programmed or otherwise specified in the animal's one-celled zygote. Nor is it easy to imagine how there could ever have been a sustained and large population of lens-injured amphibians with otherwise functional eyes — a population large enough, that is, to enable a supposedly mindless process of natural selection to evolve over great lengths of time a specific, novel solution to the problem of lens regeneration.

The problem of form exists even at the molecular level

The problem of form has long been central to biology, where each creature so notably reproduces after its own kind and according to its own form. "It is hardly too much to say," wrote geneticist C. H. Waddington, "that the whole science of biology has its origin in the study of form." In both their descriptive and theoretical activity, biologists "have been immersed in a lore of form and spatial configuration" (Waddington 1951, p. 43).

"Immersed in a lore of form" is, however, an oddly mild way of putting it. "Hopelessly adrift upon a fathomless sea of mystery" might be more fitting. An observer surveying the biological disciplines today (some seventy years after Waddington's comment) can hardly help noticing that every organism's stunning achievement of form has become an enigma so profound, and so threatening to the prevailing style of biological explanation, that few biologists dare to focus for long on the substance of the problem.

We will find it necessary in our further discussion to keep in mind that the mystery is at least as apparent on the microscopic (and even the molecular) level as it is at more easily recognizable levels. We have already seen this in earlier chapters. For example, in Chapter 2 ("The Organism's Story"), we heard the English neurophysiologist, Sir Charles Scott Sherrington, describing how a severed motor

BOX 10.1

The Miracle of Wound Healing

Here is a description offered by English biologist Brian Ford (2009):

"Surgery is war. It is impossible to envisage the sheer complexity of what happens within a surgical wound. It is a microscopical scene of devastation. Muscle cells have been crudely crushed, nerves ripped asunder; the scalpel blade has slashed and separated close communities of tissues, rupturing long-established networks of blood vessels. After the operation, broken and cut tissues are crushed together by the surgeon's crude clamps. There is no circulation of blood or lymph across the suture.

"Yet within seconds of the assault, the single cells are stirred into action. They use unimaginable senses to detect what has happened and start to respond. Stem cells specialize to become the spiky-looking cells of the stratum spinosum [one of the lower layers of the epidermis]; the shattered capillaries are meticulously repaired, new cells form layers of smooth muscle in the blood-vessel walls and neat endothelium; nerve fibres extend towards the site of the suture to restore the tactile senses ...

"These phenomena require individual cells to work out what they need to do. And the ingenious restoration of the blood-vessel network reveals that there is an over-arching sense of the structure of the whole area in which this remarkable repair takes place. So too does the restoration of the skin. Cells that carry out the repair are subtly coordinated so that the skin surface, the contour of which they cannot surely detect, is restored in a form that is close to perfect."

nerve in some animals manages to grow back, through many obstacles, toward the far-distant muscle it was originally attached to.

Somehow the minuscule nerve "knows" where it is within the vast three-dimensionality of the animal's body — knows its own place in contradistinction to that of all the other nerves in different parts of the body. It likewise "senses" where it needs to get to in the local context, and how to find its way there. It's as if it had a detailed map of the terrain.

When we consider the more general case of wound healing described in Box 10.1, we find ourselves watching how the nerves, blood vessels, muscles, and all the diverse, mangled tissues in a wound sort themselves out. It is all somehow governed by what the description's author calls "an over-arching sense of the structure of the whole area in which [the] repair takes place." The original form is restored as far as possible. But what is being sensed? how is it sensed? and "who" is doing the sensing? — these most basic questions remain unanswered.

We saw in Chapter 8 ("The Mystery of an Unexpected Coherence") that a similar problem faces us when we look at the several scores or hundreds of molecules engaged in the intricate molecular "surgery" known as *RNA splicing*. We know that all the complex, carefully sequenced, splicing interactions respect every bit of physical and chemical understanding we have amassed, and so we can go about explaining them in that sense. But a *biological* understanding — an understanding of the effective, flexible, context-dependent coordination of physical events toward a desirable result — remains indescribable in the currently acceptable terminology of biology.

And so the problem of form, even when we try to approach it at the molecular level, seems intractable from the standpoint of conventional biology. In the case of RNA splicing, we can ask how each molecule among the large crowd cooperating in the activity of splicing is synthesized in the right amount; how each one is brought to the right place for splicing, and at the right time; how it manages to interact with properly selected molecules among all the available partners in the operation, doing so in a carefully choreographed sequence; how the overall cooperation among all the molecules is achieved; and how this cooperation is properly aligned with the *needs* of the cell at a particular time — a time when one form of the spliced RNA rather than another happens to be called for, requiring the "surgery" to be performed with unique variations.

Need is not a term of physical science. Further, all this occurs in a fluid or highly plastic medium, with no crucial and precisely machined channels of communication such as those carved in silicon chips at our high-tech factories. The externally imposed mechanistic constraints, such as those required for the operation of our machines, simply are not there in the organism.

Of course, researchers have traced all sorts of molecular syntheses, movements, and interactions. We can be sure that everything in the entire picture proceeds lawfully, and in this very constricted sense every local event looks necessary. And yet we can find no combination of physical laws capable of "enforcing" the proper form of all the different parts of the body of this or that animal. In the case of a wound, there is no purely physical necessity to achieve the "proper" form in the face of wildly variable conditions.

In other words, the mere fact of physical lawfulness does not explain the *coordination of events* along an extended timeline in the narrative of healing, from infliction of the wound to the final restoration of normalcy. Nor does it explain the narrative of RNA splicing, from the occurrence of an RNA molecule in need of reconfiguration, to the final product of those scores or hundreds of participating molecular "surgeons." We can watch the molecules performing in a way that gives expression to the overall sense, or meaning, of the activity, but we do not have even the barest beginnings of a purely physical explanation for their commitment to that meaning.

Michael Levin: Revolutionary

I wrote above that every organism's stunning achievement of form has become an enigma so profound, and so threatening to the prevailing style of biological explanation, that few biologists dare to focus for long on the substance of the problem. Michael Levin is one of those few. An enthusiastic, prolific, hyper-achieving researcher, he appears to represent at least part of the future of biology. As the Vannevar Bush Professor at Tufts University near Boston, Levin is principal investigator of the Levin Lab there, director of the Tufts Center for Regenerative and Developmental Biology, and team leader of the Allen Discovery Center at Tufts. He also holds positions at Harvard and MIT. The wide-ranging work under his supervision includes pioneering explorations of the role of electrical fields in the production of biological form. (See Box 10.2.)

But what is perhaps most impressive about Levin is his willingness at least to make a start at acknowledging certain extraordinarily difficult questions biologists must raise if they want to face organisms squarely, as we actually observe them. Chief among these is the problem of organic form.

Levin is particularly explicit about this problem in a paper entitled "The Biophysics of Regenerative Repair Suggests New Perspectives on Biological Causation," published in *Bioessays* (Levin 2020). We will now give particular attention to this paper, which will be the source of all quotations unless otherwise indicated.

The way in which tissue voltage patterns prefigure the developing form of embryos has been central to Levin's thinking (Box 10.2). This prefiguring, he emphasizes, is not in the first instance a genetic event, but "a [much higher-level] physiological event ... causally responsible for a given patterning outcome" — and therefore also a cause of the gene expression required for that outcome.

BOX 10.2

Electricity in the Developing Tadpole

In the summer of 2011 a team of researchers at Tufts University produced a startling time-lapse video of a developing tadpole embryo[1] (Vandenberg et al. 2011). Due to the use of special dyes reporting the electric potentials across cell membranes, areas of the embryonic surface successively lit up brightly and then went dark. For a few seconds of the time-lapse film (representing the events of several hours), the featureless part of the embryo that would eventually become the animal's head flashed the image of a tadpole face.

But no actual face had yet developed. Nevertheless, the electrical pattern did "signal" where key elements of the tadpole's face, such as its eyes, would eventually appear. Regional changes in electric potential, these scientists concluded, "regulate expression of genes involved in craniofacial development."

According to Michael Levin, head of the laboratory where the tadpole research was performed, "Ion flows and the resulting [membrane voltage] changes are components of long-range conversations that orchestrate cellular activities during embryonic development, regeneration, and ... tumor suppression." He adds that "bioelectric cues are increasingly being found to be an important regulator of cell behavior," controlling the proliferation and death of cells, their migration and orientation, and their differentiation into different cell types. "We are," he wrote further, "just beginning to scratch the surface of the bioelectric code — the mapping between voltage properties and patterning outcomes, akin to the genetic, epigenetic, and perhaps other codes remaining to be discovered" (Levin 2012).

Levin's team quickly went on to manipulate the distribution of membrane voltages in developing embryos so as to provoke the generation of eyes in decidedly unexpected places — for example, on the back and tail, and even in the gut, of a frog embryo. The results were fragmentary and rather chaotic — the ectopic ("out of place") eyes were partial or deformed — but the result was nevertheless as startling as it was monstrous (Pai et al. 2012).

In other words — and this is where Levin particularly sees himself offering something new — there is a kind of causation, somehow active in the larger pattern, that we cannot understand by adding together the causal action of molecular-level entities upon each other. The tissue-wide electric potentials can fairly be said to play a decisive role in stimulating cascades of gene expression on the way toward formation of entire organs. But, in the reverse direction, genes cannot be said to cause, or explain, the patterns of electric potential.[2]

Similarly with the examples in the opening section of this chapter. They all raise the problem of causation from whole to part — and (although this is not a point Levin raises) they all vex our efforts at strictly physical understanding. The question we need to ask ourselves is this: "How can the physical body of a relatively undeveloped organism — a body already exhibiting coordinated physical processes perfectly adapted to its present state — redirect and transform those well-adapted physical processes so as to conform to a different and more 'mature' pattern that is not yet there?"

Whole-part, future-oriented causation

Why does holistic causation refuse strictly physical understanding? A key difficulty, as I have been emphasizing, lies in the observation that every embryo seems, in its holistic manner, to be *reliably guided* toward a future state. It is as if that future state were somehow present and influential along the entire path of its own material realization — as if the developing embryo were expressing from the very beginning its own *telos*, or the essential idea of its ultimate maturity and wholeness, as a very real and present power.

In a moment we will have to ask to what degree Levin clearly recognizes how thoroughly the problem of causation running from whole to part and directed toward the future disrupts conventional thinking. He is, in any case, fascinated by what he often refers to as "top-down causation" — "an important distinct type of causation" in which "a future state … guides the behavior of the system." He recognizes the "incredibly tangled details underlying system-level outcomes in biological systems," and instead of immediately pivoting away from the challenge of future-directed, higher-level causation in order to resume the analysis of microstates, he questions the wisdom of such a strategy:

> Embryonic patterning, remodeling, and regeneration achieve invariant anatomical outcomes despite external interventions. Linear "developmental pathways" are often inadequate explanations for dynamic large-scale pattern regulation, even when they accurately capture relationships between molecular components.

That is, even in the face of the researcher's deranging intrusions, the embryo does its best to re-organize itself in the light of a characteristic outcome yet to be fully realized — all in a way that does not seem to be explained by the activity of lower-level entities.

The common expectation, which dominated twentieth-century molecular biology, had been that we would learn to track every microstate in every cell and organism, and in doing this we would gain all the understanding of biological processes we could ask for. Levin wonders whether this expectation isn't having the unfortunate effect of "delaying the development of higher-level laws" that could advance our interests more effectively.

So, then: what might he mean by "higher-level laws"?

Michael Levin: Counter-revolutionary

Levin has seen deeply into decisive and overlooked problems of biology. It is, therefore, all the more revealing of the state of modern biology to see how conventional dogma sets bounds to the solutions he can conceive. Despite his desire to frame a new paradigm of causation in living beings, his work testifies to the deeply entrenched power of conventional biological thinking. In fact, at times he seems drawn to the most abstract and least biological aspects of this thinking.

Counter to what you might have thought based on the preceding descriptions, Levin's interests center emphatically on *machine-like* models, *control*, and *prediction*. (I count forty-eight occurrences of the word "control" in the main body of his article.) He repeatedly expresses his confidence in explanatory models based on "top-down" techniques already "exploited very successfully by control theory, cybernetics, computer science, and engineering of autonomous robotics" — and is also impressed by "new developments in information theory that help to rigorously identify and quantify tractable macrostates with maximal causal power." These top-down tools of control could now "enable transformative advances in biomedicine."

By "top-down" Levin typically means: driven by something like an engineer-designed computer program embodied in things like circuits and switches. The *new* in his "new paradigm" consists largely of the fact that the program is thought to be (somehow) distributed throughout tissues and organs, rather than encoded in the tight "logic" of the DNA sequence.

Even the bioelectric features of tissues (Box 10.2) become, for Levin, the manifestations of digital devices. When he looks at these features, he sees circuits, biolectric networks that serve as "a rich computational medium," and feedback loops "equivalent to transistors — fundamental building blocks of logic circuits and decision-making machinery."

And so, he is convinced, appropriate machine models present a wonderful opportunity: we may gain "predictive control in regenerative medicine and synthetic biology." All that is required is a high-level focus on "control mechanisms that harness cell behavior toward specific organ-level outcomes." His complaint about microstates as presumptive causes is that they do not enable us "to make quantitatively accurate predictions with respect to the complex final outcome ... which is the key property we require from a purported explanation of a biological process."

One might have thought that at least one key thing we want from biological explanations is an understanding of the unique, qualitative ways of being distinguishing the life of one organism from another — for example, the zebra from the lion (Holdrege 2020). The narrow interest in "quantitatively accurate predictions," on the other hand, stems from the long-running commitment of science to the discovery of clear and unambiguous causes of a certain sort — reliable causal factors that, within carefully controlled systems, *consistently make specific things happen*, and therefore can be used technologically.

Certainly we do want a maximally *effective* medicine, just as we want a maximally effective political or educational system. But this does not mean we can healthily understand political or educational processes by grounding ourselves in machine models of causation. And the same goes for medicine.

The main problem we have in following Levin is that we arrive at neither a revolution nor a new paradigm for causation merely by changing our level of observation from microstates to macrostates — from molecules to tissues and organs. As long as we remain committed to the same physical and mechanistic notion of causation that has dominated biology for the past few centuries, we can hardly claim to have arrived at a profoundly new understanding of biological causation.

I believe Levin has glimpsed the fact that something *can* radically change when one begins to talk about top-down causation — especially if one realizes that, in organisms, we are looking not only at causes running from the whole toward the parts, but also at a kind of future-oriented causation. But he has compromised this insight by forcibly marrying it to tired, machine-based habits of explanation that represent nothing but the old paradigm.

Of course, he might well object to this. His references to cybernetics, control theory, and computational neuroscience show that he sees himself focusing on a *distinct type* of machine — namely, those operating under some form of programmed control and feedback. Don't we see in these machines a kind of top-down and purposive causation that seems to match that of organisms? The inadequacy of current thinking about biological causation, he is suggesting, lies in biologists' failure to exploit the analogies between living beings, on one hand, and machines of this *particular sort*, on the other.

He is right — and importantly so — about biologists' failure to take seriously the fact of purposive biological processes. But does he himself fully acknowledge the purposive dimension of organic activity? Or does he instead think in terms of activity that only *looks*, rather illusorily, "as if" it were purposive? And do programmed machines point us toward a useful understanding of biological causation?

In what sense are machines end-directed?

In his paper, Levin addresses the idea of "setpoints as causes." Setpoints, he says, are not-yet-existing "future states" that somehow "guide the behavior of the system" toward a realization of those future states. As it stands — and in relation to living beings — the assertion is as vague as it is radical. But Levin makes clear the kind of thing he has in mind: it is illustrated above all by the kind of feedback and control systems we routinely rely upon in devices we use daily.

In such systems, the setpoint is embodied in a mechanism or controller that can be set to some value. In a very simple case, this could be a thermostat set to a particular temperature. That temperature is the setpoint, and the thermostat uses it to control a heating system, such as the one in many homes.

A more complex case would be a computer taking input from buttons you may have on your automobile's steering wheel, where the input represents a desired cruise control speed. Or think of a cruise missile flexibly seeking out a specified target with the help of "sensing" instruments and a complex, computerized guidance system. The target (set point) must, in one way or another, be entered into the computer.

It is obvious that we can say, abstractly and analogically, that organisms pursuing their own purposes have "setpoints." The lion (in some sense) races "like" a cruise missile toward the antelope, adjusting its course as the antelope turns this way and that. And, likewise, the lion embryo flexibly pursues a reliable "trajectory" toward its mature form. But — although Levin often seems to forget the fact — such remote analogies fail to show that the lion can in any meaningful sense be explained as the functioning of a programmed machine. This would have to be demonstrated.

Surely (to change the image) it is difficult to find much commonality between the transformation of a single zygotic cell into a mature eagle, on one hand, and the "development" of a missile, on the other. If, before venturing upon its flight, the missile were to "mature" from a single transistor (or circuit board) into the totality of a functioning, deadly efficient vehicle; and if, during its flight, all its physical constituents were metabolizing and metamorphosing as an essential part of the overall operation; and if, instead of a single "setpoint," there were a massively interwoven and nearly infinite collection of "setpoints" governing each of the missile's "organs," each "cell," the entire missile as a whole, and all its environmental relations — well, as you can see, taking the comparison with living beings seriously could get silly fast.

In any case, the decisive issue is not difficult to grasp. Cruise missiles — and, for that matter, kitchen blenders, electric hand drills, and textile looms — consist of materials we articulate together for use as tools in accomplishing our own tasks. The "top-down" ideas guiding assembly are ours; they do not come to expression through holistically active developmental processes in which all the growing parts participate. *Our* ideas are not native to the collection of parts. Our ideas are not active at the very root of material manifestation in the way that physical laws and biological principles are inseparable from the substance in which they work. We merely rearrange, in an external manner, materials already given to us. We cannot penetrate to the inherent lawfulness of physical materials with the force of our wills, except in moving our own bodies. (And even there, the doing is currently inaccessible to our understanding.)

When we want to explain the operation of a kitchen blender (or a missile), we require no reference to its intentions, or to any striving toward a future state. When we do make such reference, we are really talking about our own purposes in structuring the device for employment in service of our interests. There is no more an *immanent* end-directedness in a cruise missile than in a blender. Both devices are simply put together in accordance with *our* purposes.

By contrast, a developing organism's living "trajectory" results from its *growing* directionally into its mature functioning. We never see a designing power or force that assembles an organism from pre-existing parts in anything like the way we build tools and machines. *Organisms are not designed and tinkered with from without, but rather are enlivened from within*. The wisdom we find at play in them is intrinsic; it is their own in a sense wholly untrue of the external intelligence with which our mechanical inventions are structured.[3]

Does this not make a great difference for our thinking about causation in organisms and machines? The act of structuring and programming a physical device such as a cruise missile is our own. The missile itself has no intentions, and is not "aiming at" anything, no matter how great our role as inventors and builders. In this regard it is simply a more complex kitchen blender. We may have gotten more sophisticated in shaping tools to our own ends, but that is our development, not the machine's.

(For a fuller discussion of the machine model of organisms, see Talbott 2014.)

A deep issue, unaddressed

I have several times mentioned in these pages that all biologists do recognize the agency — the *telos*-realizing, purposive, task-oriented, and storytelling (narrative) activity — of organisms. Biological research is structured by our interest in the things organisms do and accomplish so differently from what rocks "do" and "accomplish." These organic achievements include gene expression, DNA replication and cell division, growth and development, and animal behavior.

But, as I have also mentioned, this awareness of agency remains, for most biologists, blindsighted, and therefore does not make its way into biological theory and explanation, or even into the biologist's own clear consciousness. Levin therefore provides a valuable service by encouraging a more general awareness of what he occasionally refers to as the "teleological" dimension of biology.

I do regret, however, that despite his extraordinarily wide-ranging familiarity with the technical literature, he shows no evidence of having mined the rich wisdom in the works of the organicist biologists of the twentieth century — figures such as E. S. Russell and John Scott Haldane (not to be confused with his son, J. B. S. Haldane) in Britain, and Paul Weiss in America.[4] These prominent and well-respected researchers had already grasped the centrality for biology of the coordinating ("top-down") agency at work in organisms seen as wholes.

A familiarity with this earlier work might have prodded Levin to take a more critical approach to the machine models he so insistently applies to organisms. As it is, he makes no very apparent effort to justify a substantive comparison of living activity to humanly designed machine operation. He does, however, assure us that, with respect to developing organisms, "work is ongoing to understand the molecular nature of the processes that measure the [current] state, maintain the setpoint, and implement the means-ends process to achieve the target morphology."

But, in the work he cites, I see nothing to suggest answers to the most obvious questions. Where might a machine-like setpoint be physically embodied — where might it even *conceivably* be embodied — so as to represent the entire, infinitely detailed and intricately interwoven morphology of a given animal? Once found, how might this setpoint actually direct and coordinate all the animal's living activity over a lifetime — or over a single healing episode such as described in Box 10.1? And where do we find evidence that an organism's fundamental activity of *growth*, *striving*, and *self-transformation* can be understood on the model of our technological devices?

Much of the work Levin draws upon to illustrate machine-based theorizing about the top-down performance of organisms comes from neuroscience, and especially computational neuroscience. The naïveté expressed in this work can be startling. This is illustrated by how quickly, in the dawning computer age, neuroscientists decided that neurons (the only cells in the brain taken with any seriousness at the time) were essentially binary, on-or-off devices more or less like transistors. Even today that basic mind-set seems entrenched, despite the inevitable complicating factors emerging year after year.

It all reminds me of the prominently honored theoretical neuroscientist, Larry Abbott, who, in a genuine attempt to support the prevailing mindset, wrote a book chapter about the brain entitled "Where Are the Switches on This Thing?" (Abbott 2006) There turns out to be no obvious answer.[5]

An unquestioned model

The machine model seems so deeply embedded in Levin's thinking that one can only surmise he has never thought of questioning it. He seems to think it inevitable that any analogy anyone puts forward between an organism and a machine, however remote and abstract, means the organism must work the way the machine works. He is properly struck by the remarkable achievements of development and regeneration we mentioned earlier in this chapter. But when he looks at these achievements, he immediately, and without further question, sees in them "extensive proof-of-principle of control circuits that enable efficient self-repair and dynamic control of multicellular, large-scale shape" (Pezzulo and Levin 2015).

In other words, the fact that we see the organism developmentally transforming itself and healing wounds — and doing so as a coherent whole — is already proof for him that we are dealing with large-scale "control circuits." Certainly there *is* a physical activity through which the organism's transformation and healing are realized. But nowhere in the physical lawfulness of this activity do we find the requisite principles of coordination and control. The fact that we can build machines with certain kinds of controls does not show that organisms function causally in the manner of these machines.

As for the *predictability* in which Levin sees evidence of top-down controls, his prime illustrations are the achievement of his laboratory in stimulating the development of eyes on the tails (or in the guts) of tadpoles, and in producing two-headed flatworms — all by means of bioelectric manipulations. It is true enough that when we forcibly intervene in an animal's life, giving it biological signals that would not normally occur, it can only take the signals as reality and respond holistically as best it can. Presumably, if we intervene to keep experimental conditions constant, we might (more or less predictably) expect similar insults to produce similar responses.

But it isn't clear how "throwing a wrench into the works" by deranging an animal's normal developmental processes, thereby causing the formation of dysfunctional eyes and supernumerary heads, constitutes the kind of predictability we would want from an understanding of the *true nature* of an organism. And, in any case, none of this testifies to the machine-like nature of the processes by which an organism carries out even deranged living activities.

It is precisely *because* every organism is, in a holistic sense, an agent, that it can respond to violent interventions in a meaningful and creative manner. This holistic response is what seems to entrance Levin. He wants other biologists to recognize the organism's top-down performance. But not only does he fail to reckon with the work of earlier biologists who both described such holistic agency *and* denied the machine interpretation; he sees no need to make his own case for that interpretation. He just takes it for granted.

Given his promise as a biologist, I could dearly wish that Levin would consider something like the process of RNA splicing described in Chapter 8, or DNA replication and damage repair, or cell division, or just about any other sustained biochemical or physiological activity in living beings. And then I would love to see him view this activity in light of the observation by Paul Weiss we heard above: The behavior of the whole "is infinitely less variant from moment to moment than are the momentary activities of its parts." Where are the machine models that can meaningfully elucidate the overall *coherence* of these largely fluid phenomena?

I am sure Levin would be pleased to see how Weiss' work thoroughly supports his own interest in top-down causation. And I suspect that he would recognize the wisdom in Weiss' refusal of machine-based explanation.

WHERE ARE WE NOW?

Organic form and machine models

We have been introduced to the problem of form — the problem Michael Levin so eloquently brings to the biologist's attention. How does an organism move in a persistent, adaptive, and sometimes strikingly novel way toward the realization of a living shape and functioning that are in some sense "given in advance"? Levin has clearly seen that this sort of activity, like purposive or future-oriented activity in general, requires us to recognize a kind of causation that somehow works not only from the present into the future (or, we might even say, from the future into the present), but also from the whole into its parts.

But we also see in Levin's response to this problem the remarkable and seemingly unshakable power of machine-based thinking in contemporary biology, especially as exemplified in computers. Having effectively posed questions that could radically re-shape today's biology, he is content to return to the worst tendencies of the life sciences. As I have tried to show in this and the preceding chapters, the machine model fails the organism at virtually every point of comparison. Nor is the matter particularly subtle. It does not require much insight to see that the notions of wired cells, master controllers, computer-like instructions conveyed from here to there, or inert, unliving, machine-like parts coming together to form a living cell or organism simply don't carry any convincing weight.

In sum, machine-based ideas are neither revolutionary nor particularly helpful for our approach to questions concerning the character of biological activity.

In the next chapter we will look at another take on the problem of biological form — the one offered by evolutionary developmental biologist Seain his book, *Endless Forms Most Beautiful*. He, too, is enamored of machine- and computer-based thinking. But his way of approaching the problem of form will enable us to get at a rather unexpected conclusion: form is not something we should be feeling a need to explain, least of all to explain with our familiar mechanistic notions. Once we rise above those notions, we may be able to gain our first glimpse of a game-changing question: Might it be that the proper apprehension of form is itself the understanding we were really seeking all along?

NOTES

1. Vandenberg et al. 2011. As of July 10, 2025, the video was available at https://now.tufts.edu/2011/07/18/face-frog-time-lapse-video-reveals-never-seen-bioelectric-pattern.
2. The point is that bioelectric fields across tissues are the result of physiological processes at a considerable remove from gene expression. While genes are certainly required for the production of the ion-transporting proteins that help produce electric fields, these genes can hardly be said to control the subsequent activity of these proteins. This activity includes the elaborate and sensitively shifting play of bioelectric signaling of the sort involved in craniofacial patterning of the tadpole.
3. The poet and philosopher, Samuel Taylor Coleridge, captured the distinction between machine and organism very well when he wrote:

   > The form is mechanic when on any given material we impress a predetermined form, not necessarily arising out of the properties of the material — as when to a mass of wet clay we give whatever shape we wish it to retain when hardened. The organic form, on the other hand, is innate; it shapes as it develops itself from within, and the fullness of its development is one and the same with the perfection of its outward form. Such is the life, such is the form (quoted in Guite 2017, p. 365).

   The original source is given as *Lectures 1808–1819 on Literature*, by Samuel Taylor Coleridge, edited by R. A. Foakes, vol. 1, p. 495.
4. See, for example, Weiss 1962, Weiss 1968, Weiss 1973, Russell 1930, Russell 1945, Russell 1938, Haldane 1917, and Haldane 1923.

5. The thing that stands out most egregiously in Levin's various discussions of theoretical work in neuroscience is his casual conflation of chemistry and cognition. In describing naïve, switch- and circuit-based theorizing about neurons, he seems to assume that he is also talking directly about cognitive activity such as decision-making, learning, and memory formation. "High-level mental processes," he claims, illustrate how "encoded information" possesses "causal power" (Pezzulo and Levin 2015).

   Is he actually talking about high-level *mental* processes, or instead referring to collections of neurons? The two seem indistinguishable in his thinking. And yet cognitive scientists today (generally by their own admission) do not yet have any clue as to how the meanings of the chemist and physiologist relate to the qualities and meanings at work in our mental and cognitive activity. No one doubts that, in thinking, we employ our brains (and, presumably, much else). But we can no more say that brain activity *is* our thinking than we can say, "muscle activity *is* our willing" or "neuronal activity connected to the retina *is* our perceiving."

   Surely Levin is right in arguing for the causal effectiveness of our mental activity. What is disturbing is the way this immediately translates for him into an assumption about the applicability of the models used by mechanistic- and computational-minded neuroscientists.

SOURCES

Abbott, Larry F. (2006). Where Are the Switches on This Thing?, in J. L. van Hemmen and T. J. Sejnowski, editors, *23 Problems in Systems Neuroscience*. Oxford: Oxford University Press.

Farinella-Ferruzza, N. (1956). The Transformation of a Tail into Limb after Xenoplastic Transplantation, *Experientia* vol. 12, pp. 304–05. https://doi.org/10.1007/BF02159624

Ford, Brian J. (2009). On Intelligence in Cells: The Case for Whole Cell Biology, *Interdisciplinary Science Reviews* vol. 34, no. 4 (December), pp. 350–65. doi:10.1179/030801809X12529269201282

Gilbert, Scott F. (1994). *Developmental Biology*, 4th edition. Sunderland MA: Sinauer Associates.

Guite, Malcolm (2017). *Mariner — A Voyage with Samuel Taylor Coleridge*. London: Hodder and Stoughton.

Haldane, John Scott (1917). *Organism and Environment as Illustrated by the Physiology of Breathing*. New Haven: Yale University Press.

_______ (1923). *Mechanism, Life and Personality: An Examination of the Mechanistic Theory of Life and Mind*, second edition. London: John Murray. Facsimile of second edition published in 1973 by Greenwood Press, Westport CT.

Holdrege, Craig (2020). The Intertwined Worlds of Zebra and Lion, *In Context* # 43, pp. 12–19. https://natureinstitute.org/pub/ic/ic43/zebralion.pdf

Levin, Michael (2012). Molecular Bioelectricity in Developmental Biology: New Tools and Recent Discoveries, *Bioessays* vol. 34, pp. 205–17. doi:10.1002/bies.201100136

_______ (2020). The Biophysics of Regenerative Repair Suggests New Perspectives on Biological Causation, *BioEssays* vol. 42, no. 2 (February). doi:10.1002/bies.201900146

Lewontin, Richard C. (1983). The Corpse in the Elevator, *New York Review of Books* vol. 29, no. 21 (January 20), pp. 34–37. https://www.nybooks.com/articles/1983/01/20/the-corpse-in-the-elevator/

Pai, Vaibhav P., Sherry Aw, Tal Shomrat et al. (2012). Transmembrane Voltage Potential Controls Embryonic Eye Patterning in *Xenopus laevis, Development* vol. 139, pp. 313–23. doi:10.1242/dev.073759

Pezzulo, Giovanni and Michael Levin (2015). Re-membering the Body: Applications of Computational Neuroscience to the Top-down Control of Regeneration of Limbs and Other Complex Organs, *Integrative Biology*. doi:10.1039/c5ib00221d

Russell, E. S. (1930). *The Interpretation of Development and Heredity: A Study in Biological Method*. Freeport NY: Books for Libraries Press.

_______ (1938). *The Behaviour of Animals*. London: Edward Arnold. Second edition.

_______ (1945). *The Directiveness of Organic Activities*. Cambridge UK: Cambridge University Press.

Talbott, Stephen L. (2014). Biology's Shameful Refusal to Disown the Machine-Organism. https://bwo.life/org/comm/ar/2014/machines_18.htm

Vandenberg, Laura N., Ryan D. Morrie and Dany Spencer Adams (2011). V- ATPase-Dependent Ectodermal Voltage and pH Regionalization Are Required for Craniofacial Morphogenesis, *Developmental Dynamics* vol. 240, pp. 1889–1904. doi:10.1002/dvdy.22685

Waddington, C. H. (1951). The Character of Biological Form, in *Aspects of Form: A Symposium on Form in Nature and Art*, edited by Lancelot Law Whyte, preface by Herbert Read. Bloomington IN: Indiana University Press, pp. 43–52.

Weiss, Paul (1962). From Cell to Molecule, in *The Molecular Control of Cellular Activity*, edited by John M. Allen. The University of Michigan Institute of Science and Technology Series. New York: McGraw-Hill, pp. 1–72.

_______ (1968). *Dynamics of Development: Experiments and Inferences*. New York: Academic Press.

_______ (1973). *The Science of Life: The Living System — A System for Living*. Mount Kisco NY: Futura Publishing.

CHAPTER 11

Why We Cannot Explain the Form of Organisms

QUESTIONS OF FORM have seemed oddly resistant to the biologist's quest for explanation. Darwin himself seemed to sense the difficulty in that famous instance where he recoiled from contemplating the subtle perfections in the form of the eye: "To suppose that the eye with all its inimitable contrivances for adjusting the focus to different distances, for admitting different amounts of light, and for the correction of spherical and chromatic aberration, could have been formed by natural selection, seems, I freely confess, absurd in the highest degree" (Darwin 1859, chapter 6).

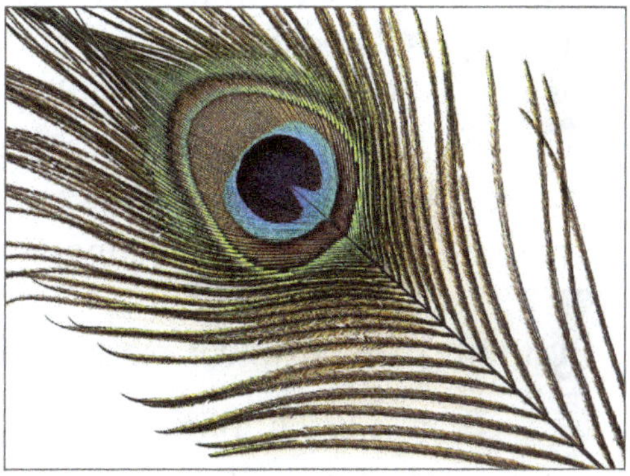

Figure 11.1. An eyespot in a peacock feather.[1]

Of course, as Darwin quickly added, his theory convinced him that he was merely suffering from a lack of imagination. All that was really needed were the creative powers of natural selection acting through eons upon an endless supply of small, helpful changes. But his underlying malaise was not so easily vanquished: "It is curious," he wrote to the American botanist Asa Gray in the year following publication of the *Origin*, "that I remember well [the] time when the thought of the eye made me cold all over, but I have got over this stage of the complaint, and now small trifling particulars of structure often make me very uncomfortable. The sight of a feather in a peacock's tail, whenever I gaze at it, makes me sick!" (Darwin 1860).

We can assume that Darwin got over that stage of the complaint as well. But, thankfully, the biologist is still now and then allowed, if not a complaint, at least an honest expression of wonder. The great twentieth-century student of animal form, Adolf Portmann, writing not of the peacock, but of another bird with a remarkable pattern of coloration on its wings, shares with us his own wonder:

> If ... we look at the speculum on a duck's wing, we might imagine that an artist had drawn his brush across some ten blank feathers, which overlap sideways — making white, bluey-green, and black lines —

> so that the stroke of the brush touched only the exposed part of each feather. The pattern is a single whole, superimposed on the individual feathers, so that the design on each, seen by itself, no longer appears symmetrical. We realize the astonishing nature of such a combined pattern only when we consider that it develops inside several or many feather sheaths completely separated from one another; and that in each individual feather it appears at an early stage while it is still tightly rolled up, the joint pattern not being produced until these feathers are unfolded. What sort of unknown forces direct the constructional work in the "painting" of these feather germs? (Portmann 1967, p. 22)

Whatever Portmann's "unknown forces" may be, they seem to work to perfection. But how are we to understand this perfection? What sort of explanation are we looking for when we want to make *sense* of form? In the case of that patch of color on the duck's wings, surely we will eventually be able to trace exhaustively the processes and connections by which each molecule of pigment seems lawfully "compelled" to take up its proper place in the various feathers. But where, amid the innumerable, widely dispersed molecular jigglings, transits, collisions, interactions, and chemical changes, will we glimpse a global *coordinating* power capable of guaranteeing the overall, aesthetically satisfying outcome in the face of all the degrees of freedom possessed by the interacting molecules in all the individual and separate feathers?

Looking for physical explanations of form

Sean Carroll thinks he has an answer to this question. A geneticist and developmental biologist, Carroll tells the story of the rising discipline of evolutionary developmental biology in a widely read and beautifully illustrated book, *Endless Forms Most Beautiful: The New Science of Evo Devo* (Carroll 2005).

Figure 11.2. A mallard duck with a speculum on each of its wings (left); and an individual speculum feather (right).[2]

Inspired by work in the relatively young discipline of evolutionary developmental biology ("evo devo"), he aims to explain "the invisible genes and some simple rules that shape animal form and evolution" (p. x).

Carroll's triumphalist narrative focuses heavily on the role of "tool kit" or "master" genes. (On "master" controllers in general, see Chapter 9, "A Mess of Causes.") Until the discovery of these genes, he tells us, biologists had known that "evolution is due to changes in genes, but this was a principle without an example. No gene that affected the form and evolution of any animal had been characterized" (p. 8).

That state of affairs apparently ended with the identification of a relatively small number of genes whose presence, absence, or mutation turned out to be associated with the formation (or malformation) of large-scale, discrete features of an organism — and they were often associated with similar features in widely differing organisms. These tool kit genes may, for example, produce proteins that are distributed in bands, stripes, lines, or spots in a young insect embryo. This geographical distribution turns out to be a kind of map of certain features that will develop later.

Carroll reproduces beautiful photographs of fly embryos showing (by means of technical manipulation) brightly colored regions, where each region — blue, green, red, yellow — corresponds to the activity of a particular collection of genes. A couple of hours after fertilization, the oblong embryo is about one hundred cells in length from end to end (or from "west" to "east," as the researchers prefer to say, with west corresponding to the future head pole). Thanks to the differentiated activity of tool kit genes, the western, middle, and eastern sections of the embryo clearly reveal themselves as separate bands.

As these bands fade, they are replaced by seven stripes over the eastern two-thirds of the embryo. Each stripe, together with the neighboring darker band, marks out a pair of future segments of the fly larva. Then these stripes, too, under the influence of yet another group of genes, give way to fourteen stripes indicating the fourteen segments of the larva individually. Most of these latter stripes persist throughout development, and they lead rapidly to actual segmentation of the embryo.

The photographs are spectacular, and leave no doubt in one's mind that the early embryo, uniform and undistinguished as it might appear under ordinary light, is in fact an embodiment of order and form. There is a head and tail, with degrees of longitude between them, and likewise a top and bottom (dorsal and ventral regions), with degrees of latitude. And different "modules" (as Carroll calls them) are already marked out for the development of specific organs and appendages.

Carroll's own work has focused on butterflies. Here again the design to come is signaled by the distribution of tool kit proteins. Carroll produces photographs showing these proteins in the developing wing, occupying exactly those locations where the beautifully decorative spots and stripes and rings will eventually appear. *It's as if the future design were in some way already there.*

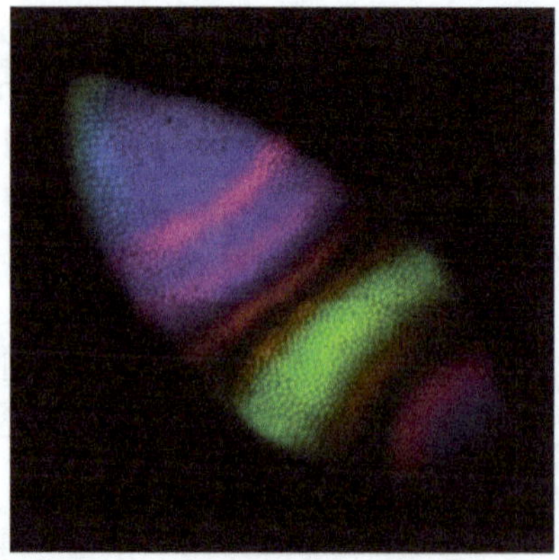

Figure 11.3. Artificially colored bands on a fruit fly embryo, showing the location of particular proteins, which in turn result from differential gene expression and signal the future location of fly segments.[3]

The mastery of genetic switches

But tool kit genes are only part of the picture. It's true that the protein bands in the early embryo are associated with genes that are activated in those bands so as to produce ("express") the proteins. Certain genes that are "on" or "off" within the band, will be in the opposite state outside the band. But what is supposed to coordinate this activation and deactivation of genes?

Carroll's answer is at the same time his central theme: the tool kit genes are systematically turned on and off by a computer-like "operating system" — a vast network of switches residing in those portions of DNA that do not "code" for proteins. Acting, according to Carroll, like a global positioning system (GPS), these switches "integrate positional information in the embryo with respect to longitude, latitude, altitude, and depth, and then dictate the places where genes are turned on and off."

Each switch, as Carroll describes it, is actually a short stretch of DNA controlling a particular tool kit gene. Often there are multiple switches controlling a single gene. Proteins (produced by yet other tool kit genes) can bind to these switches, altering their state. The overall pattern of switch states for a particular gene then determines whether that gene will be activated or repressed. This allows a single gene to be used in many different ways at different times and places — for example, in the development of our own heart, eyes, and fingers. Everything depends on the states of its associated switches. "The entire show," writes Carroll, "involves tens of thousands of switches being thrown in sequence and in parallel" (p. 114).

The governing image in all this is that of the computer. He refers to DNA switches as "fantastic devices [that] translate embryo geography into genetic instructions for making three-dimensional form" (p. 111). The computational powers of the controlling network of switches, he tells us, allow fine-grained management of the expression of individual genes. But at the same time the switches are the key to a software-like modularization of the organism, making it possible for entire features (a spot on a wing, an insect's eye, a digit on a mammal's foot) to come or go — or be modified in dramatic ways — with the flip of a few switches.

Can we trace form to something other than earlier form?

All this raises an obvious question, which, in a way, Carroll himself acknowledges. Suppose we have a fly embryo divided into three regions marked out by proteins A, B, and C.

> You might ask, where do these patterns of tool kit proteins A, B, and C come from? Good question. These patterns are themselves controlled by switches in [the associated] genes *A*, *B*, and *C*, respectively, that integrate inputs from other tool kit proteins acting a bit earlier in the embryo. And where do those inputs come from? Still earlier-acting inputs. I know this is beginning to sound like the old chicken-and-the-egg riddle. Ultimately, the beginning of spatial information in the embryo often traces back to asymmetrically distributed molecules deposited in the egg during its production in the ovary that initiate the formation of the two main axes of the embryo … I'm not going to trace these steps — the important point to know is that the throwing of every switch is set up by preceding events, and that a switch, by turning on its gene in a new pattern, in turn sets up the next set of patterns and events in development (p. 116).

Here, then, is the general thrust of Carroll's attempt to elaborate "the simple rules that shape animal form." But perhaps we may be forgiven a certain unease at this point — a discomfort, first of all, with a claim of simplicity applied to "tens of thousands of switches being thrown in sequence and in parallel." Before we can see the exquisitely detailed and aesthetically satisfying spatial pattern of pigments on the butterfly's wings (or the peacock's feathers), there must be a correspondingly exquisite and detailed pattern of flipped genetic switches. The form at the later stage must in some way be foreshadowed by the form at the earlier stage.

Figure 11.4. The moth, *Automeris janus*, illustrated by George Shaw (1751–1813) in *The Naturalist's Miscellany.*[4]

It is no wonder that Carroll says "I'm not going to trace these steps." For it appears that the tracing would not give us an *explanation* for the form of development of an organism — not in the usual causal sense of "explanation." It would simply trace the form through successive manifestations, one momentary snapshot of which might be given in an image such as that of the fruit fly embryo's colored bands.

The problem with the usual sort of causal explanation is that, as we saw in Chapter 7 ("Epigenetics: A Brief Introduction"), and will see much more fully in Chapter 14 ("How Our Genes Come to Expression"), the "causal factors" elucidated in studies of gene expression end up converging upon each other in endlessly varying patterns — patterns extending throughout the entire cell and organism.

So we might wonder whether the effort to define unambiguous biological causes always resists a final resolution in terms other than those of form — that is, resists our attempts to *explain* form. If in fact a biological performance always involves an intentional, directed (that is, *temporally formed*) coordination of physically lawful interactions, then explanation in terms of the physical interactions alone will never rise to the level of biological understanding. It is the *pattern* — the thought-infused, aesthetic, and qualitative aspect of the coordination — the meaning of it all — that we really want to lay hold of. The form of an organism's body and behavior just is this meaning put on display.

Perhaps, in other words, we never are, at any stage of our investigation, tracing physical mechanisms that *explain* observed form. Perhaps apprehending form in its own terms — and doing so as perceptively as possible — is how we make sense of biological phenomena, because form is itself the decisive explanatory principle. It seems worth considering whether form is what every material phenomenon most essentially *is* for our understanding. After all, the form of a thing is not just a particular feature that can be pasted onto the thing. It belongs to the creative, interior aspect that makes the thing this *sort of appearance* and not *that sort*.

A second source of our unease with Carroll's supposedly explanatory genetic tools and switches is the casual assumption that something in fluid, ever-transforming cells operates in meaningful analogy to a computer's precisely machined, rigidly fixed, transistor-based hardware. No specific support is offered for this critical and wholly improbable fundament of Carroll's argument.

Moreover, we do know that his language at this point is misdirected. He speaks as if particular switches "control" genes or "dictate" such-and-such an outcome. But, as we saw in Chapter 9 ("A Mess of Causes"), such straightforward, machine-like causes are foreign to the life of organisms. The ever-expanding sciences of genetics and epigenetics have shown us that influences flow toward genes from just about every corner of the cell and organism — and they do so as all those corners are themselves caught up in the overall developmental transformation of the whole organism. Contrary to any picture of neat controlling causes, we are forced to understand the entire organism as itself the fundamental, ever-flowing, metamorphosing "cause" of its own development.

Discomfort also arises when we contemplate Carroll's ever-receding series of "inputs" that, as we look further and further into the past, finally peters out in the vagueness of "asymmetrically distributed molecules" in the earliest stages of an egg's development.[5] These randomly disposed, "primordial" molecules in the egg hardly seem the ultimate, revelatory basis for explaining the ceaseless metamorphoses leading to the mature organism. Can we really believe that the asymmetrically arranged molecules at the beginning of an egg's life are patterned with almost infinite precision and delicacy so as to exercise (somehow) a physically controlling influence over the highly coordinated activities of countless trillions of molecules freely diffusing through the watery milieu of thousands or millions of dividing cells in the developing empryo? What exactly is the *explanation* Carroll claims to possess?

Such vagueness at the decisive beginning of the entire developmental process, when all the organism's still-to-unfold features lie potent in the egg, does not say much for our present understanding of the supposedly "simple rules" that explain the observed complexity and seamless unity of every unique life form.

Carroll's whole approach raises one other concern, perhaps the most fundamental of all. All form seems to be essentially qualitative. And this has to do with the fact that to apprehend an appearance is to grasp at least part of its *meaning*. We see directly, perceptually — not only through technical analysis — what constitutes it a *this* rather than a *that*, a redwood rather than a willow, a squirrel rather than a chipmunk, a virtuous act rather than a dastardly one. When we try to capture such differences in words, we always slip into a qualitative language — for example, the language of art — even if we immediately obscure that language behind the terminology of mechanism.

This brings us to the underlying difficulty that Carroll (and biologists generally) run up against. Their physical world has, in the style of nineteenth-century classical physics, finally been reduced to inert, mindless, and therefore qualityless, particles. These particles can have nothing to do with the reality of inherently qualitative form. And so, in order to make sense of Carroll's non-explanatory explanations, readers must superimpose upon his toolbox language whatever pictures of form they have gained from his illustrations. And then, as I have already pointed out, all they have is form related to form.

But *this* form — true form — qualitative as it inevitably is, remains wholly disconnected from Carroll's tools, switches, and networks. There is, from the standpoint of contemporary science, no bridge from a mechanistic to a qualitative understanding. The only bridge we have runs in the reverse direction: we make what sense we can of those switches and networks by viewing them in terms of meaninguful and purposeful form, not the other way around. There is nothing about a set of switches and networks *as such* that causally explains the purposes to which they are put — nothing that explains their formal and meaningful patterns of activity.

So, then, returning to our central question: where in the entire developmental sequence can we honestly say, "Here we are *explaining* the form itself, as opposed to simply *describing* a continuous manifestation and transfiguration of form, which must be understood in its own, immediately given terms?"

If the arrangement of an insect's body segments is prefigured by various patterns of protein deposition, and if the protein patterns are prefigured by patterns of gene expression, and if the patterns of gene expression are prefigured by precisely arranged spatial patterns of switches being turned on and off, then we may be describing a play of form over time and at specific levels of observation. But if we try to see this as an explanation of how significant form arises from the supposedly unformed, we can hardly help noticing that we have merely pushed the problem of form backward in time and downward in scale, until it vanishes from sight, still unexplained.

Every stage of the most complex animal, starting from the single-celled zygote and extending all the way to maturity, is in fact the proper and complete form of the animal at that particular stage. To understand the form of an organism is to enter as fully as possible with our imagination (which is itself a power of forming and transforming) into the power manifested in the continous metamorphosis of form from the beginning to the end of that organism's life.

Endless transformations most beautiful

All processes of development and growth are *metamorphoses*. If we were able to view, with understanding, a three-dimensional movie showing the magnified interior of our own developing bodies, the significance of the proceedings would be overwhelming. We would watch a single zygotic cell reproduce and diversify, yielding eventually a trillion or more cells proceeding along hundreds or thousands of distinct trajectories of differentiation.

It would almost be as if we were watching a vast menagerie of wildly different, single-celled organisms, multiplying, writhing, dancing, and contorting themselves in different rhythms and patterns in countless niches or compartments throughout all the tissues and organs of the body. Each of those "organisms" has its own intricate form, changing from cell generation to cell generation, and yet it all happens under the "discipline" of the larger and unfathomably complex, developing form of the whole organism.

Every organ would have its own distinct story to tell. In our developing brains, for example, we would see not only the differentiation of the many unique cellular lineages in that organ, but also the forming of significant functional connections and patterns of interaction as the brain shaped itself (or was shaped) to the form of our cognitive experience and motor activity. The lungs would likewise be shaped for and by the air and our eyes for and by the light, just as our bones are shaped for mobile support under the influence of gravity and our habits of movement.

And, of course, the picture is just as lively and striking when we step back and look at any organism as a whole. Here is the well-known description by Thomas Huxley, Darwin's pre-eminent apologist during the latter part of the nineteenth century:

> Examine the recently laid egg of some common animal, such as a salamander or newt. It is a minute spheroid in which the best microscope will reveal nothing but a structureless sac, enclosing a glairy fluid, holding granules in suspension. But strange possibilities lie dormant in that semi-fluid globule. Let a moderate supply of warmth reach its watery cradle, and the plastic matter undergoes changes so rapid, yet so steady and purpose-like in their succession, that one can only compare them to those operated by a skilled modeller upon a formless lump of clay. As with an invisible trowel, the mass is divided and subdivided into smaller and smaller portions, until it is reduced to an aggregation of granules not too large to build withal the finest fabrics of the nascent organism. And, then, it is as if a delicate finger traced out the line to be occupied by the spinal column, and moulded the contour of the body; pinching up the head at one end, the tail at the other, and fashioning flank and limb into due salamandrine proportions, in so artistic a way, that, after watching the process hour by hour, one is almost involuntarily possessed by the notion, that some more subtle aid to vision than an achromatic, would show the hidden artist, with his plan before him, striving with skillful manipulation to perfect his work (Huxley 1860).

Do we really need some still more subtle instrument that will reveal a hidden artist working from outside — which, of course, Huxley didn't believe in — or do we need rather to credit the capacity of our own, educated eyes to see, as Huxley did, the *inherent artistry* that informs the processes right there in front of us? The embryo plainly and objectively manifests a power of unified expression, of metamorphosing organic form — something a child can recognize. Why should we not accept this power exactly as and where we observe it — as a living *power*

— just as we accept the very different power of gravity in exactly the terms of *its* manifestations?

And, despite Huxley's reference to "a formless lump of clay," never in all this drama of transfiguration do we witness a cell or any other element being constructed from formless substance (if such substance could even be imagined) — or being built from preexisting, "plug-and-play" parts. The parts undergo transformation simultaneously with the whole, and only as expressions of the whole.

The starting point of it all is the living zygote, and in its flourishing and wonderfully structured context-embeddedness, its life "overflows" and multiplies. The zygote's original, one-celled unity is never lost, but rather is subdivided and differentiated. It is worked on from within and influenced from without (that is, from the environment), according to the unfolding of its governing principles of form.

These principles — those of the type, or species — are regarded by every embryologist as telling *one, unified story* from zygote to maturity and senescence. Further, the informing power that is characteristic of that story remains "in force," as far as circumstances allow, regardless of drastically different nurturing environments, and even in the face of disfiguring insults inflicted by laboratory technicians. The organism responds to every insult by bending it, as far as possible, toward the normal pattern of development.

The existence of this governing pattern, or form, in every different sort of organism is a decisive truth of biology. No matter how far down toward the molecular we go in trying to explain form, we find our explanations themselves, so far as they are biological and not merely physical or chemical, to be always based on considerations of form. We never seem able to get beneath or behind these considerations so as to grasp something more fundamentally explanatory than form itself.

Even the classic efforts to explain everything based on genes have now become ever more vividly an elucidation of form — form that is already in play at the level of genes and chromosomes. For example, some geneticists speak of "genomic origami," while others refer to the three-dimensional "dance" of chromosomes in the nucleus — a spatially significant performance essential to the expression of the right genes in the right amounts at the right times (Chapter 3, "What Brings Our Genome Alive?").

Apparently Carroll, and all the other biologists who in one way or another employ the same language, have come to the (perhaps unconscious) conclusion that we really do need to find Huxley's "invisible artist" — but that we must do so mechanistically, re-imagining the artist as a designer-engineer (often working in the guise of natural selection). It somehow seems too distasteful to take seriously the transformative artistry we can observe actively at work in the organism itself.

This is a good place to return to the wisdom of the twentieth-century cell biologist, Paul Weiss, who once remarked:

> There would be less room for misconception if instead of referring to developmental dynamics as "formative," we were to designate them as "transformative," for then the notion that order or organization as such are created *de novo* [anew] within a totally random pool of unit elements could not arise (Weiss 1971, p. 39).

Finally, form as a principle of explanation is found throughout the sciences. For example, the mathematically formulated laws of force in physics are, like all principles of form, ideal (conceptual or thought-based), even if they are very different from the formative principles we find in biology. These physical laws can remind us how readily scientists have for centuries accepted the play of formative thought both in the unfolding of material phenomena and in the explanation of those phenomena. Actually, the very idea of explanation already seems rooted in a conviction that understanding is a matter of thought, and that this thought explains by showing us the characteristic form of a phenomenon. A mathematical equation does not so much explain the form as it provides an extremely condensed description of it. Confusion about this – and perhaps an unwillingness to accept the explanatory primacy of form – is perhaps why the law of gravity has historically been rejected by some physicists and philosophers as "unscientific" and even "mystical."

WHERE ARE WE NOW?

Is form a primary and irreducible feature of the organism?

In the chapter introduction, I asked where we might glimpse the global, coordinating power that guarantees the infinitely detailed and aesthetically satisfying form of organisms — for example, the pattern of color in a duck's speculum — given that physical laws by themselves know nothing of the sustained and meaningful coordination required.

In both Chapter 10 ("What Is the Problem of Form?") and this one I have argued that mechanisms do not give us workable models for the play of form in organisms. In this chapter I have suggested further that the attempt to explain form seems misconceived in the first place, since we can never get "behind" form to an explanatory principle more basic. I have also pointed out that an appeal to form is usually an appeal to some part of the qualitative thinking through which we discover a phenomenon to be understandable.

If the effort to explain form is misdirected, does this mean that the idea of explanatory causes has no place in our understanding of biological form? Not at all. Maybe we will be reminded here of the fact that *formal causes* have long been recognized as essential for our understanding, going back to Aristotle. Perhaps the apprehension of principles of form yields understanding precisely because those principles are themselves causal, although in a crucial sense differing from our usual understanding of causes.

The conclusions of this chapter will, I suspect, seem rather anemic to conventionally minded biologists. This is because we still need to illustrate as vividly as possible what it means to gain a profound grasp of an organism's form, and also because we need a fuller reckoning with the *causal* role of form. These topics will be touched upon in Chapter 12 ("Is a Qualitative Biology Possible?"), where we will look at three research efforts aimed at elucidating form in the phenomena of life. These efforts are irreducibly qualitative, and in their light we may begin to understand that form is not so much a result of causes (as causes are commonly understood in biology), but rather is itself a too-long-neglected kind of cause.

NOTES

1. Figure 11.1: free-use image (https://www.pxfuel.com/en/free-photo-joqxe).
2. Figure 11.2 credits (left): Krista Lundgren, United States Fish and Wildlife Service, (https://www.flickr.com/photos/usfwsmtnprairie/34431869295), CC BY 2.0; (right:) Courtesy of The Feather Factory (https://thefeatherfactory.co.uk/products/duck/31/butchers-blue-mallard-duck-feathers-pack-of-10-/1224/).
3. Figure 11.3 credit: FlyEx data base.
4. Figure 11.4 credit: See here: https://www.rawpixel.com/image/386761/free-illustration-image-moth-butterfly-george-shaw.
5. The origin of this asymmetry is often assigned by biologists to the "random movements" of some number of molecules. But such randomness does not contribute much, if anything at all, to scientific understanding. If we consider the eggs, or germ cells, of species with radically different forms — say, anteaters and eagles — random movements in the developing germ cells cannot meaningfully explain the specific and differing character of those forms.

SOURCES

Brady, Ronald H. (1987). Form and Cause in Goethe's Morphology, in *Goethe and the Sciences: A Reappraisal*, edited by F. Amrine, F. J. Zucker, and H. Wheeler, pp. 257–300. Dordrecht, Holland: D. Reidel. Available at https://www.natureinstitute.org/ronald-h-brady/form-and-cause-in-goethes-morphology

Carroll, Sean B. (2005). *Endless Forms Most Beautiful: The New Science of Evo Devo*. New York: W. W. Norton.

Darwin, Charles (1859). *The Origin of Species*. Available online at https://archive.org/details/harvardclassics11eliouoft

_______ (1860). Letter to Asa Gray (April 3). Available at https://www.darwinproject.ac.uk/letter/DCP-LETT-2743.xml doi:10.1016/j.jvcir.2020.102758

Huxley, T. H. (1860). *The Origin of Species*, Collected Essays vol. II. Available online: https://mathcs.clarku.edu/huxley/CE2/OrS.html

Portmann, Adolf (1967). *Animal Forms and Patterns: A Study of the Appearance of Animals*, translated by Hella Czech. New York: Schocken Books. Originally published in 1952.

Weiss, Paul A. (1971). The Basic Concept of Hierarchic Systems, in *Hierarchically Organized Systems in Theory and Practice* by Paul A. Weiss, H. K. Buechner, J. S. Coleman et al., pp. 1–43. New York: Hafner.

CHAPTER 12

Is a Qualitative Biology Possible?

THE PHILOSOPHER, Ronald Brady, once wrote about his undergraduate experience this way:

> When I began college as a chemistry major my enthusiasm for science was somewhat dampened by meeting a professor of chemistry who pointed out the difference between my own goals and those he, as an experienced professional, would call mature. My passion, he noted, was entirely focused on direct experience — my sense of chemical change was invested in sensible qualities: in smells, colors, the effervescence of liquids, the appearance of precipitates, the light and violence of flame, etc. But, he countered, this was probably closer to medieval alchemy than to chemistry. The latter is really a matter of molecular and atomic events of which we can have only a theoretical grasp, and the sensible experience on which my excitement centered was secondary ... I was reminded of him when I spoke to a morphologist at Berkeley about my interest in Goethe's attempt to approach science by keeping to direct experience. The morphologist responded: "You are interested in this approach because you are a Nature appreciator, while I am a productive scientist." It is always nice to see where one stands.[1]

Ever since the Scientific Revolution, physical scientists have held to the conviction that, whereas nature speaks decisively in the language of mathematics, the qualities *of nature* are not actually qualities of nature, but rather additions provided "from outside" by human subjectivity. And where physical scientists have led, biologists have done their best to follow.

If, as is commonly thought, qualities reside outside the bounds of any rigorous science, including biological science, then the very idea of a qualitative biology is self-contradictory. There should be no such science. Since this entire book is founded on the contrary assumption — an assumption explicitly defended in Chapters 13 and 24 — it feels obligatory to provide some particularly instructive examples of what a qualitative biology might look like.

In what follows I offer three such examples of widely differing sorts. The first involves the study of a single animal, the second a study of leaf sequences along the stems of certain plants, and the third a study of systematic morphological, behavioral, and other patterns recognizable in evolved groups of organisms, yet inexplicable in terms of present evolutionary theory.

An animal expressing the character of the tropical forest

Craig Holdrege is a biologist with a predilection for what he calls "whole-organism studies." In his research he struggles to arrive at a unified picture of an organism by approaching it from all sides: morphological, physiological, behavioral, ecological, and more. The knowledge he arrives at in this way is irreducibly qualitative. But what does that mean? Before we look at Holdrege's sketch of the three-toed sloth, let's take in some basic information about this creature of the South American rain forest:

- The sloth spends much of its active life clinging to or hanging from the branches of trees. It sleeps or remains inactive for the greater part of every day.
- The sloth has proportionately less muscle mass than most mammals. It also has a higher percentage of retractor (pulling) muscles, and its muscles react more slowly than those of other mammals.
- The sloth makes use of smelling more than seeing or hearing.
- Its body temperature varies more with the ambient temperature than in most other mammals.
- The fur coat of the sloth is often covered with algae. Also, beetles, moths, and various other insects, as well as mites, may inhabit the fur, sometimes with the individuals of a particular insect species numbering a hundred or more.
- Gestation period: four to six months.
- Teeth: continually growing; not pre-formed, but shaped by use.
- Eyes: can retract into their sockets.
- The sloth descends from the trees to the ground about once per week to defecate. Its feces are only slightly decomposed after six months. In defecating and urinating, the animal may lose more than a quarter of its bodily weight.
- The sloth is relatively non-reactive to pain and injury.

There you have a collection of facts about the sloth. But you hardly have a coherent *picture* of the sloth. Based on these facts, viewed in mutual isolation, you

Figure 12.1. A three-toed sloth flowing up a tree limb.[2]

can say little about the distinctive *qualities* of the animal. But now let me briefly summarize a part of Holdrege's discussion of the sloth as a "whole organism." (The balance of this section is drawn from Holdrege 2021.)

What first of all strikes one about the sloth is, of course, its "slothfulness." It is indeed a slow creature, spending the greater part of the day sleeping or otherwise inactive. It will sometimes cling so stubbornly to a given position that a tree limb must be sawed off in order to remove it. When it does move of its own accord, it pulls itself slowly along the tree branches from which it hangs "by all fours," drawing leaves to its mouth with its front limbs and eating them. When it descends from the tree to urinate and defecate on the ground, the process is so deliberate and gradual that the wingless moths who have taken up residence in the sloth's fur have plenty of time to crawl off the animal, lay their eggs in the fresh dung, and return to their furry habitat.

But "slothfulness" is much more than mere speed of movement. It qualifies every aspect of the animal. For example, the sloth's digestive processes, about which its life seems to be centered, are remarkably slow. According to one researcher, "after three or six days of fasting, the stomach is found to be only slightly less full." The stomach is four-chambered like the cow's, but digestion takes about ten times longer than in the cow.

With its reduced muscle mass, the sloth generally performs about ten percent of the physiological work typical of similar-sized mammals. "All metabolic processes are markedly measured in tempo. Sloths use little oxygen, breathe slowly, and the respiratory surface of their lungs is small." Further, a four-to-six-month gestation period compares to a little over two months for the similar-sized cat. And even the sloth's dung may be only slightly decomposed after six months — this amid the intense decompositional processes of the rain forest. This is thought to help slow down the high nutrient recycling rates for certain trees, helping to stabilize some components of the ecosystem. In sum,

> The sloth brings slowness into the world. This is not only true of its reactions, movements and digestion. It also develops slowly in the womb and has a long life span for a mammal of its size.

Clearly the sloth is not a creature of rapid or pronounced change. In this it expresses features of its environment. The tropical rain forest is a place of great constancy — days of equal length throughout the year, the air warm and humid with little seasonal variation, the light levels always low beneath the dense forest canopy, afternoon rains every day.

> The uniformity of light, warmth, and moisture — in intensity and rhythm — mark the rain forest. And it is hard to imagine a rain forest dweller that embodies this quality of constancy more than the sloth. From meters below, the sloth is sometimes described as looking like a clump of decomposing leaves or a lichen-colored bough.

But there are many ways an animal can reflect its environment. The sloth exhibits a certain passive, yielding character so that it is, in a sense, "formed from the outside." For example, in a way that is extremely unusual for warm-blooded animals, the sloth's internal temperature varies considerably — and does so less in accord with its own activity than with the ambient temperature. (Unlike other mammals, the sloth cannot actively raise its temperature through the muscular activity of shivering.)

Similarly, the sloth does not so much overcome gravity as yield to it. With its skeletal structure loose and flexible rather than fixed, and with retractor (pulling) muscles dominant, it lacks the ability to push against gravity and raise itself up. Placed upright on a smooth, flat surface, its legs will splay out and it will be helpless to move unless it can find toeholds (clawholds) for pulling itself along. It spends much of its life either curled up in a ball or hanging by its hook-like claws from tree branches.

Figure 12.2. A three-toed sloth trying to cross a road. Sloths are rather helpless on a flat surface unless they can find toeholds to pull themselves along.[3]

In maintaining the balance of its life, the sloth does not strongly counter external forces and conditions with its own activity.

This, perhaps, makes it less surprising that the sloth is so oddly nonreactive to experiences of pain or injury. Pain occurs where the boundary between self and world is violated, but the sloth seems to have no vivid sense of this boundary. It will cling stubbornly to the very object that is injuring it. One researcher who kept sloths in his home tells of an animal burning and smoking as it sat on a light bulb in a lamp. But upon being rescued, it only protested and tried to cling to the lamp. Another researcher describes a sloth that acted "normally for a long time after it had received a wound which practically destroyed the heart." As part of its receptivity to the world, the sloth (Holdrege writes) "seems not to live as intensely in its body as other mammals, being quite insensitive to pain."

Even in its digestion the sloth shows its passive and nonreactive character. Although its stomach is four-chambered like the cow's, this stomach "is more like a vessel that needs to remain full than a place of intensive muscular activity, secretion, mixing and breaking down, as it is in the cow." Or, again, the sloth's teeth are not pre-formed with crown cusps and ridges as in other mammals (and especially grazers); rather, they emerge as simple cones and are shaped through their engagement with food. In this sense, the sloth's teeth are formed from the outside.

So we see that in many ways the sloth does not so much respond to the rain forest environment as receive its imprint. Even the sloth's fur, which soaks up water "like a sponge," is often green-tinted from the growth of algae. So it assumes some of the appearance and character of its surroundings. And this fur provides a little rain forest habitat of its own, being the home, as we have noted, for numerous beetles, moths, and other insects, as well as mites.

Like most mammals, sloths do occasionally groom themselves. But, as one pair of researchers reports, the grooming effort is so sluggish that moths "may be seen to advance in a wave in front of the moving claws of the forefoot, disturbed, but by no means dislodged from the host."

Fully consistent with this image of an animal that receives the environment into itself rather than actively projecting itself outward, Holdrege recognizes in the sloth "a primary gesture ... of pulling in or retracting." We have already noted the predominance of retractor muscles along with the manner in which the sloth pulls itself along a branch and brings leaves to its mouth. The head itself is a picture of this withdrawn and in-drawing manner of being. Lacking the protruding snout of most grazers, the skull is extraordinarily round and the head is not clearly separated from the rest of the body. The sloth's ears are tiny and do not project out into the environment. Its eyes can retract into their sockets. Both sight and hearing are, in the sloth, quite weak; smell — a sense whereby part of the environment is drawn deeply into the organism — is the primary sense. Imagine yourself living in a world of wafting smells: no distinct boundary between self and other is given through this sense.

Slowness and constancy; receptive openness to the environment; a passive, somewhat withdrawn character; a gesture of pulling in or retracting rather than projecting outward; being formed from the outside — each of these phrases emphasizes a slightly different side of a unitary way of being. We can, with inner effort, bring all the sloth's traits into a coherent picture that holds together. And when we do this, claims Holdrege, we find that "every detail can begin to speak 'sloth'." That is, we can recognize a quality of "slothness" that shines through all the details and makes them into a single, expressive whole.

Of course, Holdrege's own description is much more organic than this haphazard, fragmented, and incomplete summary. But, in comparing the list of facts offered at the beginning of this section with the attempt to weave these facts into at least the bare beginning of a connected fabric, perhaps you can begin to glimpse the meaningful unity that a qualitative approach to the sloth might make available. The qualities are, so to speak, recognized *between* the isolated facts. Only by virtue of this bridging function of qualities through which diverse features are seen in a common light can we apprehend the unity of an organism.

It is impossible to comprehend this unity when we approach an organism in the usual terms of evolution and natural selection — that is, when we approach it as a collection of independently arising traits, each of which offers its own selective advantage. There is, in that, no principle of unity. We see the unity only in terms of the organism itself, viewed as a whole, expressing itself out of its own nature. And if typical evolutionary explanations give us no approach to this readily observable unity, then clearly something fundamental is missing from our evolutionary understanding.

The problem of organic form

Johann Wolfgang von Goethe (1749–1832), who pioneered morphological studies (and gave us the word "morphology"), wanted his readers to understand about the new science that "its intention is to portray rather than explain." At the same time, however — and rather mysteriously for most modern ears — he emphasized that the portrayal was itself all the explanation we needed: "Everything in the realm of fact is already theory ... Let us not seek for something behind the phenomena — they themselves are the theory."[4]

This is the puzzle that the philosopher Ronald Brady undertook to elucidate in one of the most important (and most widely unread) papers of the twentieth century: "Form and Cause in Goethe's Morphology" (Brady 1987):

> Any modern reading of Goethe's morphological writings must struggle with the author's apparent satisfaction that his "morphology" ... was both a descriptive science and a causal one. This unlikely attitude is made all the more difficult by Goethe's suggestion that form — at least in the sense of "archetypal" form — is itself causal ... I shall argue

> in this paper that Goethe's notion of archetypal form represents an important advance in the phenomenology of organic form, and that it does indeed have causal implications.

(All quotes will be from Brady's paper unless otherwise indicated.)

We are assessing form when we judge, for example, whether two trees — one short, thin, and spindly, growing at the alpine tree-line, and one tall and lush, growing at sea level — are both Norway spruce. Likewise, we are assessing form when we ask whether the human arm is *homologous* with the fin of a whale and wing of a bird. That is, can we say that arm, fin, and wing are in some sense the *same* limb, whatever transformations may have differentiated one from another? And a similar question arises when we consider the succession of vertebrae along the spine of a human being or other vertebrate. Can they be seen as transformations of a single entity or idea?

In studying plants during the later eighteenth century, Goethe recognized a commonality uniting such diverse features as the seed leaves, foliage leaves, sepals, petals, pistils, and so on. All these organs in any particular plant, he claimed, are transformations of a single archetypal form, a form he chose to call the *leaf*. The foliage leaves are just one set of embodiments of this archetypal *leaf*. But while the validity of Goethe's discovery has been widely accepted within biology, the nature of that discovery, according to Brady, has been just as widely misinterpreted.

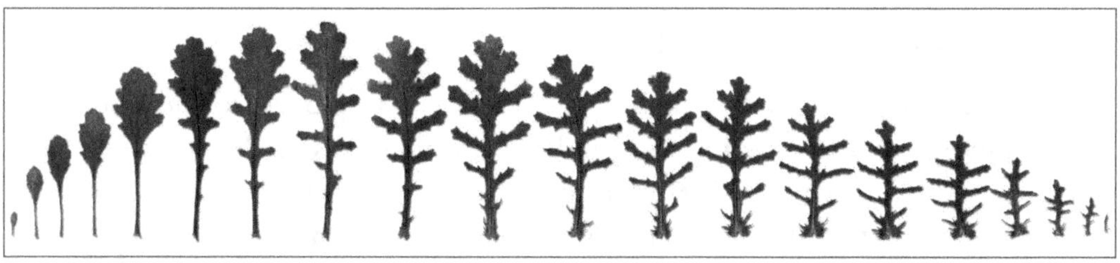

Figure 12.3. Leaf sequence of common groundsel (*Senecio vulgaris*). The figure shows all the foliage leaves of the main stem of a single plant from bottom to top (in the figure from left to right). The first (at left) is one of the two cotyledons of the plant. The sequence is otherwise complete; no leaf is missing.[5]

Goethe was not simply abstracting a set of common features from a diverse set of forms, yielding a fixed schema. As we will see, no such schema can make sense of the processes of becoming we observe, for example, in the leaves appearing in sequence along the stem of a plant that grows anew from ground level every year (whether as an annual or perennial — see Figure 12.3):

> Goethe's common organ, or *leaf*, is not a simplification of foliar members. All empirical forms are, for him, equally particularized, and

> his general organ can be general only by lacking such particularity. His *leaf* accomplishes this requirement by *having no form at all.*

In other words, the archetypal leaf of the plant has no form in the usual sense — no *static* material form — but rather is a special dynamic sort of form that is *generative* of particular, sensible forms. We recognize it as a formative power or potential. This is where *something like* causation (for which there is no clear concept in modern science) enters the picture, and it is also where the modern reader stumbles. However, Brady takes great pains to make the point accessible. We will follow his line of thought in some detail.

How to generalize upon a transformational series

The figure below shows, from bottom-left and clockwise around the circle to bottom-right, a sequence of leaves taken in ascending order along the stem of a single meadow buttercup (*Ranunculus acris*). For pedagogical purposes the sequence is somewhat simplified, with some leaves omitted. Also, for some of the leaves only part of the leaf stalk is shown.

You will note that the attempt to abstract a list of features common to all the leaves might yield something more or less like the simple form at the end of the clockwise movement (bottom-right).

Figure 12.4. A leaf sequence from a buttercup plant (*Ranunculus acris*). See text.

But suppose we declared this one leaf to be the "Gestalt" underlying all the leaves in the sequence. This would be of no value, because the simplified leaf, from which so much detail has been removed, fails to provide a principle for recognizing the fit (or lack of fit) of the other forms — or of any new leaves we might be shown. We can imagine countless different ways for a leaf to be tripartite without at all conforming to the pattern that distinguishes *Ranunculus acris* from other species.

No features abstracted from all the forms so as to yield a single form or schema can generalize upon a series of organically related forms. Such a schema, as Brady remarks, will always be "closer to one stage of the series than it is to the others." It cannot be equally related to them all. Yet the history of biology is replete with attempts to identify fixed schemas and to make them determinative for various biological "kinds."

If we want to understand the relations between these leaf forms, we cannot begin with any single and definite form, whether that form be given by nature, abstracted from various exemplars, or invented by ourselves as a mediating design. Rather, "we must begin our study of the series *from the progression itself*":

> Let the reader imagine, for a moment, how one could decide whether an additional form, not included in the series as yet, could be placed within it. By what criterion could the judgment be made? (Since I have performed the experiment with luckless classrooms of students — mostly ignorant of biology — I can report that the solution is almost immediate for most observers.) The forms of a graded series have the peculiar property of appearing to be arrested stages — we might call them "snapshots" — of continuous "movement." If we begin with the first leaf (lower left) and follow the transformation to the last (lower right), we have the sense that we are in fact watching the form on the lower left turn into the form on the lower right. Because we "see" the series in the context of this imagined or "intended" movement (to use the phenomenological term), an adequate criterion for accepting or rejecting a new member is near at hand.

Understanding what is meant here by "movement" is the decisive thing. Brady helps us along with a series of succinct observations.[6]

The movement is continuous and ideal. The formative movement from leaf to leaf in an organically ordered series becomes more vivid to the degree that more transitional forms are supplied between the shapes we already have:

> The movement we are *thinking* would, if entirely phenomenal, be entirely continuous, leaving no gaps. Thus as gaps narrow[,] the impression of movement is strengthened, and the technique by which a new form can be judged consists in placing that form within one of the gaps or at either end of the series and observing the result. When the movement is strengthened or made smoother the new form may be left in place. But if the impression of movement is weakened or interrupted, the new form must be rejected. Thus the context of movement is itself a criterion by which we accept or reject new forms.

But note: while the movement may be said to *produce* sensible forms, the movement itself is neither sense-perceptible nor physical. Yes, each physical leaf goes through its own unique and continuous development, as does the plant as a whole. But the unifying movement, or "gesture," we recognize in passing from one leaf to the next is apprehended only in thought and imagination. One leaf does not physically metamorphose into the next leaf. So our practical and *objective* criterion for recognizing candidate leaves and correctly placing them in the sequence is an *ideal* movement.

The formative movement requires both difference and sameness. A critical point: "The impression of 'gradual modification' cannot depend any more on what each form has in common with its neighbors [such as an overall tripartite form] than upon what it *does not share* with them. Change demands difference, and continuous change, continuous difference." That is, a transformational series is united as much by differences as by similarities. We cannot have transformation without differences, and the nature of the differences tells us about the nature and distinctive unity of the transformation. One sort of transformation will require very different differences compared to another sort of transformation.

And so we are able to "see" the movement from form to form "only by a distribution of sameness and difference between them." We test in our imagination the dynamic context — the smooth movement that expresses a differential within the context of a unifying gesture — because by this movement "*the lawful relation between the forms is made manifest.*" All this normally happens without our noticing it. But if we want to understand biological form, it can be well worth noticing what we usually ignore.

An awareness of the movement changes our perception of the leaves. In seeing the movement that unites the forms, we shift our intentional focus "from text to context, from the individual particulars to the unifying movement." This necessarily changes the way we see the individual leaf, which now becomes merely an arrested stage of the movement — a momentary expression or visible trace of a passage — rather as we can isolate a series of still shots from a movie.

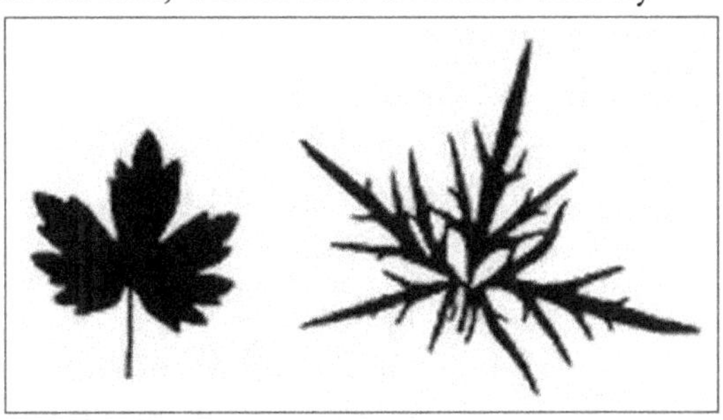

Figure 12.5. Two leaves from the buttercup leaf sequence. See text.

We can see how this works by considering an extreme case where we are given just the two leaves of the sequence (Figure 12.5). If we were seeing these for the first time, we could hardly regenerate an entire series of buttercup leaves from them. But if we first live with the more complete series, entering into the implied movement, and if we then look again at the two isolated leaves shown here, they will "no longer seem unlike. They will, in fact, bear a distinct resemblance to each other, and bear it so strongly when the trick is learned that the impression arises that they are somehow the *same form*. *Here is the intuited 'single form' of the series, but it cannot be equated with anything static*" (emphasis added).

By expressing not just the sameness of one or more features abstracted from all the individual forms, but also a differential running throughout the series, the movement "specifies the forms possible to the series." Here, in this movement, we have truly generalized upon the entire series in a way that no abstract and simplifying reduction to common features and no fixed image would allow. What a common schema fails to provide is "the differential that runs throughout."

If there were no differential — if the image above showed two identical forms — then we would have no way to identify any sort of transformative movement. But because there is a differential, our enlivened and mobilized imagination can recognize in each sensible leaf the one true "leaf" capable of generating it. This is not a physical leaf, but rather the *single movement* out of which the sensible leaves have "fallen." Thanks to our apprehension of this movement, the sensible leaf is no longer perceived as merely itself, but as a manifestation of a gesture.

The movement manifests itself through the particulars. It might seem odd to speak of movement rather than a *thing moving*. But what seems odd for contemporary habits of understanding may be exactly what's required for overcoming the limitations of our understanding. After all, we have no difficulty speaking of the "movement of thought" — which may, in fact, be an aspect of what we are talking about here. And, in any case, it is not so difficult to see that no static form or particular thing can capture the quality of a movement *between forms*.

And let's note also that we *are* talking about qualities here. We cannot grasp whatever is distinctive and significant about a gesture of any sort without a qualitative movement of our own thought. Only in qualities do we find the kind of multivalent potentials that can unify different (and otherwise disconnected) expressions or forms. Where a quantitative science might see in qualities only vagueness, a qualitative science gains access to dimensions of reality hidden to quantitative approaches.

The conclusion of all the foregoing (which will require further elaboration) is that "The movement specifies forms … by generating them."

> The movement ... is a continuity which must contain, in order to be continuous, multiple *Gestalts*. Thus the movement is not itself a product of the forms from which it is detected, but rather [it is] the unity of those forms, from which unity any form belonging to the series can be generated. Individual forms are in this sense "governed" by the movement of the series in which they are found — their shape and position in that series are both functions of the overall transformation.
>
> At this point in the argument, the project of description must permanently shift from static to mobile form, for the latter generalizes upon the former.

So how do we come to terms with a generative movement that is not a material thing? This will bring us to the culmination of Brady's presentation.

The question of causality

We have been trying to understand the "movement" at work in the leaf sequence as a *form-making* principle. But, Brady remarks, if we are justified in speaking of this movement as a *making* principle rather than a *thing made*, then we seem to be attributing causal efficacy to the movement.

But we have so far been engaging in a purely descriptive project. Can such a project, however accurate, thorough, and fruitful for understanding, yield a causal principle? Brady's concern was to identify the characteristic features of the leaf sequence produced by a particular plant — the features by which we recognize "this plant is a specimen of *Ranunculus acris*." What's decisive, it has turned out, is not a particular static form or material entity, but an imaginal movement with its own distinctive qualities.

Having enabled us to recognize this movement for ourselves, has Brady also given us a causal understanding? Do we now see an *enabling power* by which the leaves manifest as they do?[7]

Current mental habits make it easy for us to picture *things* producing a well-formed movement, but very difficult to believe an ideal movement could somehow govern the production of things. This is the mental block Brady would have us overcome, and in a section of his article entitled "Form and Potency" he proceeds by refining his analysis of our experience with the buttercup leaves. Again we look at key points in his discussion.

Each individual leaf is "coming from" and "passing to." When we grasp the unity of the leaf sequence, we have recognized the differential underlying the transformation of one leaf to the next. The experience is dynamic, and this changes our perception of the individual leaves. As a result, as we discovered above, even two leaves from different parts of the sequence can strongly suggest the character of the overall transformation. The individual leaf at this point is not perceived as a mere fixed form, but rather as a movement "caught in the act."

> As our familiarity with transformation sequences increases ... so does the capacity of a single form to bring other forms to mind, or of two forms to build a connecting bridge between them. The morphologist not only "sees" that two distinct configurations are still "the same," but is made aware, by the same faculty, of nascent potentials that seem to arise from every juxtaposition. This peculiar potency of organic form has acted as a constant spur to thought, and a fair amount of theory — including speculations on "vital force" and "final cause" — has responded to it.

As for vital forces and final causes, we can perhaps understand how easy it has been for observers to imagine them. What is *making* each leaf conform to the pattern revealed by its predecessors and successors, if not some special sort of force? And doesn't the directionality of the overall sequence suggest a goal that can be thought of as the final cause?

But Brady, as we will see below, finds no justification for vital forces or final causes. He wants nothing more than to clarify observation, and his fascination is with the way a perceived form relates to potential forms, given the right sort of transformative context. In this way the individual forms lose their independence. As an "arrested movement" — as a phenomenon *arising from* a predecessor and *developing toward* a successor — each leaf is inseparable from a before and after. This is how it so powerfully suggests the "missing pictures" of the transformation. "The single image now becomes transparent to the whole 'gesture' — which it now seems to express ... Potential forms come to mind because they are contained in the whole we are trying to see."

The point is crucial enough to bear repeating in slightly different words. Once we have established the context of movement, each individual leaf — by *coming from* something and *passing to* something — "represents, to our mind, more than itself — it can no longer be separated from its before and after. Indeed, its only distinction from these moments lies in the conditions of arrest — i.e., we see it 'caught in the act' of becoming something else."

The sensible (visible) form shows itself to be but a partial disclosure of a forming activity. The instant it loses (due to our weakened perception) the *coming from* and *passing to*, it ceases to offer this disclosure. It then appears cut off from its own fuller reality — cut off from the reality and the whole in which it essentially participates, from the reality where we must look for causal relations, from the whole that is "somehow all the forms at once." *So the recognizable truth of the individual leaf is lost when it is detached from the ideal movement, the dynamic context, out of which it arises.*

Whatever specifies the appearance of forms in time has causal significance. We come, then, to the heart of the matter.[8] Just as, in space, we can represent a set of distinct loci as a spatial unity (whether the unity of an imagined triangle or a single tree), so, too, we can represent successive manifestations in time as a unity. "*A principle by which we represent the distinct moments of time as a unity, even as we represent the loci of space as a unity, is a principle of form. But this sort of form must be a causal principle as well*" (emphasis added).

When we have a principle that tells us, consistently and correctly, something about *what we can expect to happen next* — what will follow a preceding event, so that the two events can be understood in terms of a single patterning idea — such a principle accords with what we usually think of as causal explanation.

As we have noted, the individual leaf form, insofar as it discloses a larger context of movement, contains within itself a "*felt potency to be otherwise*" — the sort of felt potency that leads some people to speak of a vital force. But the essential thing to realize, according to Brady, is that "*the sensed power is at the same time logical necessity.*" We are aware of this necessity when, presented with a buttercup leaf not currently shown in the leaf sequence, we find that it *must* be placed at only one location; otherwise, it will violate the living dynamism of the sequential movement.

The idea of logical necessity here tells us that the "sensed power" is not just brute, formless power, but a specific shaping power with its own character, or necessity. This dynamic principle remains itself only through its ability continually to become other in its successive incarnations, thereby maintaining its identity as a consistent principle of transformation. If the generative principle (or archetypal idea) were not determining a successor in this way, it would no longer be the unifying truth we have objectively recognized in the leaf sequence. We discover in it the necessity and power of change — and do so without adding any prejudicial theoretical structure to what observation yields.

> It is clear that the sense of power is part of the logical structure of the form, and not a subjective reaction on our part ... once we have accepted the dynamic context the rest follows of its own necessity rather than by any further choice on our part.[9]

In general, we recognize causation when we see one event following another in what we think of as a lawful manner — that is, according to a discernible pattern that reflects one or another sort of ideal necessity. Nevertheless, I suspect that some readers may still have difficulty believing that the kind of ideational or archetypal formative movement Brady has identified in the buttercup leaf sequence plays a role we can properly think of as *causal*. In the next section I offer further supporting commentary.

A clarification of dynamic form as cause. The idea that the dynamic, generative form we've recognized in the leaf sequence — a form or potency we've been calling a "movement" — should be viewed as causal immediately raises a question for most people: "But what is *making things happen*?" Where is the necessary material influence, the matter impinging on matter, the coercive gears and levers that bring something about? How can an immaterial form, however dynamically we imagine it, causally intervene in the growth of a plant?

The questions are understandable in the light of contemporary thinking. But this does not absolve them of extreme naïveté — a naïveté that Aristotle had already overcome when he recognized what he called "formal causes" at work in material interactions.

The fact is that *all* material causation is an expression of immaterial (ideal) relations. Bodies moving at random in the solar system would tell us nothing about causes or laws; but if we observe certain geometric regularities — movements, for example, tracing the forms of conic sections such as ellipses or parabolas — or if, in investigating the fall of objects toward the earth, we eventually arrive at the formulas, $F = ma$ and $F = Gm_1m_2/d^2$, then we have discovered a certain lawfulness. We can talk about material objects acting on material objects, but without conceptual relations of *some sort* — and conceptual relations (including those expressed in equations) are *not* material things — we have no lawful regularity and therefore no causation at all in any defensible sense.

Biologists today remain determinedly focused on material manifestations rather than on the living activity through which the material organism takes shape. Their conviction is that what has already become determines what *will* be. Brady's discussion of leaf sequences shows how wrong this is. The already manifested leaves, as material achievements, do not cause or explain the form of the next leaf. Rather, they, along with all the forthcoming leaves, testify to the ideal movement that has given rise to them and rules them.

We can say much the same thing about the developmental processes we observe in complex, multicellular animals. Nowhere along the path from the zygote to the mature form is the future form determined by what has already come to manifestation. That's why (to take a more extreme example), if we were seeing an insect larva or a tadpole for the first time, we would have no purely physical ground for sketching a clear picture of the butterfly or frog to come.

And, as we saw in the discussion of RNA splicing and Paul Weiss' work in Chapter 6, even at the molecular level the freedom of movement ("degrees of freedom") possessed by molecules in a fluid medium makes it impossible to treat the outcome of elaborate molecular operations as if each step were strictly determined by the material result of the previous step.

So the rule in biology is this: while the previous material achievements may be required (as preconditions) for whatever comes next, they do not *bring forth* the next steps. The failure of a materialist biology to reckon with this truth has distorted the entire science. And the failure extends all the way to evolution. As my colleague, Craig Holdrege, has summarized it, "You can't grasp evolution by staying with its material products."[10]

Threefoldness, the biology of form, and evolution

In 1977 the German biologist, Wolfgang Schad, published a substantial volume called *Man and Mammal: Toward a Biology of Form*. The richly illustrated book was founded so fully upon direct observations, and these observations required so thorough a reconceptualization of the foundations of biology and evolution, that biologists could scarcely afford to take note of it. So (for the most part) they didn't.

But Schad never ceased his undertakings, and in 2020 there appeared in English translation a vast, 1300-page, two-volume, hardcover expansion of the original work, with hundreds of color plates and with a new title: *Threefoldness in Humans and Mammals: Toward a Biology of Form*. It's all there, ready to be taken in by any with the requisite interest and willingness to see the biological world with new eyes. Here I can offer only an inadequate sort of abstract merely gesturing toward the broader themes of this work. It will be enough, I think, to suggest how little the problem of biological form in relation to evolution has yet been recognized by the biological community as a whole.

Schad acknowledges how much biologists have learned about "genetic factors, basic physiological processes, predictable instinctive reactions, and the social behavior of animals." But we can in this way learn a great deal about the physical parts and developmental processes of an animal without coming to a recognition of the formative ideas governing these processes. Explaining any organism in terms of genotypes subject to natural selection under the pressure of environmental conditions still leaves us wondering *what the organism has to say about itself through its own unique form*.

About the beaver, Holdrege remarks that its "teeth are good for gnawing wood, the large flat tail for swimming and as a paddle to slap against the water to alert other beavers about the presence of potential predators, and the high-set eye sockets for swimming inconspicuously with its head only slightly above the water surface." All this sounds good in terms of fitness and survival strategies. And yet what does it tell us about why the beaver took on its own specific, unified character and fashioned its own special niche within a larger environment inhabited by so many other organisms that traversed very different adaptive pathways? The conventional approach

> leads us to mentally dissect the animal into different traits, each of which has its own type of survival value. The coherence and integrity of an animal dissolves into a collection of traits, and all its characteristics are considered solely as adaptations that secure survival (Holdrege 2019).

But we have already seen that, if we looked at the leaf sequence discussed above without attending to its unifying idea, we would miss a decisive causal truth about its character. Similarly, by limiting our attention to the survival value or "fitness" of different animal traits — traits that *could* have taken countless other forms — we remain ignorant of the expressive unity of the specific animal. "No one," Schad says, "can tell us why well-known hoofed mammals, like cattle, deer, and rhinoceroses, have head protuberances, while horses, donkeys, tapirs, and camels do not. Neither molecular biology nor behavioral research concerns itself with the significance of an animal's form" (Schad 2020, p. 2).

David Seamon, editor of the journal, *Environmental and Architectural Phenomenology*, and co-editor of the book, *Goethe's Way of Science*, wrote of Schad's work:

> In the holistic biology that Schad presents, each feature of an animal is seen as significant because the whole is reflected in each part. The aim is to recognize the inner organic order in an animal in such a way that its individual features can be explained by the basic organization of the animal itself (Seamon 2020).

A starting point: living polarity in the human being

Schad attempts to appreciate organisms in the living terms that have long been recognized by the best biologists as essential to any profound understanding. I mean the terms of a dynamic interweaving of activities whereby parts come into being and gain their specific identity, not as independent elements, but rather as integrated expressions of a pre-existing whole.

The key to Schad's approach lies in his understanding of the organism as a being organized according to principles of polarity. By "polarity" he does not at all suggest what is usually meant by "polar opposites." There is no absolute opposition or incompatibility of parts. Rather, we see a mutual participation of parts within an integral whole that lives by reconciling the creative tension between opposing tendencies.

A merely static image of polarity is given by a bar magnet, each of whose poles extends as an active principle all the way into the opposite pole. Cut a small slice off one end of a bar magnet, and you have a second, smaller bar magnet with the same "opposition" of two poles. Each pole's character not only penetrates all the way to the opposite pole, but can exist only in conjunction with the activity of that opposite pole.

Our own human organization is a good place to start in seeking a more living example of this "unity of contrary tendencies." But here we discover, as with organisms generally, that this unity is not merely bipolar in the manner of an inert bar magnet. Rather, the active, living interpenetration of the two poles points to a third aspect of our being — a rhythmic and harmonizing activity that mediates between the poles, effectively raising the contrary tendencies to a higher level where unity is achieved.[11]

And so, looking at the human being, Schad sees three functioning systems. One is centered in the head, one in the abdominal cavity and limbs, and one — mediating between the other two — in the chest region.

In the head we find gathered together most of our sense organs, through which we more or less consciously relate ourselves to the "outer" world — for example, through sight (eyes) and our sense of balance (inner ear). The center of our nervous system (brain) is enclosed in a protective, globe-like exoskeleton, comprised of bones largely fused together.

At the other pole we find our limbs, with their endoskeleton. Far from globe-like and fused together, the parts are linear. The bones are connected by elaborate joints allowing the relatively independent movement of parts. The puzzle here might be that Schad conflates the limbs with the abdominal cavity and its intense metabolic processes, as if they comprised a single, coherent system.

What the conflated functional aspects have in common is a power of movement, where "movement" is used in an older (Aristotelian) sense, overlapping with the sense of "change" (*metabolē*). Motion, according to Aristotle, can be of several types, involving change in identity, quality, quantity, or place. "The last named is the primary kind of motion but involves the least change, so that the list is in ascending order of motions but descending order of changes" (Sachs 1998, p. 249). Of course, our movement in space makes intense demands upon our metabolism for energy. We can also say that both the metabolism and the limbs serve to maintain an animal's autonomy from its environment. They do this physiologically — through the digestion and assimilation of "alien" food into the structure of one's own body — and in terms of the ability to relocate oneself in space (p. 16).

The organs of digestion in the abdomen are not invested with, or protected by, a bony structure, but are an altogether soft part of the body. Their activity, contrasted with the almost "inert" quality of the brain and its nerves — and also contrasted with the functioning of the limbs — consists of intense *internal* movement. This includes the muscular and mechanical movement of the digestive organs, but also, and most prominently, the transformation — breaking down and building up — of substances.

In this way Schad refers to the *nerve-sense system* on the one hand, and the *metabolic-limb system* on the other. In between, in the chest area, is the *respiratory-circulatory system*, or the *rhythmic system*, centered in the activity of lungs and heart.

This middle region of the body is marked by a transition from the character of the head region to that of the abdomen and limbs. It is surrounded by the partly open rib cage, where the relatively immobile bony structure toward the head is more closed-in, with the ribs circling all the way around from the backbone to the sternum. But lower down the ribs become gradually shorter, straighter, and more mobile, "the last two pairs remaining close to the spinal column, where they 'float' freely and point downward. The sternum is broadest near the head and relatively narrow where it ends only part way down the chest cavity" (p. 18).

Here in the middle region we do not see a battle between the two poles, but rather a harmonization of them. The rhythms of breathing and heartbeat bring the breath and oxygenated blood to every part of the body, maintaining complex processes of balance or homeostasis.

> Lungs and heart are rhythmically pulsating organs. In each, contraction and expansion, tension and relaxation, compression and dissolution alternate constantly. The polarities of the organism, therefore, are always present in this region, but here they do not maintain their spatial separateness; rather, they actively complement one another through their rhythmical alternation in *time* (p. 16).

That is, this middle system is itself a manifestation of polarity, but through rhythmic alternation the poles are fully reconciled with each other. We see this polarity expressed in the relation between lungs and heart, where

> the lungs tend more toward the upper processes of the body that are centered in the head. Through the trachea, the lungs reach up into the head and establish a direct connection with the outside world ... Their passivity, much like that of the head, appears also in the fact that they are incapable of self-initiated motion and are moved by the thorax and diaphragm (pp. 16–17).

The heart, on the other hand, "initiates its own movement" and is "closed off from the outside world":

> The largest of the arteries originating in the heart, the aorta, turns downward toward the lower part of the body, where the blood relates directly with the processes of the metabolism. Only through the circulation of the blood do the lungs have access to the dominant processes of the metabolism; conversely, the blood gains contact with the outer atmosphere only through the lungs (p. 17).

Interpenetration of the three aspects

Referring to the relative immobility of the head, Schad writes: "Above the runner's flailing limbs and panting chest, the head quietly keeps the goal in view" (Schad 2020, p. 15). But here we need to keep in mind that the threefold aspects of the human being are neither abstract principles nor the material end-products of activity. What we find are qualities of character that continually interact and mutually influence each other, much as motifs, themes, and harmonies may play into each other throughout a musical composition.

It is clear enough that the nerves are not only contained in the brain, but also extend throughout the body, just as do our senses, which give us awareness of many internal processes of our body. Likewise, our circulation and breathing do not exist only in the heart and lungs. The circulating blood, with its finely balanced gases, flows throughout the body, and the breathing function includes the nose, mouth, and vocal organs. And so, too, metabolic activities proceed not only in the digestive organs, but in every cell of the body.

There are other ways we can look at this functional interplay. An example is given by the way our own head organization not only represents one of the poles of our being, but also bears within itself a somewhat muted image of our whole, threefold being. That is, the head has its own opposed (upper and lower) poles as well as a reconciling middle. The nervous system comes to a clear focus in the immobile, bone-enclosed brain. At the opposite pole we have the "limb" system manifesting in the movable, hinged, lower jaw. With its chewing motions to grind food and the digestive processes initiated by saliva, the jaw brings metabolic-limb activity to our heads, where this activity "establishes direct contact with the outside world" (p. 19). And through our breathing and speech we see the performance of the middle (rhythmical) system. One way Schad makes this latter point is by referring to the air-filled cavities "found in the middle section of the cranium, between the sensory area of the face and the braincase itself":

> They include the larynx, the cavities of the throat and nose, and the more ossified air-filled cavities in the upper jaw (maxillary sinuses), middle ear, and frontal and sphenoid bones. Here, in a delicate way, the head's own respiration takes place. When the lungs exhale, air is pressed into the head's cavities; when the lungs inhale, the cavities of the head exhale. These cavities are lined with a moist inner layer that allows for gaseous exchange. Thus the middle region of the head also participates in respiration and in the organism's rhythmic functions. (p. 19)

Furthermore, even when we look only at the mouth and throat we find all three aspects of the organism coming into play. The forward part of the mouth, with the sensitive lips and tip of the tongue, manifests the dominance of the conscious nerve-sense pole, whereas the middle system comes to the fore in the rhythmic chewing activity. Finally, the food is (with a diminished role for consciousness) moved to the rear of the mouth, swallowed, and passed down into the unconscious, metabolic center of the body (p. 40).

Despite all this mutual interpenetration of functional characteristics, we can certainly say that the nerve-sense system is *centered* in the head, just as the rhythmic system is centered in the chest and the metabolic processes in the abdomen.

A great part of Schad's research consists of a kind of "musical" analysis whereby he traces the endless lawful interplay within the threefold organization of the body. We will see more hints of this subtlety before we finish our discussion of his work.[12]

Threefold organization in mammals

It is commonplace to note that many animals possess specializations that make them, in one regard or another, superior to humans. The sight of an eagle, the dog's power to follow a scent, the gnawing ability of a beaver, the sonar-like hearing of the bat, the digging skill of a gopher — we could scarcely hope to match these abilities with our own natural equipment.

Humans, we might say, specialize in non-specialization. Our hands and arms, good for neither digging nor flying, neither swimming nor swinging from tree branches, can employ an endless range of tools of our own devising, from computer keyboards to the knitting needles through which we have long fashioned clothing adapted to numerous environmental conditions. While we lack the well-developed instincts that fit animals for particular environments, our brains remain plastic throughout our lives in decisive regards. "Most of the regions of our neocortex have to be differentiated through active learning. We can change established habits and continue learning indefinitely without ever exhausting the functional potential especially of the right hemisphere" (p. 10).

A central truth found in Schad's work is that the various mammalian groups develop the threefold organization of their lives with different balances among the three functional systems. In other words, they can "specialize" not only in specific behavioral traits or morphological features, but also in one or another of the three functional systems.

For simplicity, Schad's work is often presented initially by focusing on three groups of mammals. One group shows an especially strong development of the nerve-sense system, one emphasizes the polar opposite metabolic-limb system, and one reflects especially well the principles of rhythm, harmony, and balance characteristic of the middle system. These are, respectively, the rodents, the ungulates (hoofed mammals), and the carnivores.

Mice, with their nervous sensitivity and refined sense organs, exemplify the emphatic nerve-sense development typical of rodents. This is evident even in the limbs of rodents, which tend to be small and delicate, with long and narrow fingers and toes, and nails shaped like tiny claws. "The forepaws of squirrels, for example, are adept at grasping, handling, and feeling. Their limbs have acquired a sensory function. Long sensory facial hairs (whiskers), and shorter ones over the entire surface of their body including their bushy tail, project beyond their warm coat and enable squirrels, fitfully twitching and hopping, to find their way in the surrounding world ... Agile and quick in its reactions, a rodent lives in constant agitation, alarmed pauses, and rapid flight. Even in sleep, nervous spasms periodically run over its small body" (p. 38).

This differs greatly, for example, from the powerful digestive processes and strong, hoofed limbs of the ungulates such as the dairy cow. "In contrast with the five-digit type of limbs of the less specialized mammals, the ungulates' feet have regressed to a few bones, which, however, are very strongly formed. This specialization of the limbs extends even to the powerful enlargement of the nail into a hoof ... The limbs of horses and cattle support massive bodies and, in stamping and galloping, horses express the powerful, animating forces within them" (p. 38).

Whereas a mouse must eat frequently, preferring energy-rich, easily digested foods and leaving behind dessicated droppings with little fertilizer value, the ruminants (which Schad considers the "most characteristic" group of ungulates) are well-known for their four-chambered stomachs, their extremely long intestines, and their ability to digest cellulose.

> Contented peace and restfulness suffuse the cow's placid gaze, especially when, ruminating for hours, she devotes herself entirely to her food. Her eyes, and the eyes of all ruminants, lack the yellow spot, the *macula lutea*, which is the part of the retina with clearest sight. To the ruminants, the outside world appears diffuse. They have a stronger experience of smell and taste, senses more connected with the inner working of the metabolism than the eyes and ears. A cow is never as completely awake as a mouse; the unconscious processes of digestion predominate even in the ruminant's state of half-wakefulness. (pp. 38–39)

The carnivores, with their intermediate character, which lacks the distinctive and one-sided development of the rodents and ungulates, are less easy to describe. Schad spends a good deal of time working out the sometimes subtle ways in which different groups of carnivores lean slightly more toward the nerve-sense pole or the metabolic-limb pole, while generally falling in the broad middle area between the two poles. (See his discussion of dogs and cats below.)

It happens that organisms in each of the three major groups tend to fall in different size categories. Rodents are smaller, ungulates larger, and carnivores take up a position between them. And there is an inverse relation between size and the quality of the food each group favors. Rodents prefer highly nutritive, energy-rich foods — fats, oils, and starches. Breaking these down for immediate use, they tend to store very little in the way of bodily reserves of energy. Ungulates, on the other hand, eat poor-quality food, and build up from it great energy reserves — illustrated by the hump of a camel or the subcutaneous tissue (ham) in pigs. And so,

> While nervous constitutions characteristically break down substances, metabolic ones rebuild and augment them. The nutritive processes of the carnivore represent an intermediate state. When a leopard devours a gazelle, a true change of substances does of course take place during digestion, but the change from one form of protein to another hardly alters the chemical energy level (p. 40).

Figure 12.6. Taxidermied lesser Egyptian jerboa at the Natural History Museum in London.[13]

Schad notes what might almost seem a counter-intuitive relation between, on one hand, the nerve-sense or metabolic-limb emphasis and, on the other hand, the overall form of the animal. The rodent, with its strong nerve-sense orientation, tends toward an accentuation of the posterior end of its body, with long tail and the hind legs longer and stronger than the forelegs. The head is not dramatically separated from the rest of the body (think of the mouse). For a rather extreme example of this posterior emphasis, see Figure 12.6.

By contrast, the American bison, with its highly developed metabolic-limb system, presents an anterior emphasis, with its powerful neck and head, and the great hump above its shoulders (Figure 12.7). The giraffe, with its long neck and forelegs and its even more "shrunken" hindquarters is an extreme example of this tendency.

Figure 12.7. An American bison.[14]

The carnivores in general occupy the middle ground, where balance is achieved between the posterior and anterior ends of the animal (Figure 12.8). Or one can picture the chase, where a lion pursues its prey with a burst of energy, its forelimbs and hindlimbs contributing equally to the task. And then, in the natural rhythm of its life between sudden exertion and inactivity, the great predator, having eaten its fill, is overtaken by lassitude. Its rest and sleep are the very picture of flexible bodily relaxation.

Figure 12.8. Bengal tiger (*Panthera tigris tigris*) female, Kanha National Park, India.[15]

Compare that with the ungulates:

> [Speaking at first of the bison:] The front pole of the body with its morphological over-accentuation constitutes the animal's center of gravity. When cattle stand up, they first straighten their less heavily burdened hind legs; only then do they raise the heavier, front part of their body. They lie down, too, in a way that seems strange to us: First, they bend their front legs, laying the main burden of their body down upon the ground, and then the hindquarters follow effortlessly …
>
> In a rodent — a squirrel, for example — the posterior limbs and the tail are over-accentuated. The lighter front part of its body and its smaller forelegs are less ponderous than most ruminants', enabling it to sit up on its haunches and raise its head, which is quite typical for all mice, hamsters, dormice, chipmunks, ground squirrels, marmots, beavers, etc. This is quite the opposite of the buffalo, whose mighty head is bowed down by heaviness (pp. 294–95).

Subtle interweaving

I mentioned above that a certain threefoldness manifests within the "one pole" of the human head — and again within just one part of the head, the mouth-throat area. (Schad also discusses at length how the teeth alone strongly manifest a threefold nature.) This illustrates the general principle of "the whole within the part."

This kind of interweaving is in fact evident everywhere. But it occurs in a continually different expressive fashion. Schad subtly traces the differing relative prominence of the three functional systems not only in the three major groups (rodents, carnivores, and ungulates), but also within many of the subgroups as well as entirely different major groups. He shows, for example, how, in two subgroups of carnivores — felines and canines — we see a degree of leaning toward either the nerve-sense or the metabolic-limb pole. This is despite the fact that both groups clearly exhibit, overall, the rhythmic or middle emphasis of the carnivores.

Cats, with their highly developed senses of sight and hearing, and their sensitive whiskers, tend toward the nerve-sense pole. The dog's primary orientation is toward that of the rather duller sense of smell. "The cat's sensitive constitution is also revealed in its paws, with their retractile claws, so different from the dog, whose limbs have become tools for running, with immovable claws" (p. 48).

As for the cat, "even its method of hunting is in keeping with its strongly developed senses: it prowls stealthily, then crouches motionless with all its senses focused on its prey, and finally pounces with lightning speed. By contrast, wolves, as well as their descendants, the dogs, hunt by pursuit. Tirelessly, they drive their victim until it is exhausted and must surrender. Cats hunt primarily with their senses, thus avoiding great physical exertion; dogs hunt with their limbs, powerfully activating their metabolism. Dogs and cats have thus developed polar modes within the 'attack' behavior of all carnivores ... And yet in their supple agility, well-proportioned form, and moderate size, both are typical carnivores, shaped primarily by the rhythmic system" (p. 48).

Briefly: we see a similar polarization within the ungulate group. With its strongly developed digestive system, the bison (or cow) exemplifies the least development of the nerve-sense pole, while the giraffe, with its more refined head raised high above its digestive organs and alert to the larger environment, shows a relatively strong nerve-sense emphasis. And just as the lion and wolf occupy a middle place among the mammals generally, the deer, with its nerve-sense and digestive functions more or less in balance, holds a middle ground among the ungulates.

I hope all this illustrates a crucial truth. We are not talking about fixed schemas and opposing structures, but rather about qualitative tendencies that can play into each other with infinite subtlety and variation. Clearly, as with all qualitative science — and as illustrated by our discussions of the sloth and plant leaf sequences above — science must engage with art in the effort to apprehend the morphological and functional characteristics of animals. The faithful grasp of polarity requires a lively imagination immersed in the rich world of phenomena.

To a bench scientist in the laboratory, bent on uncovering unambiguous causes and "master molecular regulators," it may seem that a qualitative science is no science at all. And yet, to anyone profoundly attuned to the living world, it is within the laboratory that, all too often, the organism disappears and biology comes to a dead end.

Can evolutionists escape responsibility for explaining these patterns?

It's worth mentioning that the polarity we are speaking about here bears strongly on evolution. The differing but ordered qualitative emphases among the different groups of mammals are invisible to current evolutionary theory. So we are forced to ask, "What is missing from this theory?"

The observed patterns, according to Schad, include this one: Within any group there is an evolution from smaller, more active, nerve-sense-dominated animal forms toward larger, less active, metabolically oriented forms. The latter represent a kind of conclusion, after which evolution takes up a fresh start from another point, just as, upon the extinction of the dinosaurs, the tiny mammalian forms existent at the time became the basis for a new evolutionary thrust.

Such regularities of form can hardly be understood in terms of conventional evolutionary theory. Someone else who appreciated the difficulty of the problem of form relative to contemporary concepts of evolution was the widely respected twentieth-century Swiss zoologist, Adolf Portmann, as evidenced by his observations in a book that has been translated into English as *Animal Forms and Patterns — A Study of the Appearance of Animals*.

By paying attention to form, Portmann recognized trends and relationships overlooked in standard approaches to evolution. He found the external appearance of animals to be the self-presentation of creatures with an inner way of being — a presentation that includes ways of perceiving, moving, behaving, and all forms of color and shape expression in space and time. This expressiveness with its meaningful patterns, he claimed, goes far beyond what might be advantageous relative to natural selection.

Portmann offered a simple, but useful reminder of the expressive luxuriance of nature when he mentioned in passing how plants present us with "a variety of leaf shapes in a profusion of unsuspected magnitude" despite their not being "favoured by any animal selection." That is, the leaf shapes cannot be explained by selective pressures such as those that might involve the leaves and the insects or other animals that feed on them. Similarly with "the many varied types of design on the shells of snails and bivalves," and also with "the whole world of astonishing shapes found in the shells of the microscopically small, one-celled Radiolarians of the open sea" (Portmann 1967, pp. 114–5, 124).

Portmann's excellence as a zoologist is undisputed. But he had the misfortune of pursuing the main body of his work on the eve of the all-out triumph of molecular and genetic approaches to the organism. Few wanted to look at the animal in the qualitative manner he did, so they did not see what he saw.

Among those who do look at animal form in its own terms — and who have extended Schad's work by applying it to their own research — we should at least take note of biologist Mark Riegner and his investigation of the plumage patterns and coloration in birds.[16] Then there is the Welsh dinosaur expert, Martin Lockley. As a paleontologist and professor of geology for thirty-two years at the University of Colorado, Denver, Lockley wrote his popular 1991 book, *Tracking Dinosaurs: A New Look at an Ancient World*, as well as numerous technical publications.

In his 2007 paper, "The Morphodynamics of Dinosaurs, Other Archosaurs, and Their Trackways: Holistic Insights into Relationships between Feet, Limbs, and the Whole Body," Lockley wrote that within the two main dinosaur groups,

> one can detect a spectrum of form between small, long-tailed, narrow-bodied, bipedal species (posterior emphasis) and large, wide-bodied, shorter-tailed, quadrupedal species with various cranial processes (crests

> and horns), indicating an anterior emphasis. These same or similar formative movements reiterate at many different taxonomic levels, and even reiterate within organs of the whole organism such as skulls and feet (Lockley 2007).

Noting that these morphodynamics of dinosaurs can also be recognized among birds and ungulates, Lockley recommended that paleontologists pay much more attention to such patterns of form, and he suggested that the "traditional emphasis on Darwinian functionalism will assume less importance, while the significance of inherent morphodynamics becomes more fully appreciated." After all, we can't be so easily satisfied with the explanation that some particular dinosaur developed a large head "for use in combat" when we find that the movement toward larger heads happens repeatedly and lawfully — in harmonious relation to many other morphological trends — within every group of dinosaurs.

This, of course, amounts to a startling rejection of conventional evolutionary reasoning. The typical causal, deterministic language of biology is simply ill-suited to an understanding of changing patterns of form. Lockley formulates this rule: "*rather than single organs changing for specific adaptive purposes, all organs may change simultaneously as part of a shift in a complex, highly dynamic organic system*." And what is true of the individual organism seems to be true also of the way species are ordered within higher taxonomic groups, so that "*the evolution of species may be, at least in part, an inherent biological dynamic associated with large-scale evolutionary shifts affecting multiple species*" (emphases in original).

This dramatic claim leads to another one of those bombshells Lockley rather casually drops throughout his paper as he unfolds the implications of wide-ranging, repeated patterns of form in the animal kingdom. He notes that different animal groups show three sequential cycles, first of posterior, then of balanced, and then of anterior development, and that the great dinosaur extinction terminating the Mesozoic era came at "*precisely* the end" of the third cycle, "when horned dinosaurs (like *Triceratops*) had developed maximum head size ... This seems to be a rather remarkable coincidence, in which a large-scale, inherent biological cycle coincides so precisely with a purported extrinsic cause (meteorite or comet impact)." He goes on:

> If a significant number of morphodynamic cycles, culminating in anterior (metabolic) specialization, also result in, or coincide with, extinction, the implication is that extinction, at least to some degree, is an inherent, biological dynamic analogous to a large-scale "life cycle."... Therefore, efforts to seek external causes may be unnecessary and result in misleading, or at best incomplete, explanations and correlations.

And in yet another jettisoning of standard evolutionary thought, Lockley questions whether evolution proceeds "by some process of random mutation." After all, given repeated and dynamic morphological tendencies exhibited widely among different animal groups and manifesting their own relational lawfulness, it is hard to reconcile these with the supposedly random generation of variation. If, as Lockley suggests, "it may be possible to predict the general form and physiology of the whole animal from an analysis or understanding of the parts," and if a similar coherence of form exists within the "superorganisms" comprising the various taxonomic groups, then we are a long way from both the usual adaptationist explanations of the features of animals, and also from chance as the primary generator of variation for natural selection to act upon.

More generally, Lockley argues for a holistic approach to animal morphology, rather than an attempt at feature-by-feature explanation. The latter focuses upon adaptive function (horns are used for butting) whereas the former reckons with the fact that in any organism the modification of one part "will lead to a compensation or ripple effect throughout the whole" organism.

Lockley's work on dinosaurs is vastly more complex and subtle than I could possibly indicate here (or properly understand in my own right). But, following Schad, he is clearly suggesting the need for radical new perspectives on evolution. Yes, we must investigate how the various features of an organism help to make it fit for the requirements of its life within a particular environment. That's part of getting to know what sort of organism (and environment) we're dealing with. But when this investigation is narrowed down to a search for survival mechanisms offering a competitive advantage — when the explanatory significance of every feature is reduced finally to the terms of a quantitative judgment about fitness to survive, so that the feature itself is not taken to express anything significant apart from its contribution to survival — when the artful pattern on the butterfly's wing becomes no more than, say, a deceit aimed at birds to avoid being eaten — then we lose the organism as such.

We lose it because we're not really *seeing* it; we're not allowing its features to speak for themselves, in their own expressive terms. Everything has to be reduced to fit an interpretation that says a feature is *for* some particular survival benefit rather than for the entire, uniquely formed way of being of the organism itself. We thereby ignore the lawful patterns visible in the way an evolving species picks one path rather than another through the infinite landscape of survival possibilities

Naturalists may develop a profound sense for the inherent lawfulness of a particular organism's way of being. But, unfortunately, naturalists do not have much standing in the age of molecular biology. As Schad puts it on the opening page of his two-volume work: "The immediate observation of nature and the study of natural science as commonly practiced today have generally become different activities." A profound truth whose disturbing implications are not often considered.[17]

If, however, it is true that the organism is a recognizable unity; if there are consistent harmonies sounding through its various "survival mechanisms," bringing them together in one song — a song as distinct from those of other organisms as a lullaby is from a patriotic march; if the organism, not only as a product but also as a shaper of its environment, takes up its creative opportunities and employs them with all the coherence and expressive focus we find in the work of a perceptive artist; if, in sum, there shines consistently through all the morphological, physiological, and behavioral details of an organism a character declaring something more than "I have survived," but also "I am my own sort of being, unified, bound by a lawfulness not only of matter but also of form, and this lawfulness is accessible to those who approach me respectfully" — well, then, the supposedly solid foundations underlying contemporary evolutionary theory will have crumbled beneath us.

WHERE ARE WE NOW?

We have seen what a qualitative biology can be.

Despite the several-century effort within science to formulate a quantitative discipline without any explicit acknowledgment of the role of qualities, the goal is impossible to achieve, and we always in fact have an "invasion" of qualities in our science. But because the qualities go unacknowledged and are rarely if ever consciously taken up as an issue for scientists to come to terms with, the invasion generally takes unhealthy form — something I will hint at in Chapter 13 ("All Science Must Be Rooted in Experience").

In the present chapter I have drawn on the work of three researchers in order to present diverse examples of biological work where qualities are not only front and center, but also where the qualitative nature of the work is readily recognized as decisively important for scientific understanding. With these examples we have addressed the following questions: (1) How can we characterize the way of being of a specific kind of organism (the sloth)? (2) What sort of immaterial and qualitative understanding gives us our basis for recognizing the material, species-specific, leaf sequence patterns in certain plants? (3) Do we discover distinct and lawful relations between the forms of the various mammalian groups, and do these relations present problems for current evolutionary theory?

What then are qualities? It will be evident from the discussion in this chapter that there is no great mystery here. A qualitative language describes what a thing is in its own, observable and sense-perceptible terms — the terms that are a prerequisite for our having a conviction that anything material is actually *there*, anything from which we can, if we wish, proceed to abstract mathematical relations.

In Chapter 24 I address the broadest and most fundamental question of all: whether, within science or outside it, we can speak coherently of a material world without first taking qualities seriously.

NOTES

1. (Brady 2006). The last sentence of this quote ("It is always nice to see where one stands") is not present in the current cited source, which reads instead: "I left his office feeling very deflated. Again a representative of science had put his finger on my immaturity." The quote I have reproduced in the main text was from an earlier version of Brady's book chapter.
2. Figure 12.1 credit: Courtesy of Craig Holdrege.
3. Figure 12.2 credit: Courtesy of Craig Holdrege.
4. Goethe 1995, pp. 57, 307. On the relevance of Goethe's scientific work to today's science, see Craig Holdrege's "Goethe and the Evolution of Science" (Holdrege 2014), an expansion of a talk given in October 2013 to an interest group at the New York Academy of Sciences.
5. Figure 12.3 credit: Courtesy of Craig Holdrege.
6. Keep in mind that, without our active participation in the leaf progression — without experiencing qualitatively through our own willed inner movement the character of the transition from leaf to leaf — we will not come to any full appreciation of Brady's discussion. It is, in any case, not an unhealthy exercise to bring about through our own effort the transformation of one leaf form to the next, an exercise requiring a muscular and fluid imaginal activity that habits of abstraction easily bypass.
7. It is important to realize that the movement Brady speaks of cannot by itself *wholly* determine leaf forms:

    The movement of the series cannot, of course, demand that any particular potential will be realized, but it does give the range of potential forms — those which would become actual were the imagined continuous transformation to become actual. Whether an actual leaf will realize this or that potential is determined by something else, but it is the movement which defines the potential forms.

    Only a few out of a continuous series of possibilities are actually realized on a single plant, and the form of those few will be influenced by environmental factors. Unusually cold or dark or dry weather will have its effect — but always consistent with the recognizable potentials of the species we are looking at.
8. Brady considers form and cause in the context of Immanuel Kant's treatment of organic form in the *Critique of Judgment*. My present purposes forbid extending the discussion in this direction. But see the following footnote.

9. For the philosophically minded, Brady offers the following aside:

> I am aware of course that the coincidence of logical necessity and causality is something that one does not think to see after the work of Hume and Kant. With regard to Kant I can only point to the potential breakdown of his system that threatens to emerge from the *Critique of Judgment*. Goethe may be understood as exploiting the seeming contradiction that we can intend what we cannot understand. Of all our experiences, intentionality is potentially the most clear, for what we do ourselves is open to our intimate gaze. Kant did not attempt to observe his own intentional acts, and thus never investigated this possibility. Goethe, coming to Kant when he was already engaged in this project, was simply made more conscious of it. He read Kant as if Kant were proposing a similar "adventure of reason."

With regard to Hume we must return to the problem of causality in general. It should be clear to us that however we normally think of causal necessity, we must intend it as a necessity that stretches over different moments in time, and it is the ultimate exclusion of one moment from the next that defeats Hume's attempt to think it out in terms of logical necessity. An identity that bridges that exclusion would also solve the logical problem, and just such an identity is intuited in the observations described. It should be of some interest to rethink Hume's problem on these grounds, for it rests upon the assumption that the distinctions of time are primary. If, on the other hand, the time-form is primary, we should discover that we must intend this unity in order to perceive the "movement of time" itself. The project is too fundamental to consider any further in this discussion.

Regarding Kant, Brady provides a succinct summary of the issues in another luminous article (Brady 1998) not currently available online:

> Kant made science into a study of appearances aimed at bringing them under rational law, that is, if we could understand and predict appearances, our inability to understand their ultimate source would not be a serious debility.
>
> Unfortunately there were some appearances that resisted this project. Kant was acutely aware that our notion of life was formed by the sense of inward unity, an agency that produced and governed the organism from within. This inner agency could not be brought to the understanding by a conceptual summary of its parts, as is the case with inorganic compositions. In its earliest stages, in fact, the organism had yet to develop the organs by which its later existence would be supported, making the inward unity *antecedent* to the developing parts, a whole which makes its own parts necessary rather than a result of the combination of the parts. To the degree that the combination of parts may be said to be causal, each part aided in the production and maintenance of all the others, and all the others did the same for each. As a result, the physical organs had to be recognized as both cause and effect of themselves. The linear chain of causes by which mechanical events were understood here curled up into a circle, depriving the chain of explanatory power.
>
> If one reflects more deeply, it seems obvious that the mechanical laws do not show the requisite logical structure to explain life. Inert objects were moved from without by impressed forces. Laws governing their movement, therefore, are also "external" to the things moving, that is, the laws of mechanics sum up the interactions of objects while being perfectly indifferent to the individual natures of

> those objects. The organism, however, could not be known in this abstract manner, and predictions concerning its changes were dependent on a knowledge of the species. Even the sort of materials out of which it was constructed are an expression of species identity, and thus the governing laws had to be identified with the object they governed, that is, such laws not only governed, but also produced, their objects. Or, at least, Kant argued, these results express the way things *appear* to immediate perception …
>
> Analytic thought, which understands the whole through summing the effects of the parts, could not comprehend a whole that preceded the parts or accomplish a path of thought that moved from the general to the particular. Such a movement, Kant argues, would be that of an *intuitive* intellect, which humanity does not possess.

Brady goes on to say that, while Kant never made a project of actually testing whether a properly developed scientific understanding could embrace the becoming of an antecedent, organic whole, this was exactly the test that Goethe did make — and made successfully, as shown by his work on plants.

10. Craig Holdrege, personal communication.

11. Schad's "threefold" understanding of the human being draws from the threefold picture first offered by Austrian philosopher Rudolf Steiner in his 1917 book, *Von Seelenrätseln*. That work has been translated into English under various titles, including "The Riddles of the Soul."

12. Schad's description of the relationship between the organism and the world shows the impossibility of any rigidly schematic notion of threefoldness:

> Initially, I characterized the whole upper system as *directed outward toward the world*, and the lower system, by contrast, as *self-enclosed*, with the rhythmic system mediating between the two. Though this relationship is an essential characteristic of the threefoldness of the bodily organism, we can, as we have also seen, come to a more nuanced understanding of this pattern in its particular characteristics. On the one hand, the organism communicates with the outer world in three quite different ways: primarily via the *sense organs*, but also through *breathing* and through *limb activity*. On the other hand, it establishes its *specific physiological competency*, its independence [or autonomy relative to the world], chiefly through the *digestive organs*, but also through its relatively closed *circulatory system* and its almost wholly encapsulated *nerve center* (p. 23).

He immediately adds that the human organism "is as much a member of the surrounding world as it is an independent world of its own; and by mediating between these two kinds of existence, between its biological self and the surrounding world, it creates an active interplay between the two. It always gives the lie to any one-sided explanation of its reality, which we can approach only by adopting multiple perspectives" (pp. 23–24). So we find a unity of polar opposites, not only within the organism, but also between the organism and the world.

13. Figure 12.6 credit: Emöke Dénes (https://commons.wikimedia.org/wiki/File:E_-_Jaculus_jaculus_2.jpg), CC BY-SA 4.0.

14. Figure 12.7 credit: RedGazelle123 (https://commons.wikimedia.org/wiki/File:American_bison_DZS.jpg), CC BY-SA 4.0.

15. Figure 12.8 credit: Charles J. Sharp (https://commons.wikimedia.org/wiki/File:Walking_tiger_female.jpg), CC BY-SA 4.0.
16. See Riegner 2008. Riegner's work on birds would take us too far afield to allow for coverage here. He has also written an important paper in the philosophy of biology, dealing with the "new archetypal biology and Goethe's dynamic typology as a model for contemporary evolutionary developmental biology" (Riegner 2013).
17. Schad goes on to remark on the second page: "My purpose is to place in the absolute center of inquiry the direct perception of the animals most closely related to us — the mammals — as they live in their natural environment. We shall approach them with the confidence that their lives openly and plainly convey what is essential for our understanding of them. As we recognize the unique quality of each animal form, it poses a much neglected question whose answer ... can be supplied only by the living form of the animal itself."

SOURCES

Brady, Ronald H. (1987). Form and Cause in Goethe's Morphology, in *Goethe and the Sciences: A Reappraisal*, edited by F. Amrine, F. J. Zucker, and H. Wheeler, pp. 257–300. Dordrecht, Holland: D. Reidel. Available in the Nature Institute's Ronald H. Brady Archive: https://www.natureinstitute.org/ronald-h-brady/form-and-cause-in-goethes-morphology

_______ (1998). The Idea in Nature: Rereading Goethe's Organics, in *Goethe's Way of Science: A Phenomenology of Nature*, edited by David Seamon and Arthur Zajonc, pp. 83–111. Albany NY: State University of New York Press.

_______ (2006). Direct Experience, Chapter 1 in *Being on Earth: Practice in Tending the Appearances*, by Georg Maier, Ronald Brady, and Stephen Edelglass. Full text is available at https://www.natureinstitute.org/book/being-on-earth. Hardcopy book is available from Logos Verlag Berlin (2008): https://www.logos-verlag.de/cgi-bin/engbuchmid?isbn=1887&lng=eng&id=

Goethe, Johann Wolfgang von (1995). *Scientific Studies*, edited by Douglas Miller. Vol. 12 in Goethe: Collected Works. Princeton NJ: Princeton University Press.

Holdrege, Craig (2014). Goethe and the Evolution of Science, *In Context* #31 (spring), pp. 10–23. https://natureinstitute.org/article/craig-holdrege/goethe-and-the-evolution-of-science

_______ (2019). A World of Dynamic Connections: A Commentary on Wolfgang Schad's *Understanding Mammals*. *In Context* #41 (spring), p. 9. https://www.natureinstitute.org/s/In-Context-41.pdf

_______ (2021). What Does It Mean To Be a Sloth?, Chapter 1 in *Seeing the Animal Whole — And Why It Matters*. Great Barrington MA: Lindisfarne, pp. 19–40. An earlier version of the chapter (only slightly revised for the book) is available at https://natureinstitute.org/article/craig-holdrege/what-does-it-mean-to-be-a-sloth

Lockley, Martin G. (2007). The Morphodynamics of Dinosaurs, Other Archosaurs, and Their Trackways: Holistic Insights into Relationships between Feet, Limbs, and the Whole Body, SEPM Special Publication No. 88 (published by the Society for Sedimentary Geology), pp. 27–51. doi:10.2110/pec.07.88.0027

Portmann, Adolf (1967). *Animal Forms and Patterns: A Study of the Appearance of Animals*, translated by Hella Czech. New York: Schocken.

Riegner, Mark F. (2008). Parallel Evolution of Plumage Pattern and Coloration in Birds: Implications for Defining Avian Morphospace, *The Condor* vol. 110, no. 4, pp. 599–614. doi:10.1525/cond.2008.8503

_______ (2013). Ancestor of the New Archetypal Biology: Goethe's Dynamic Typology as a Model for Contemporary Evolutionary Developmental Biology, *Studies in History and Philosophy of Biological and Biomedical Sciences* vol. 44, no. 4, Part B (December), pp. 735–44. doi:10.1016/j.shpsc.2013.05.019

Sachs, Joe (1998). *Aristotle's Physics: A Guided Study*. New Brunswick NJ: Rutgers University Press. Originally published in 1995.

Schad, Wolfgang (1977). *Man and Mammals: Toward a Biology of Form*, translated by Carroll Scherer. Garden City NY: Waldorf Press.

_______ (2020). *Threefoldness in Humans and Mammals: Toward a Biology of Form*. Harlemville NY: Adonis Press. This is a substantially revised and expanded, two-volume edition of Schad 1977. An early, limited printing of the expanded book (200 copies) was titled, *Understanding Mammals*.

Seamon, David (2020). "Book Note," review of Wolfgang Schad's *Threefoldness in Humans and Mammals* (under the title, *Understanding Mammals*). *Environmental & Architectural Phenomenology* vol. 31, no. 1, p. 15.

CHAPTER 13

All Science Must Be Rooted in Experience

IN PREVIOUS CHAPTERS we have seen how organisms, as centered agents, present us with rich, narrative contexts — mortal performances that proceed, with characteristic expressiveness and intention, through the stages of a unique life drama qualitatively recognizable as belonging to a particular species. And yet, as we have also seen, a powerful urge drives biologists to ignore, as far as they can, every *living* feature of those performances. The aim is to employ strictly physical, inanimate, unliving terms of description.

They ignore, for example, what it must really mean when they say that animals "strive" to maintain their life, or that a wound "heals" itself, or that an organism "adapts" to its environment, or that an animal "perceives" a threat and "responds" to it. (Inanimate physical objects — stones, clouds, whirlpools, and dust storms — do not, in the biological sense, strive, heal, adapt, perceive, or respond.) But it is all too easy for any scientist to side-step such meanings and analyze the organism's story into lifeless sequences of precisely lawful molecular interactions. And since there appear to be no gaps in the molecular-level picture, the resulting explanations seem complete. Only the organism is missing.

In other words, seamless as they may be in their own impoverished terms, such explanations are not in fact complete. They miss the simply observed fact that molecular-level interactions in an organism are always caught up in, and governed by, the higher-level pattern of a life story. We always find ourselves watching the meaningful *coordination of causal processes in an extended narrative* — an end-directed coordination that cannot be explained by the processes being coordinated. This is why explanations that never move beyond physics and chemistry stop short of biology.

Non-living explanations do, however, have one advantage: they conveniently avoid all those troublesome words I use throughout this book in discussing organic contexts and life stories — words such as *intention* and *purposiveness*, *idea* and *thought*, *agency* and *end-directedness*, *interests* and *meaning*. Most biologists prefer to have nothing to do with such terms.[1]

One stumbling block associated with those words is that they relate to features of our own inner lives — our human *experience*. It is, of course, healthy to avoid an anthropomorphic projection of human experience upon other organisms, where it does not belong. But we, too, are organisms, and therefore we have no cause to question whether conscious human experience belongs in our biological science. Instead we can only ask, "*Where* does this experience belong in our biological science?"

If we ignore the character of our own life and experience, can we fully understand a world that mustered its resources, material and otherwise, in human form — a world that ultimately came to present itself in the form of human understanding? And how can any biologist today make the evolution-denying assumption that our own experience has absolutely nothing to do with our evolutionary ancestors — was in no way pre-figured in them? Further, how can we gain legitimate scientific understanding, if it is not *empirical* — if it is not an expression of our most rigorously considered *experience*?

Perhaps first of all we need to ask what is meant when we refer in this way to our own experience.

Two distinguishable but indivisible aspects of human experience

It is clear enough — trivially clear, it seems — that we cannot conceive any material phenomenon, or any reality at all, that is inconceivable. If an object or phenomenon did not lend itself to our conceptualizing — if nothing of its true nature could be captured in thought — we would never know it because we would not even be able to think it. If we cannot conceive something, it cannot appear as a definite and coherent fact of our experience. Either the world's character is at least partly given in thought, or else it is altogether alien to our understanding.

Some truths are so obvious and foundational that we easily forget them in our quest for new knowledge. The fact that anything we can understand must share in the nature of thinking — must in one way or another be meaningful — may be one such truth. If a thing cannot present itself to us as thought, it may as well not exist as far as we are concerned.

But our conceptualizing or thinking capacity is only one of the contributors to our experience, and therefore to an empirical science. Our senses also contribute. And here, too, we can say that, without the qualities of sense, we have no material world to talk about. If you open yourself to any phenomenon whatever and then (in imagination) remove all sensible qualities from it — all the given colors, sounds, touch sensations, smells, and so on — nothing will be left. You are confronting an absolute void.

Not even the most rigorous mathematics can give us a world, since nothing in mathematical thought itself tells us what the mathematics is about. We must *apply* the mathematics to sensible experience if we want to see how mathematical ideas are expressed in material reality. But the same applies to all thinking, not just to the purely quantitative ideas of mathematics: only by bringing our thought into relation with what comes through our senses do we find the world taking shape around us. This is a key idea that we will flesh out below.

There seems to be no basis for assigning priority either to our sensing or our thinking. These are two inseparable yet distinguishable aspects of a single reality. Our access to them via the more or less distinct human functions of thinking and sensing happens to be a characteristic of the structure of our being as cognitive agents, not a bifurcation in the world itself. This structure of our cognitive experience, we will see in Chapter 23 ("The Evolution of Consciousness"), has changed over the course of human evolution.

For the moment, it is enough to ask ourselves: Do we have any knowledge of the material world that is not an intimate marriage of sense and thought?[2] It will not require much work to realize that the answer is "No."

Without relations of thought, we have only chaos

Many of us (especially as we grow older) have had the experience of "losing our bearings" while driving or riding in normally familiar territory. Suddenly a powerful sense of disorientation takes hold of us, and the entire landscape becomes a disconcerting question mark. For a moment we have no sense for where we are or where we are headed, so that our usual feeling of comfort with our surroundings is lost. The confusion that sets in, however short-lasting, is one of profound lostness. The connections linking where we are at the moment to a wider, coherent context have gone missing. The conceptual map through which we grasp the meaningful arrangement of the larger landscape is no longer anchored to our current location.

One might think that the problem here applies only to matters of spatial location. After all, when I become disoriented while driving, all the particular objects around me — houses, trees, road surfaces, animals — continue to make perfectly natural sense. My disorientation applies only to a certain contextual aspect of my environment.

But the fact is that all the particular things around me also depend on the thinking that weaves parts into a meaningful whole — not only spatially, but also, for example, functionally.

Suppose I were to lose all conceptual grasp of the relations governing the scene outside the window where I am now writing — a scene with a great white pine tree standing just a few meters beyond my desk. I would then have no more reason to connect the particular branch I am now looking at with the trunk of the tree than I would have for connecting it with the contiguous patch of blue sky. The idea that the sky is *up there* while the tree is *here*, or that the pressure of the wind against the branches is responsible for their waving, or that the roots in the ground provide mechanical strength in support of the tree's uprightness, or that the entire tree as an integral unity is *growing* — these would no longer serve to hold the tree together in my understanding as the unitary kind of thing, or being, it really is.

A wholly unformed content of sense perception is something we presumably never experience as such — because it would not yet be *experience*. It can assume meaningful, experiential form only so far as it is informed by thought. In routine perception, this informing is already accomplished before we are aware of it. Through long training, our senses are educated by our thinking, so that we do not need to reflect anew, over and over again, upon familiar contents in order to form them into elements of our experience.

We can see the issues more clearly when we consider cases where the normal education of the senses has been partly lacking.

We do not see with our eyes alone

It can be hard for us to recognize all the thinking that is woven into our perceptual experience, much of it originating far back in childhood. But there are now well-studied cases where some aspects of the usual marriage of sense and thought never occurred in the first place — not until adulthood. I have in mind those individuals, born blind, who were much later given sight through operations.[3] Here we find vivid evidence for the insufficiency of mere sense impressions, and for the role of thinking in giving us lucid, intelligible experience of the world.

The British neuropsychologist, R. L. Gregory, describes the case of "S. B.," who received donated corneas to replace his own congenitally opaque ones at age fifty-two. After the operation, the bandages were removed from his eyes, and

> he heard the voice of the surgeon. He turned to the voice, and saw nothing but a blur. He realised that this must be a face, because of the voice, but he could not see it. He did not suddenly see the world of objects as we do when we open our eyes.

He made progress while still in the hospital, but it all involved learning *how to understand what he was looking at* so as to bring it to coherent and meaningful form. At first he judged that he could hang from the window ledge of his room with his feet touching the ground when the distance was in fact at least ten times his own height. When, on the other hand, he had had previous touch experience with objects, he could estimate visual distances much more realistically.

S. B., like many such patients, found it stressful to adjust to his new powers of sight. For example, he had difficulty "in trusting and coming to use his vision when crossing a busy road":

> Before the operation he was undaunted by traffic. We were told that previously he would cross roads alone, holding his arm or his stick stubbornly before him, when the traffic would subside as the waters before Christ. But after the operation it took two of us, on either side, to force him across: he was terrified, as never before in his life.

Following his operation, S. B. fell into an increasingly deep depression. Making sense of things was hard work, and he would often prefer to encounter new objects with the familiar sense of touch alone. "Some of these people," Gregory writes, "revert very soon to living without light, making no attempt to see. S. B. would often not trouble to turn on the light in the evening, but would sit in darkness." Over time "he gradually gave up active living, and three years later he died" (Gregory 1978, pp. 193–98).

Such cases highlight for us the extent of work required to make rational sense of the unformed content supplied by our senses. This depends a great deal on the availability of relevant prior experience — that is, experience that results from already having made sense of prior perceptions. But the unnatural work of suddenly having to cope in adulthood with an overwhelming mass of unfamiliar sensations so as to find the connecting thoughts that form them into a coherent and satisfying picture can clearly prove exhausting.

How do things around us become what they are?

We have all been exposed to so-called ambiguous figures — images cunningly contrived like those of Figure 13.1 so that they can come to meaningful appearance with the aid of at least two altogether different and conflicting conceptions of their governing relations. While the "image on our retinas" remains the same, the way we *think* the image makes a huge difference in what we see.

Figure 13.1. Three examples of ambiguous figures: Necker cube (left); a duck or rabbit (middle); young woman or old hag (right).

This usefully draws attention to how we must *participate with our thinking* in the appearance if in fact its potentials as an actual phenomenon are to be realized. However, the fact that the ambiguous figure allows different interpretations does not mean that the contribution of thought is arbitrary or merely subjective. If we try to think the Necker cube with the idea of a sphere, we will not come to a meaningful image. Our thoughts must be those already implicit in the sensible aspect of the appearance.

Ambiguous figures are an unusual case. What remains true even in the more general case of great art is that we can always deepen our thoughtful understanding of it. Anyone as artistically unaware as the present writer may have the experience of hearing an art historian lecture about a particular painting or a particular cultural tradition of painting, and then find that he looks at certain works with newly and refreshingly informed eyes. The picture he sees now is not the same one he saw before.

But this is true also of natural scenes. Confronted by a violent thunderstorm, Stone Age man did not actually *see* the same atmospheric phenomenon we see today. *Our* "art lecturer" in this case has been the scientist, whose conceptualizations have been assimilated by the entire culture of the last few hundred years. The lecture has ceaselessly entered our ears through the words and meanings we have learned from childhood onward. We see with the perceptual and conceptual resources of our own era. (As for Stone Age man and ourselves, it may be that we *both* miss important aspects of the thunderstorm. But that is a point for Chapter 23, "The Evolution of Consciousness.")

I would be saying nothing unusual if I were to remark that we have no *theories* except by virtue of the thinking that constitutes them as what they are. It is a vastly more difficult matter, however, to realize, as we surely must, that we have no *things* to theorize about in the first place except by virtue of the thinking that constitutes them as the things they are. So if we are content merely to accept things as they are now given to us, then before we even begin our scientific work, we have already committed ourselves to the particular, culturally influenced thoughts that bring to appearance the objects and phenomena currently available to our scientific curiosity.[4]

There remains the question, then: "How adequate are the thoughts through which our natural surroundings have gained whatever meaningful form they now have for us?" Every bit of nature can be seen more or less profoundly, with deep insight or a superficial glance; with an intense, trained perception, or a lazy attention that merely glides over surfaces; with loving, qualitative detail or with remote abstraction. We may not easily misconceive a cube as a sphere, but we *can* be content to see far less of the world than is actually available to a more penetrating vision. From force of habit we of the past few centuries may, for example, see merely "mindless objects," despite the fact that it is our own, culturally informed minds through which the objects come to the only appearance we are given.

Recognizing the element of our own thinking in the data of science — both the truth of this thinking and its limitations — would seem to be a prerequisite for any rigorous scientific understanding.

Do we really want an empirical science?

The fact that thinking is already present in the only phenomena available to scientific investigation is one of those fundamental truths, easily recognized yet widely ignored, that can change everything. It tells us something about how intimately we as thinking beings are woven into a universe that invested in us powers of thought coordinate with the thinking already inherent in that universe. Or, in the lower-level (molecular) context of the preceding chapters: it reminds us how intimately the world's wisdom has been woven into the directed activity through which our bodies, including our brains, have been formed (Chapter 8, "The Mystery of an Unexpected Coherence").

But, important as thinking is, we have seen that it cannot by itself give us a world. There is also the "something" that thinking illuminates — the unformed contents provided by our senses. If, as we saw above, our senses cannot give us identifiable or nameable or recognizable *things* without first being informed by thinking, neither can thinking give us any such *things* without there first existing a sensible content capable of being so informed.

A new kind of attention to the senses was the glory of the Scientific Revolution — a revolution that was felt to be, in part, a reaction against the empirically untethered intellectual flights of the medieval doctors. The pioneers of modern science sought to bring their thinking into disciplined connection with careful observation and manipulation of the world around them. Thus was born the ideal of an *empirical* science — a science of practical experience rather than speculation. To this day the ideal remains sacrosanct among scientists.

But here a curious contradiction emerges. For, the ideal is directly belied by an entrenched conviction (elaborated in the following section) that human sense experience is irreducibly subjective and illusory. If this is true, how is an empirical science supposed to give us an objective understanding of the world? Doubt on this score has been met by an ever greater reliance on the extremely thin "experience" of instrument dials, gauges, and read-outs.

The idea behind this reliance is that the quantitative rigor and sensitivity of the instruments can compensate for the limitations of the human senses. But whatever those limitations might be, the senses are what give us access to the world. Numbers are not material entities. They are conceptual, and the fact remains that thinking alone — including, as I have already indicated, mathematical thinking — cannot give us a world. We must *apply* the mathematics to sensible experience if we want it to tell us something about material reality. Where are we to gain that experience (so as to have actual things to talk scientifically *about*), if not through our supposedly unreliable senses?

Our contradictory attitude toward human experience — hailing it as the foundation of any true science, while denigrating it as the source of confusing subjectivity — has long been an open wound in the body of science. Yet the issue is rarely given thought by the working scientist. Philosophers, meanwhile, continue picking at the wound as they have for the past few centuries, to little avail.

Nevertheless, the entire problem, having been falsely posed, can be simply resolved.

It is careless thought that deceives us, not our senses

Who has not heard the various clichés about how our senses "lie" to us. Try immersing one hand in a bowl of hot water, and the other in a bowl of crushed ice, holding them there for a while. Then remove them both and place them together in lukewarm water. Initially, one hand will feel the water as cool and the other as warm. So goes the "proof" that the felt qualities of things are subjective and misleading compared to the objective report of a thermometer.

The conclusion is wrong. If you follow an identical procedure with two thermometers, you get a similar result: the two columns of mercury initially show different temperatures. Over time they move in opposite directions until, as happens with our hands, equilibrium is reached. Nor does hand or thermometer offer false reports during the period of adjustment. At every moment the reading correctly reflects the changing *relations* between water and measuring instrument. Such relations must be grasped in thought, which is the only way we ever make sense of our senses.

How many school children have been given an experience of these bowls of water! And how many have been taught the lesson that their experience is worthless and deceptive! All the better, I suppose, to prepare them for further misconceptions of the sort we will now consider.

Earth and sun

Another classic example of our "lying" senses has to do with an appearance we witness every day: it *looks*, we are told, as if the sun goes around the earth, not as if the earth is rotating as it goes around the sun. In his play, *Jumpers*, Tom Stoppard skewered this particular claim by having one of his characters ask: "Well, what would it have looked like if it had looked as if the earth was rotating?"

Surely it *should* look exactly as it *does* look; any other appearance would have been false to the fact of rotation. It's just that we have to employ our thinking in order to make sense of *any* appearance. Once we grasp this truth, we cannot help realizing how wrong it is to declare the appearances from earth to be false. We are free to take up any vantage point we choose. Copernicus chose to look, in imagination, from the vantage point of the sun. This was a decisively important step. But surely we have no more right to absolutize that perspective than we do the one from earth. The heliocentric view is as "parochial" as the geocentric view compared, say, to a galactocentric view, where observations over time would make it clear that both the earth and the sun engage in a complex dance around an ever-changing point that is neither at the center of the earth nor the center of the sun — a dance that is influenced by all the other planets.

Figure 13.2. Nicolaus Copernicus.[5]

Scientists, in their research, do in fact routinely and justifiably employ purely local coordinate systems for their immediate purposes wherever they happen to be on earth. It would make no sense to use a heliocentric coordinate system when mapping out the placement of plants in an experimental garden. And neither scientists nor the rest of us have any particular difficulty holding all the various possible perspectives harmoniously together. When standing in a group around a tree, we all perceive the same tree, even if no two of us see exactly the "same" image of it. Our senses must be informed by our thinking. Only then does a coherent appearance — as opposed to a chaotic aggregation of disconnected sense impressions — present itself.

The atom and beyond

Here is another scientifically sanctioned "old wives' tale," taken from a PBS television special written by science journalist, Timothy Ferris:

> The baseball and the bat are mostly empty space. Their solidity is an illusion created by the electromagnetic force field that binds their atoms together ... We credit the home run to the batter, but the fundamental force responsible is electromagnetism (Ferris 1985).

The picture we are invited to contemplate is one of atoms. Each atom is said to consist of minuscule particles packed into an infinitesimally small nucleus. Added to these are even more minute electrons traversing enormous tracts of empty space as they orbit the nucleus at a vast distance. It is, we are told, the electromagnetic force binding the particles together that deludes us into losing sight of all that empty space comprising nearly the whole of the individual atom, and therefore also nearly the whole of the bat and ball.

But notice: "empty space" gains its meaning here only when we picture the nucleus and the orbiting electrons as a collection of nicely solid particles — solid like little space-occupying bits of the actually experienced world. We are then supposed to contrast these particles in our minds with the great expanses occupied by no particles at all.

But this is the picture that physicists labored throughout much of the twentieth-century to eradicate from our imaginations. For good reason: they well know that the erstwhile "particles" of atomic theory do not exist — not as bits of material stuff occupying discrete volumes of space that we can contrast with empty space. The only *material stuff* we are given in the universe is the sensible content of our perception.[6]

Look at what is happening here. Ferris is trying to get us to doubt our perception of the material world. Yet he is doing so by asking us to imagine imperceptible "particles" as if they were little bits of perceived material stuff. Rather than discrediting our perception, he is in fact illustrating the impossibility of imagining a world otherwise than in terms of perception. The only illusion is on his part: he is projecting the contents of perception into a theory-laden, falsely imagined, submicroscopic realm where in fact no perceptible content is given to us.

To reinforce the point, listen to neuroscientist and philosopher, Paul Churchland, assuring us that our various forms of observation — sight, hearing, touch, and so on — are not to be trusted:

> The red surface of an apple does not *look* like a matrix of molecules reflecting photons at certain critical wavelengths, but that is what it is (Churchland 1988, p. 15).

Our senses, in other words, are said to fail us because they do not show us the red surface of the apple as *really* consisting of unimaginably small "billiard balls" or "wave packets" reflecting other balls or packets. And so, again, apart from such sense-based imagery — the very thing that physics today forbids us from projecting into atomic theory — Churchland's argument would be wholly unpersuasive.

The point is decisive. Only by picturing particles (or waves) as little bits of the qualitatively experienced world can the reader fill in Churchland's description in a way that makes it sound meaningful. But this sensible perception of the world's qualities is exactly what Churchland is trying to dismiss. While telling us that the familiar qualities of the world are illusions, he invites us to project these same qualities into the sub-microscopic realm. That realm then becomes proof that the familiar qualities aren't to be taken seriously. Apparently sensory qualities, such as the firmness and solidity of material things, are illusions *here* (where we can experience them), but real *there* (where we cannot).[7]

The moral of the story? Even when we are trying to talk about a world without the qualities of our senses, we end up talking about the qualities of our senses — but in a nonsensical way.

Our "missing" bat sense

One last example. Those who disparage our experience love to point to creatures who perceive things we cannot. Wouldn't we live in a different reality if, say, we had the infrared vision of some snakes or the "sonar" (echolocation) sense of a bat? Of course we would — but only in the way those who are deaf or blind would live in a different reality if their senses were unimpaired. Perhaps the most striking thing about our perceptual worlds is their continuity and coherence, despite the supposedly discrete nature of the sense data and of the different senses themselves. Adding a new sense gives us a richer picture, but it is a richer picture of the unified world we already know.

We heard above in the case of S. B. that it can be difficult, as an adult, to cope with an overwhelming content of sense perception through organs of sense that have not, in the normal course of things, already been educated by thinking. But the fact remains that the normal course of education presents no particular difficulty at all.

If the bat's echolocation were suddenly and miraculously added to our own array of senses, we would presumably suffer some disorientation, just as S. B. did when the bandages were removed from his eyes. Like all our other senses, our new sense would need to be educated by our thinking. But we would have no reason to think that our new world stood in contradiction to our previous experience.

Nor is there any reason to think that a person naturally born with a capacity for echolocation would find his world *conflicting* with that of the rest of us. The two worlds would certainly vary in the richness of the contributions made by the different senses, but they would no more disagree with each other than the truly vast difference between the most sensitive musician's ear and the dullest, least attentive ear among the rest of us would spell a disagreement of sense perception.

To believe that we can truly know the world is not to believe that our present knowledge is exhaustive, or that the world cannot present itself within many modes of consciousness, or that our present powers of perception cannot be deepened beyond anything we can now imagine.

One reason we can be confident that newly developed senses — whether those of a bat or otherwise — would harmonize perfectly with our previously existing senses is that the harmony results from the thinking aspect of things. It seems safe to say that the education of our senses by thinking is essential to the unity of our experience of the world. Thinking has the quality that all thoughts can enter into harmonious relation with all other thoughts. The thought-world knows nothing of absolute disconnection or contradiction. To take a trivial example, we bring "truth" and "falsehood" into meaningful and harmonious relation when we say, "Truth and falsehood are contradictories."

The world of thought is, in a profound sense, *one*, and this is what enables us to have *one tree* despite the fact that we view the tree from many sides and never have two identical visual impressions of it. This unity of the thought-world also explains how it can be that, in any text or speech, individual words can be informed by their context. Their meanings are shaped by the thought of the context as if they were essentially of one substance with it. They merge their own identity into the integral and coherent unity of the whole.

Even the recognition of a logical contradiction requires a perspective wherein we can see particular thoughts joined together by a relation of sameness as well as significant difference. There can be no *absolute* opposites, for if they had nothing at all in common, there would be no way for us to think them together in order to compare them or pronounce them "opposite." We can have contrary things to say only about ideas participating in a common realm of meaning.

Closing thoughts

The bare contents of our senses — if we could somehow know them before they were illuminated and given form by thinking[8] — could not possibly lie to us. Our senses just are not in the business of being either true or false. Truth and falsehood are features of thinking. In fact, as the ambiguous images in Figure 13.1 were meant to illustrate, particular sensible contents are not even *there* for us in any meaningful sense — and certainly not in a manner we could call "true" or "false" — until the illumination by thinking has occurred. It is only this thinking that can be more or less faithful to whatever comes through our senses.

You may recall a few occasions when you saw a slowly moving object high in the sky that might have been a bird or a plane. If you initially and unquestioningly took it for a bird, then that was the appearance you saw. But if, due to a sound reaching you or the hint of a contrail, you eventually realized that it was a plane, then the appearance changed and now became stable in the way that your first impression was not. Your initial judgment was, you might want to say, false.

But even though our perceptual judgments may need to be corrected, they are rarely if ever *absolutely* false. Even when you falsely thought you were looking at a bird, you correctly believed you were looking at a moving object in the sky — unless, perhaps, you later discovered that it was neither a bird nor a plane, but a floater in your eye. Even so, you would at least have been correct to note a real object in your field of vision — unless it turned out that you were mentally disturbed and starting to hallucinate.

The thought-aspect of perception — the bringing of perceived contents into this or that form — is a complex matter, sometimes requiring high skill and practiced judgment. And it is only this thought that we can evaluate for adequacy or inadequacy, truth or error. Eventually — and with reinforcement from our social surroundings — we do reach a stable judgment of some sort about most objects in our immediate environment. But even so, we can always deepen those judgments. I do not know the woods through which I walk with anything like the insight of a life-long forester. Which is to say that I have not yet begun to bring the woods to profound appearance in the way the forester has.

And then there is the problem inherent in what we might call "collective illusions" or "historical aberrations." Such might be, for example, the contemporary experience of the world's objects as wholly "out there," separate from ourselves in a mind-independent way, even though we know very well that we play a role in how they come to appearance. This, in fact, is a confusion intimately related to our present discussion. So let's pause for a moment to ask ourselves, "What if the world is not mind-independent? What if its essential nature lies in its power of *manifestation* — its power of coming to *real and substantive appearance* for percipient beings precisely in terms of their various cognitive capacities, including their sense of touch and other senses, as well as their power of making sense of things?"

This is an issue we will look at more closely in Chapter 24, "Is the Inanimate World an Interior Reality?" For the moment it is perhaps enough to remark that I have never seen evidence produced in favor of the view that the mind-dependence of our cognition somehow demonstrates the unreliability of our knowledge. It would be just as logical to conclude that the interior qualities of our cognition are exactly the right prerequisites for our understanding a mind-dependent world. A mind-dependent world, after all, has the advantage that it would presumably be a mind-accessible world. Granted, this accessibility may in certain matters require a rigorous, skillful, and highly developed cognitive activity. But just about all worthwhile achievements on earth similarly demand effort and skill.

It is easy to forget that the cognitive capacities by which we bring the world to appearance are the cognitive capacities the world itself has brought to fruition within us. We are certainly free to doubt them, but we ought at least to ask whether the capacities by which we daily judge ourselves to be knowers of the world — capacities born of the world and through which we make sense of the world — might be just what they seem to be and ultimately perfectible without limit. Nothing about the development of human cognitive capabilities in almost every direction, from science to the arts, seems to suggest that we face narrow constraints, or that our minds are fundamentally alien to reality. Every time we gain understanding of anything to any degree, we seem to find ourselves invited "further up and further in." That is, we find ourselves becoming ever more familiar with a mind-soaked reality.

Actually, a mind-soaked reality is the only sort of reality we could ever hope to know. In fact, it's the only sort of world to which the idea of *knowing* could be applied. So even to ask whether we can know reality *in truth* may already be to assume that something knowable, something mind-soaked, is potentially waiting to be known — ready to be embraced, mind to mind.

There seems something rather odd about the turn in thought of the past few centuries whereby we have come to assume, without evident reason, that the world's knowability is somehow compromised by the fact that we happen to know things by means of our own capacities and from our own vantage point. Who else's vantage point would we want to assume? Is it even conceivable that any phenomenon of the world should present itself "neutrally," as if from no vantage point at all? What could this mean? Could a real, material ice cube present itself other than from a particular point of view? Perhaps the unthinkability of what we might consider the strictly "objective" view only confirms that it is in the nature of the world to be a content of particularized experience.

In any case, for the student of the evolution of consciousness (Chapter 23), the question is not, "How can anyone arrive at the 'crazy' idea that thinking belongs to the warp and woof of the world?" but rather, "How did it happen, in this last brief, historical moment, that we have come, 'crazily,' to doubt a world humming with the high tension of creative thought?"

WHERE ARE WE NOW?

Is our way of knowing the world truly revelatory?

Biologists have studiously applied themselves to continual reinforcement of a materialist attitude that aims to ignore everything *living* about organisms. And this attitude is most intense when it comes to ignoring the reality of human experience — human interiority — through which alone we can have an empirical science. It all makes for a science that is extraordinarily inattentive to the ground upon which it stands.

This ignoring of the ground on which we stand is a strange thing, and (as I have tried to show in this chapter) has led to all sorts of self-contradictory claims about the uselessness of direct human experience for science. We have considered arguments such as, "It doesn't look as though the earth rotates on its axis and revolves around the sun"; or "Putting our hands in separate bowls of cold and hot water, and then putting them both in a bowl of tepid water proves the fallibility of our sense for warmth"; or "The red surface of an apple does not *look* like a matrix of molecules reflecting photons at certain critical wavelengths, but that is what it is."

The proper conclusion is that our senses, considered by themselves (and it takes some critical work to consider them that way) *never* lie to us. They're not in the business of being true or false; they just are what they are. Truth and falsehood are features of thinking, not of the raw givenness of sense. They apply, for example, to the thinking that, joined to the reports of our senses, brings the world to more or less coherent and revealing appearance.

Humans belong to the world, are nurtured by the world, and are naturally given means to know the world in which we are so intimately immersed and from which our own substance and capacities are derived.

Putting it in slightly different terms: If we believe in practice, as virtually everyone does, that we can know the world, we must believe that, by nature, it lends itself to our understanding. It "speaks" to us in the language of our own experience, which is to say that its native language is also our language. The language of the world's expression is the language of our experience.

In Chapter 23 ("The Evolution of Consciousness") we will look at the powerful historical evidence grounding this understanding of the relation between the world's speaking and our own speaking. And in Chapter 24 ("Is the Inanimate World an Interior Reality?") I will attempt to carry the considerations of this present chapter as far toward a conclusion as I can.

NOTES

1. There is also the phenomenon I have referred to as biological blindsight. Biologists certainly do recognize an end-directed coordination of events in organisms. They want to understand how cells, by means of almost unthinkably complex organizational activity, prepare for and go through cell division. Or how predators strategically mobilize all their physical resources in order to capture prey. It's just that the *explanations* for such coordinated activities are, for artificial reasons, required to consist, at bottom, of causal processes that make no reference to the fact of higher-level coordination.
2. I take the phrase, "marriage of sense and thought," from a wonderful book of that title (Edelglass et al. 1997).
3. The classic study is that of M. von Senden. See also the discussion of "S. B." in Gregory 1978 and that of "Virgil" in Sacks 1995.
4. Anyone who would like a fuller exposition of the role of thought in what we perceive might want to read the three chapters by philosopher Ronald Brady in the freely available online book, *Being on Earth: Practice In Tending the Appearances*. See also Chapter 23 ("The Evolution of Consciousness") and Chapter 24 ("Is the Inanimate World an Interior Reality?") of the present book.
5. Figure 13.2 credit: District Museum in Toruń (https://commons.wikimedia.org/wiki/File:Nikolaus_Kopernikus_MOT.jpg), Public Domain via Wikimedia Commons.
6. If you wanted to speak in terms of physics, you would have to talk about forces *entirely filling* the space of the atom (and extending far beyond it). Such forces can be measured, but bits of atomic or subatomic "stuff" are *never* seen. The "pictures of atoms" we are sometimes shown are in fact graphs — for example, graphs of measured forces — designed to look like material objects. And if the space of the atom is wholly permeated with forces, *that* fact gives us no basis for contrasting substantive particles with empty space. It merely shows that particle physicists have abstracted their understanding so far from the perceptible world that many of their theoretical constructs do not refer to anything like familiar elements of experience. These constructs are undoubtedly rooted in meaningful structure at the submicroscopic level — structure such as that given in a pattern of forces — but this is not yet to be speaking about *things* in the sense of material reality. As we have seen above, such *things* are always products of the "marriage of sense and thought." Without both of these together, nothing is *there* for us.
7. Physicists, having learned long ago not to assert the existence of real particles and waves in the sub-microscopic realm, came to speak instead of mathematical probabilities corresponding to various instrumental read-outs. What material reality these probabilities correspond to cannot be meaningfully discussed — so they often tell us. And this should be no surprise, given that the only reality we have is a reality of *experience*. Talking about contents of experience that we cannot actually experience leads to gibberish.
8. To know a content of our senses would, of course, already be to have illuminated that content with thinking.

SOURCES

Churchland, Paul (1988). *Matter and Consciousness*. Cambridge, MA: MIT Press.

Edelglass, Stephen, Georg Maier, Hans Gebert and John Davy (1997). *The Marriage of Sense and Thought: Imaginative Participation in Science*. Hudson, NY: Lindisfarne.

Ferris, Timothy (1985). *The Creation of the Universe*, PBS television special, which Ferris authored and narrated.

Gregory, Richard L. (1978). *Eye and Brain: The Psychology of Seeing*, 5th edition. New York: McGraw-Hill.

Maier, Georg, Ronald Brady and Stephen Edelglass (2006). *Being on Earth: Practice In Tending the Appearances*. Full text is available at https://www.natureinstitute.org/book/being-on-earth. Hardcopy book is available from Logos Verlag Berlin (2008): https://www.logos-verlag.de/cgi-bin/engbuchmid?isbn=1887&lng=eng&id=

Sacks, Oliver (1995). An Anthropologist on Mars: Seven Paradoxical Tales. Vancouver, WA: Vintage Books.

von Senden, M. (1960). *Space and Sight: The Perception of Space and Shape in the Congenitally Blind Before and After Operation*. London: Methuen.

TECHNICAL SUPPLEMENT
to Part 1

CHAPTER 14

How Our Genes Come to Expression

(It Takes an Epigenetic Village)

IF YOUR UNDERSTANDING of genetics comes from your newspaper's science section, or a popular science magazine, or any other source intended for the general public, then you probably will not have been given the remotest glimpse of what actually goes on with the genes in our bodies. In fact, geneticists themselves have been known to lament how limited their knowledge of gene-related activity is, simply because the demands of professional specialization scarcely allow a wide field of view.

But it turns out that a wide field of view is the one critical prerequisite for any adequate understanding of genes. Only a broad survey can illustrate how every gene, like a significant word in a text, receives its full meaning only through the interweaving and converging influences issuing from all the elements of its context.

My aim here is to offer such a wider, "epigenetic" view — and to do so in the briefest space possible. If I succeed, you will begin to sense a biological landscape that reconfigures many long-standing assumptions, not only about genetics itself, but also about the character of all living activity.

High expectations: the promise of molecular biology

After the discovery of the structure of the DNA double helix in 1953 and the elaboration of the "genetic code" during the early 1960s, the *expression* of a gene was thought of as the production of a functional protein corresponding precisely to *instructions* in the gene — coded instructions that were spelled out in the gene's *sequence* of DNA "letters," *or nucleotide bases*. The protein's production, based on this sequence, was routinely described as a cut-and-dried, fully determined, rather mechanistic affair. The larger picture was sometimes summed up in this formula:

DNA makes RNA, RNA makes protein, and protein makes us.

A few key terms may help to flesh out the formula as it was then understood. (All the special vocabulary is elaborated in an online glossary at https://bwo.life/mqual/glossary.htm.)

The first step in gene expression was thought to be the binding of a protein *transcription factor* (one of many such factors existing in the cell) to DNA at or near a target gene. This led to the adjacent binding of a complex protein called *RNA polymerase* (often described as a "molecular machine"), which then *transcribed* the DNA sequence of the gene into an RNA molecule closely mirroring the DNA sequence.

Finally, the RNA was exported from the cell nucleus into the cytoplasm, where it was *translated* into a specific protein. The translation was carried out by another complex "molecular machine," known as a *ribosome*. The sequence of amino acids in the resultant protein was said to have been *coded for* by the sequence of nucleotide bases in the gene and the similarly coded sequence of the RNA. A parallel was sometimes drawn with Morse code, in which a sequence of dots and dashes codes for a sequence of alphabetic letters.

The discovery of the entire scheme, so neat and tidy, testified to the impressive technical sophistication of the researchers, and was universally acclaimed.

But there was already a curiosity. Consider the picture. The production of a protein from DNA was initiated by a protein transcription factor. The "molecular machines" doing the work of transcription and translation consisted, in whole or in part, of proteins. Moreover, it was recognized that proteins were decisive for the very existence of DNA, as well as its replication, maintenance, and repair. So not only were proteins required in order to explain their own synthesis, but they were also required in order to explain the existence of DNA.[1] At the same time, DNA was clearly required for the existence of proteins.

You might think the chicken-and-egg problem here would have given the scientific community pause during its single-minded, twentieth-century rush toward a gene-centered view of life. Was it really genes that made the organism, including its proteins? Or was it proteins that made the organism, including its genes? Or were both points of view terribly flawed and unbiological, so that we were being asked to rise to a more living and integral level of understanding where it is impossible to say that one thing unambiguously "causes" another?

Complications

Fast forward to today, and consider just one of the terms mentioned above: "transcription factor." A riddle posed by many such protein factors involves their "promiscuous binding." Transcription factors, of which there are over a thousand in the human genome, are not targeted to specific DNA sequences by some iron necessity. Most of them are quite capable of binding at thousands of locations throughout the genome — that is, at far more loci than they are actually found at in typical assays of living cells. In other words, we have to look for much more than a definitive, sequence-based targeting logic if we want to understand how transcription factors activate (or inhibit) specific genes in this or that specific kind of cell and context.

So the question arises, How does a transcription factor "know" which gene or genes to interact with? If its specificity — its ability to bind where it is needed — is not dictated by a simple and determinative match between its own binding domain and the DNA sequence it binds to, then how do we make sense of its well-directed activity? Is this activity merely expressing something like the logic at work in a humanly devised mechanism? Or is it more like a living language, where words can have diverse expressive potentials that are in part lent to them by their context?

The answer — or, rather, the many answers — are still unfolding today. The one indisputable truth is that it takes a molecular "village" — a vigorous and entire cellular context — to establish the correct and ever-changing relations between a transcription factor and the genes it helps bring to expression. The old idea that the relations among transcription factors, genes, and gene products are unambiguous — are governed by a fixed, necessary, and easily comprehended logic — is no longer tenable.[2]

Transcription factors and DNA engage in a complex play of form

To begin with, not just the DNA sequence, but also the moment-by-moment sculptural form, or conformation, of a DNA locus affects the binding potential of a transcription factor. This dynamically imposed form reflects the cellular environment. Also decisive are the plastic conformational potentials of the transcription factor itself. And then there are the many other essential molecules ("co-factors") that may not even have the ability to bind to DNA, but which are nevertheless essential co-participants, along with transcription factors, in an interactive community through which a gene, or set of genes, is made ready for transcription.

For example, one way a transcription factor can contribute to the expression of a gene is by bending a short stretch of DNA into a shape conducing to further interaction. (For a striking, if highly schematic, illustration of this, see Figure 14.5 below.) By this means the initial presence of a transcription factor can make it easier than it would otherwise be for a second protein to bind nearby. In the case of one gene relating to the production of interferon (an important constituent of the immune system), "eight proteins modulate [DNA] binding site conformation and thereby stabilize cooperative assembly [of gene-regulating proteins]" (Moretti et al. 2008).

And so, despite the fact that "DNA is often mistakenly viewed as an inert lattice" onto which proteins bind in a sequence-specific way (Chaires 2008), the fact of the matter is altogether different. Proteins and DNA are caught up in a continual conversation of mutual influence and shifting form. It becomes obvious, then, that "No simple code combines all the various determinants of transcription factor binding specificity" (Slattery et al. 2014).

In other words, a transcription factor's "recognition" of a DNA binding site is not a digital, yes-or-no matter, but a community judgment. And how could it be otherwise, given that no cell in our bodies (and no collection of molecules) lives merely for itself? Our activities always involve vast, cooperating communities of various sorts. Every cell and cellular organelle is caught up in a larger context of meaning and must be capable of adapting itself to, and supporting, virtually any of the infinitely varying activities we find ourselves engaging in.

A living flexibility is therefore crucial. So it is no surprise when one pair of researchers, studying a group of transcription factors in the genomes of animals, report "a dazzling array of strategies employed by [these] transcription factors to control gene expression." The "emerging, unifying theme," they say, is the ability of these transcription factors "to interact with many diverse partners. This high connectivity is probably crucial to assemble highly context-specific, transcriptionally active complexes at selected sites in the genome" (Bobola and Merabet 2017).

Genes and proteins interact in tangled causal webs

It is hard to take in the full significance of this "high connectivity," which is typical of so many biological processes. One way to visualize the complications is to consider the fact that some transcription factors can target genes for other transcription factors. And, of course, this second group of transcription factors might target the genes for still other transcription factors as well as the genes or regulatory sequences associated with the first group. We can easily imagine the tangled causal webs resulting from this kind of inter-connectivity, where causal "arrows" can eventually circle back to their starting point. Unsurprisingly, there are entire fields of research today given over to complex gene and regulatory networks such as shown in Figure 14.1.

Returning to the puzzle of transcription factor "promiscuity": this word reflects neither undisciplined profligacy nor uncertainty of function. Rather, it points to the unbounded, context-specific potentials of transcription factors. Their contribution to essential cellular processes, after all, is properly focused and far from promiscuous. They are caught up within a wisdom that seems to "know" exactly what it is doing. It's just that this doing is complex and *living* — flexible and adaptive — far beyond what a simple, definitive, one-dimensional mapping between DNA sequence and a rigidly complementary protein shape would allow.

This flexibility is what allows community-tuned activity in the larger surround to influence local goings-on in endlessly nuanced ways — all so as to satisfy the needs of the current context.

It is important to underscore here a fact we have found ourselves coming up against throughout this book: the tangled causal web we discover in organisms is not merely a matter of complexity. There are many nonliving physical contexts so complex that, as a practical matter, we cannot easily trace precise lines of cause and effect. This is true of eddies in a great river or in the atmosphere, and it is even true of some kinds of computer program. And yet no one would doubt in these cases that the relevant causes *could* be traced, at least in principle, or that the tracing would give us what is usually (if erroneously) considered to be a full accounting of what we are looking at.

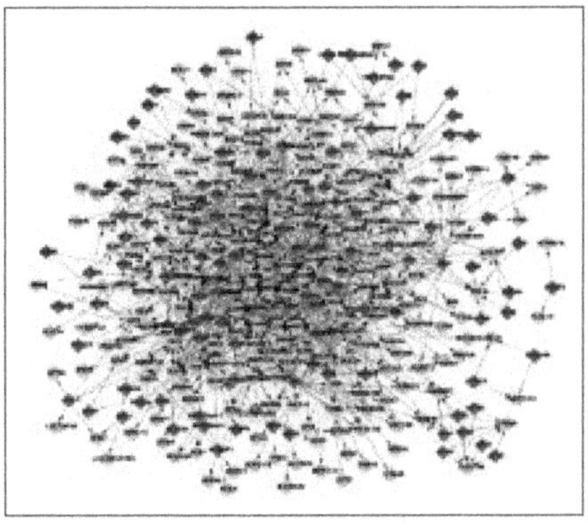

Figure 14.1. Transcription regulation network of Parkinson's disease, showing differentially expressed genes (pink) together with some of the transcription factors (blue) playing a role in regulating those genes. The figure is too small to read — purposely. Researchers sometimes lightheartedly refer to such diagrams as "hairballs," which is about all you need to know.[3]

But, as I began explaining in Chapter 2, the purposive behaviors of organisms exhibit a kind of coherence and meaning that is not satisfactorily explained when we look only at principles of physical causation. The "causal confusion" in the organism's case is not due merely to the complexity of the always lawful and harmonious physical relations, but rather to the fact that purposive and *narrative explanation* must be found at a "higher" level of meaning than physical lawfulness. The significance of what is going on is recognized only when we consider the insistent coordinating principles through which physical events are caught up in serving the needs and interests of organisms. Because concepts such as "need" and "interest" are incommensurable with the accepted principles of physical explanation, they demand recognition as explanatory principles in their own right.

The cell holds DNA in an intimate and instructive embrace

Our brief discussion of genes and transcription factors has, so far, been hopelessly simplistic. The chromosomes in our cells do not consist of a naked DNA double helix sporadically bound at particular sequences by this or that transcription factor. The picture is wholly different. Our DNA is intimately bound up with a massive, intricate, and dynamic protein-RNA-small molecule complex that, together with the DNA, is called *chromatin*. "Chromatin," in other words, can pass as simply a name for the full substance of chromosomes. The proteins in this complex are as weighty as the DNA itself — and much more active and directive when it comes to gene expression.

Some of the protein constituents of this chromosomal substance — both the longer-term and the many transient constituents — can bind directly to DNA, thereby facilitating, blocking, or modifying the transcription of this or that gene. But other elements of chromatin, while not directly bound to DNA, nevertheless contribute crucially to the regulation of gene expression. Overall, the molecular factors associated with chromatin play roles such as the following:

- they help to condense or decondense the packing of the DNA (more tightly condensed DNA tends to be less accessible to activating factors);
- they move chromosomes or parts of chromosomes to different regions of the cell nucleus (the interior of the nucleus tends to be more transcriptionally active than the periphery);
- they attach parts of chromosomes to the nuclear envelope (many factors at or near the envelope bear on gene expression);
- they interweave and (almost miraculously, it might seem) disentangle chromosomes, while also forming decisively important chromosome loops (such as those we heard about in Chapter 3, "What Brings Our Genome Alive?") — all so as to form various-sized "communities" of functionally related chromosomal loci;
- they untwist (loosen) the two strands of the double helix in some places and twist them more tightly in others, which can make the difference between a gene's accessibility or inaccessibility to transcription factors;
- and they alter the electrical characteristics of particular loci (yet another feature bearing on the expression of affected genes).

As you may surmise, then, it's not as if the power to determine gene expression outcomes is one-sidedly delegated to any genetic sequences, any transcription factors, or any other entities. It is rather as if the result arises in the way a musical performance is evoked from a jazz ensemble. A distinct locus of DNA certainly offers its own expressive potentials, but there is no telling — no predicting solely from an analysis of the sketchy DNA "musical score" — how the locus may be employed within the improvised, multi-cellular performance leading from a single fertilized egg cell to the mature human being.

But perhaps we would do better to imagine an exquisitely detailed, never-ending, self-assured, yet highly improvisational dance involving billions of molecular dancers within a cell — all coordinated with the choreography in neighboring cells and with the ongoing story of the organism as a whole. The performance, involving the fluid identity of countless players, is a long way from that of calculating or information-processing hardware and software.

In any case, the present point is that our DNA is thoroughly "wedded" to — bound together with — an almost unfathomably intricate arrangement of protein, RNA,[4] and small molecules. The protein and RNA constituents of this chromatin complex are fully as "information-rich" as the DNA. Genes, as such, cannot *do* anything, and certainly cannot transcribe themselves. The information-rich, if unquantiable, doing is in large part a function of the associated proteins, which, among other things, thereby participate in their own genesis. Alongside them are many other molecules, including water molecules (Chapter 5, "Our Bodies Are Formed Streams"), all of whom give collective expression to the purposive coherence of the cell as a whole.

I have so far offered only a rather vague and general description of the highly effective embrace in which DNA is held. In later sections we will look further at some of its key features.

Getting started is hard to do

Meanwhile, leaping tall edifices of thought in a single bound, we will pass over the question how cells "know" which genes need to be expressed within the current context of a person's activity and within the trillions of cells constituting our bodies. We will also avoid asking how any single cell — which can play only a spatially minute part within an organ such as the liver or within a process such as wound healing — finds its own proper role in whatever the current larger performance happens to be. And so, assuming all the necessary contextualization and direction to be somehow wisely taken care of,[5] we will imagine just one cell embarking on a single task: to give expression to one among its 20,000 or so genes. How might this cell proceed?

Our imaginative exercise will necessarily be more than a little artificial. That's because we need to think one thing at a time, whereas in the cell countless mutually entangled things are all happening at once. But we will try to make the best of it.

You may recall from Chapter 3 ("What Brings Our Genome Alive?") that packing DNA into a typical human cell nucleus is like packing about 24 miles of very thin, double-stranded string into a tennis ball, with the string divided into 46 separate pieces, corresponding to our 46 chromosomes.

To locate a modest-sized protein-coding gene within all that DNA is like homing in on a half-inch stretch within those 24 miles.[6] Or, rather, two relevant half-inch stretches located on different pieces of string, since most of our cells have two copies of any given gene, residing on different chromosomes. Except that sometimes one copy differs from the other and one version is not supposed to be expressed, or one version needs to be expressed more than the other, or the product of one needs to be modified relative to the other. So part of the job may be to distinguish one of those half-inch stretches from the other, and to act differently in the two cases. "Decisions" everywhere, it seems.

As a functional unit, a gene must participate in a performance appropriate to the current cellular and extra-cellular context, and the highly distributed activity responsible for its function must be cobbled together by the cell according to the needs of the moment. There is no *predefined* path to follow once the cell has located the "right" half inch or so of "string," or once it has done whatever is necessary to bring that locus into proper relation with other chromosomal loci participating in, and essential to, a joint performance.

One issue has to do with the fact that there are two strands of the double helix, and (in a chemical sense) these complementary strands "point" in opposite directions. In humans, protein-coding sequences can occur on both strands. Likewise, transcription (of both protein-coding and regulatory sequences) occurs on both strands, which is to say that the transcribing enzyme (RNA polymerase) can move in either direction along the double helix. The direction chosen — that is, the strand along which the RNA polymerase will move — depends on the meaning within the current context of the sequences that exist at the current locus. Somehow, acting within and guided by its present context, RNA polymerase must have the "good sense" to choose the appropriate activity from among the various possibilities.

And even when the cell "knows" to initiate transcription in one particular direction, it must "choose" the exact point in the genetic sequence at which to begin. Different starting points can yield functionally distinct results. "Many studies focusing on single genes have shown that the choice of a specific transcription start site has critical roles during development and cell differentiation, and aberrations in… transcription start site use lead to various diseases including cancer, neuropsychiatric disorders, and developmental disorders" (Klerk and 't Hoen 2015).

Intertwined with all the preceding issues is the cell's task of assembling a *pre-initiation complex* (PIC). This variable arrangement of regulatory elements typically sets the stage for the transcriptional activity to follow. Figure 14.2 is a cartoon figure that merely names some of the protein PIC constituents that arrange themselves on DNA (shown as a black line) near locations where gene transcription is to begin. You needn't concern yourself with names and meanings, beyond the general description I am offering now.

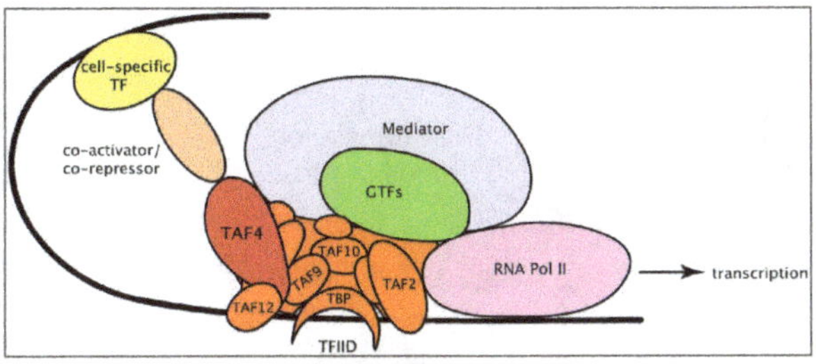

Figure 14.2. The pre-initiation complex (cartoon representation).[7]

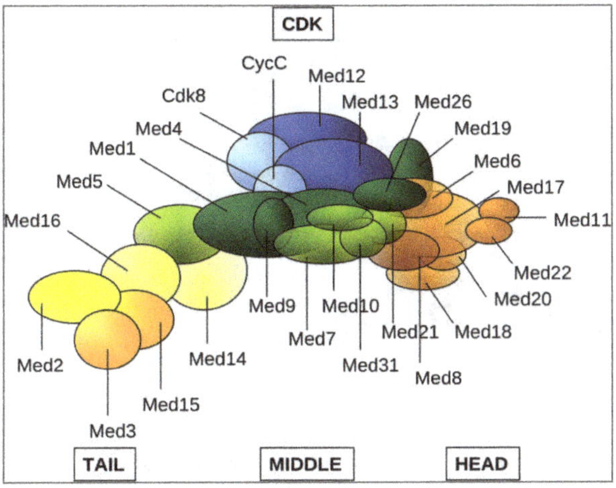

Figure 14.3. Some subunits of the Mediator complex (cartoon representation) captured at the CDK gene locus.[8]

The cell's narrative at this point could hardly be more dramatic — or more subtle. The largest oval in Figure 14.2, named "Mediator," is a massive molecule consisting (in humans) of 26 protein subunits (Figure 14.3) arranged in modules and interacting in numerous ways among themselves, as well as with other PIC constituents and "visiting" molecules. Depending on context, Mediator can vary endlessly in both subunit composition and function. Its effects upon gene expression are many, and still only fragmentarily grasped.[9]

Figure 14.4 shows the known interaction partners for the Mediator subunits in just one cell type — mouse neural stem cells. The figure omits the numerous interactions among the Mediator subunits themselves. It also omits the interactions among the molecules shown in the surrounding circle. And, perhaps most importantly, it omits the interactions those molecules have with still others not shown in the diagram. For it is just a fact that each of these molecules shown in the outer circle could be made the center of its own diagram. Reflecting on this can usefully remind us of what it means to say that all biological activity in a cell, no matter how micro-focused our vision, turns out upon broader inspection to be an almost impossibly intricate and coordinated activity of the whole.

And, of course, Mediator is just one element of the PIC. Each of the other elements has its own story to tell. The entire PIC was once regarded as a rather mechanical, routine, and mostly unvarying assembly of "parts" whose unproblematic duty was to initiate gene transcription in a standard way. But, of course, that was to overlook how thoroughly every aspect of gene expression *must* vary if it is to serve the needs of a living being. The PIC is now seen to be an infinitely modifiable, highly dynamic complex, responding both to the immediate DNA context and to influences arriving from distant reaches of the cell. Its overall

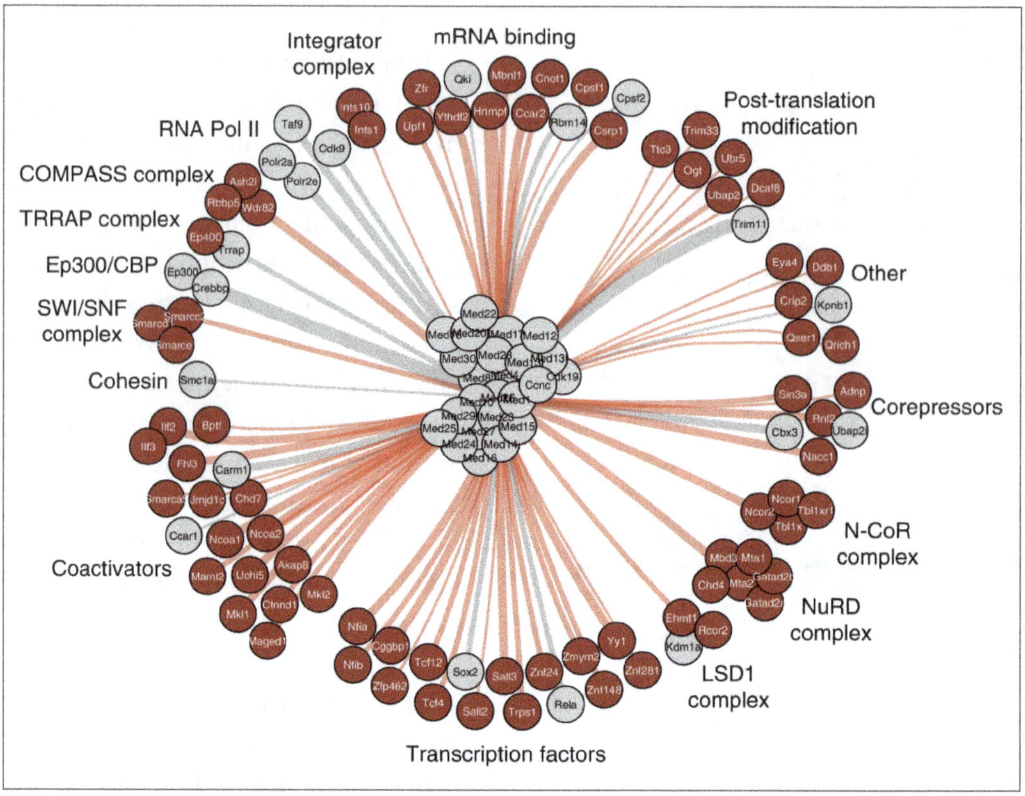

Figure 14.4. Interactions of the Mediator complex in mouse neural stem cells. Mediator subunits are shown in the middle. Gray circles and lines represent interaction partners that were already known at the time (2019) when the research was carried out. Red circles represent interactions newly discovered by the authors of the paper from which this figure was taken.[10]

"decision-making" role, which can differ from one gene to the next, is hardly the functioning of a routinely analyzable mechanism.

It doesn't require of the reader a technical penetration of these figures to get a sense for the kind of thing that is going on — especially if one keeps in mind that we are talking, not about rigid machinery of the sort we are familiar with in our daily lives, but rather about *molecular interactions* within a highly fluid context where machine-like constraints to forcibly channel the interactions are altogether absent.

I will mention here just one other element of the pre-initiation complex. Figure 14.5 shows DNA (in a wholly artificial, simplistic, and impossibly rigid, concrete representation) being "gripped" by the tata-binding protein (TBP), shown in blue. (TBP is also seen as the crescent-moon shape at the bottom of Figure 14.2.) The protein "clasps" the DNA in an intimate and rather tortuous manner — a clasp that might remind one of the forcible interaction between two human wrestlers.[12]

A severe bend of about eighty degrees is thereby applied to the double helix. This bend, which also tends to pull the two strands of the helix apart, is a general prerequisite for the assembly and activity of the rest of the PIC. As always, the cell is doing something *sculptural*, not narrowly informational in the usual sense.

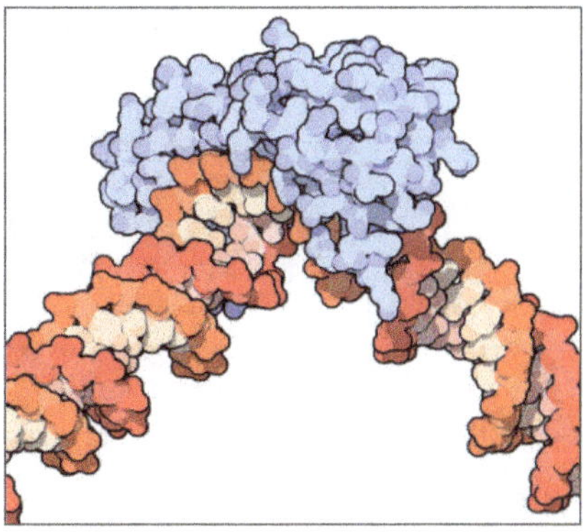

Figure 14.5. DNA (red) in the grip of the tata-binding protein (blue).[11]

Carrying on

As we heard at the outset, the (protein) enzyme that transcribes DNA into RNA is RNA polymerase.[13] The enzyme certainly does not work alone, however, and its task is by no means automatic. To begin with, its critical interactions with various elements of the pre-initiation complex help determine whether and exactly where transcription will begin. Then, after those "decisions" have been made, RNA polymerase moves along the double helix transcribing the sequence of genetic "letters" into the complementary sequence of an RNA.

Throughout this productive journey, which is called *elongation*, the RNA polymerase still keeps good and necessary company. Certain molecular co-activators modify it during its transit of a gene's sequence, and these modifications not only enable transcription elongation to begin, but also provide binding sites for yet other proteins that will cooperate throughout the transcription journey. The collective interaction here, as in the activities discussed above, can vary in many details from one context to another — all in order to contribute to a meaningful narrative that could hardly repeat itself in exactly the same way.

The table below offers some perspective on the number and variety of protein factors influencing elongation. You need not puzzle over the details. A quick browse of this incomplete listing (as of 2013) will give you at least an inkling of the kind of intricate complexity the cell must organize in order to carry out transcriptional elongation. As always, it is important to realize that each of the factors listed here enters the picture out of its own world of regulation. At the molecular level of the organism we are always looking at ever-widening circles of interaction, without limit. It's just a question of how narrowly we choose to focus our attention — and how much of the context we consequently block from view.

Table 14.1. Some factors regulating RNA polymerase elongation (copied from Kwak and Lis 2013)

DON'T READ THIS TABLE! (JUST FEEL IT.)

CLASS	FACTOR NAME	FUNCTION	RELATED FACTORS AND NOTES
GAGA factor	GAF	Generates nucleosome-free region and promoter structure for pausing	NURF
General transcription factors	TFIID	Generates promoter structure for pausing	
General transcription factors	TFIIF	Increases elongation rate	Near promoters
General transcription factors	TFIIS	Rescues backtracked RNA polymerase II	RNA polymerase III
Pausing factors	NELF	Stabilizes RNA polymerase II pausing	
Pausing factors	DSIF	Stabilizes RNA polymerase II pausing and facilitates elongation	
Positive elongation factor	P-TEFb	Phosphorylates NELF, DSIF, and RNA polymerase II CTD for pause release	
Processivity factors	Elongin	Increases elongation rate	
Processivity factors	*ELL*	Increases elongation rate	AFF4
Processivity factors	SEC	Contains P-TEFb and *ELL*	Mediator, PAF
Activator	c-Myc	Directly recruits P-TEFb	
Activator	NF-κB	Directly recruits P-TEFb	
Coactivator	BRD4	Recruits P-TEFb	
Coactivator	Mediator	Recruits P-TEFb via SEC	
Capping machinery	CE	Facilitates P-TEFb recruitment, counters NELF/DSIF	
Capping machinery	RNMT	Methylates RNA 5' end to complete capping	Myc
Premature termination factors	DCP2	Decaps nascent RNA for XRN2 digestion	Dcp1a/Edc3
Premature termination factors	Microprocessor	Cleaves hairpin structure for XRN2 digestion	Tat, Senx
Premature termination factors	XRN2	Torpedoes RNA polymerase II with RNA 5'-3' exonucleation	
Premature termination factors	TTF2	Releases RNA polymerase II from DNA	
Gdown1	GDOWN1	Antitermination and stabilizes paused RNA polymerase II	TFIIF, Mediator
Histone chaperone	FACT	H2A-H2B eviction and chaperone	Tracks with RNA polymerase II
Histone chaperone	NAP1	H2A-H2B chaperone	RSC, CHD
Histone chaperone	SPT6	H3-H4 chaperone	Tracks with RNA polymerase II
Histone chaperone	ASF1	H3-H4 chaperone	H3K56ac
Chromatin remodeler	RSC	SWI/SNF remodeling in gene body	H3K14ac
Chromatin remodeler	CHD1	Maintains gene body nucleosome organization	FACT, DSIF
Chromatin remodeler	NURF	ISWI remoderling at promoter	GAGA factor

Poly(ADP-ribose) polymerase	PARP	Transcription independent nucleosome loss	Tip60
Polymerase-associated factor complex	PAF	Loading dock for elongation factors	SEC, FACT
Histone tail modifiers	MOF	Acetylates H4K16 and recruits Brd4	H3S10ph, 14-3-3
	TIP60	Acetylates H2AK5 and activates PARP	
	Elongator	Acetylates H3 and facilitates nucleosomal elongation	Also in cytoplasm
	Rpd3C (Eaf3)	Deacetylates and inhibits spurious initiation in gene body	H3K36me3
	SET1	Methylates H3K4	MLL/COMPASS
	SET2	Methylates H3K36 and regulates acetylation-deacetylation cycle	Rpd3C
	PIM1	Phosphorylates H3S10 and recruits 14-3-3 and MOF	
	RNF20/40	Monoubiquitinates H2BK123 and facilitates nucleosomal DNA unwrapping	UbcH6, PAF

I will mention here only one aspect of this cooperation of multiple factors. Transcription is an essentially rhythmical performance, with various sorts of pauses along the way. (Again, dynamic sculpture, or dance!) One pause of great significance occurs after RNA polymerase has just begun transcribing DNA but before it has fully separated from the pre-initiation complex. The factors that influence whether transcription will continue at this point — or remain paused for an extended period — play a large role in the regulation of gene expression.

But once that first pause is ended, the elongation journey often continues to be marked by a series of further, generally briefer pauses. These have to do, at least in part, with the need to disengage DNA from its intimate mutual embrace with certain constituents of chromatin (histone complexes, about which we will learn more below). The polymerase has various assistants to aid in this disengagement, which may involve disassembly of the protein complexes. Typical of chromatin in general, these histone complexes are rich repositories of regulatory information, so they will need to be reassembled behind the transcribing complex, and the remarkably nuanced meanings embodied in their composition and structure will somehow have to be preserved, reestablished, or modified.

So the rhythm of pauses depends, at least in part, on the polymerase's helper molecules and on the positioning of certain protein complexes along the double helix, both of which will vary from one gene to another and even from one time to another. All this, and not just the so-called genetic code as such, shapes the functional significance of the DNA sequence within its chromosomal context. As we will see shortly, different versions of a protein may be produced, depending on the timing of the pauses.

Shaping a significant end

Finally — and mirroring all the possibilities surrounding initiation of gene transcription — there are the issues relating to its termination. Again, they are far too many to mention here. Transcription may conclude at a more or less canonical terminus, or at an alternative terminus, or it may proceed altogether past the gene locus, even to the point of overlapping what, by usual definitions, would be regarded as a separate gene farther "downstream." The cell has great flexibility in determining what, on any given occasion, counts as a gene, or transcriptional unit.

The very last part of the transcribed gene is generally non-protein-coding, but nevertheless contains great significance. Examining this region in a single gene, one research team identified "at least 35 discrete regulatory elements" to which other molecules can bind (Kristjánsdóttir, Fogarty and Grimson 2015). Importantly: additional dramatic and diverse regulatory potentials arise from the customized "tail" that the cell commonly adds to the end of an mRNA *after* its transcription from DNA. The regulatory processes called into play by this tail can affect everything from the stability of the mRNA to its cellular localization and the efficiency of its translation into protein. It can even play a role in determining exactly what protein will ultimately be produced. And the patterns of these added tails tend strongly to differ from one tissue type to another. "Decisions" yet again.

Much of this post-transcriptional regulation is accomplished by proteins and other molecules that bind, not only to the end, but also to the various regulatory sequences at the *head* of the RNA transcript. It all occurs in a context-sensitive manner, where cell and tissue type, phase of the cell cycle, developmental stage, location of the transcript within the cell, and converging environmental factors, both intra- and extra-cellular, may all play a role.

But it's not only the RNA *sequence* that provides opportunities for management by the cell. The three-dimensional, folded *structure* of the RNA molecule offers boundless occasion for further regulation. So here, as with DNA, we find gene expression to be in part a matter of sculptural performance. And, again, it is not just a matter of static form, but of movement. According to molecular biologists at the University of Michigan and Duke University, "RNA dynamics play a fundamental role in many cellular functions":

> [There are] many structural maneuvers that occur over timescales ranging from picoseconds to seconds ... These transitions include large-scale secondary-structural transitions at [greater than tenth-of-a-second] timescales, base pair/tertiary dynamics at microsecond-to-millisecond timescales, stacking dynamics at timescales ranging from nanoseconds to microseconds, and other "jittering" motions at timescales ranging

from picoseconds to nanoseconds. RNAs often harness multiple modes to achieve complex "functionality" (Mustoe et al. 2014).

From genetics to epigenetics

"Epigenetics" refers to that which is not genetics as such, but rather is "added to," or "on top of" genetics. You might therefore think that the transcription factors, RNA polymerases, and other proteins mentioned above, which are not themselves *genetic* elements, would be treated under the heading of epigenetics. Oddly, however, this has not been the case. Presumably, the reason is that these factors have for so long been taken for granted as if they were mere adjuncts to the "controlling logic" of DNA sequences.

But this never made much sense. What I have tried to suggest in my descriptions above is that these "mere tools" are more and more being recognized as participants in a dynamic communal context out of which alone our genes come to disciplined expression according to the needs of each cell.

Now, however, it is time to approach — albeit with painful brevity — what is generally considered the epigenetic mainstream. After all, we now know that gene transcription is merely a small part of all the activity shaping gene expression. The many processes "on top of" transcription are fully as rich and multifaceted as the various features of transcription itself.

We have already heard about *RNA splicing*, which we looked at in Chapter 8, "The Mystery of an Unexpected Coherence." As we learned in that chapter, cells don't just passively accept the RNAs that emerge from the transcription process, but rather "snip" them into pieces and "stitch" (splice) some of the pieces back together, while leaving others aside for purposes both known and unknown. It happens that these operations typically begin before the RNA is fully transcribed, and the rhythm of pauses by RNA polymerase during elongation influences which pieces are chosen for the mature transcript.

For the vast majority of human genes the splicing operation can be performed in different ways, yielding distinct protein variants (often called *isoforms*) from a single RNA. It would be hard to find any major aspect of human development, disease etiology, or normal functioning that is not dependent in one way or another on the effectiveness of this liberty the cell takes with the products of its gene sequences.

But RNA splicing is hardly the end of it. Through *RNA editing* the cell can add, delete, or substitute individual "letters" of the RNA sequence.[14] Or, leaving the letters in place, the cell can apply over 170 distinct chemical *modifications* to them.[15] Both the editing and the modifying are major topics in themselves, but not ones we can linger on here.

MicroRNAs: a large world of tiny regulatory factors

An entire, diversified area of research involves small, non-protein-coding RNAs. There are many different kinds of noncoding RNAs, but the only ones we will discuss here are known as *microRNAs* (miRNAs), which are generally derived through the cleaving and processing of longer RNAs. A microRNA commonly joins forces with a large protein complex, called the *RNA-induced silencing complex* (RISC). The microRNA guides the RISC to specific mRNAs by means of (sometimes only rough) base pair complementation. (See "base pair complementarity" in the online glossary at https://bwo.life/mqual/glossary.htm#base_pair.) Once a target mRNA is located, the RISC can cleave or otherwise degrade it, or else block its translation. In this way a typical microRNA can degrade or tune the amounts of a considerable number of different mRNAs.

Such degradation is an example of *RNA decay* in general, for which there are many different, interwoven pathways in cells. It is easy to overlook the fact that decay is fully as important — and fully as much in need of careful regulation — as the production of the RNA in the first place. During development, for example, cell differentiation would be impossible if the RNAs and proteins appropriate for an earlier form of a cell could not be recycled. In this way their constituent nucleotides or amino acids can support synthesis of new RNAs and proteins necessary for the cell's forthcoming, more differentiated form. Such a refocusing of energies may be required by any changing conditions that require fresh responses from the cell.

MicroRNAs are key fine-tuners of the relative numbers of mRNAs in a cell under any given circumstances — and therefore also of the relative numbers of various proteins. We can only wonder how the microRNAs are "instructed" by the larger context so as to "know" what these relative numbers ought to be. But we do know some of the means employed.

One of the current stories about the role of microRNAs in regulating gene expression points to a complexity almost beyond all hope of detailed understanding. Evidence suggests that just about any RNA in the human body can help to regulate any number of other RNAs, just as it in turn is regulated by them. This intertwining of fates is due not only to the competition for resources (an extremely abundant RNA, by monopolizing the available amino acids in a cell, can make it more difficult for other RNAs to be translated into protein), but also to the impact of microRNAs. Here's one way it works:

Many protein-coding RNAs are densely covered with binding sequences for microRNAs, so that a typical microRNA will find about 200 different RNA species it can target for decay or modification. This means that if a particular RNA is being highly expressed — and all the more if it is a "microRNA sponge" possessing multiple binding sites for a specific microRNA — it can have the effect of up-regulating other RNAs that are targets for the same microRNA. It "soaks up" most of the microRNAs that might otherwise degrade those targets. The RNAs that in this way regulate other RNAs by competing for shared microRNAs are known as "competing endogenous RNAs" (ceRNAs).

One research group (Tay, Rinn and Pandolfi 2014) traced the relations among a small network of twelve ceRNAs, which included the RNAs, PTEN (derived from the *PTEN* gene) and PTENP1 (derived from the *PTENP1* gene). PTEN, when translated, yields a protein that is, among other things, a tumor suppressor. (It also appears to facilitate cell migration, and to play a part in the adhesion of cells to each other.) PTENP1, on the other hand, is an RNA derived from a so-called "pseudogene," assumed to result evolutionarily from a mutational duplication of the *PTEN* gene, followed by further mutations compromising its protein-coding function. Pseudogenes are one more example of those many DNA elements, once written off as nonfunctional "junk," which are now being "caught in the act" playing important roles.

In the present case, we know at least one role for *PTENP1*. Its RNA may be incapable of being translated into protein, but it nevertheless shares many microRNA binding sites with the PTEN RNA. By sequestering those microRNAs away from PTEN, PTENP1 allows the tumor-suppressor to be expressed at proper levels. If, on the other hand, the pseudogene becomes dysregulated for some reason so that PTENP1 is not produced, then microRNAs that would otherwise bind to PTENP1, end up instead binding to, and repressing, PTEN, which reduces its tumor-suppressing activity. It has in fact been shown that PTENP1 functioning is selectively lost in certain human cancers, consistent with its importance as a microRNA sponge.[16]

And yet, the situation is actually much "worse" than I have so far indicated. MicroRNAs can also regulate other microRNAs, whether by direct targeting or, indirectly, by targeting transcription factors or regulators of those other microRNAs. For example, one particular microRNA (known as miR-499) was shown not only to regulate target genes (via their mRNAs) in the usual way, but also altered the expression of 11 other miRNAs. These changes resulted in 969 down-regulated genes, only 7.8 percent of which were directly targeted by miR-499. In other words, "hundreds of genes may be altered in expression" via these indirect pathways radiating from a single microRNA (Hill and Tran 2021).

Here we see the same obstacle to any straightfoward causal understanding that we encountered above regarding transcription factors activating or repressing other transcription factors. Tracking the mutual, broad-scale, and often subtle interactions where "everything seems to be affecting everything else" will presumably challenge researchers for a very long while. It looks like a classic picture of the unanalyzable holism of all cellular processes. All the other interwoven aspects of gene regulation discussed in this chapter, when added together, only further complicate the problem of unanalyzability.[17]

DNA methylation

Some epigenetic processes profoundly implicated in gene expression transform the DNA sequence itself. That is, they modify the nucleotide bases ("letters") of the so-called "genetic code." One of these processes, known as *DNA methylation*, is extremely important for gene regulation.

DNA methylation is the addition of a methyl group (with chemical formula –CH3) to certain DNA bases. There are four different bases in DNA, and the one most commonly methylated is cytosine. In its methylated form, this has been referred to as the "fifth base of DNA." Millions of bases throughout the genome are selectively and dynamically methylated in the cells of normal human tissues. The difference between a methylated and unmethylated base is hardly less significant, in its own way, than the difference between one base and another. But, unlike the general rule for the "raw" sequence of DNA bases, the methylation of those bases can be altered during development and in response to environmental influences. In this sense, much of our DNA inheritance is not at all the fixed-once-and-for-all destiny it is so often taken to be.

An "attached" methyl group is said to "tag" or "mark" the affected base. However, words such as "attach," "tag," and "mark" are grossly inadequate, suggesting little more than an annotation in the margin of a text, or a digital label on an otherwise unchanged entity. But in fact what DNA methylation gives us is chemical transformation — the metamorphosis of many millions of letters of the human genome under the influence of pervasive and incompletely understood cellular processes. And the altered balance of forces — the modulation of chemical, electrical, and sculptural qualities of chromosomes — resulting from all these chemically transformed bases, certainly plays with endless possible nuances into the expression of our genes.

We have been learning about the extreme consequences of these metamorphoses. In the first place, the transformations of structure brought about by methylation can render DNA locations no longer accessible to the protein transcription factors that might otherwise bind to them in order to activate nearby genes. On the other hand, by changing the local physical properties of the double helix, methylation "is observed to either inhibit or facilitate [DNA] strand separation, depending on methylation level and sequence context" (Severin et al. 2011). This has a direct effect on gene expression — for example, because strand separation is essential for the work of the polymerase that transcribes DNA.

Many proteins that recognize and bind specifically to methylated sites are then able to recruit other proteins that restructure and functionally alter the chromatin — for example, condensing it in a manner conducing to gene repression throughout an entire chromosomal region.

It would be difficult to overstate the pervasive role of this epigenetic factor in the organism. Stephen Baylin, a geneticist at Johns Hopkins School of Medicine, says that the silencing, via DNA methylation, of tumor suppressor genes is "probably playing a fundamental role in the onset and progression of cancer. Every cancer that's been examined so far, that I'm aware of, has this [pattern of] methylation" (quoted in Brown 2008). In one study among various others — a study of colorectal cancer tissues — the researchers identified 1549 genomic regions with methylation patterns differing from the patterns in similar, non-cancerous tissues (Wei et al. 2016). There are often many more methylation anomalies in cancerous tissues than there are mutated genes.

In an altogether different vein, researchers have found that "DNA methylation is dynamically regulated in the adult human nervous system." Distinctive patterns of DNA methylation are associated with Rett syndrome (a form of autism) and various kinds of mental retardation. Changing patterns of methylation also figure in aging, and constitute a "crucial step" in memory formation (Miller and Sweatt 2007).

Among many other things, DNA methylation appears to play a key role in tissue differentiation; in the activation (rather than only the repression) of gene transcription; and in the regulation of alternative RNA splicing. And, as by now we might expect, DNA methylation itself is regulated by processes converging from all corners of the cell and larger context.

The nucleosome: a complex marriage of DNA and protein

Nothing more vividly illustrates the cell's dynamic and transformational "embrace" of its DNA than the thirty million or so *nucleosomes* that form the main bulk of human chromosomes. Each nucleosome consists of several histone proteins complexed together in a *core particle*, around which various other proteins help to bend and wrap the rather stiff DNA double helix. The DNA circles the core particle approximately twice and is (more or less) held in place there, largely by means of electrostatic forces and hydrogen bonding. It is time to focus on this remarkable protein-DNA complex — a complex that, for all its centrality, scarcely figures in the broader public understanding of genetics.

Figure 14.6 is an electron microscope-derived image obtained in the 1970s by the discoverers of the nucleosome, Ada and Donald Olin, who were then researchers at the University of Tennessee and Oak Ridge National Laboratory.

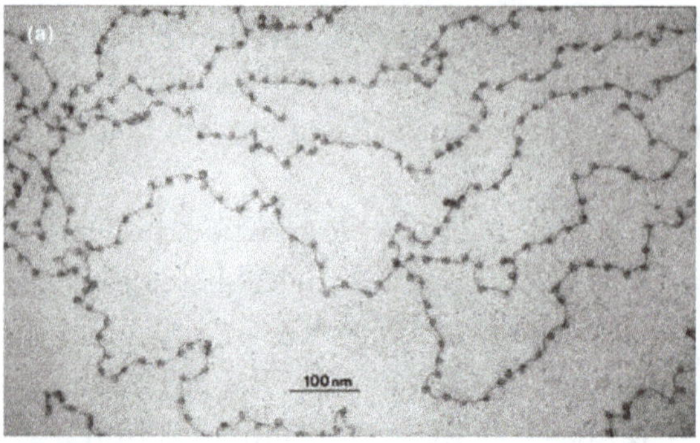

Figure 14.6. DNA (black "string") and nucleosomes ("beads" on the string), as imaged by an electron microscope.[18]

You can see the nucleosomes as "beads" along the string-like DNA.

A nucleosome most commonly consists of eight histone proteins (two copies of each of four histones, known as H2A, H2B, H3, and H4). The two stretches of *linker DNA* at the entry and exit points of the nucleosome are often held together by a *linker histone* (H1). The latter plays a role both in influencing how the DNA is bound to the core particle and also in managing the packing together of neighboring nucleosomes.[19] (See the cartoon representation in Figure 14.7.)

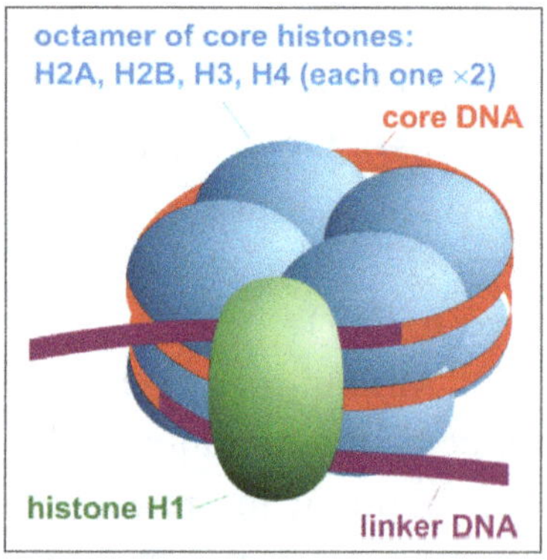

Figure 14.7. A schematic representation of a nucleosome, together with the linker histone (H1) and the encircling DNA.[20]

I referred earlier to the challenge of packing all the DNA of a cell into the space of the nucleus. As it happens, nucleosomes play a large role in this packing. Depending on their arrangement, which varies with the context, they help to organize the DNA molecule into a fiber that is said to be anywhere from (roughly) 1/5 to 1/50 of the uncondensed length. Something like 75 percent of our genome is wrapped up in nucleosomes, and a typical gene will have scores of nucleosomes within its body. This radically alters the popular image of a chromosome as a vast, uninterrupted length of the spiraling double helix.

Figure 14.8 shows (again in cartoon form) nucleosomes with and without linker histones, as well as the varying degrees of DNA compaction that can be achieved with the aid of nucleosomes.

"Ribbon" images of the nucleosome core particle, as in Figure 14.9, though highly schematic, are intended to signify certain abstract features of the histone protein structure. The DNA encircling the histones is shown, cartoon-like, in purple.

And yet again, though still with extreme artificiality in terms of the visual image, we have representations such as Figure 14.10, which are generated using data from sophisticated molecular imaging techniques. The red, white, and blue stick figure represents the DNA encircling (about one and two-thirds times) the histone core particle. Red and blue patches on the core particle represent acidic and basic areas, respectively. These, via their effect on the distribution of electrostatic charge over the surface of the histones, have a bearing on many of the functional aspects of the nucleosome discussed below.

Here it is well to remember one of the primary lessons of twentieth-century physics: we are led disastrously astray when we try to imagine atomic- and molecular-level entities as if they were tiny bits of the stuff of our common experience. It would be far better to think of the core particle's "substance," "surface," "contact points," and "physical interactions" as forms assumed by mutually interpenetrating forces in their intricate and infinitely varied play.

In particular, as geneticist Bryan Turner of the School of Cancer Sciences at the University of Birmingham (UK) reminds us, the nucleosomal core particle "is much more flexible than the crystal structure [which is the basis for images like Figure 14.10] might lead us to believe," and our current understanding of it "does

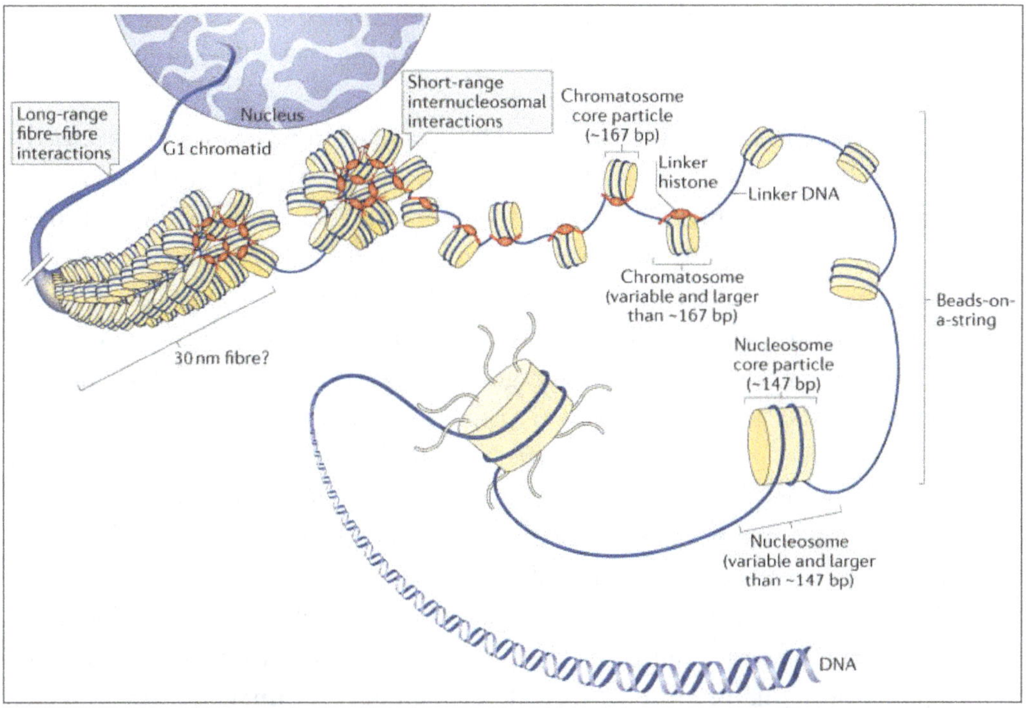

Figure 14.8. Levels of chromatin folding and compaction. Here the "chromatosome core particle" refers to the nucleosome core particle with linker H1 added. (However, all such histone-plus-DNA configurations can still be referred to as "nucleosomes.") The abbreviation "bp" refers to nucleotide base pairs, so that "167 bp" and "147 bp" refer to the approximate length of DNA wrapped around nucleosomes with and without linker histones, respectively. DNA is ever more fully compacted as the nucleosomes are packed more tightly together. For simplicity, DNA-bound proteins other than histones are not shown. Also, only histone-DNA associations on a single chromatin fiber (chromosome) are depicted here, not associations among different chromosomes.[21]

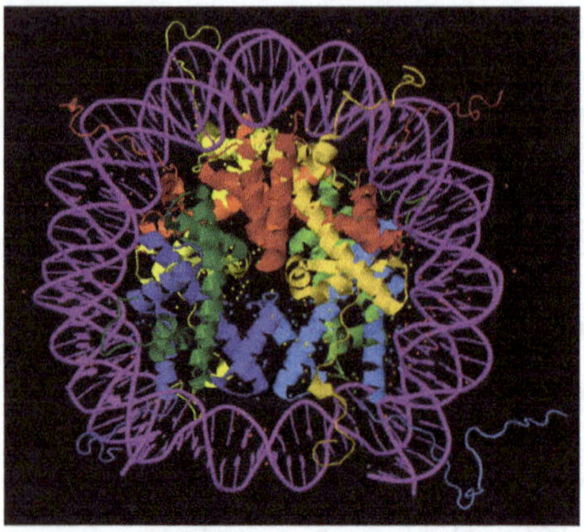

Figure 14.9. A "ribbon" representation of nucleosome structure.[22]

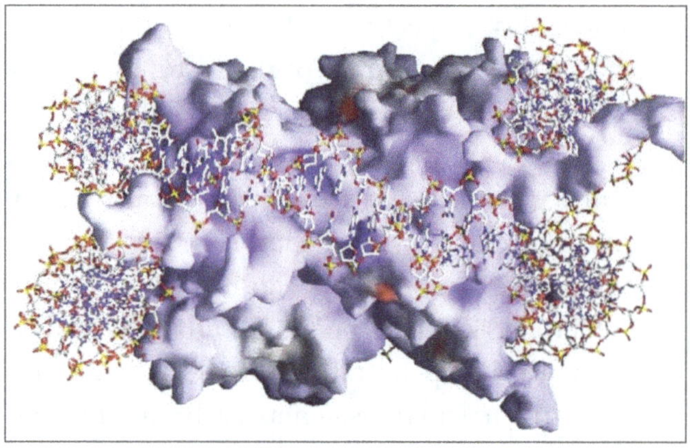

Figure 14.10. Yet a different way to represent the structure of a nucleosome. See main text.[23]

not lend itself to simplifying generalisations" (Turner 2014). As we will see, the impressive enactments of form and force about the nucleosome are central to any understanding of gene function.

Every "thing" in biology is really an activity, or is caught up in activity, and the extraordinarily dynamic nucleosome is no exception. For example, nucleosomes are the primary feature of chromatin that, as we noted earlier, must be disassembled, or at least "remodeled," during gene transcription, and then restored to a fully functional state after the transcribing enzyme (RNA polymerase) has passed by.

More generally, the individual histones in a nucleosome can come and go at an almost alarming rate — with an average exchange time of just a few minutes for many nucleosomes. And in some situations the histones exchanged in this way can be *different* histones — known as "histone variants" — with each variant exerting its own distinct sort of influence on gene expression and chromatin dynamics. Individual histones can even be removed from a core particle altogether, leaving it "incomplete" and now with seriously altered function.

Further: in the course of its life the cell can, and does, reposition huge numbers of nucleosomes along the double helix, bringing to bear upon them a whole galaxy of regulatory interactions. The positioning of nucleosomes — which may be achieved by protein complexes that slide the DNA around the core particle — matters at a highly refined level: a shift by as little as two or three bases (two or three "letters" of the "genetic code") can make the difference between an expressed or silenced gene (Martinez-Campa et al. 2004). (Individual genes typically contain thousands of bases.)

Still further: not only the exact position of a nucleosome along the double helix, but also the precise *rotation* of the helix in its embrace of the histones is important. "Rotation" refers to which part of the DNA double helix faces toward a histone surface and which part faces outward. Depending on orientation, the nucleotide bases will be more or less accessible to the various gene-activating and repressing factors that recognize and bind to specific sequences.

This in turn relates to the fact that there are two grooves (the *major* and *minor* grooves) running the length of the double helix (Figure 14.11). Proteins that recognize a particular sequence of nucleotide bases typically do so in the major groove, where the sequence is most readily accessible.

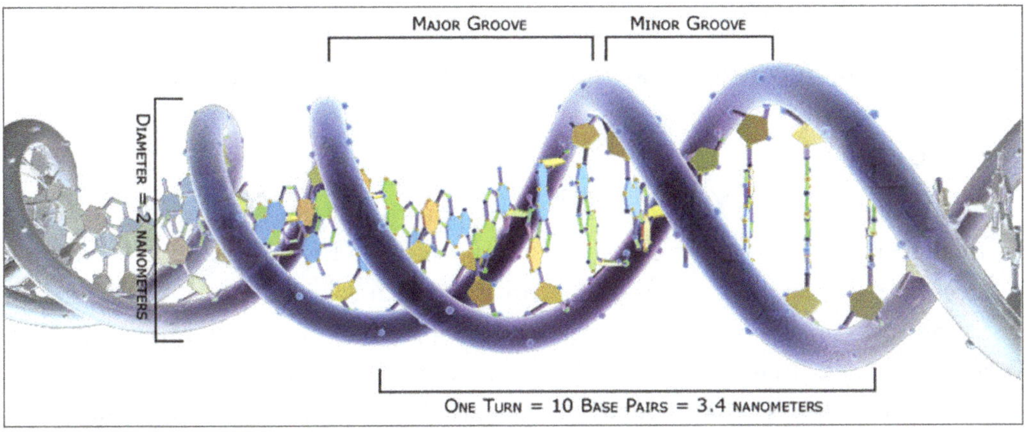

Figure 14.11. A schematic representation of the DNA double helix, showing the major and minor grooves.[24]

However, many proteins bind to DNA in highly selective ways that can be determined by factors other than the exact DNA sequence. For example, investigations have shown that the minor groove may be compressed so as to enhance the local negative electrostatic potential. Regulatory proteins "read" the compression and the electrostatic potential as cues for binding to the DNA. The "complex minor-groove landscape" (Rohs et al. 2009) is indeed affected by the DNA sequence, but also by associated proteins. Regulatory factors "reading" the landscape can hardly do so according to a strict digital code. By our musical analogy: it's less a matter of identifying a precise series of notes than of recognizing a melodic and harmonic motif performed by a full orchestra.

You can see, then, why one molecular biologist has referred to the "bewildering array of molecular mechanisms that have evolved to alter the physical properties of nucleosomes" and thereby to play a role in gene regulation (Cosgrove 2012). Also consider this:

> Influences such as DNA methylation, posttranslational modifications of the core histone proteins, histone variants, [histone gene] mutations

> and the level of chromatin compaction may each contribute to a multitude of additional energy states within the chromatin network. All these factors can potentially alter intra- and internucleosomal forces and establish a different or more extended ensemble of nucleosome conformational states, and therefore further fine-tune the functional activities. This is consistent with the notion of a heterogeneous population of nucleosomes within chromatin, all in a dynamic state and able to respond to continuous changes from environmental ques [sic] (Joshi et al. 2012).

But our story of nucleosome-based regulation has so far been radically incomplete.

A tale of tails

We will now look more closely at those parts of the nucleosome where it may be that the most dramatic story unfolds. Refer back to Figure 14.9, representing a nucleosome. The eight histones of the core particle are shown as a ribbon diagram, with the DNA double helix (schematically depicted in purple) wrapped around it somewhat less than two times. You will note a number of thin yellow, red, blue, or green "pig's tails" extending outward from the core histones. These are the thin, flexible, and mobile *histone tails*, ten of which are present in the typical core particle. There are hundreds of distinct chemical modifications of these tails (referred to as *post-translational modifications*), and the countless resulting patterns of modification within any given nucleosome or group of nucleosomes are intimately bound up with the expression of genes. In fact, there is little relating to gene regulation, DNA replication, chromatin structure and dynamics, or the overall functional organization of the nucleus that is not correlated in one way or another with patterns of histone tail modifications.

Learning about these tails, we may be reminded (albeit in a highly fanciful manner) of both the sensory functions of insect antennae and the motor functions of limbs. On the "sensory" side, the tails are receivers of molecular signals coming from all directions in the form of post-translational modifications. The nucleosome provides a context where the integrated significance of these signals can be "read off" (to use the standard phrase) by the gene-regulatory proteins that are sensitive to them. These readers may then "recruit" (again standard usage) various other proteins that either help to restructure chromatin in one way or another, or more directly regulate the expression of genes.

There are in fact many protein "readers" that interact with single modifications, or with groups of them, or with the asymmetrically modified tails of a histone pair, or with a histone modification in proximity to a site of DNA methylation. Every such reader protein acts out of its own world of biochemical genesis, folding, post-translational modification, and conformational plasticity, and together these proteins tell an important part of the story of gene regulation.

Finally, the tails can also act with a kind of brute force as "muscular" effectors. They can, for example — no doubt depending at least in part on their various modifications and protein associations — insinuate themselves into one of the grooves of the double helix, thereby loosening the DNA from the nucleosomal core particle (and making it more available for transcription), or else binding it more tightly. In both cases, one way this is accomplished is by altering the electrical interaction between histone and DNA.

Some of those tails are also thought to establish nucleosome-to-nucleosome contacts, helping to compact a stretch of chromatin. How and whether this is done can make genes either more or less accessible for transcription and various forms of regulation.

Perhaps you can now see why the members of one research team, writing about histone tail modifications, find themselves reflecting upon

> the incredibly intricate nature of the chromatin landscape and resultant interactions. The biological consequences of [interactions between histone tail modifications and regulatory proteins] are highly context dependent, relying on the combinatorial readout of the spatially and temporally fluctuating local epigenetic environment and leading to a highly fine-tuned [regulation] of particular genomic sites (Musselman et al. 2012).

A still closer look

We have progressively magnified our field of view by shifting from the overall structure of chromatin, to the nucleosome with its histone core, and then to the individual histone tails. Important principles of gene regulation operate at each different level. Now, magnifying our view one last time, we will home in on a single histone tail modification. The most commonly discussed modifications are the *acetylation* and *methylation* of certain lysine amino acids in the tails, but there are many other kinds of modification. Here I will focus on the modification called *ubiquitination* simply because its gene regulatory roles do not seem quite as extensive (or just are not as well investigated) as those performed by some other tail modifications. This makes their description here a little more manageable.

Monoubiquitination is the "attachment" (a poor word, as I indicated above) of a single ubiquitin chemical group to a lysine amino acid of a protein. In the case of histone tails, this can be done at more than one lysine, but we will look only at the monoubiquitination of lysine 120 on a tail of the histone known as H2B (that is, the lysine at the 120th sequential position along the tail), all of which can be designated *H2BK120ub1* (where 'K' is the symbol for lysine), but which will be abbreviated here as H2Bub1.

So what is the significance of this modification at a single histone tail location? Here's one summary:

> H2Bub1 takes part in almost every molecular process associated with chromatin biology. H2Bub1 has been shown to regulate transcription initiation and elongation, DNA damage response and repair, DNA replication, nucleosome positioning, RNA processing and export [from the nucleus], chromatin segregation and maintenance of chromatin boundaries. Given the large number of molecular processes regulated by H2Bub1, it is not surprising that H2Bub1 plays a vital role in some of the most fundamental biological processes that occur within multicellular organisms. [Loss of an enzyme responsible for ubiquitination] results in very early embryonic lethality. Furthermore, aberrant H2Bub1 levels can affect cell cycle progression, apoptosis ["programmed cell death"], stem cell differentiation, development, viral infection outcome and "tumorigenesis" (Fuchs and Oren 2014).

(I draw largely on the paper by these authors in the remainder of this section.)

Of course, H2Bub1 does nothing "in general"; results are always specific and context-dependent. For example, blocking this modification in a particular human cell line was found to upregulate some genes, downregulate others, and leave a great many unchanged. Under some circumstances, H2Bub1 is particularly needed for the transcription of relatively long genes. And the modification also plays an important role in histone "crosstalk," helping to regulate other crucial modifications within the same or on different histones.

A search for "effector" molecules that, singly or cooperatively, associate and interact with the H2Bub1 modification led to the identification of more than ninety proteins, many with known functions in gene regulation consistent with those known to be "effects" of H2Bub1. This points us to what could be a still further extension of our survey, whereby we might analyze one or more of those proteins. We would then have to trace the modifications *they* undergo, and the larger regulatory world in which they are caught up. But there would be no end of this, since following up any particular line of inquiry in a cell or organism sooner or later leads to everything else.

I have made repeated reference to these ever-widening circles of causal influence. Here I will just momentarily hint at this broader reality in relation to the histone tail modifications called "methylation" (not to be confused with DNA methylation). A methyl group is added to various histone amino acids by enzymes called "methyltransferases," and is removed by other enzymes called "demethylases." The mammalian genome is said to encode thirty five histone methyltransferases and twenty three demethylases. This is where the complications enter.

In an article entitled "Controlling the Controllers," the authors discuss how these methylating and demethylating enzymes are themselves modified and regulated by the addition of phosphoryl groups, with "diverse effect" on enzyme function. Further, the phosphorylation of the enzymes is in turn "regulated by upstream signalling pathways." And, still further, "different histone methyltransferase and demethylase enzyme families are connected to upstream signalling pathways in different ways" (Separovich 2020). And so the circles widen. But now we must return to our narrower focus.

It remains to mention only that, with ubiquitination as with so many other molecular biological investigations, researchers are vexed by an imagined "need to establish causality more unequivocally" (Fuchs and Oren 2014) — a need that never seems fully satisfied as our understanding grows. This search for unambiguous causes is a fruitless one (Chapter 9, "A Mess of Causes") because the kinds of causes being looked for don't exist in organisms.

As for the relations that do exist in organisms, just reflect for a moment. Think, for example, of the transcription network vaguely depicted in Figure 14.1. Then think of the networks of hundreds of mutually regulating mRNAs and microRNAs also discussed above. And now consider the virtually infinite combinations of histone tail modifications and their endlessly elaborated meanings and pervasive "crosstalk." Many other domains of gene regulation have been alluded to in preceding sections, and untold others *could* have been mentioned. And now ask yourself what all this must mean. There seem only two possibilities: complete bedlam and chaos of causes working at cross-purposes, or else the play of a coherent, unified, and encompassing wisdom whose all-embracing effectiveness and power of coordination we can hardly yet even begin to conceive.

Movement and rhythm

Few if any details of nucleosome structure and dynamics are fixed and constant. Nothing illustrates this more vividly than the fact of *DNA breathing* on the nucleosome surface. This refers to the partial and rhythmical unwrapping and re-wrapping of the double helix, especially near the points of entry and exit on the nucleosome. This provides what are presumably well-gauged, fractional-second opportunities for gene-regulating proteins to bind to their target DNA sequences during the periods of relaxation:

> Some transcription factors (TFs) only recognize nucleosomal DNA when nucleosome "breathing" occurs, that is when the DNA is partially and temporarily unwrapped from the nucleosome surface ... histone post-translational modifications facilitate DNA breathing. TF binding facilitates further nucleosome unwrapping by promoting the binding of additional TFs, and/or in coordination with chromatin remodelers. Some TFs can bind their cognate motifs on fully compacted nucleosomal DNA and initiate ATP-independent DNA unwrapping or even histone eviction. However, outcomes in which TF binding stabilizes nucleosomes are also possible (Makowski et al. 2020).

This breathing also relates to the transcriptional pausing by RNA polymerase (discussed above). The polymerase appears able to take advantage of the breathing in order to move, step by step and with significant pauses, along the genes it is transcribing. In this way the characteristics of nucleosomes — how the DNA breathes, and whether it is firmly or loosely anchored to a histone at any particular moment and place — can affect the timing and frequency of pauses. And, as we saw earlier, the rhythm of pauses and movements then affects the splicing and folding of the RNA being synthesized, which in turn bear on how the RNA can be regulated as well as the structure and function of the protein molecule produced from the RNA. A proper "music" is required for the overall performance to be successful. So it appears that the references to "choreography" and "dance" one sometimes encounters in the literature may be more than mere poetic niceties.

With a different sort of rhythm nucleosomes will sometimes move — or be moved (as I have remarked before, the distinction between "actor" and "acted upon" is forever obscured in the living cell) — rhythmically back and forth along the DNA, shifting between alternative positions in order to enable multiple transcriptional passes over a gene by RNA polymerase.

Stem cells exhibit what some have called "histone modification pulsing," which results in the continual application and removal of both gene-repressive and gene-activating modifications of nucleosomes. In this way a delicate balance is maintained around genes involved in development and cell differentiation. The genes are kept, so to speak, in a finely poised state of "dynamic and balanced readiness," so that when the decision to specialize is finally taken, the repressive modifications can be quickly lifted, leading to rapid gene expression (Gan et al. 2007).

This state of suspended readiness in stem cells also seems to be served by a rhythmical (10–100 cycles per second), back-and-forth spatial movement, or vibration, of chromatin within the cell nucleus. Associated with "hyperdynamic binding of structural proteins" mediated by nucleosomes, this vibration is thought to help maintain the largely open chromatin state characteristic of stem cells. The movement depends on the metabolic state of the cell and is progressively dampened as the stem cell differentiates into a specialized cell with substantial portions of its chromatin in a condensed state (Hinde 2012).

BOX 14.1

From Static Mechanism to Dynamic Regulator

In an article entitled "Understanding Nucleosome Dynamics and Their Links to Gene Expression and DNA Replication," Pennsylvania State University molecular biologists William Lai and Franklin Pugh concluded their review of nucleosomes this way:

> Originally viewed as a rather static mechanism of chromatin packaging, the nucleosome core complex is now well recognized as one of the key regulatory components of the genome. We also now see that instead of static protein complexes, nucleosomes are in fact exceptionally dynamic and that their positioning and composition are crucial for genome regulation. As such, the study of nucleosome dynamics is essentially the study of genome regulation. The complex interaction between nucleosome occupancy and positioning allows the cell to properly regulate accessibility of various proteins and their complexes to DNA and thus to regulate gene expression programmes. A variety of regulatory cofactors such as chromatin remodellers, chaperones and general regulatory factors operates both independently and synergistically to maintain the precise organization and composition of nucleosome arrays at specific genomic loci. This dynamic environment probably exists so that the genome may respond and adapt quickly to both external stimuli as well as be able to quickly recover from chromatin-disruptive activities such as transcription and replication (Lai and Pugh 2017).

With reference to that last sentence, it needs adding that what "responds and adapts quickly" to external and internal stimuli is not really the rather passive genome so much as the entire, all-encompassing regulatory environment, of which the nucleosome is a neat picture and summary.

But quite apart from stem cells, it is increasingly appreciated that nucleosomes play a key role in holding a balance between the active and repressed states of genes in many cell types. As the focus of a highly dynamic conversation involving histone variants, histone tail modifications, and innumerable chromatin-associating proteins, decisively placed nucleosomes can (as biologist Bradley Cairns writes) maintain genes "poised in the repressed state," and "it is the precise nature of the poised state that sets the requirements for the transition to the active state." Among other aspects of the dynamism, there is continual turnover of the nucleosomes themselves — and of their separate components — a turnover that allows transcription factors to gain access to DNA sequences "at a tuned rate" (Cairns 2009).

It is perhaps worth mentioning here that in certain bacteria a 24-hour (circadian) rhythm correlates with the changing state of DNA supercoiling — that is, with a tighter or looser twisting of the double helix. It appears that something similar may be going on in higher animals, where DNA supercoiling is so closely "wrapped up" with nucleosomes. In these organisms one of the factors involved in the extremely complex processes by which genes are regulated in a circadian fashion is the rhythmic application of histone modifications to selected nucleosomes (Woelfle et al. 2007), presumably with direct implications for chromatin structure and DNA supercoiling.

The nucleosome, we can fairly say, is a ceaselessly transforming matrix and organizational hub whose structure and pattern of activity is never exactly duplicated anywhere in the genome. It is where the infinitely ramified interface between the larger cell and its DNA comes to its most focal expression. And that expression turns out to be livingly nuanced activity, dynamic beyond what anyone imagined during the age of the double helix as the one-dimensional "secret of life."

And so, seemingly in the grip of the encircling DNA with its relatively fixed and stable structure, yet responsive to the ceaselessly varying flows of life around it, the nucleosome holds a muscular and intelligent balance between gene and context — a task requiring flexibility and a play of appropriate rhythm (Box 14.1).

Such, then, is the intimate, intricate, well-timed choreography through which our genes come to their proper expression. And the plastic, shape-shifting nucleosome in the middle of it all provides an excellent vantage point from which to view the overall drama of form and movement.

A story mostly untold

We have, in our review, only sparsely sampled the overwhelming number of causal factors participating in gene expression. The topics not touched upon here — the unmentioned domains of regulatory, or epigenetic, activity affecting what the cell makes of its genes — would extend the presentation vastly beyond the topics I have briefly alluded to here.

There is, for example, the recently intensifying exploration of the importance of modifications, not only on the histone tails, but also on the histone cores. These also are proving relevant to gene expression, and in complex ways, both direct and roundabout.

We could also have talked about the entire universe of regulation governing the translation of mRNA molecules into protein after they have been exported from the cell nucleus into the cytoplasm. The task is accomplished by complexes of protein and RNA known as "ribosomes." The diverse factors the cell gathers together for translation rival those we see in gene transcription.

And once a protein is generated, there is the problem of its folding (and re-folding), often with the help of "chaperone" proteins. Many proteins can potentially fold in an almost unlimited number of ways, yet achieving the "right" folds is crucial for protein function. This folding of a protein can begin already as it is being *translated* from RNA. Moreover, the folding outcome may be affected by the innumerable factors playing into the activity of translation. We do not often find just one thing at a time being accomplished by any biological process. (Something similar is true of RNAs. We have seen that both alternative splicing and folding of an RNA can occur — with major functional implications — during its *transcription* from DNA.)

Then, still further downstream from gene transcription, there are the various post-translational modifications (PTMs) that may be applied, removed, and re-applied to any gene-regulatory protein (transcription factors, co-activators, co-repressors, chromatin remodelers, and so on), just as we saw with the histone proteins belonging to nucleosomes. These again shape the molecule's function, often in a dynamic, ever-shifting way as the modifications come and go. Together, the many thousands of proteins subject to PTMs, and the diverse effects of these modifications, make for a vast regulatory landscape almost impossible to encompass in thought. The resulting regulatory activity is always context-dependent, relating to larger, governing purposes rather than being the mere effect of a local physical necessity.

We could also talk about what is, in one sense, the most fundamental biological activity of all — metabolism. After all, every performance of our body derives in one way or another from the food we eat. Metabolites and the organization of metabolic processes play critical roles in many aspects of gene expression related to everything from circadian rhythms to cancer.

Or we could talk about how some RNAs, especially non-protein-coding RNAs, form a "scaffolding" that gives structure to the cell nucleus and therefore plays a fundamental role in just about all nuclear functions. Except that words such as "scaffolding" and "structure" can be very misleading, as two researchers point out in a paper entitled "Role of Nuclear RNA in Regulating Chromatin Structure and Transcription." We should expect, they write, that "any nuclear structure that is assembled employing RNA cannot be static but [must be] constantly recycling degraded RNA with newly synthesised ones." So "the original concept of a static nuclear matrix must be re-evaluated in terms of a dynamic scaffold" (Michieletto and Gilbert 2019).

Perhaps the most intense and significant, newer field of research bearing on gene regulation in recent years relates to *phase transitions* in the cell, and especially in the nucleus. (See Chapter 5, "Our Bodies Are Formed Streams.") Like ice crystals forming and dissolving in water held near the freezing point, or like oil droplets in some other liquid (or like water droplets in oil), complex combinations of proteins, RNAs, and other molecules can form separated-out liquid or semi-solid aggregates (droplets) within the cellular plasm. The dynamic functional role of these aggregates in bringing molecular communities together at the right place, in the right amounts, and at the right time is now a prime topic relating to just about everything discussed in this chapter. The new understanding we are gaining in this field makes a mechanistic or deterministic interpretation of cellular physiology even less tenable than it already was.

And if any new topic of research ranks second to phase transitions in importance, it surely must be the one focusing on the role of the *microbiome.* The total DNA sequence of all the microorganisms in our bodies exceeds that of the trillions of cells in our bodies. The processess rooted in this "foreign" DNA can affect our biology, much as can the processes stemming from our own DNA. And the effects extend to regulation of our genes.

But surely it is time for us to stop. Anyone desiring a glimpse of the wider range of topics relating to gene expression might wish to scan the expanded outline of topics near the beginning of the article, "How the Organism Decides What to Make of Its Genes" (Talbott 2021).

Concluding thoughts

A decisive problem for the classical view of DNA is that a human cell employs its 20,000 or so genes to generate an estimated 250,000 to 1 million distinct proteins (Klerk and 't Hoen 2015). The activities shaping these abundant outcomes are not strictly determined by DNA. Rather, they arise from all corners of the cell and larger organism, just as the outcomes themselves — all those distinct proteins — are ushered to their proper places in every cell of every tiniest niche throughout the whole. We are always watching integral and unified performances. The idea that genes are *originating* causes that make everything else happen is grotesquely wrong-headed.

Mina Bissell, a researcher who has received many recognitions, has, along with her co-author, put the matter this way: "The sequence of our genes are [sic] like the keys on the piano; it is the context that makes the music" (Bissell and Hines 2011). We might add that the raw DNA sequence does not even contain all the keys; let's say: just the white keys. The flats and sharps, without which the music would lose its savor, are provided by DNA methylation, RNA editing, and so much more.

And Shelley Berger, the Daniel S. Och professor of cell and developmental biology at the University of Pennsylvania School of Medicine's Wistar Institute — after noting that a single histone tail modification "recruits numerous proteins whose regulatory functions are not only activating but also repressing," and that "many of these marks have several, seemingly conflicting roles" — summarized the situation this way:

> Although [histone] modifications were initially thought to be a simple code, a more likely model is of a sophisticated, nuanced chromatin "language" in which different combinations of basic building blocks yield dynamic functional outcomes (Berger 2007).

What she says about histone tail modifications could just as well be said, as we have seen, about the entire universe of gene regulation. We are looking at a meaningful, qualitative, and thoughtful language through which living narratives are constructed. In slightly different terms, Berger envisions histone modifications as participating in "an intricate 'dance' of associations."

In the plastic organism, what goes on at the local level is always shaped and guided by a larger, coherent context — a context that surely has meaning, but (as in natural languages) never an absolutely fixed grammar or logic. And, in fact, while overwhelming evidence for a meaningful, gene-regulatory conversation involving histone modifications has emerged, there is little to suggest a rigid code — this despite the strong urge in molecular biologists to find one.

The overall picture of gene expression is one of unsurveyable complexity in the service of remarkably effective living processes. What all the foregoing shows is that the whole cell and the whole organism are forever carrying out *narrative* tasks. We have no explanatory coherence so long as we are following individual chains of molecular causation. The mutually interpenetrating lines of influence converging upon and issuing from our DNA reveal their full meaning only when we consider what needs and interests are reflected in the overall, coordinated pattern of causes — *what the organism is doing and why*.

WHERE ARE WE NOW?

Gene expression: A long and winding journey

If you feel exhausted at this point, I will understand. So do I. Any effort to fully take hold of life, at any scale of observation and activity, can prove exhausting. The way in which gene expression arises from, or is disciplined by, or is made to serve, all aspects of an organism's life may be tiring to explore, even in the sorely incomplete manner of the foregoing. But taking note of the basic fact of the matter is well worthwhile. I am not at all tempted to try to summarize anew here the ground we have covered. But I will extract two statements from the text above suggesting one way to view the significance of everything we have looked at:

> (1) Given the play of infinite, interwoven influences at the molecular level, where non-mechanical fluidity rules and the number of actors relevant to just about any function of the cell or organism is unlimited, there seem only two possibilities: complete bedlam and chaos of causes working at cross-purposes, or else the play of a coherent, unified, and encompassing wisdom whose all-embracing effectiveness and power of coordination we can hardly yet even begin to conceive.
>
> (2) In the plastic organism, what goes on at the local level is always shaped and guided by a larger, coherent context — a context that surely has meaning, but (as in natural languages) never an absolutely fixed and determining grammar or logic.

These conclusions could hardly be more upsetting for a molecular biology centered on theoretical notions of code, informational logic, and discrete causes. We need not only a tracing of physical and chemical lawfulness, but also an understanding of the meaning, end (*telos*), and purposiveness of things — a hard pill to swallow for the conventionally trained biologist. But it's not as if much imagination is required in order to see which way the current is pulling us in today's deep-diving explorations of molecular biology.

We had an introduction to epigenetics (as genetics seen in context) in Chapter 7. That, together with this current chapter, as well as much else in the first half of the book will need to be kept in mind as we pass on to the discussion of evolution in the second half of the book. We will see that the main point of the older, outmoded concept of gene expression was to eliminate the life of the organism from evolutionary theorizing. If you remember what you have read here, you will have much less difficulty thinking about how organisms themselves — collectively organized in a species or population — might be the real drivers of evolution, much as the cells and microbiome, collectively in each of us, are so organized as to give adaptive expression to the life of the individual.

NOTES

1. In Chapter 8 ("The Mystery of an Unexpected Coherence") we looked at how proteins can rescue completely shattered DNA.
2. The "promiscuity" of binding — that is, binding in the absence of definitive binding sequences — is a problem relating to protein-nucleotide interactions in general. For example, 55 percent of RNA-binding proteins "do not contain any known RNA-binding domain at all" (Editors of *Nature Structural & Molecular Biology* 2021).
3. Figure 14.1 credit: from Li Wei et al. (2021), CC BY-NC-SA 4.0.
4. I will not discuss the RNA portion of chromatin here. But its functions, which researchers are now struggling to unravel, look as though they may rival the diverse functions of the protein portion.
5. No contemporary biologist has a sound basis for assuming "necessary contextualization and direction," because the idea of wise direction is foreign to the current presuppositions of biology. But every biologist, in talking about specific molecular processes, nevertheless *does* make the assumption — and makes it for the simple reason that there is no alternative. We either assume the wisely guided context or our immediate work becomes meaningless. It loses its whole point, which is to explain how one or another process contributes to a *function* or *task* — that is, to an effectively directed, purposive activity (Chapter 2, "The Organism's Story"). So biologists are forever implicitly placing themselves within a theoretical framework that, from their own standpoint, is indefensible.
6. By "modest-sized" I mean: about 2000 nucleotide bases in length.
7. Figure 14.2 credit: From Kazantseva and Palm 2014, CC BY-SA 3.0.
8. Figure 14.3 credit: From Tóth-Petróczy et al. 2008, CC BY-SA 4.0.

   The article from which the figure was taken concerns the propensity of Mediator proteins to contain "intrinsically disordered" regions. The authors conclude that "conserved intrinsically disordered regions contribute to the gene-specific regulatory function of the Mediator. Intrinsically disordered regions with weak sequence restraints can provide an evolutionarily economic solution for the Mediator to handle a steadily increasing amount of complex regulatory signals."
9. Here is one paragraph from a paper on the Mediator complex:

   > The Mediator is an evolutionarily conserved, multiprotein complex that is a key regulator of protein-coding genes. In metazoan cells, multiple pathways that are responsible for homeostasis, cell growth and differentiation converge on the Mediator through transcriptional activators and repressors that target one or more of the almost 30 subunits of this complex. Besides interacting directly with RNA polymerase II, Mediator has multiple functions and can interact with and

> coordinate the action of numerous other co-activators and co-repressors, including those acting at the level of chromatin. These interactions ultimately allow the Mediator to deliver outputs that range from maximal activation of genes to modulation of basal transcription to long-term epigenetic silencing (Malik and Roeder 2010).

Mediator also has tissue-specific aspects:

> Adding yet another degree of complexity, members of the same transcription factor family can target different Mediator subunits to activate transcription of the same gene, through the same promoter elements, in different cell types (Conaway and Conaway 2011).

10. Figure 14.4 credit: From Quevedo et al. (2019), CC BY 4.0.
11. Figure 14.5 credit: RCSB Protein Data Bank (http://www.rcsb.org), courtesy of David S. Goodsell.
12. The Wikipedia article, "Tata-binding protein" (accessed on April 1, 2019), offers a succinct description of part of this interaction: "When TBP binds to a [particular sequence] within the DNA, it distorts the DNA by inserting amino acid side-chains between base pairs, partially unwinding the helix, and doubly kinking it. The distortion is accomplished through a great amount of surface contact between the protein and DNA. TBP binds with the negatively charged phosphates in the DNA backbone through positively charged lysine and arginine amino acid residues. The sharp bend in the DNA is produced through projection of four bulky phenylalanine residues into the minor groove. As the DNA bends, its contact with TBP increases, thus enhancing the DNA-protein interaction."
13. There are actually three RNA polymerase enzymes in humans: RNA polymerase I, II, and III. I will be speaking of RNA polymerase II, which transcribes the great majority of our genes. Also, "RNA" in the following descriptions will refer either to messenger RNA (mRNA), which can be translated into protein, or else to RNA more generally. References to specific non-protein-coding RNAs such as microRNAs (miRNAs) will be flagged as such.
14. Just about any functional significance of an RNA — from what protein it produces, to its stability and cellular localization, to the various roles of its three-dimensional structure — can be affected by this editing. One kind of editing (known as A-to-I editing) "is extremely abundant in primates: over a hundred million editing sites exist in [RNAs derived from] their genomes" (Levanon and Eisenberg 2014). However, biologists have only begun to explore the functional significance of most of this editing, and there remains among the majority of researchers today a tendency to dismiss as "random noise" whatever their current methods and concepts cannot presently illuminate.
15. Frye 2018. Regarding one of these modifications, known as mRNA adenosine methylation (m6A), Timothy Nilsen, a molecular biologist at Case Western Reserve University in Cleveland, has written:

> A series of papers have appeared in rapid succession, together providing a wealth of unequivocal evidence for m6A function. But these findings still have not led to a coherent picture of the number and variety of functions of the m6A modification (Nilsen 2014).

In the years since he wrote that, the picture has, bit by bit, been filled in, and continues to be filled in. But there is a long way to go.

16. The ceRNA network we're discussing is extremely simple. The authors of the paper presenting it refer to a study of brain cancer (glioblastoma) where "the analysis was significantly extended beyond the binary ceRNA associations described in most other studies," and "the PTEN ceRNA interactions were found to be part of a post-transcriptional regulatory layer comprising more than 248,000 microRNA-mediated interactions."
17. Of course, anything can be analyzed in one way or another if we narrow our vision sufficiently and disregard, for example, the purposive (*telos*-realizing) aspects of what is going on. The question is whether analyzing living activity by breaking it into physically explicable part-processes yields an explanation or understanding of its *telos*-realizing character. Throughout this book I have been pointing out the incommensurability between a strictly physical analysis of biological phenomena and the recognizable *meaning* of those phenomena.
18. Figure 14.6 credit: Courtesy of Donald Olins.
19. An example of the functioning of linker histones: "Our results establish H1 as a critical regulator of gene silencing through localized control of chromatin compaction, 3D genome organization and the epigenetic landscape" (Willcockson et al. 2020).

    The functions of the linker histone are also indicated by the fact that "mutations in H1 drive malignant transformation primarily through three-dimensional genome reorganization, which leads to epigenetic reprogramming and derepression of developmentally silenced genes" (Yusufova et al. 2020). And then there is this: "The biochemical functions of H1 in the regulation of nuclear DNA metabolism should not be limited to a single, one-size-fits-all DNA compaction paradigm. Rather, H1 appears to be an active biochemical player in chromatin and a potent effector of multiple aspects of chromosome structure and chromatin functions" (Fyodorov 2018).
20. Figure 14.7 credit: Darekk2 (https://commons.wikimedia.org/wiki/File:Nucleosome_organization.png), CC BY-SA 3.0.
21. Figure 14.8 credit: From Fyodorov et al. 2018.
22. Figure 14.9 credit: Darekk2 (https://commons.wikimedia.org/wiki/File:Nucleosome_core_particle_1EQZ_v.5.jpg) based on data from the Protein Data Bank, CC BY-SA 3.0.
23. Figure 14.10 credit: From Luger 2006.
24. Figure 14.11 credit: Zygote Media Group (https://commons.wikimedia.org/wiki/File:3DScience_DNA_structure_labeled_a.jpg), CC BY 2.5.

SOURCES

Berger, Shelley L. (2007). The Complex Language of Chromatin Regulation during Transcription, *Nature* vol. 447 (May 24), pp. 407–12. doi:10.1038/nature05915

Bissell, Mina and William C. Hines (2011). Why Don't We Get More Cancer? A Proposed Role of the Microenvironment in Restraining Cancer Progression, *Nature Medicine* vol. 17, no. 3 (March), pp. 320–29. doi:10.1038/nm.2328

Bobola, Nicoletta and Samir Merabet (2017). Homeodomain Proteins in Action: Similar DNA Binding Preferences, Highly Variable Connectivity, *Current Opinion in Genetics and Development* vol. 43 (April), pp. 1–8. doi:10.1016/j.gde.2016.09.008

Brown, Valerie (2008). Environment Becomes Heredity, *Pacific Standard* (July 14). https://psmag.com/nature-and-technology/environment-becomes-heredity-4425

Cairns, Bradley R. (2009). The Logic of Chromatin Architecture and Remodelling at Promoters, *Nature* vol. 461 (September 10), pp. 193–98. doi:10.1038/nature08450

Chaires, Jonathan B. (2008). Allostery: DNA Does It Too, *ACS Chemical Biology* vol. 3, no. 4 (April 18), pp. 207–9. doi:10.1021/cb800070s

Conaway, Ronald C. and Joan Weliky Conaway (2011). Function and Regulation of the Mediator Complex, *Current Opinion in Genetics and Development* vol. 21, pp. 225–30. doi:10.1016/j.gde.2011.01.013

Cosgrove, Michael (2012). Writers and Readers: Deconvoluting the Harmonic Complexity of the Histone Code, *Nature Structural and Molecular Biology* vol. 19, no. 8 (August), pp. 739–40. doi:10.1038/nsmb.2350

Editors of *Nature Molecular & Structural Biology* (2021). The Nucleotides That Bind, *Nature Structural & Molecular Biology* vol. 28, no. 1 (January). doi:10.1038/s41594-020-00552-8

Frye, Michaela, Bryan T. Harada, Mikaela Behm and Chuan He (2018). RNA Modifications Modulate Gene Expression during Development, *Science* vol. 361, no. 6409 (September 28), pp. 1346–49. doi:10.1126/science.aau1646

Fuchs, Gilad and Moshe Oren (2014). Writing and Reading H2B Monoubiquitylation, *Biochimica et Biophysica Acta* vol. 1839, pp. 694–701. doi:10.1016/j.bbagrm.2014.01.002.

Fyodorov, Dmitry V., Bing-Rui Zhou, Arthur I. Skoultchi and Yawen Bai (2018). Emerging Roles of Linker Histones in Regulating Chromatin Structure and Function, *Nature Reviews Molecular Cell Biology* vol. 19, no. 3 (March), pp. 192–206. doi:10.1038/nrm.2017.94

Gan, Qiong, Tadashi Yoshida, Oliver G. McDonald et al. (2007). Concise Review: Epigenetic Mechanisms Contribute to Pluripotency and Cell Lineage Determination of Embryonic Stem Cells, *Stem Cells* vol. 25, no. 1, pp. 2–9. doi:10.1634/stemcells.2006-0383

Hill, Meredith and Nham Tran (2021). Global miRNA to miRNA Interactions: Impacts for miR-21, *Trends in Cell Biology* vol. 31, no. 1, pp. 3–5. doi:10.1016/j.tcb.2020.10.005

Hinde, Elizabeth, Francesco Cardarelli, Aaron Chen et al. (2012). Tracking the Mechanical Dynamics of Human Embryonic Stem Cell Chromatin, *Epigenetics and Chromatin* vol. 5, no. 20. doi:10.1186/1756-8935-5-20

Joshi, Sachindra R., Yaw C. Sarpong, Ronald C. Peterson and William M. Scovell (2012). Nucleosome Dynamics: HMGB1 Relaxes Canonical Nucleosome Structure to Facilitate Estrogen Receptor Binding, *Nucleic Acids Research* vol. 2012, pp. 1–11. doi:10.1093/nar/gks815

Kazantseva, Jekaterina and Kaia Palm (2014). Diversity in TAF Proteomics: Consequences for Cellular Differentiation and Migration, *International Journal of Molecular Sciences* vol. 15, no. 9 (September 19), pp. 16680–97. doi:10.3390/ijms150916680

Klerk, Eleonora de and Peter A. C. 't Hoen (2015). Alternative mRNA Transcription, Processing, and Translation: Insights from RNA Sequencing, *Trends in Genetics* vol. 31, no. 3 (March), pp. 128–39. doi:10.1016/j.tig.2015.01.001

Kristjánsdóttir, Katla, Elizabeth A. Fogarty and Andrew Grimson (2015). Systematic Analysis of the *Hmga2* 3' UTR Identifies Many Independent Regulatory Sequences and a Novel Interaction Between Distal Sites, *RNA* vol. 21, no. 7 (July), pp. 1346–60. https://rnajournal.cshlp.org/cgi/doi/10.1261/rna.051177.115

Kwak, Hojoong and John T. Lis (2013). Control of Transcriptional Elongation, *Annual Review of Genetics* vol. 47, pp. 483–508. doi:10.1146/annurev-genet-110711-155440

Lai, William K. M. and B. Franklin Pugh (2017). Understanding Nucleosome Dynamics and Their Links to Gene Expression and DNA Replication, *Nature Reviews Molecular Cell Biology* vol. 18, pp. 548–62. doi.org/10.1038/nrm.2017.47

Levanon, Erez Y. and Eli Eisenberg (2014). Does RNA Editing Compensate for Alu Invasion of the Primate Genome?, *Bioessays* vol. 37, pp. 175–81. https://doi.wiley.com/10.1002/bies.201400163

Luger, Karolin (2006). Dynamic Nucleosomes, *Chromosome Research* vol. 14, pp. 5–16. doi:10.1007/s10577-005-1026-1

Makowski, Matthew M., Guillaume G. Aullier, and Karolin Luger (2020). Picking a Nucleosome Lock: Sequence-and Structure-Specific Recognition of the Nucleosome, *Journal of Biosciences* vol. 45, article 13. doi:10.1007/s12038-019-9970-7

Malik, Sohail and Robert G. Roeder (2010). The Metazoan Mediator Co-Activator Complex as an Integrative Hub for Transcriptional Regulation, *Nature Reviews Genetics* vol. 11 (November), pp. 761–72. doi:10.1038/nrg2901

Martinez-Campa, Carlos, Panagiotis Politis, Jean-Luc Moreau et al. (2004). Precise Nucleosome Positioning and the TATA Box Dictate Requirements for the Histone H4 Tail and the Bromodomain Factor Bdf1, *Molecular Cell* vol. 15 (July 2), pp. 69–81. doi:10.1016/j.molcel.2004.05.022

Michieletto, Davide and Nick Gilbert (2019). Role of Nuclear RNA in Regulating Chromatin Structure and Transcription, *Current Opinion in Cell Biology* vol. 58, pp. 120–25. doi:10.1016/j.ceb.2019.03.007

Miller, Courtney A. and J. David Sweatt (2007). Covalent Modification of DNA Regulates Memory Formation, *Neuron* vol. 53 (March 15), pp. 857–69. doi:10.1016/j.neuron.2007.02.022

Moretti, Rocco, Leslie J. Donato, Mary L. Brezinski et al. (2008). Targeted Chemical Wedges Reveal the Role of Allosteric DNA Modulation in Protein-DNA Assembly, *ACS Chemical Biology* vol. 3, no. 4, pp. 220–29. doi:10.1021/cb700258r

Musselman, Catherine A., Marie-Eve Lalonde, Jacques Côte and Tatiana G. Kutaleladze (2012). Perceiving the Epigenetic Landscape Through Histone Readers, *Nature Structural and Molecular Biology* vol. 19, no. 12 (December), pp. 1218–27. doi:10.1038/nsmb.2436

Mustoe, Anthony M., Charles L. Brooks and Hashim M. Al-Hashimi (2014). Hierarchy of RNA Functional Dynamics, *Annual Review of Biochemistry* vol. 83, pp. 441–66. doi:10.1146/annurev-biochem-060713-035524

Nilsen, Timothy W. (2014). Internal mRNA Methylation Finally Finds Functions, *Science* vol. 343 (March 14), pp. 1207–8. doi:10.1126/science.1249340

Quevedo, Marti, Lize Meert, Mike R. Dekker et al. (2019). Mediator Complex Interaction Partners Organize the Transcriptional Network That Defines Neural Stem Cells, *Nature Communications* vol. 10, no. 2669. doi:10.1038/s41467-019-10502-8

Rohs, Remo, Sean M. West, Alona Sosinsky et al. (2009). The Role of DNA Shape in Protein-DNA Recognition, *Nature* vol. 461 (October 29), pp. 1248–53. doi:10.1038/nature08473

Separovich, Ryan J., Chi Nam Ignatius Pang, and Marc R. Wilkins (2020). Controlling the Controllers: Regulation of Histone Methylation by Phosphosignalling, *Trends in Biochemical Sciences.* https://doi.org/10.1016/j.tibs.2020.08.004

Severin, Philip M. D., Xueqing Zou, Hermann E. Gaub and Klaus Schulten (2011). Cytosine Methylation Alters DNA Mechanical Properties, *Nucleic Acids Research* vol. 39, no. 20, pp. 8740–51. doi:10.1093/nar/gkr578

Slattery, Matthew, Tianyin Zhou, Lin Yang et al. (2014). Absence of a Simple Code: How Transcription Factors Read the Genome, *Trends in Biochemical Sciences* vol. 39, no. 9 (September), pp. 381–99. doi:10.1016/j.tibs.2014.07.002

Talbott, Stephen L. (2021). How the Organism Decides What to Make of Its Genes. https://bwo.lifc/org/support/gcncrcg.htm

Tay, Yvonne, John Rinn and Pier Paolo Pandolfi (2014). The Multilayered Complexity of ceRNA Crosstalk and Competition, *Nature* vol. 505 (January 16), pp. 344–52. doi:10.1038/nature12986

Tóth-Petróczy, Ágnes, Christopher J. Oldfield, István Simon et al. (2008). Malleable Machines in Transcription Regulation: The Mediator Complex, *PLOS Computational Biology* (December 19). doi.org/10.1371/journal.pcbi.1000243

Turner, Bryan M. (2014). Nucleosome Signalling: An Evolving Concept, *Biochimica et Biophysica Acta* vol. 1839, pp. 623–26. doi:10.1016/j.bbagrm.2014.01.001

Wei, Li, Fei He, Wen Zhang, Wenhua Chen, and Bo Yu (2021). Analysis of Master Transcription Factors Related to Parkinson's Disease Through the Gene Transcription Regulatory Network, *Archives of Medical Science* vol. 17, no. 5.

Wei, Jiufeng, Guodong Li, Shuwei Dang et al. (2016). Discovery and Validation of Hypermethylated Markers for Colorectal Cancer, *Disease Markers* vol. 2016. doi:10.1155/2016/2192853

Willcockson, M. A., S. E. Healton, C. N. Weiss et al. (2020). "H1 Histones Control the Epigenetic Landscape by Local Chromatin Compaction, *Nature* (December 9). https://doi.org/10.1038/s41586-020-3032-z

Woelfle, Mark A., Yao Xu, Ximing Qin et al. (2007). Circadian Rhythms of Superhelical Status of DNA in Cyanobacteria, *PNAS* vol. 104, no. 47, pp. 18819–24. doi:10.1073/pnas.0706069104

Yusufova, Nevin, Andreas Kloetgen, Matt Teater et al. (2020). Histone H1 Loss Drives Lymphoma by Disrupting 3D Chromatin Architecture, *Nature* (December 9). https://doi.org/10.1038/s41586-020-3017-y

CHAPTER 15

Puzzles of the Microworld

ALL PHYSICAL SCIENTISTS, in an effort to understand reality, take their stand upon a tiny island of knowledge, surrounded by an immense, fathomless sea of ignorance. The island is forever threatened and re-shaped by revelatory eruptions from the surrounding deep.

According to the celebrated physicist, Richard Feynman, "we have no knowledge of what energy *is*" (Feynman et al. 1963). Nor, for that matter, do we know what a *force* is. And the same is true of all the foundational terms of physics. *Matter*, the supposedly solid ground of material reality, remains an enigma that has only grown more perplexing along with advances in quantum physics. Other basic terms such as "space," "time," and "field" — while perfectly workable as conceptual black boxes in the context of the physicist's narrow mathematical aspirations — are not themselves so much elements of adequate explanation as they are perplexities in need of understanding.

The general fact of *chemical transformation*, by which, for example, hydrogen and oxygen gases can be made (in the right proportions and in the presence of a flame) to "explode" into water, remains for our present understanding something like a miracle. The supposedly explanatory "particles" involved — which we know only as abstract, mathematized constructs altogether lacking sensible qualities — are said to rearrange themselves in an instant. According to the standard picture, the rearrangement of the qualityless particles somehow yields a radical transformation in the qualities of the reacting gases, releasing in the process a great amount of Feynman's unknown energy. And so, gaseous elements of the atmosphere, flown through by birds, transmute before our uncomprehending eyes into a fluid element of the sea, swum through by fishes.

The mysteries we confront are as great as the universe itself. Physical laws — and, in general, the rational coherence and order of the world — remain puzzles for us at least as profound as they were for Galileo and Newton. And so also with the ever-growing conundrum of human thought — that "unnatural quirk" in the universe in terms of which, it just so happens, the universe bares for us something of its most intimate nature. As for the "Big Bang," it brings no more lucidity to the question of origins than "God made it so."

It would be a stretch to think that any of our "settled" science is immune to serious reconsideration, depending on whatever revelations eventually illuminate these fundamental questions. I am not talking about a need to recalculate, say, the numerical value of the gravitational constant, but rather our understanding of the character of the physical cosmos and the manner of our participation in it as knowers. Or the significance, among scientists, of their paradoxical commitment to a materialist dogma at a time when no one can define "material" but all do consider themselves *thinkers* who take their own scientific descriptions to be both *meaningful* and true to the world's reality.

In sum: our accumulating grasp of (mostly technological) know-how, stunning as it is in practical terms, is nevertheless a power enveloped by profound ignorance. What little understanding we have of the world we so skillfully manipulate is at every moment subject to modification by whatever yet-unimagined insights may eventually bring clarity to this or that enigmatic term at the root of our science.

And yet — isn't it odd? — we find it so natural and easy to forget all this! In our primary cognitive enterprises — science, education, religion — training for the young focuses on what we already know, or think we know, rather than on our ignorance and the corresponding promise of new understanding. On my own part, I feel an obligation at least to acknowledge the largely unaddressed mysteries shadowing our understanding.

This is not your familiar Aesop

A mouse and an elephant live in fundamentally different physical worlds. The fact is evident enough in the way mice scurry around, darting this way and what, while the elephant carries its weight more slowly and deliberately. Or, to approach the matter from a very different direction: if you dropped a mouse from seven meters (twenty-three feet) above a meadow, it would likely right itself after landing and scamper away. If you dropped an elephant from that height, it would die from massive internal trauma. And if you simply left a beached blue whale where it lay, it might die from any of several different causes, one of which is being crushed under its own weight. All this has to do with the changing relation between the weight of an animal and the surface area of its body as its overall size changes.[1]

So when we talk about the diverse environments in which organisms live, one aspect of the diversity has to do with their varying experiences of the force of gravity in relation to the dimensional aspects of their lives. To be a different size is already to live in a different world.

Einstein, so it is said, was led to his theory of special relativity due in part to his having imagined what it would be like to "ride on a light beam." Might we possibly discover equally strange things if we tried to imagine what it would be like to dwell within an individual living cell?

Unlike Einstein with his task, ours would be much simpler. It would not require bold new understandings in physics, but simply a willingness to imagine the changing play, at different dimensions, of already formulated physical laws. And, fortunately, we have at least one scientific paper, written over thirty years ago, that has already done much of the work of imagining the startlingly different conditions of life at the scale of the cell.

That 1990 paper was written by Guenter Albrecht-Buehler of the Northwestern University Medical School in Chicago. He began his professional life as a physicist before moving into cell biology. However, unlike what you might expect of a physicist, one of his larger concerns was rooted in the conviction that we cannot build up an understanding of organisms by starting from the molecular level. His paper, titled "In Defense of 'Nonmolecular' Cell Biology," has not, in my judgment, received the attention it deserves. The present chapter represents my effort to summarize only that part of the paper dealing with the wildly unexpected consequences of differences of scale, and then to offer a few additional comments of my own.

Unless otherwise indicated, quotes in the following section are drawn from Albrecht-Buehler's paper.

Warning: This chapter is a bundle of contradictions. In fact, that is more or less its point. The ways we think and speak about the submicroscopic world are almost guaranteed to be impossibly off the mark, and yet anyone who would point this out has no choice but to use the established, off-the-mark language, which is the only language we currently have available. So if you begin to notice a jarring dissonance between the intended meaning and the actual language of particular statements — and I hope you will — you can take it as a sure sign that you are getting the point of the chapter.

For example, you will hear me saying that "If you considered two isolated electrons to be point masses and placed them 1 meter apart ..." You will likewise hear me talking about the "collisions" of "particles," and you will listen to a prominent cell biologist remarking how the 5 billion proteins in a cell are "jammed shoulder to shoulder, [while] also charging past one another at insanely high speeds." These references to "isolated electrons," "point masses," "collisions," and "proteins charging past one another" all seem to demand that we imagine particular *things* acting in the manner of the familiar objects of our experience.

But, as I hope you will realize by the end of the piece, there are no *things* "down there" of the sort we almost inevitably find ourselves imagining. What *is* down there is a very good question. And if you are asking it by the time you finish reading this, then the chapter will have accomplished its purpose.

From here to there – or, down the rabbit hole?

Albrecht-Buehler begins his main discussion by remarking that the size of cells "is so dramatically much smaller than the macroscopic objects we are accustomed to judging, that it is fair to say they live in an utterly alien world." The surface-to-volume ratio of a cell — a crucial consideration underlying the mouse–elephant comparison above — is 100,000 times greater for a typical cell-sized sphere than for an everyday-sized sphere with a diameter of 50 centimeters (about 20 inches). But the "alien" character we discover by imagining the life of a cell at its own dimensions goes far beyond the principle we learn by dropping mice and elephants to the ground. Nevertheless, that principle isn't a bad place to start.

From wine to jelly

Suppose we shrink a bottle full of wine to one-tenth its normal size, reducing the 2-centimeter diameter of its neck to 2 millimeters. If we now turn the bottle upside down, nothing pours out. This is, again, due to the changing surface-to-volume ratio as the size of an object (wine bottle) decreases. Given the shrinkage of the bottle, the volume (and therefore the weight) of the wine has decreased much more than the surface area of the air-wine boundary in the bottle's neck. The shaping forces[2] that hold the wine together in one compact mass at that boundary are now too strong for the reduced gravitational weight of wine in the bottle to overcome.

Figure 15.1. A dew droplet on a leaf. The droplet is about one millimeter in diameter.[3]

We see the natural tendency of such shaping forces in water when we observe tiny droplets of dew on a waxy leaf. Instead of spreading out over the leaf, the water draws itself into a roughly spherical shape. But if we instead had a ball of water 10 centimeters (4 inches) in diameter and could manage to place it on a flat surface, the water's much greater weight would overwhelm its shaping forces, so that the liquid would flow out in all directions. Only in the tiny droplets that might remain here and there would we again see the spherical, dew-drop shape we are familiar with on leaves, grass blades, and so on.

The point to attend to, then, is that change of size can result in dramatic differences in the play of forces. Of course, our wine bottle's reduction in size was not very great. Reflect now upon the fact that the volume of water in a typical cell is not 10 times, but rather 28,000 times smaller than the volume of a wine bottle. Albrecht-Buehler remarks of the non-flowing wine in the neck of the shrunken wine bottle that it appears to have become rigid, "like jelly." Indeed, "wine can turn into jelly just by existing in smaller amounts." Try to imagine the implications of that statement in light of a scale reduction by a factor of 28,000!

But your imagination would probably be seriously errant. This is because, as we will see below, there are other forces involved, and they, too, can baffle our understanding at smaller scales.

Viscous drag

A fluid's *viscosity* is a measure of its "thickness," or its internal, frictional resistance to free flow. Molasses is more viscous than water. And the more viscous the fluid, the greater the *drag*, or resistance, it presents to an object moving through it.

Albrecht-Buehler compares the effects of viscous drag upon two objects moving through water — a spherical cell, and a sphere with a 50-centimeter diameter. Both spheres are assumed to consist of the same protein matrix. He asks: If an initial movement of one diameter per second is imparted to both of them, how quickly would they come to a stop due to the resistance of the water? It turns out that the larger sphere will travel long enough to traverse many diameters. By contrast, the cell-sized sphere will stop within about a millionth of a second, during which it will have traveled about a millionth of a diameter — which is more or less to say that it stops immediately and doesn't travel at all.

This might seem to suggest that if you or I lived at the size of a cell — or, worse, a molecule within a cell — and if we wanted to take a swim, we might just as well try swimming inside a large block of concrete. But this can't really be the case, and only illustrates the difficulty of transporting ourselves in imagination to a different scale of existence. Objects like you and me — or pebbles and flowers, or the gears and levers of a machine — could not be scaled down to a sub-cellular level and still remain what they were in any meaningful sense. They would become objects of an entirely different character.

Further, molecules "live" at a radically reduced scale compared to the cell, so in moving from the whole cell to the molecular level (what I will call the "microworld"), we see the various lawful relations changing yet again. In reality, molecules move through their cellular environs (as we will see below) with remarkable speed. Moreover, despite the example above, even cells move quite well in their viscous environment. So still other factors must come into play.

Brownian movement

In 1827 the Scottish botanist, Robert Brown, used a microscope to observe tiny pollen granules, about 5 microns (5 millionths of a meter) long, suspended in water.

(For comparison, the diameter of a typical human cell nucleus is about 10 microns.) He observed a continuing series of movements — a "rapid oscillatory motion" — in what appeared to be random directions. Such movements, apparently coming from nowhere, were a considerable mystery at the time.

The motion, which gained the name "Brownian," was further characterized by later investigators. Their work confirmed three features of the movements: they were indeed random in the sense that all directions were "equally likely"; "further motion seemed totally unrelated to past motion"; and "the motion never stopped." In addition, "small particle size and low viscosity of the surrounding fluid resulted in faster motion" (Encyclopedia Britannica editors 2024).

In the early twentieth century the French physicist, Jean Baptiste Perrin, recorded the positions of three particles in water at 30-second intervals, as viewed through the microscope. His representation is shown in Figure 15.2.

Today Brownian movement is commonly visualized, however problematically,[5] as being due to random collisions ("random thermal fluctuations") of a liquid's molecules with a very small suspended object. In this sense, writes Albrecht-Buehler, the contents both within a cell and in its external, watery environment are "jerking violently." Moreover, these effects outweigh those of gravity to such an extent that collisions with just two to three molecules in a cell's environment are enough to counterbalance the gravitational weight of the cell, keeping it from sinking in water. Given the countless trillions of such impacts coming from all sides, "another way of formulating this result is to say that gravity is an entirely irrelevant force in the violently chaotic world of cells."[6]

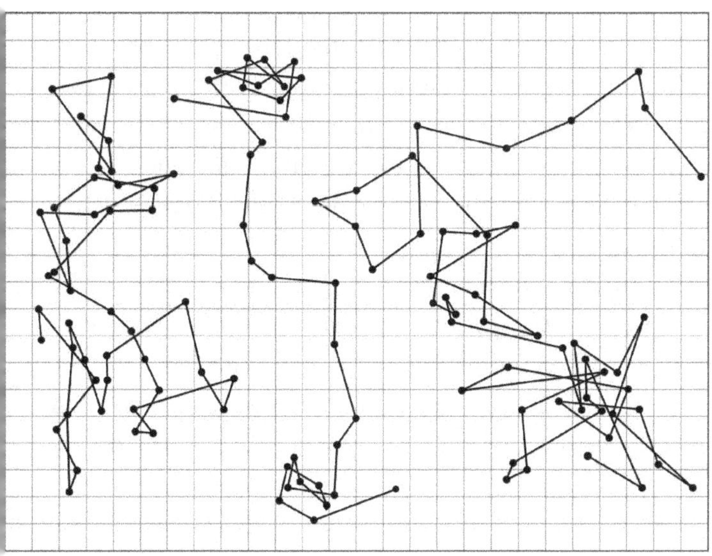

Figure 15.2. Tracings of the motions in water of three colloidal particles of radius 0.53 microns, as seen under the microscope. Successive positions every 30 seconds are joined by straight line segments. The grid lines are 3.2 microns apart. Note that the straight lines are artifacts of the fact that positions were recorded at 30-second intervals. More frequent measurements would have yielded smoother curves (but the overall movement, with its directional changes, might still be termed "jerky").[4]

Chemical energies

A cell, turbulent as it may seem from some standpoints, is actually far from being an "out-of-control" world. One good reason for this has to do with the chemical bonds between atoms and molecules. Even the weakest (hydrogen) bonds are strong enough to remain stable in the presence of Brownian fluctuations. So the making and breaking of these bonds involves the ordered direction and redirection of vast amounts of energy.

Here is one example of the use of chemical energy. A single muscle cell contains hundreds of subunits ("sarcomeres") whose dimensions are less than 3 millionths of a meter. They contract by converting chemical energy into mechanical energy. The force delivered by one sarcomere, as Albrecht-Buehler remarks, is such that "it can lift 60 entire cells! In other words, the cells submersed in violently jerking molasses of their surrounding aqueous media have literally gigantic forces at their disposal."

Electrical forces

If gravitational forces tend toward complete insignificance at the cellular level, the same can hardly be said of electrical forces. The first thing to realize is how much more powerful than gravity is the electrical force. Here is one way to think about it. If you considered two isolated electrons to be point masses and placed them 1 meter apart, there would be a certain force of gravitational attraction between them. Suppose, then, that you wanted to know where you should place them in order for the magnitude of the electrical force between them (a force of repulsion rather than attraction in this case) to be of the same magnitude as the gravitational force at 1 meter.

The answer is that you would have to separate the electrons by approximately 200,000 light years. This hardly seems believable but is, I am assured on good authority,[7] the correct answer. Two hundred thousand light years amounts to more than 34 billion times 34 billion miles. This is too much to get one's head around, so the take-home point is simply that the electrical force is inconceivably stronger than the gravitational force.

The remarkable thing is that, in most of our routine experience of the world around us, we would hardly suspect the ubiquitous presence of such monstrous forces relative to our experience of gravity. This has to do with the fact that, in the world we normally experience, the bearers of negative electrical forces, such as electrons, are more or less counterbalanced by bearers of positive electrical forces, such as protons.

The way in which charged particles naturally tend to distribute themselves gets very complex, but the upshot of it all is the following: while the electrical forces between cellular constituents are unthinkably more powerful than the gravitational forces, they don't simply rip the cell to smithereens. Here, too, negative and positive charges tend to balance each other out, but the operative word is "tend." The imbalances that do exist are enough to help account for a lot of what goes on.

Albrecht-Buehler puts the matter this way: in the molecular collectives of cells, "[charged] molecules do not notice each other until they come closer than about one-third of their diameter. Once they are that close, however, they are attracted or repelled with almost irresistible electrical forces." And again: a single electron charge within the typical electric field spanning a nerve membrane "can balance the weight of an entire cell." He goes on to mention that "cell surfaces contain thousands of electron charges."

We might also consider, not just static electrical forces, but electrical currents. Michael Persinger, the late Laurentian University (Canada) neuroscientist who investigated bioelectric phenomena in both the brain and the earth's atmosphere, was looking, not for great differences, but for close parallels between the two widely varying scales. And he found them. But even here the parallels show how differently we must think, for example, of the brain compared to our routine picturing of physiological processes.

For example, the electrical impulse traveling along the axon of a neuron is driven by what might seem to be a trivial action potential of 0.09 volts. But this voltage applies across a 10 nanometer neuronal membrane, which means that it amounts to millions of volts per meter. This is on the order of the action potential of an atmospheric lightning bolt. And the density (amperes per square meter) of the current traveling along the neuronal path is, according to Persinger, "remarkably similar" to the density of the electric current flowing in a lightning bolt.

So the reality looks rather as if our brains are continually "lit up" by countless cascading, lightning-like discharges — perhaps on the order of a billion discharges per second (Persinger 2012).

Polymerization

"One of the strangest forces that we can encounter in the world of cells that has no counterpart in our world are the forces of polymerization." We came up against polymerization in Chapter 4 ("The Sensitive, Dynamic Cell"), where we talked about the various thin filaments forming the cellular cytoskeleton. The filaments are polymers, composed of repeating protein subunits that can be added or removed at the ends of filaments in a dynamic fashion. The process of adding subunits to a polymer is called "polymerization." When a cell is migrating, some of these filaments are being extended forward (by means of polymerization) in the direction of the migration, thereby facilitating the cell's movement.

This can happen because the chemical addition of another subunit to a polymer of the cytoskeleton is an energetic process. "The force of the addition of only one [protein] subunit is ten times larger than the weight of a cell!" In theory, therefore, "adding one subunit to a polymer could lift ten cells by the thickness of the subunit." This tells us a good deal about how cells can move. At the normal scale of our lives we see nothing like this ability of a tiny unit of matter to be chemically joined to others of its kind and thereby to shift material objects (cells) that happen to be billions of times more massive than that tiny unit. (A typical human cell has been estimated to contain several billion protein molecules, in addition to water, lipids, carbohydrates, and all its other contents.)

You will recall from our earlier discussion that a dew drop on a leaf is "pulled" into a sphere by its shaping forces. (See Figure 15.1.) Further, we heard that these forces, relative to the gravitational force that might break the droplet's form and cause it to flow over the flat surface of the leaf, become vastly greater at very small scales. At the level of a cell, one of these shaping forces (surface tension) is "several thousand times larger than the weight of the cell, and we should expect the surface force to shape the cell as a perfect sphere."

The question, therefore, is why a cell is not held rigidly in the shape of a sphere (Figures 15.3 and 15.4). Cells often have all sorts of non-spherical protrusions, and some kinds of cell readily flatten themselves against a surface and slide over it. In

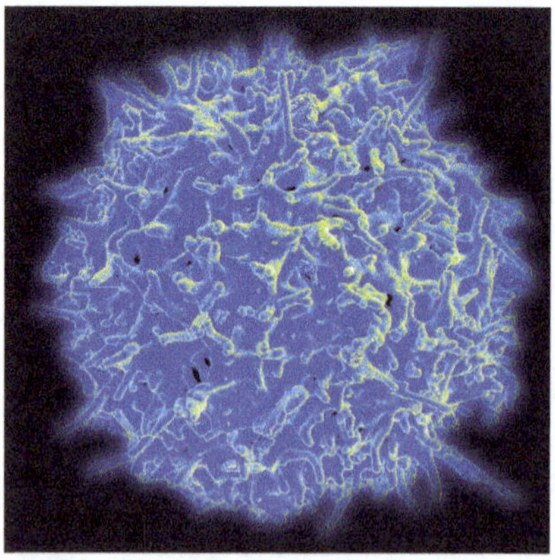

Figure 15.3. Colorized scanning electron micrograph of a human T lymphocyte (also called a T cell) from the immune system of a healthy donor.[8]

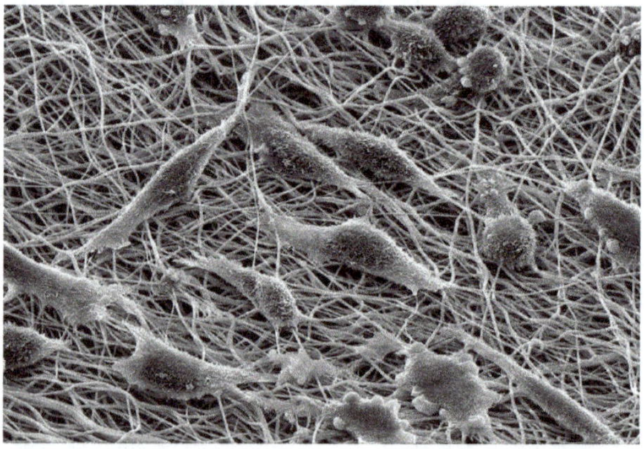

Figure 15.4. Scanning electron microscopy image of mouse fibroblasts cultured on artificial filamentous material.[9]

doing so, they are overcoming the hugely powerful shaping forces just mentioned. Part of the answer to this particular puzzle is, in Albrecht-Buehler's words, that "the surface forces are no match for the strong polymerization forces." Bundles of cytoskeletal filaments extending in a common direction have no difficulty reshaping a cell and helping to bring it into movement.

A world hard to get a grip on

How all these unfamiliar elements of the cellular world add up is not easy to picture. And it becomes even less easy when we look at some of the apparent dynamics of cellular life. "Imagine packing all the people in the world into the Great Salt Lake in Utah — all of us jammed shoulder to shoulder, yet also charging past one another at insanely high speeds. That gives you some idea of how densely crowded the 5 billion proteins in a typical cell are."[10]

Those "insanely high speeds" in crowded places are thought to explain how, as a standard textbook puts it, "a typical enzyme will catalyze the reaction of about a thousand substrate molecules every second" — meaning that the enzyme must bind to a new substrate in a fraction of a millisecond. This happens despite the fact that there tend to be relatively few substrate molecules per cell. If, for example, there is only 1 substrate molecule for every 100,000 water molecules, "nevertheless, the active site [the place where catalysis occurs] on an enzyme molecule that binds this substrate will be bombarded by about 500,000 random collisions with the substrate molecule per second." At the same time, "a large globular protein [like many enzymes] is constantly tumbling, rotating about its axis about a million times per second" (Alberts et al. 2002, pp. 77–78).

As if everything we have heard so far is not difficult enough to comprehend, the problem of imagining microworlds truthfully is greatly magnified by emerging technologies that generate seductive images. When biologists speak so casually of atoms and molecules as *things*, and when engineers then present us with "pictures" of them, we can hardly help taking the pictures as images of actual phenomena. And so they are. But the phenomena we are dealing with are not "down there." They are "up here," where we are experiencing our instruments. Those instruments may be telling us *something* truthful about the microworld, but we have to figure out what that something is.

What we derive from "down there" (at the atomic and molecular levels) is mostly mountains of data produced by our instruments. The pictures we look at are representations of that data. If we take these pictures at face value — if we unthinkingly accept them *in the same way* we accept the terms of our visual engagement with the familiar world — then we are projecting into the microworld phenomena that are not actually there.

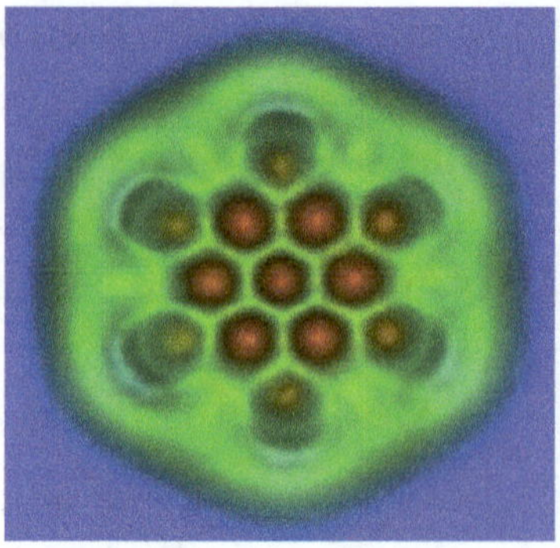

Figure 15.5. An image produced by the interaction of a non-contact atomic force microscope with graphene (a lattice of carbon atoms), in an IBM laboratory. The bright green lines forming approximate hexagons are taken to represent the molecular bonds between carbon atoms.[11]

This is a problem. If images like the one in Figure 15.5 truly represented anything like the physical objects around us, merely reduced to very small dimensions, and if billions of such objects (commonly, if nonsensically, referred to as "molecular machines") were racing around inside the cell at "insanely high speeds," tumbling around while rotating a million times per second, they would presumably achieve nothing but rampant destruction within the cell.

Figure 15.5, which is said to represent carbon atoms, is not in any normal sense a photograph of atoms, as the scientists and engineers who produce such images well know. There is no "thing" anywhere in the world that looks like this, except the picture itself. Responsible physicists do not talk about *things* at this level of observation at all. In this particular case we are looking at a kind of colored graph of a data set produced by an atomic force microscope. The spatial distribution of the artificial colors represents the relations between the highly refined measuring instrument, on one hand, and forces at an extremely small (atomic) scale, on the other. It is a picture of a distribution of forces. Forces are not things.

What then?

So what do we make of all the foregoing? It's hard to say — and maybe that itself is the important point. It is clear enough that when we imagine the world of atoms and molecules in terms of our familiar experience, we are far from truth. If we want some sort of picture, it will hardly do to conjure images of robots or sewing machines or pliers, merely reduced in size ("nanorobots") and spinning around a million times per second, or a brick beneath a skyscraper receiving an electrical charge and thereby raising the building off the ground, or molecules looking like brightly colored baubles.

The one thing we can be sure of is that the cellular realm is not composed of anything like our familiar objects, only made smaller. The really foundational question is whether, and at what scale of observation, we are justified in talking about "things" at all, as opposed to forces or potentials.[12] This question certainly bears on the common appeal by molecular biologists to machine and computer models. Regarding computers, Albrecht-Buehler has written:

> To my knowledge, there is not even a clue as to how to build a liquid miniature computer that would function despite thermal fluctuations and other turbulences in the liquid that would disrupt the circuitry (Albrecht-Buehler 1985).

There is, quite simply, nothing there that could remotely qualify as "circuitry" in the sense of familiar "machine parts."

One might have thought that the puzzling revelations from our indirect, instrument-mediated encounters with the microworld would have opened up a space for free inquiry as we considered the nature of perceptible, material appearance versus the theoretical constructs and misleading imagery through which we try to picture a realm of which we have no direct experience. One might indeed have expected that — given a microworld considered fundamental to our understanding yet inaccessible to the direct activity of our senses — we might have warned ourselves about the temptation to project falsely imagined perceptual contents into what is an experiential blank for us.

And, given the scientific commitment to empirical (experience-based) evidence, what are we to make of a microworld characterized almost solely in terms of thought-models, mathematical formulae, and theoretical constructs, with no sense experience to ground us? All we have really asked of our models and theoretical constructs — from the solid, indivisible atom of Dalton to the "solar-system" atom of Bohr to the probability clouds of our own day is that they be successful technologically, enabling us to make things (including scientific instruments) that work. And each model has in fact helped us to make things — until it didn't. This practical goal has always been the essence of trial-and-error methods, and our models have become little more than helpful tools for wonderfully expediting such methods.

But, so far anyway, it is still hard to repress the occasional question: "How much do we actually *understand*?"

WHERE ARE WE NOW?

What does the microworld have to do with us?

You may be thinking that the topic of this chapter is off the main track of the book. Perhaps so. But then again, just about the entire book is off the main track of the biological sciences today, including molecular biology and (as we will see) evolutionary biology. It is not a bad thing if this chapter encourages us to take note of the limits of our understanding. Nor is it a bad thing if scientists put themselves into a questioning mode rather than the "we understand perfectly well" mode. And certainly it is not a bad thing if, when we look at the inevitable schematic "pictures" of cells in textbooks, we realize how little we understand what we are looking at.

What we don't understand goes far beyond the issues discussed in this chapter. The question I have been posing throughout the book is this: Given the wise and well-directed coordination of all the physically lawful life processes we have ever observed, and given the fact that physical lawfulness alone provides no accounting for either the wise coordination or its end-directedness, how can we arrive at a new way of thinking about the problems of life? It hardly seems justified to ignore this question simply because it too easily invites answers that go contrary to our existing intellectual commitments.

In light of such a profound question, it is surely healthy to acknowledge how little our normal habits of thinking allow us to picture what is actually going on at the molecular level where so many have been determined to find their answers. For example, regarding what we have learned in the preceding chapter ("How Our Genes Come to Expression") and Chapter 8 ("The Mystery of an Unexpected Coherence") about the regulation of gene expression by countless molecules interacting in a fluid medium, it seems impossible to believe that we currently understand even the most basic truths about how the meaningful coordination of seemingly independent events actually occurs.

If one thing is clear, it is the implausibility of the usual fantasy of molecular-level "machines." At the very least, we can say that these could have virtually nothing in common with the machines we know of. This means that the most common way of imagining the wise and well-directed coordination of events in the microworld of the organism — by picturing something like the intelligently designed machines of our own making — is a non-starter. Of course, we have already seen (for example, in Chapter 10, "What Is the Problem of Form?") many reasons for dismissing the machine-model of organisms, quite apart from those of the current chapter.

Let this be a time for opening our minds rather than sealing them shut. And if the present chapter encourages such opening, so much the better. In any case, we will now move on to evolution. If there is any topic that demands of us an open and questioning mind, it is this one.

NOTES

1. As the size of an animal decreases, its volume (and therefore its weight) decreases much more rapidly than its surface area. In other words, as any given object is reduced in size, its surface-to-volume (surface-to-weight) ratio rises. The increased surface-to-weight ratio of the mouse is why its rate of fall is reduced by air resistance more than the elephant's rate of fall. A falling leaf is a more extreme example.

   More significantly for the fate of the mouse and the elephant in our rather twisted thought experiment, the different surface-to-weight ratios mean that the weight of the mouse per square centimeter of its body surface striking the ground is minuscule compared to the weight overlying the elephant's area of contact with the ground. So the crushing effect of the impact is much greater for the larger animal.
2. Among the interrelated shaping forces of a liquid such as water are internal cohesion and surface tension.
3. Figure 15.1 credit: Michael Apel (https://commons.wikimedia.org/wiki/File:Dew_2.jpg), CC BY-SA 3.0.
4. Figure 15.2 credit: Original observations made by Jean Baptiste Perrin. Digital rendering by MiraiWarren (https://commons.wikimedia.org/wiki/File:PerrinPlot2.svg), Public Domain via Wikimedia Commons.
5. The word "collisions" suggests an activity of *particles* conceived in the manner of our everyday experience of tiny bits of matter. Thinking of water molecules in this way is not something any physicist today would want to defend.
6. It is worth remembering that the lives of large, multicellular organisms — ourselves, for example — are not centered upon the cellular and molecular level. As we walk, run, and otherwise pursue our lives on earth, our bodies must work against the pull of gravity. If we do not sufficiently perform that work — if we are bedridden or live a sedentary lifestyle — our bodies suffer ill effects.

   We know further that the weightlessness endured by astronauts on long missions results in significant loss of bone mass, density, and strength (Keyak 2009). Likewise, lions raised in zoos, apart from the rigors and stresses of hunting and the need to patrol large territories, have a bone structure differing from lions raised in the wild (Holdrege 1998).

   So Albrecht-Buehler's assertion that "gravity is an entirely irrelevant force in the violently chaotic world of cells," while it may be true when we are looking at the interplay of forces in the decontextualized cell, can hardly be true for cells in the context of our bodies. If someone experiences changes in bone mass and muscle strength while living in a gravity-free environment, this implies radical changes in cells, including the loss (death) of cells. The fact that, when a person stands upright on earth, the weight of a 150-pound body comes to bear upon the small surface area of two feet certainly makes gravity a "relevant force" for the tissues and cells on the bottoms of our feet. And much the same can be said about the distribution of weight and weight-bearing surfaces throughout our bodies.

   Actually, the importance of a larger context was very much part of Albrecht-Buehler's argument in his paper. He was claiming, quite rightly, that we cannot explain either cellular or organismic behavior by trying to ground our picture upon decontextualized molecular-level analyses.

7. I have this answer courtesy of the physicist, George Burnett-Stuart.
8. Figure 15.3 credit: NIAID/NIH (https://commons.wikimedia.org/wiki/File:Healthy_Human_T_Cell.jpg), CC BY 2.0.
9. Figure 15.4 credit: Judyta Dulnik (https://commons.wikimedia.org/wiki/File:Fibroblasts_2.jpg), CC BY-SA 4.0.
10. Callier 2021, citing a comparison offered by Anthony Hyman, a British cell biologist and a director of the Max Planck Institute of Molecular Cell Biology and Genetics in Dresden.
11. Figure 15.5 credit: IBM Research–Zurich, CC BY 2.0.
12. Some experimental techniques do give us a form of sense-perceptible report from the microworld. For example, the relatively small "green fluorescent protein" (GFP) can be fused to particular molecules of interest in a cell. When the cell is irradiated with blue or ultraviolet light, the protein fluoresces, revealing under a light microscope the distribution of the target molecules in a cellular location. Again, however, blobs of fluorescent light, while informative of location, do not give us pictures of molecular "objects" residing at that location.

    When a student collects a quantity of DNA on a glass rod, she is not looking at DNA molecules, but rather at a white, sticky substance. Similarly, a prospector may be looking at a chunk of iron ore, but he is not examining iron atoms. To say that our instruments, by eliciting responses at an atomic scale, can trace significant structure at that scale, is not to answer in any meaningful experiential sense, "structure of what?" — not if by "what" we refer to objects of the microworld possessing sense-perceptible, material descriptions. We can legitimately relate the structure to white, sticky substance or to iron ore, but not to atomic particles imagined in the mode of that substance or that ore.

    As a hypothetical question: what would we "see" if, through some sort of inner work, we should develop in the future a cognitive (clairvoyant?) capacity to experience — bring to appearance — whatever can be found, say, at the quantum level? This is, of course, pure speculation. But my surmise is that we would discover an intricately structured play of "forces" of will. We would discover, that is, a field of potential that, when probed in appropriate ways, can be brought to manifestation as materially engaged force. The fact that our own wills (in a manner of which we are completely ignorant and unaware) can take form in the enfleshed mechanical forces of our arms and legs might be suggestive in this regard.

SOURCES

Alberts, Bruce, Alexander Johnson, Julian Lewis et al. (2002). *Molecular Biology of the Cell*, 4th edition, pp. 75–76. New York: Garland Science.

Albrecht-Buehler, Guenter (1985). Is Cytoplasm Intelligent Too?, chapter 1 in *Cell and Muscle Motility*, edited by Jerry W. Shay, pp. 1–21. Boston: Springer.

_______ (1990). In Defense of "Nonmolecular" Cell Biology, *International Review of Cytology* vol. 120, pp. 191–241. doi:10.1016/S0074-7696(08)61601-0

Callier, Viviane (2021). A Newfound Source of Cellular Order in the Chemistry of Life, *Quanta Magazine* (January 7). https://www.quantamagazine.org/molecular-condensates-in-cells-may-hold-keys-to-lifes-regulation-20210107/

Encyclopedia Britannica editors (2024). Brownian Motion (July 25). https://www.britannica.com/science/Brownian-motion. Downloaded July 29, 2024.

Feynman, Richard, Robert B. Leighton and Matthew Sands (1963). *The Feynman Lectures on Physics*, three volumes. Reading MA: Addison-Wesley.

Holdrege, Craig (1998). Seeing the Animal Whole: The Example of the Horse and Lion. In *Goethe's Way of Science*, edited by David Seamon and Arthur Zajonc. Albany: SUNY Press, pp. 213–32.

Keyak, J. H., A. K. Koyama, A. LeBlanc, Y. Lu and T. F. Lang (2009). Reduction in Proximal Femoral Strength Due to Long-Duration Spaceflight, *Bone* vol. 44, no. 3, pp. 449–53. doi:10.1016/j.bone.2008.11.014

Persinger, Michael A. (2012). Brain Electromagnetic Activity and Lightning: Potentially Congruent Scale-Invariant Quantitative Properties, *Frontiers in Integrative Neuroscience* vol. 6, article 19 (May). doi:10.3389/fnint.2012.00019

Part 2:
EXTENDING THE ORGANISM'S STORY
Toward an Evolutionary Narrative

CHAPTER 16

Let's Not Begin with Natural Selection

EVOLUTIONARY THEORISTS TEND to become frustrated when many of the rest of us fail to "get" the revolutionary and convincing simplicity of natural selection, the supposedly primary engine of adaptive evolution also known informally as "the survival of the fittest." For example, Niles Eldredge, a paleontologist and, for several decades, a curator at New York's Museum of Natural History, has wondered, "Why do physicists, who have the reputation of being among the best and the brightest, have such a hard time with the simple notion of natural selection? For simple it is." He then quotes the familiar passage from Charles Darwin:

> As many more individuals of each species are born than can possibly survive; and as, consequently, there is a frequently recurring Struggle for Existence, it follows that any being, if it vary however slightly in any manner profitable to itself, under the complex and sometimes varying conditions of life, will have a better chance of surviving, and thus be *naturally selected*.

"The concept," Eldredge writes, "is definitely simple enough. This description of natural selection may be a bit longer than the elegantly brief F=MA [force equals mass times acceleration — Newton's second law of motion]. Conceptually, however, it is hardly more complicated" (Eldredge 2000, pp. 89–90).

The simplicity of what is being promulgated as "natural selection" can hardly be doubted. In his landmark book on *The Nature of Selection*, the philosopher of evolutionary theory, Elliott Sober, considered it "remarkable that a hypothesis of such explanatory power could be so utterly simple conceptually: If the organisms in a population differ in their ability to survive and reproduce, and if the characteristics that affect these abilities are transmitted from parents to offspring, then the population will evolve" (Sober 1984, pp. 21–22).

Figure 16.1. Charles Darwin.[1]

The idea of natural selection seems to many so straightforward and conclusive that it forces its way into the receptive mind without much need for evidence. August Weismann, whose importance for nineteenth-century evolutionary theory has been considered second only to Darwin's, rather famously wrote in 1893 that we must accept natural selection as the explanation for the wondrous adaptation of organisms to their environments "because it is the only possible explanation we can conceive."

Further, according to Weismann, "it does not matter" whether we can demonstrate the role of natural selection in particular cases. "Once it is established that natural selection is the only principle which has to be considered, it necessarily follows that the facts can be correctly explained by natural selection" (quoted in Gould 2002, p. 202).

The compelling simplicity of natural selection, according to Ernst Mayr, is so pronounced as to have proven a stumbling block for many. Mayr, whose influential career spanned the entire twentieth-century history of the modern evolutionary synthesis, proposed that "startling simplicity was the most formidable obstacle that the selection theory had to overcome. Students of the phenomena of life found it undignified to explain progress, adaptation, and design in nature in so mechanistic a manner" (Mayr 1964, p. xviii).

Brief summary statements of the simple logic of natural selection abound. In philosopher Daniel Dennett's succinct formulation, "evolution will occur whenever and wherever three conditions are met: replication, variation (mutation), and differential fitness (competition)" (quoted in Lenski et al. 2003). Or, expanding the idea just a little, we might say that evolution is guaranteed to occur under three conditions:

- There must be trait *variation* among individuals in a breeding population. Without variation, nothing new could ever come about.
- This variation must to some degree be *heritable*, so that offspring generally resemble their parents more than they resemble others. (This is Dennett's principle of *replication*.) If offspring didn't tend to resemble their parents, then it's not clear how variants, even if they occurred in specific members of a population, could spread through the population as a whole.
- Individuals possessing different variants of a trait must, at least in some cases, exhibit *differential fitness* (or differential survival) — that is, they must produce, on average, different numbers of offspring, whether immediate offspring or later descendents. This is often referred to as the principle of *competition* or *survival of the fittest*. The advantageous adaptation of the fittest organisms is what gives them a better chance of surviving and contributing their fit genes to the descendent population.

> Similarly, less fit organisms will have a reduced chance of surviving and passing on their genes.

With various terminological variations, that is the textbook presentation of natural selection. According to the influential popularizer (and noted theorist) of evolutionary theory, Stephen Jay Gould, the basic idea has the simplicity of a syllogism. He referred to it as the "syllogistic core" of natural selection (Gould 2002, pp. 125–26n). For Dennett, this core is a "mindless" recipe, or *algorithm*, — one so obvious and universal that it could be derived even without reference to organisms, while nevertheless offering "guaranteed results" in biology. The algorithm is "Darwin's dangerous idea," and it is the key to making sense of everything from the simplest irritable cell to human meaning, cognition, culture, and morality (Dennett 1995, pp. 51, 163–81).

Variation, inheritance, and survival of the fittest: for a certain mindset (well-established in our day), something does indeed seem irresistible and self-evident about the way these conditions testify to the idea of change. And — Eldredge's obtuse physicists apart — more than enough students of evolution do seem smart enough to "get" the extraordinary power and simplicity of natural selection. The widely read British psychologist and science writer, Susan Blackmore, speaks for many when she says that "evolution is inevitable — if you have information that is copied with variation and selection then you *must* get [quoting Dennett] 'Design out of chaos without the aid of mind'." Blackmore goes on almost rapturously: "It is this inevitability that I find so delightful — the evolutionary algorithm just *must* produce design, and once you understand that[,] you have no need to *believe* or not believe in evolution. You see how it works" (Blackmore 2014).

This cocksureness about the simplicity, universality, and persuasive force of the evolutionary algorithm as an explanation for the complex forms of life we observe seems to know no bounds. It extends, for example, even (or especially) to computationally oriented researchers. In 2003 Christoph Adami, who was then head of the Digital Life Laboratory at the California Institute of Technology, defended the value of trivially simple and non-living "digital organisms" — bits of computer code standing in for genes and living processes — as teachers instructing us about evolution. What we learn, he said, is that the principles of evolutionary theory are "very, very general, and very simple," so that our predictions "don't depend on these little details of molecular biology" (quoted in O'Neill 2003).

It is, we may sense with a certain unease, almost as if actual phenomena become irrelevant to the researcher, who needs only to work out a simple logic.

And our sense of unease only grows when we hear Richard Dawkins discussing how some animals cleverly coerce the behavior of others. For anyone skeptical of his explanation, Dawkins had this word of encouragement: "With natural selection working on the problem, who would be so presumptious as to guess what feats of mind control might not be achieved?" (Dawkins 2008, p. 71). One almost hears an echo of the parent trying to soothe a child's perplexity about some puzzle of creation: "Surely, with God working on the problem …."

And, indeed, over-estimation of the explanatory power of natural selection may be why Darwin's contemporary, the geologist Charles Lyell, accused him of "deifying" the theory.[2] A century later, in 1971, Lila Gatlin, a biochemist and mathematical biologist who figured centrally in developing the conception of life as an "information processing system," could summarize contemporary usage by saying "the words 'natural selection' play a role in the vocabulary of the evolutionary biologist similar to the word 'God' in ordinary language" (quoted in Oyama 2000a, p. 31). Such is the power of logical constructions over the contemporary human mind.

No doubt the "evolutionary algorithm" truly is simple, and its logic, as far as it goes, is self-evident. But we might want to keep in mind how thin and unstable is the strip of intellectual real estate between "self-evident" and "vacuous" — especially when we find ourselves preferring abstract logical necessity and simplicity to superfluous "little details" (Adami), such as the difference between a computer program and the life of a tiger or octopus.

When we allow ourselves to ignore material particulars, we depart from science. We allow ourselves to be moved by the force of a discarnate logic whose "bodily substance" derives from vague and abstract mental constructs rather than careful observation of the world. These constructs reflect the presuppositions and biases of our own untethered minds much more than any truths of the organisms whose lives have disappeared from our thinking.

What are the "guaranteed results" of natural selection?

It may take a while, and it may be rather uncomfortable, to digest the anti-scientific attitudes we have just heard from leaders of science and its philosophy. But recall the substance of it. Apparently dismissing as unnecessary the role of painstaking observation in science, August Weismann declared in "medieval" fashion that what his mind had been able to conceive was "the only possible explanation" for the still largely unknown facts of evolution. Daniel Dennett was sure that it didn't require any knowledge of organisms in order to see the unquestionable truth of organisms under natural selection, an attitude seconded by Christoph Adami, with his blithe disregard of the facts ("little details") of molecular biology.

The decisive truth of the theory of natural selection, it seems, could be spun as pure thought-stuff out of the minds of its devoted theorists, and what their minds conceived in this way was, as Susan Blackmore put it following Dennett, a process that just *must* produce "Design out of chaos without the aid of mind." The hypothesis of natural selection, according to Elliott Sober, married great explanatory power to utterly simple conceptuality. Putting all this together, we might conclude that, with "startling simplicity" and independence from all the complexities of biological fact, natural selection somehow explains just about all the great issues of biology — "progress, adaptation, and design in nature" (Ernst Mayr).

It's as if we were celebrating a recipe by marveling at the extraordinarily simple procedure by which one can proceed through Steps 1 to 5 and, suddenly, behold! — automatically, simply, mindlessly, and guaranteed — a mouth-watering *chicken cacciatore*! And why not? Why not ignore the bothersome details about how all the ingredients came about — how the herbs and spices were grown, where the mushrooms and onions came from, the skills of the chef, how fresh or rotted the ingredients were, and, above all, the life, growth, health, and care (or lack of care) of the chicken. The five steps are all we need to focus on, because we just *know* they must already be the adequate explanation of a delicious chicken dish.

And, of course, they are, depending on how we contextualize them. But, considered independently from their context — for example, from the availability and quality of all the ingredients — we also recognize that they tell us very little about what, if anything, awaits our discriminating palettes.

In order to assess the foregoing claims about natural selection, I would begin by inserting two question marks in the testimonials we have heard. First, there is Darwin's statement: "Any being, if it vary however slightly in any manner profitable to itself, under the complex and sometimes varying conditions of life, will have a better chance of surviving, and thus be naturally selected."

I would insert my first question mark after the word "vary." Notice how easily the word is slipped into the flow of thought, as if it were wholly unproblematic. *Of course, organisms do vary. We know that.* How easy to forget (like forgetting the provenance of a recipe's ingredients) that the variation referred to is never the result of anything other than the unsurveyable complexity of organic processes — processes that just happen to be *intensely organized* toward certain ends. Crucially (and despite their complexity), they always manage to be consistently expressive of a particular, qualitative way of life, while differing from the expressive character of other kinds of organism. (A sloth is not like a monkey.) "Gee," we might bring ourselves to wonder. "How do they do that?"

The main thing we should notice, especially in an evolutionary context, is that all this organic activity is characteristically *future-oriented and directive* in nature (Chapter 2, "The Organism's Story)." For example, a single-celled zygote, compensating as best it can for all disturbances along the way, determinedly pursues its unique path toward the intricate, billion-celled, not-yet-realized form of a mature trout or mountain goat.

Until we understand how an organism manages to exhibit these *intensely organized, future-oriented, and directive processes*, inexplicable as they are solely in terms of their undoubted physical lawfulness, it seems irresponsible to formulate an evolutionary logic that ignores the distinctive character of these processes. After all, it is through them that viable traits and their variations come about. Do the evident creative forces in the life of organisms have nothing to do with evolution?

Nothing in Darwin's statement leads one to dwell even for a few seconds on the infinitely complex, living realities underlying the word "vary." It is as if variation were something that "just happens" to organisms for no particular reason. Merely assuming this happening in an unreflective way ("*Of course* organisms vary!") is a prerequisite for our interpreting Darwin's words as an explanation of evolution. We don't need to ask ourselves, "Out of what sort of a life is an organism enabled to manifest those extraordinary and directive powers of development, physiology, and behavior, through which variation comes about?" Nevertheless, we *can* ask the ignored question, and we can go on to wonder, "What might these powers tell us about evolution?"

My second question mark applies to Elliott Sober's comment when he is marveling at the "explanatory power" of a simple proposition: "if the organisms in a population differ in their ability to survive and reproduce, and if the characteristics that affect these abilities are transmitted from parents to offspring, then the population will evolve."

Sober's claim is strange, given that it is flatly false — false in the sense that nothing in the bare logic of the theory tells us that populations *must* evolve in a manner that yields the new species or fundamental changes of "type" that the theory is intended to explain. It's true that healthy populations exhibit plasticity, variation, and adaptability — a spruce tree growing in the lowlands will differ greatly from one growing near the alpine treeline, and in any given location one tree will differ from its neighbor. But none of this explains the evolutionary origin of the diverse forms of life on earth.

In fact, despite such in-species variation, for millennia *all* species were widely assumed to remain constant according to their "essential" nature. Yes, untypical variation, including "monstrosities," could occur, but this only reminded our ancestors that defective organisms tended to be removed — part of the means by which the character or type of the species was preserved. So how did we learn that the situation was quite otherwise, and that species *did* evolve?

Surely a major factor was the discovery and systematic investigation of fossils. Seeing was believing. It was the apparent historical record, not the logic of natural selection, that most persuasively settled the question for us. Look at it this way: the logic can hardly be decisive because it leaves out what organisms actually *do* — and, as has long been recognized, one of the most remarkable things they are capable of doing amid all sorts of variation is to give consistent, generation-by-generation expression to the character of their own kind. Whether that kind needed to be understood as a static or dynamic reality could only be resolved through empirical investigation — and the bones continually being unearthed by paleontologists proved eloquent in this regard.

As for the character of the dynamism at issue, even today there are debates about whether evolutionary change tends to occur during relatively brief and scattered periods of intense transformation, or is instead a more or less constant phenomenon. It seems that questions about what organisms actually do can't be avoided by a mere appeal to logic. And this is true even before we take into account the evidence produced in the first half of this book for the directive nature of all biological activity.

Wholeness, unity, type: how not to over-estimate genes

My strong surmise is that a simple conviction lay behind Sober's conclusion that heritable variation bearing on fitness necessarily implies Darwinian evolution. This was the conviction that genes explain organisms, and that changes in the genome just *are* evolution. So we have no need to look at whole organisms in order to understand their evolution. Genetic variation was already enough to convince Sober that a profoundly transformative sort of evolution must be under way.

But this was to overlook the lesson we have already approached from many sides in earlier chapters: the organism as a whole exhibits a unity prior to all its parts, including its genes. This unity, though manifested in the material, is not itself a physical principle. No physical laws demand or account for an organism's holding together as the kind of whole that it is — the whole that comes to such well-directed expression throughout its development. Genes are *caught up* in the organism's unity (which we might want to call its *type* or its *kind*), so that we have to understand the causal basis[3] of the immaterial principle of unity before we can say anything about the relation between genes and evolution.

The principle of unity and wholeness applies to every level at which we analyze the organism. It was, for example, the principle that Paul Weiss was getting at when, with particular reference to the cell, he spoke of the whole being more than the sum of its parts: "certain definite rules of order apply to the dynamics of the whole system ... reflected [for example] in the orderliness of the overall architectural design, which cannot be explained in terms of any underlying orderliness of the [molecular] constituents" (Chapter 6, "Context: Dare We Call It Holism?)."

The mystery of the unity of the whole may be related to a deeply problematic aspect of all contemporary thought about evolution. How can evolution via natural selection explain any feature of an organism at all if we insist on the usual, physical, bottom-up style of explanation rather than approaching the matter from the vantage point of an immaterial principle of unity (or the type)? We have observed in the first half of this book that cellular activities in general are impossible to explain mechanistically on a genetic basis. As in our consideration of the mRNA splicing activity (Chapter 8, "The Mystery of an Unexpected Coherence"), we are always looking at fluid, complex interactions among numerous molecules in a watery medium, where the physically expected degrees of freedom of those molecules are disciplined, not by genes, but "from above" by the meaning and unity of the larger context.

More particularly, we find ourselves looking at so-called plastically "disordered" proteins with highly flexible functional structure that is not fixed by the genetic sequence; continual, on-the-fly but essential modifications of proteins by other molecules — modifications required for the ever-changing tasks to be accomplished; phase changes involving the formation and dissolution of functionally distinct droplets (collections of specialized, cooperating molecules) within the watery medium; gene-regulatory processes through which genes are maintained and variously expressed by many of the very molecules whose activity the genes supposedly account for; healing of wounds never exactly like any previous injuries in the history of the species; electrical fields that signal major changes of form before genes are summoned to their roles in those changes; and, in general, interaction networks of virtually infinite complexity whereby causal sequences circle around so as to make causes into causes of themselves.

As I have pointed out previously, we see no levers, gears, wires, or conductive channels like those in silicon by which, in any conceivable way, genes could meaningfully oversee and direct all this activity. If we only think of topoisomerases untangling knots in chromosomes, we quickly realize that no genetic mutations *in the past* can underwrite their uncoerced, contextual, never precisely repeated, moment-by-moment, purposeful activity in service of the *ever-changing, present needs* of cell and organism — and, indeed, no past genetic mutations can underwrite the *recognition* of those needs.

In sum, organisms are not the machines demanded by the logic of natural selection. So, then: in the absence of gears, levers, or logic circuits of a machine enduring from generation to generation, what exactly is the renowned evolutionary

Tinkerer[4] supposed to be tinkering with? There is, in fact, nothing we can describe as "tinkering" going on, and there is no way for the past history of a species to determine physically the exceedingly intricate, unpredictable, and fluid physiological activity through which particular traits are realized at the cellular and molecular levels, which are supposedly the most fundamental levels for biological explanation (Chapter 8, "The Mystery of an Unexpected Coherence").

Unanswered questions

It is therefore hardly surprising that natural selection tells us little or nothing about how species have evolved:

- What sorts of directionality, if any, will we discover in evolutionary change? For example, might change be directed toward more complex or less complex forms of life? Toward greater individuality or more collective interdependence? Toward increase or decrease in size? Toward the realization of human potentials? Toward competitive superiority or superior cooperation? Toward some sort of diversity, balance, and qualitative completeness upon the earth as a whole?
- What pathways of change are open to any given species at a particular time, and what pathways are closed off by the dynamic character (type) of the species itself or by the surrounding world?
- In what ways will genes or molecular and physiological processes be conserved in different organisms during evolution, and in what ways will they diverge?
- How much convergent evolution should we expect? ("Convergent evolution" refers to the independent development of similar features in distinct branches of the "tree of life" — something now known to be strikingly common, as when the "camera-eyes" of the octopus and of humans developed independently of each other.)
- How much diversity of life should we expect, and how radically disparate are the possible forms of life?
- Is evolutionary change more or less possible today than at various times in the past?
- Do populations evolve sporadically or continuously, and why?
- What accounts for the uncanny *qualitative* unity of an organism — a unity leading one observer to say of the sloth, for example, that "every detail speaks 'sloth'" (Chapter 12, "Is a Qualitative Biology Possible?").

I can think of no fundamental question about evolution whose answer *is* suggested by the advertised formula for natural selection. Everything depends on what the amazingly diverse sorts of organism actually do as they respond to and shape their environments. Contrary to Susan Blackmore's exultant insight, nothing in the "algorithmic logic" of natural selection tells us that evolution *must* have happened — and, given that it certainly has happened, the logic by itself tells us little about what we should expect to find in the fossil record. We may ask then, "What, in truth, is being celebrated as the revolutionary principle of natural selection?"

None of this is to deny the trivial validity of the idea of natural selection. *Of course* organisms that are "fitter" will generally do better in life than "unfit" organisms. That's just what "fitter" means. And *of course* a record of the winners and losers in the "struggle for survival" will tell us something about evolutionary processes. Or *could* tell us if we understood all that happened in order to establish this particular record. It is hardly unreasonable to point out that we will gain a profound understanding of evolution only when we know a fair amount about *how* it has happened among actual organisms and along its broad course down through the ages.

Every organism's life and death encompasses and, so to speak, "sums up" a vast range of purposive activities, not only on its own part, but also on the part of many other organisms — including, to begin with, its mating partners and would-be predators. One might feel, therefore, that the "theory" of the survival of the fittest can explain just about everything. Certainly the overall pattern of births and deaths must yield the observed evolutionary outcome! Actually, it just *is* that outcome, which may be why the theory strikes so many as powerfully explanatory. But an outcome — the pattern we need to explain — isn't yet the explanation.[5]

What, for example, if "fit" meant "able to thrive as such-and-such a kind of organism," where "kind" was understood statically rather than dynamically? Some species certainly do show something like this sort of constancy during lengthy periods of relative stasis. Do we understand this ability to maintain stasis well enough to say that the organism's directive powers *must* result in an evolution of kinds in the modern sense, as opposed to an active maintenance of an already existing kind?

Surely we *can* arrive, and arrive rightly, at the modern notion of dynamically evolving kinds. But the Darwinian theory of natural selection is not how we get there, because it offers no understanding of the highly directive processes through which variation arises. Nor does it offer understanding of the meaningful wholeness that is faithfully preserved from generation to generation. It doesn't tell us whether the organism's directive processes are directed at the maintenance of type, or also (at least at times) directed at the transformation of type.

Actually, the theorists of natural selection have not been interested enough in the idea of a *type* (or *archetype*, or *kind*, or *unity*, or *whole*) even to address these questions. And so the qualitative and expressive unity of every kind of organism — scientifically baffling as it is — has been given little attention. This is an egregious oversight I have tried to go some distance toward remedying in Chapter 12 ("Is a Qualitative Biology Possible?").

The "algorithm" of natural selection is widely treated as if it were an agent

The miracle of it all is that, if current evolutionary rhetoric is to be believed, the empty formula of natural selection explains just about everything you could imagine — all based on some form of "blind" agency. Despite this celebrated blindness, natural selection, we're told, is always managing to do great things, as if it were an active, well-directed power. And so we hear about the *mechanism* of selection, as well as the *forces* or *pressures* that operate in it. We learn that natural selection *shapes* the bodies and behaviors of organisms, *builds* specific features, *targets* or *acts on* particular genomic regions, *favors* or *disfavors* (or even *punishes*) various traits or behavioral strategies, *operates* in this way or that, *maintains* DNA sequences, *promotes* adaptation of populations to local environments, *polices* mutations, and, in general, *causes* an endless variety of effects. Darwin himself spoke about how

> natural selection is daily and hourly scrutinising, throughout the world, every variation, even the slightest; rejecting that which is bad, preserving and adding up all that is good; silently and insensibly working, whenever and wherever opportunity offers, at the improvement of each organic being in relation to its organic and inorganic conditions of life (Darwin 1859, p. 84).

This sort of language is now all but universal. I think it is safe to say that relatively few references to natural selection by biologists fail to assert or imply that we are looking at something like a humanly contrived *mechanism* with an effective power to do the things it was designed for, beginning with the activity of *selecting*. If what biologists say has any significant bearing on what they mean, then they are telling us, emphatically, that they believe natural selection to be an efficacious, mechanistic *agent* — an agent of evolutionary change.

And perhaps we would be forced to agree with them if in fact it was all just a matter of the living and dying of organisms whose lives meant nothing in particular — organisms in the hands of an endlessly subtle mechanism capable of performing an evolutionary work closely akin to the work organisms do in their individual development. And this seems to be the picture that is being forced upon us. All we need to do in order to accept it is to forget *how* organisms live and die — forget, for example, that we know little about what their life actually means, or what the Darwinian striving for life is a *striving* toward for each particular kind.

Developmental systems theorist Susan Oyama was fingering this forgetfulness when she reminded us that

> Nature is not a deciding agent, standing outside organisms and waving them to the right or the left. However much we may speak of selection "operating" on populations, "molding" bodies and minds, when the metaphorical dust has settled, what we are referring to is still the cumulative result of particular life courses negotiated in particular circumstances (Oyama 2000b, p. 81).

Some evolutionists are uncomfortably aware that their use of a phrase intentionally evoking the breeder's "artificial selection" invites mystical belief in a breeder-like agent supervising adaptive evolution. And so they assure us that "natural selection," despite its explicit suggestion of a selecting agent, is "just a metaphor."

The prolifically blogging defender of evolutionary orthodoxy, University of Chicago geneticist Jerry Coyne, spells it out this way: natural selection "is neither a 'law' nor a 'mechanism'." If we explain the evolution of coat color in polar bears as "'natural selection acting on coat color,' that's only our shorthand ... There is no external force of nature that 'acts' on individuals. There is only differential replication of genes" (Coyne 2010).

In other words, as Coyne goes on to say, the language of agency really refers to a mundane process — "a process that is inevitable," he adds — and here, as expected, he launches into the familiar logic of natural selection.

But it is hard to see this as anything but subterfuge. There is a reason why no effective verbal alternative to the painfully tendentious "metaphor" of selection has taken hold. After all, there really is a transforming agency at work, and no evolutionary theory of the transformation of species can prove persuasive without acknowledging it, if only in a roundabout and deceptive way. We can't have a theory of transformation without the idea of transformative power appearing *somewhere* in it.

The idea of a selecting power is deeply rooted and seemingly ineradicable from the modern biologist's thinking about evolution. Yes, we can redefine the metaphorical selecting agent as a process. But if it's the kind of process that inevitably yields exactly the results we could previously ascribe only to an intelligent agent

— yields what can be viewed as the *policing*, *targeting*, *sculpting*, and *creating* of organisms and their features — we are not getting rid of the agent. We are merely giving it a different name and comforting ourselves by calling it a metaphor. The hope that we will eventually be able to substitute a blind, mindless process for the metaphor is a hope for which we have been given no encouragement.

As far as Coyne is concerned — and this is the conventional view — natural selection gives us only a substitution of some genetic particles for others. Unfortunately, however, he offers no explanation for how the mere substitution of particles actually explains the formation of viable traits harmoniously integrated into an organism's holistic way of being and its present and future life potentials. That is the part of the "recipe" of natural selection — the living part — where he is perfectly content to let ignorance reign. So here, in this space of ignorance, is where the real fact of transformation is, for now, hidden away.

Instead of saying, "There is only differential replication of genes," Coyne should have said, "There is only the development of specific form and the creation of viable traits harmoniously integrated into the unity that is the organism as a whole (but we have no idea how this happens — and our understanding of evolution thoroughly depends on the answer we ultimately find)."

Do not underestimate the difficulty biologists have in seeing this matter clearly. Regarding the "syllogistic core" of natural selection, Gould wrote that "nearly all textbooks and college courses present the 'bare bones' of natural selection in this fashion (I have done so in more than 30 years of teaching)." After suggesting that this presentation "does not permit a teacher to go beyond the simplest elucidation of selection as a *genuine force* that can produce adaptive change in a population," he goes on to say: "In other words, the syllogistic core only guarantees that selection can work … [it] can only rebut charges of hokum or incoherence at the foundation" (Gould 2002, p. 126n; emphasis added).

It would be truer to say that the famously simple and compelling logic of natural selection, misconceived as a "force" and as the "foundation" of a powerful theory, has itself become a primary source of hokum in evolutionary thinking. It is a kind of blank template upon which overly credulous biologists and lay people can project their faith. As for the "genuine force" and causal power of the syllogistic core that Gould refers to, it is a magical invention born of the refusal to recognize agency in the only place where we ever observe it, which is in the lives of organisms.

This is not to deny that we have learned a great deal — for example, from paleontology and molecular studies — under the banner of "natural selection." After all, despite the fact that the generality and emptiness of the logical template allow the biologist to use it as a frame for just about any investigative work, the work itself often has value. Whatever it is that actually happens (which is the valuable part), we can always say (without adding anything to our understanding) that the surviving organisms were somehow or other "selected."

Certainly all extant organisms have in some sense been selected as expressions of whatever future is now being realized. The question is "What has been done, concretely, to get them here?" and the algorithm of natural selection — the idea that organisms have in fact lived and died precisely in the pattern that has landed them and us where we are now — adds little if anything beyond a certain illusion of explanation.

The inadequacy of the theory of natural selection has long been noticed

It happens that the explanatory vacuity of the logic of natural selection has been recognized by some of the most prominent and reputable evolutionary biologists for more than 150 years. They have been concerned about how complex adaptive innovations are achieved, and how, in general, we can make sense of the evident creativity in evolution. The question that nagged at them can be put this way: What does natural selection *select* — where does selectable variation come from — and why should we think that the mere preservation of variants that have already been achieved, rather than the creative production of those variants in the first place, accounts for the "accomplishments" of evolution?

The influential Dutch botanist and geneticist, Hugo de Vries, framed the matter this way during the first decade of the twentieth century:

> Natural selection is a sieve. It creates nothing, as is so often assumed; it only sifts. It retains only what variability puts into the sieve. Whence the material comes that is put into it, should be kept separate from the theory of its selection. How the struggle for existence sifts is one question; how that which is sifted arose is another (quoted in Gould 2002, p. 428).

Figure 16.2. Hugo de Vries.[6]

It was de Vries who gave currency to the catchy phrasing that has since been repeated many times: "Natural selection may explain the survival of the fittest, but it cannot explain the arrival of the fittest" (de Vries 1906, p. 826). The concern is not easily dismissed. Other biologists have added their own accents, and it is worth pausing a few moments to trace a theme that some might see as a kind of subterranean and ignored history of evolutionary thought — a history beginning no later than the year after the original publication of *The Origin of Species* in 1859:

> "If we take the three attributes of the deity of the Hindoo Triad, the Creator, Brahma, the preserver or sustainer, Vishnu, and the destroyer, Siva, Natural Selection will be a combination of the two last but without the first, or the creative power, we cannot conceive the others having any function" (Sir Charles Lyell [1860], Scottish geologist who laid the crucial uniformitarian foundation for Darwin's theory).

> "It is exceedingly improbable that the nicely adapted machinery of animals should have come into existence without the operation of causes leading directly to that end. The doctrines of 'selection' and 'survival' plainly do not reach the kernel of evolution, which is, as I have long since pointed out, the question of 'the origin of the fittest' ... The law by which structures originate is one thing; those by which they are restricted, directed, or destroyed, is another thing" (Edward Drinker Cope [1887, p. 225], noted American paleontologist and formulator of "Cope's Rule," which proposed that the organisms of an evolutionary lineage tend to increase in size over time).

> "Selection permits the viable to continue and decides that the non-viable shall perish ... Selection determines along which branch Evolution shall proceed, but it does not decide what novelties that branch shall bring forth" (William Bateson [1909, p. 96], a founder of the discipline of genetics).

> "The function of natural selection is selection and not creation. It has nothing to do with the formation of new variation" (Reginald Punnett [1911], British geneticist who cofounded the *Journal of Genetics*; quoted in Stoltzfus 2006).

> "The actual steps by which individuals come to differ from their parents are due to causes other than selection, and in consequence evolution [by natural selection] can only follow certain paths. These paths are determined by factors which we can only very dimly conjecture. Only a thorough-going study of variation will lighten our darkness" (J. B. S. Haldane [1932, pp. 142–43], a major contributor to the twentieth-century consensus theory of evolution).

> Regarding specific traits, natural selection "might afford a reason for their preservation, but never provide the cause for their origin" (Adolf Portmann [1967, p. 123], preeminent zoologist of the middle of the twentieth century).
>
> "Natural selection is the editor, rather than the composer, of the genetic message" (Jack King and Thomas Jukes [1969], key developers of the idea of "neutral evolution").
>
> "In evolution, selection may decide the winner of a given game but development non-randomly defines the players" (Pere Alberch [1980], Spanish naturalist and embryologist, sometimes spoken of as the founder of Evo-Devo — evolutionary developmental biology).
>
> "Natural selection eliminates and maybe maintains, but it doesn't create" (Lynn Margulis [2011], microbiologist and botanist, pioneer in exploring the role of symbiosis in evolution, and co-developer of the Gaia hypothesis).

Misplaced agency

We began this chapter by listening to Darwin saying that "any being, if it vary however slightly in any manner profitable to itself, under the complex and sometimes varying conditions of life, will have a better chance of surviving, and thus be *naturally selected*." And we heard much the same from contemporary philosopher, Elliott Sober: "If the organisms in a population differ in their ability to survive and reproduce, and if the characteristics that affect these abilities are transmitted from parents to offspring, then the population will evolve."

I am not sure why the void at the heart of these statements was so long invisible to nearly all biologists, or why, even where it has in one way or another been recognized, it has not fundamentally changed the dominant thinking about the theory of evolution by natural selection.

In any case, what these statements by Darwin and Sober (and just about every other evolutionary thinker) necessarily and unthinkingly assume is the existence of beings capable of sustaining their own lives and development, and also capable of producing variation. This production of variation, like development, inheritance, and organic activity generally, is always a reckoning with the present even as it is in some way oriented toward the future.

But the living beings to whom this truth applies, in the very process of being assumed as the starting point for a compelling bit of evolutionary logic, have fallen out of the picture. They play no role in the elaboration of the logic (which is why that logic can be so vacuously compelling, rather than complex and difficult to understand). Rather, organisms become "black boxes" out of which variation fortuitously appears without need for explanation.

In this way, life vanishes from the theory taken as foundational for all the life sciences, and therefore the theory's explanatory power shrinks toward nothingness. The very real knowledge of evolution we have gained so far is not owing to the theory of natural selection, even if we have learned a great deal about a subordinate topic — how gene distributions change within populations. Genetics is not evolution. Only whole organisms evolve, and genetics contributes to the picture only as one part of the whole.

Some readers will have been wondering whether I haven't fatally overlooked the latter-day turn of some biologists toward "evo-devo" — evolutionary developmental biology. And it does need saying that this turn has often included a renewed focus on organisms. Unfortunately, however, the still-uncompromised and strictly enforced materialism of the biological community has meant that the actual *life* of organisms cannot be fully acknowledged.

And so the decisive limitation of all biology remains: organisms are conceived as if they could be understood in purely physical terms — as if they were mind-independent machines existing in a mind-independent world thanks (ironically) to a God-like Designer, the mechanism of natural selection. This means that the increasing references to "purpose," "consciousness," and "agency" in certain circles around the fringes of biology do not point to genuinely interior activity, but are only circumlocutions for various patterns of machine-like interaction. And this in turn means that the unity of the organism — an immaterial, purposive, and wisdom-infused reality that must be considered *before* we can make sense of genes and all the rest — cannot be taken seriously.

It seems to have been the task of biology over the past couple of centuries to reconceive living things without their life — to see the world of organisms, not through their own eyes, but through ours, which are as if hypnotized by the well-designed automatisms that now shape every dimension of our existence. It is not often that the spell is momentarily broken, as when the philosopher of biology, Denis Walsh — after noting the indisputable yet ignored truth that "organisms are fundamentally purposive entities" — expressed his perplexity by asking, "Why should the phenomenon [of agency] that demarcates the domain of biology be off-limits to biology?"[7]

It is now my intention in further chapters to discuss evolution by articulating a different point of view, taking life in its own terms. And I see no reason to exclude what we know most directly — and in a higher key, so to speak — through our own existence as organisms. This higher key of consciousness or awareness offers us many possibilities for an immediate, inner understanding of our experience, which is hardly grounds for excluding ourselves, or our understanding of the meanings of life, from a science of organisms.

WHERE ARE WE NOW?

Sweeping out the cobwebs is good to do.

We can hardly hope to engage profitably the many puzzles and perplexities of evolutionary theory without first "cleaning out the attic of our minds," where we find stored the heritage of the past century's theoretical refusal of the life of organisms. I suppose just about everything in this book requires — and is intended to encourage — such a cleaning out in one way or another.

I have, in the above discussion, attempted to show how conventional evolutionary theory has eliminated the organism as the one available source of, or channel for, the kind of adaptive, transformational agency required by evolutionary theory. This ignoring of the organism, together with the prevailing reluctance among evolutionists *explicitly* to acknowledge that they have effectively reassigned the organism's agency to the "mechanism" of selection, has resulted in a bland formulation of natural selection as if it were the "obvious" operation of an abstract and empty logic — a dematerialized logic that somehow pretends to be causally effective despite its being abstracted entirely away from organisms. One tries not to speak *openly* of agency at all.

The logic is empty because (1) it refuses to account for the variation that is one of its core presuppositions — refuses to particularize this variation as an expression of the creative life and activity of incarnate living beings. But if we do not understand how organisms creatively produce the material of evolutionary change, then we do not understand evolution.

As we will see in later chapters, (2) the advertised logic of natural selection also fails to reckon with the organism's reliable capacity to produce an inheritance after its own kind. And, as we saw in many of the earlier chapters, (3) today's biology has also failed in understanding the relation between genes and the organism's fitness for survival — a relation whereby the organism governs its genes much more than its genes can be said to govern the organism.

So the banishing of the organism from evolution occurs in the conceptualization of all three stated requirements for natural selection to occur — (1) variation; (2) inheritance; and (3) differential fitness.

We also noted how the idea of natural selection, as it is used today, takes the organism to be a durable machine that evolution can tinker with at the cellular and molecular levels. The tinkerings are supposed to be preserved stably as fixed mechanisms that can be further tinkered with down through the geological ages so as to shape the capacities of future organisms. But as soon as we drop the fallacious machine idea and acknowledge the fluid, watery, moment-by-moment context of the internal workings of the cell, everything changes. In particular, one belief loses all credibility — namely, the belief that the evolutionary past, mediated by genes, somehow physically determines the trillions of molecular interactions every second of the cell's life so as to support the ever-changing life fuctions of the organism. I develop this point at somewhat greater length in Chapter 18 ("Teleology and Evolution").

In the next chapter we will look more particularly at the evolutionarily relevant, adaptive, and transformative powers of individual organisms, revealed especially in their development. After that, it will be necessary to look more directly at the evolutionary process itself.

NOTES

1. Figure 16.1 credit: Rakesh.infosys (https://commons.wikimedia.org/wiki/File:Charls_Darwin.jpg), CC BY-SA 4.0.
2. This according to philosopher of biology John Beatty (2010, p. 23), citing correspondence between Darwin and Lyell.
3. The kind of causation we're talking about —*formal causation* — was introduced in our discussion of Ronald Brady's analysis of leaf sequences in Chapter 12 ("Is a Qualitative Biology Possible?)."
4. The idea of tinkering — that evolution is a tinkerer rather than an engineer — traces back to an influential article by the French biologist, François Jacob (1977). "Tinkering" is now one of the clichés of evolutionary theory.
5. The American philosopher, Susanne Langer, said of natural selection that

    this constant interplay of forces, which makes shifting obstacles and openings for each individual so that variously equipped organisms are differentially brought to grief, is not a mechanism; the frequent references, in the literature, to the "mechanism of selection" bear witness to the beguiling influence of the term "natural selection," which seems to refer to an act, or at least a function, of some specific power. "Natural selection" is a historical pattern, not a mechanism; it is the pattern of the natural history of life (Langer 1967, p. 394).

6. Figure 16.2 credit: Store norske leksikon (https://commons.wikimedia.org/wiki/File:Portrait_of_Hugo_de_Vries.jpg), Public Domain via Wikimedia Commons.
7. Walsh 2015, p. ix. And yet, even Walsh, wonderfully insightful as he is, proceeds to characterize the organism's agency in a strictly materialistic manner, as if it could be understood without accepting at face value the inner dimensions of life. These dimensions include the organism's living (not camera-like or instrument-like)

perception of its surroundings (Chapter 24), the evident wisdom at work in its instincts and behaviors, and the intention and volition evidenced in its persistent and well-directed efforts to satisfy its own needs and interests. We are instead given agency without agency, life without life. Such is our way today. Biologists and philosophers call it "naturalizing" agency and purposiveness, as if even our human agency, taken at face value, were decidedly "unnatural." (This point of view is happily contradicted by the decision of those many worthy individuals who are inspired to exercise their "unnatural" capacities by dedicating themselves to the tasks of science and philosophy.)

SOURCES

Alberch, Pere (1980). Ontogenesis and Morphological Diversification, *American Zoologist* vol. 20, pp. 653–67. doi:10.1093/icb/20.4.653

Bateson, William (1909). Heredity and Variation in Modern Lights in *Darwin and Modern Science*, edited by A. C. Seward, pp. 85–101. Cambridge UK: Cambridge University Press. Text available at https://www.gutenberg.org/ebooks/1909

Beatty, John (2010). Reconsidering the Importance of Chance Variation, in *Evolution: The Extended Thesis*, edited by Massimo Pigliucci and Gerd B. Müller, pp. 21–44. Cambridge MA: MIT Press.

Blackmore, Susan (2014). A Hundred Walked Out of My Lecture. Available at https://www.richarddawkins.net/2014/08/a-hundred-walked-out-of-my-lecture/

Cope, Edward Drinker (1887). *The Origin of the Fittest: Essays on Evolution*. New York: D. Appleton and Company.

Coyne, Jerry A. (2010). The Improbability Pump, *The Nation* (April 22). https://www.thenation.com/article/archive/improbability-pump/

Darwin, Charles (1859). *On the Origin of Species by Means of Natural Selection*, first edition. London: John Murray. Available at https://archive.org/details/onoriginofspecie00darw_0

Dawkins, Richard (2008). *The Extended Phenotype*, revised edition (1999), reprinted with corrections (2008). Oxford: Oxford University Press. Original edition published in 1982.

Dennett, Daniel C. (1995). *Darwin's Dangerous Idea: Evolution and the Meanings of Life*. New York: Simon and Schuster.

de Vries, Hugo (1906). *Species and Varieties: Their Origin by Mutation*. Lectures delivered at the University of California, edited by Daniel Trembly MacDougal, second edition. Chicago: Open Court. Originally published in 1905. Full text available at https://gutenberg.org/ebooks/7234

Eldredge, Niles (2000). *The Pattern of Evolution*. New York: W. H. Freeman and Company.

Gould, Stephen Jay (2002). *The Structure of Evolutionary Theory*. Cambridge MA: Harvard University Press.

Haldane, J. B. S. (1932). *The Causes of Evolution.* London: Longmans, Green and Co.

Jacob, François (1977a). Evolution and Tinkering, *Science* vol. 196, no. 4295 (June 10), pp. 1161–66. doi:10.1126/science.860134

King, Jack Lester and Thomas H. Jukes (1969). Non-Darwinian Evolution, *Science* vol. 164, no. 3881 (May 16), pp. 788–98. doi:10.1126/science.164.3881.788

Langer, Susanne K. (1967). *Mind: An Essay on Human Feeling* vol. 1. Baltimore MD: Johns Hopkins University Press.

Lenski, Richard E., Charles Ofria, Robert T. Pennock and Christoph Adami (2003). The Evolutionary Origin of Complex Features, *Nature* vol. 423 (May 8), pp. 139–44. doi:10.1038/nature01568

Lyell, Sir Charles (1860/1970). *Sir Charles Lyell's Scientific Journals on the Species Question*, edited by Leonard G. Wilson. New Haven CT: Yale University Press. Yale Studies in the History of Science and Medicine, no. 5.

Margulis, Lynn (2011). Discover Interview: Lynn Margulis Says She's Not Controversial, She's Right, *Discover* (April). Available online at https://www.discovermagazine.com/the-sciences/discover-interview-lynn-margulis-says-shes-not-controversial-shes-right

Mayr, Ernst (1964). Introduction, *On the Origin of Species: A Facsimile of the First Edition*, by Charles Darwin, pp. vii-xxvii. Cambridge MA: Harvard University Press.

O'Neill, Bill (2003). "Digital Organisms," *PLOS Biology*. doi:10.1371/journal.pbio.0000018

Oyama, Susan (2000a). The Ontogeny of Information, 2nd edition, foreword by Richard C. Lewontin. Durham NC: Duke University Press. First edition published in 1985 by Cambridge University Press.

________ (2000b). *Evolution's Eye: A Systems View of the Biology-Culture Divide*. Durham NC: Duke University Press.

Portmann, Adolf (1967). *Animal Forms and Patterns: A Study of the Appearance of Animals*, translated by Hella Czech. New York: Schocken.

Sober, Elliott (1984). *The Nature of Selection: Evolutionary Theory in Philosophical Focus*. Cambridge MA: MIT Press.

Stoltzfus, Arlin (2006). Mutationism and the Dual Causation of Evolutionary Change, *Evolution and Development* vol. 3, no. 3, pp. 304–317. doi:10.1111/j.1525-142X.2006.00101.x

Walsh, Denis (2015). *Organisms, Agency, and Evolution*. Cambridge UK: Cambridge University Press.

CHAPTER 17

Evolution Writ Small

ONE MIGHT THINK that the natural place to look for an understanding adequate to the evolutionary history of life would be the powers of self-transformation we observe in the evolving organisms themselves. But it can be dangerous to look in a clear-eyed manner at the creative potentials of living beings. One risks having to acknowledge the evident wisdom and agency so vividly on display. In an era of institutionalized materialism, any suggestion that these inner powers are vital to the entire evolutionary story can only produce the sort of discomfort associated with a taboo.

On the other hand, Stephen Jay Gould ran afoul of no taboo when he effectively ascribed this same wisdom and agency to natural selection. Countering the questions we heard voiced in Chapter 16 about what sort of creative principle could explain the "arrival of the fittest," he asked (referring to several giants of twentieth-century evolutionary biology), "Why was natural selection compared to a composer by Dobzhansky; to a poet by Simpson; to a sculptor by Mayr; and to, of all people, Mr. Shakespeare by Julian Huxley?"

The answer, Gould said, is that the allusions to poetry, musical composition, and sculpture helpfully underscore the "creativity of natural selection":

> The essence of Darwinism lies in its claim that natural selection creates the fit. Variation is ubiquitous and random in direction. It supplies the raw material only. Natural selection directs the course of evolutionary change. It preserves favorable variants and builds fitness gradually.[1]

On its face, this argument for Darwinism was a puzzling one. Its answer to the question how variation arises amounted to saying nothing more than "It is everywhere" ("variation is ubiquitous") — which, one might have thought, only added urgency to the need for an explanation. The suggestion seems to be that, because organisms are so expert and prolific at producing new possibilities of life, the evolutionist can simply take their powers of achievement for granted without actually *looking* at them. Because *organisms* so abundantly provide what is needed ("raw materials") for the transformation of life, we are somehow free to declare *natural selection* the transforming agent. It need only preserve all those wonderfully effective new variants, and they will somehow integrate themselves into the almost infinitely differentiated unity of a living being. We need not concern ourselves with those powers of integration and unity. After all, what could they have to do with evolution?

How easy it is, apparently, to forget that the so-called "raw materials" being preserved are never merely raw materials! At the first appearance of any substantive change, the creative work has already been accomplished — if indeed the change is truly beneficial to a living being. We find ourselves looking, not at random raw materials, but at a whole harmoniously transforming itself *as a whole*, where everything tends to affect everything else.

In this way, whatever we may have falsely isolated in our minds as a "new feature" is incorporated into the tightly interwoven complexity of an organism's life. The only power we know to be capable of such incorporation is that of the organism telling its own story, a story always reflecting the qualitative, unified character and dynamic developmental potentials of a particular species.

This harmonious incorporation of new features, founded upon whole-cell inheritance and manifested in whole-organism processes of development, is where we see creative evolutionary change originating. The spreading of an already-existing change through a population is an entirely different matter.

So Gould's response shows us that one of the evolutionist's strategies for coping with taboo agency is immediately to turn the question, "How does creative change arise?" into the different question, "How does creative change, once arisen, spread through a population?" The switch of topics is not hidden, but occurs in plain sight. Only a habit of blindsight relative to the organism's agency seems able to explain such an obvious evasion of a real biological question.

None of this means we need to doubt whatever is true in the idea of natural selection (which may be very little that bears directly on evolution). Selective mortality certainly occurs throughout all domains of life. Not every organism lives out a full life. But the mere elimination of problematic traits (or defective organisms) through mortality is not the same thing as positively and viably transforming the integral unity that a particular organism is.

The point is not terribly subtle. There is simply nothing in the idea of natural selection itself that points to the creative capacities necessary for producing new adaptive features — for producing, say, a four-chambered heart (with all its organism-wide implications) from a three-chambered one. There is only the living being whose agency and activity natural selection necessarily assumes and which evolutionists have unconsciously transferred to a mystical "mechanism" of selection somehow operated by the inanimate world.

So, if we do not accept this subterfuge, we are left with the main question for this chapter: What do organisms show us, directly, compellingly, and uncontroversially, about their own powers of organic transformation? Much of the first half of this book contributes to an answer, especially at the physiological and molecular levels of observation. But in the present, evolutionary context, it will be well to look at the organism from a new angle.

A "magical" power of self-transformation

If I were to tell you that scientists have sequenced the genomes of two entirely distinct organisms — say, a flying creature such as a bird or bat, and a crawling one such as an earthworm or snake — and had found the two genomes to be identical, you would probably think I was joking. Surely such differently structured forms and behaviors could not possibly result from the same genetic instructions! A genome, we've been told time and again, comprises a blueprint for, or otherwise corresponds to, a phenotype — that is, the manifest form and functions of an organism. And what could be more different than the phenotypes of a snake and bird?

And yet a good reason for jettisoning the entire notion of a genetic "blueprint" is that there *are* flying and crawling creatures with the same genome. A monarch butterfly and its larva, for example. Nor is this kind of thing rare. A swimming, "water-breathing" tadpole and a leaping, air-breathing frog are creatures with the same inherited DNA. Then there is the starfish: its bilaterally symmetric larva swims freely by means of cilia, after which it settles onto the ocean floor and metamorphoses into the familiar form of the adult. This adult, carrying the genome passed on from its larval stage, exhibits an altogether different, radially symmetric (star-like) body plan.

Millions of species consist of such improbably distinct creatures, organized in completely different ways at different stages of their lives, yet carrying around the same genetic inheritance. (See Box 17.1.) This is something to reflect on. How could the transformation possibly be orchestrated, and where lies the power of orchestration?

To speak of the "power of orchestration" will perhaps trigger accusations of "mysticism." And yet the expression of *some* power is right there before our eyes. It is hardly anti-science to let ourselves come up against questions we cannot yet answer. They are what science is for.

One way or another we must come to terms with the fact that the organism and its cells actively *play off* the genomic sequence and all the other available resources within a huge space of profoundly creative possibility. No identifiable physical force compels or directs the cell-by-cell and molecule-by-molecule dissolution and refashioning described in Box 17.1. It is only healthy that such difficulties for our understanding should be acknowledged.

BOX 17.1

Metamorphosis of an Insect

The British physician and evolutionary scientist, Frank Ryan, described the goliath beetle's metamorphosis this way:

"Rather than a den of repose, we see now that the enclosed chamber of the goliath's pupa really is a crucible tantamount to the mythic pyre of the phoenix, where the organic being is broken down into its primordial elements before being created anew. The immolation is not through flame but a voracious chemical digestion, yet the end result is much the same, with the emergence of the new being, equipped with complex wings, multifaceted compound eyes, and the many other changes necessary for its very different lifestyle and purpose.

Figure 17.1. The goliath beetle (*Goliathus goliatus*), larva and adult. The male adult runs 2.4–4.3 inches in length. The photos are not printed at the same scale.[2]

"The emerging adult needs an elaborate musculature to drive the wings. These muscles must be created anew since they are unlike any seen in the larva, and they demand a new respiratory system—in effect new lungs—to oxygenate them, with new breathing tubes, or tracheae, to feed their massive oxygen needs. The same high energy needs are supplied by changes in the structure of the heart, with a new nervous supply to drive the adult circulation and a new blood to make that circulation work.

"We only have to consider the dramatic difference between a feeding grub or caterpillar and a flying butterfly or a beetle to grasp that the old mouth is rendered useless and must be replaced with new mouthparts, new salivary glands, new gut, new rectum. New legs must replace the creepy-crawly locomotion of the grub or caterpillar, and all must be clothed in a complex new skin, which in turn will manufacture the tough new external skeleton of the adult. Nowhere is the challenge of the new more demanding than in the nervous system— where a new brain is born. And no change is more practical to the new life-form than the newly constructed genitals essential for the most important new role of the adult form—the sexual reproduction of a new generation.

"The overwhelming destruction and reconstruction extends to the very cells that make up the individual tissues, where the larval tissues and organs are broken up and dissolved into an autodigested mush ... To all intents and purposes, life has returned to the embryonic state with the constituent cells in an undifferentiated form." (Ryan 2011, pp. 104–105)

Looking at the pupal case of a fly, the developmental biologist and evolutionary theorist, Wallace Arthur, asked: "What on earth is going on in there to turn one animal into another? If we didn't know better, we might venture 'magic' as our best attempt at an answer" (Arthur 2004, p. 45). Arthur's wonder is justified. And he surely expects, as we must, that a more satisfactory answer than "magic" will be forthcoming. Meanwhile, it is worth keeping in mind that the "magical" impression made by a phenomenon increases in direct proportion to the inadequacy of our current explanatory resources.

Metamorphosis of cells

Frogs and beetles aside, we are brought up against the same perplexities even when we consider the more "routine" developmental processes in complex organisms. Take, for example, the radical cellular transformations following from a single, fertilized human egg cell. As adults, we incarnate ourselves in trillions of cells, commonly said to exemplify at least 250 major types. And when we count subtypes and transient types, we may well find that — as cell biologists Marc Kirschner and John Gerhart tell us — there are "thousands or tens of thousands of kinds representing different stable expression states of the genome, called forth at different times and places in development" (Kirschner and Gerhart 2005, pp. 179–81).

As researchers hone their ability to investigate single cells, they are finding that even neighboring cells, "identical" in type and occupying the same tissue or niche, reveal great heterogeneity. Every cell is, in whatever degree, "doing its own thing."

Strikingly, however, the cell is not *only* doing its own thing; it is also heeding the "voice" of the surrounding context, which is in turn an expression of the unity of a particular kind of organism. So each cell is disciplined by the needs of its immediate cellular neighborhood as well as those of the entire developing organism, which in turn is conditioned by the larger environment. Every organism — even a single-celled one — is a remarkable diversity within an overall, integral unity.

In humans there are, for example, cells (neurons) that send out extensions of themselves up to a meter or more in length, while being efficient at passing electrical pulses through the body. There are contractile cells that give us our muscle power. There are the crystalline-transparent fiber cells of the lens of the eye; their special proteins must last a lifetime because the nucleus and some other subcellular entities (prerequisites for protein production) are discarded when the fibers reach maturity. There are cells that become hard as bone; as easily replaceable as skin; as permeable as the endothelial cells lining capillaries; and as delicately sensitive as the various hair cells extending into the fluids of the inner ear, where they play a role in our hearing, balance, and spatial orientation.

Many of these cells are as visibly and functionally different, in their own way, as the phenotypes of any two organisms known to us. This, you might think, would interest the evolutionary biologist. It *has* drawn the attention of a few who — if they do venture to comment about it — tend to be widely ignored on the point. One such is the much-awarded biochemist, cell biologist, and cancer researcher, Mina Bissell, who remarked in an interview with *Cell* magazine, "Your nose and mouth are completely different and yet they have the same DNA. So what on earth is telling the DNA what to do?" (Bissell 2020).

The question is as old as it is decisive. A hundred years ago, as we heard in Chapter 7 ("Epigenetics: A Brief Introduction"), the pre-eminent biologist, Frank Lillie, who served as president of the National Academy of Sciences, said that "those who desire to make genetics the basis of physiology of development will have to explain how an unchanging complex [DNA] can direct the course of an ordered developmental stream" (Lillie 1927, pp. 367–68). I can't say there's much evidence yet that evolutionary biologists feel they should bother with the question.

Organisms manage their own germlines expertly

Of all the cellular phenotypes, it would be hard to find one whose differentiation and specialization is more distinctive, or more expertly and intricately contrived, or more purposively managed, than the germ cells of sexually reproducing organisms. We can hardly help acknowledging that parental organisms, in carrying out meiosis, genetic recombination, and mating, play a massive role, not only in preserving the genome, but also in re-purposing and transforming it. Deeply embedded in time like all organisms, and therefore *always facing the future* in every aspect of their being, sexually reproducing animals express their future orientation most immediately and vividly in the gametes whose full "self-expression" belongs to the next generation.

A gamete is at least as specialized as any other cell of the body. At the same time, this gamete, along with the entire lineage leading up to it, must retain the potential to yield the totipotent zygote. That is, despite its commitment to a highly specialized, reproductive function unlike that of any other cell type in the body, the germline cell must at the same time preserve within itself the flexibility and freedom that will be required for its role as progenitor for every cellular lineage of a new organism.

It is an extraordinary mandate, and our bodies must focus extraordinary powers of development upon it. For example, the chromosomes of both sperm and egg will have been modified by epigenetic "marks" (Chapter 7, "Epigenetics:

A Brief Introduction"), ensuring that certain genes in the offspring will be active, or repressed, depending on which parent the gene was inherited from. Other widespread marks imposed by the parents will (for the most part) be erased immediately after fertilization. This leaves space for the new organism to structure the spatial, electrical, and chemical characteristics of its chromosomes (and therefore also its gene expression) according to its own way of being and developmental potentials.

And, of course, there is the elaborately orchestrated "meiotic ballet" (Page and Hawley 2003) that produces both sperm and egg, each with only half the number of chromosomes found in somatic cells, and with those chromosomes reshuffled and otherwise modified according to a logic and via activities that are still largely beyond any comprehensive understanding. But one thing is sure: the body's rearrangement ("recombination") of its germ-cell chromosomes during meiosis is now showing itself to be highly regulated. Multiple protein complexes and epigenetic modifications of chromosomes function combinatorially, with synergism, antagonism, and redundancy: "The new-found multiplicity, functional redundancy and [evolutionary] conservation" of these regulatory factors "constitute a paradigm shift with broad implications" (Wahls and Davidson 2012).

So we are given no choice but to think of the germline as an expression of that same agency — that same, end-directed transformative power — through which our bodies subtly, elaborately, and adaptively direct each of their other cell lineages toward a distinctive form and functioning within the unity of the whole. We have seen that this power of transformation comes to intense expression in entire differentiating cells, quite apart from any mutations in their DNA. And it is just a fact that an entire cell is what each parent passes on as an inheritance to its offspring.

It would be strange indeed if the organism's ability to proceed adaptively and creatively along paths of whole-cell developmental transformation were to become frozen at the very point where, via the most sophisticated activity imaginable, it prepares its whole-cell bequest for the next generation. Can we reasonably claim that this is the one cell lineage in which the organism's normal, future-oriented activity goes silent? Or that, with all the organism's expertise at producing and stably maintaining diverse phenotypes even without changes in DNA sequence, it "refuses" to employ this expertise when it comes to the preparation of inheritances? Or that the power with which the organism adapts all its cells, tissues, and organs as far as possible to new or unexpected conditions is a power lost to it in the management of its own germline?

If every organism is a living agent and power of becoming, as we know it to be, then surely that power of agency — whatever its nature, and however conditioned and constrained by the material results of its previous activity — is the decisive thing preparing the way for a new life. And yet our science has not even addressed the problem of this species-specific formative power, let alone asked about its source or about what role its unfolding expression and its development of its own potentials might play in evolution.

The questions we do ask — and ask compulsively — have to do with how an organism's genes mutate, not how, say, a mammal directs its single, inherited genome toward the radically different fates of a lens cell and a liver cell. Such cellular fates (not unlike the whole-organism fates of larva and beetle or tadpole and frog we also discussed above) are repeatedly and stably achieved before our eyes and with apparently casual ease, despite their being more complexly divergent over the space of a few weeks or months than the changes accomplished in a million years within many an evolutionary lineage.[4]

Another question we *could* wonder about is how all this creative potential of the organism bears upon the question of genetic mutation itself. And here we would have to reckon with the same future-oriented aspect we see in every cellular lineage during an organism's development, and indeed in all biological activity. Which is to say that the real question hasn't yet even been posed.

That question is not "How does a mutation affect this organism's fitness," but rather "How does it relate to where the species is going evolutionarily?" It cannot be emphasized enough that this latter question differs radically from that of fitness. After all, a tadpole in the process of transforming into a frog — having lost its tadpole organs for feeding and digestion, but not yet having completed the formation of the corresponding frog organs — is presumably not as fit as the fully mature frog. But this temporary "unfitness" is exactly what is required for the sake of a good future.

The unasked question: What can we make of the wisdom of the organism?

I have been speaking primarily about the organism's remarkable evolutionary potential quite apart from gene mutations. And for good reason, since the picture we've been given of genes and their mutations has undercut any interest the biologist might have had in that wise and inherently directive evolutionary potential. But, of course, the genome, too, belongs — and belongs importantly — to the whole organism. Everything we have learned in previous chapters about the purposiveness and end-directedness of organisms with respect to the management of their physical resources certainly holds true of their management of their genomes.

In her 1983 Nobel address, geneticist Barbara McClintock cited various ways an organism responds to stress by, among other things, altering its own genome. "Some sensing mechanism must be present in these instances to alert the cell to imminent danger," she said, adding that "a goal for the future would be to determine the extent of knowledge the cell has of itself, and how it utilizes this knowledge in a 'thoughtful' manner when challenged" (McClintock 1983). Subsequent research has shown how far-seeing she was.

James Shapiro, a leading microbiologist and geneticist at the University of Chicago, worked for a considerable time with McClintock, and has himself gone a good way toward achieving her "goal for the future." In his impressive and sprawling book, *Evolution: A View from the 21st Century*, he has painstakingly documented innumerable ways that organisms carry out what he calls "natural genetic engineering" on their "read-write genomes." The relevant molecular biological research is rapidly intensifying today and throwing ever more light on the subject. We can be quite sure that Shapiro's understanding will become more and more the "view from the 21st century." However, there is not much reason for me to recapitulate any of Shapiro's massive work here, and I wish only to add one line of thought of my own.

As long as there has been a modern biological science, it has been common for biologists to mention in passing the "wisdom of the organism." But this has hardly been a theme seriously influencing their scientific understanding, and I imagine that McClintock's rather more tendentious references to the cell "sensing" danger, and to the use of its "knowledge" of itself in a "thoughtful" manner, has raised more than a few skeptical eyebrows in the years since her Nobel address. But what, really, is the issue here?

Do we not know that the DNA of a human cell suffers tens of thousands of lesions ("mutations") per day, and that without the cell's *skillful* and well-informed repair of nearly all this damage we would not long survive? Or that when germline cells undergo the cell divisions producing gametes, they routinely and competently restructure their genomes via a process known as "genetic recombination"? Or that all dividing cells pass through "checkpoints" at which they *assess* whether they have accrued enough unrepaired DNA damage to require a *decision* in favor of cellular suicide? Or that immune cells creatively reconstitute their genomes so as to enable the potential production of many millions of distinct proteins required for immune activity — proteins that could not have been produced before the elaborate reconfiguration? Or that certain one-celled organisms (*Deinococcus radiodurans*) are capable of reassembling a functional genome after their chromosomes have been shattered into more than a thousand fragments by radiation (Chapter 8)? Or that topoisomerases — enzymes that cut one or both strands of a DNA molecule and then stitch them back together so as to release twisting tension or undo knots — do so with uncanny *knowledgeability*, so that their discoverer, Harvard molecular biologist James Wang, after calling the feat "amazing," explained:

> An enzyme molecule, like a very nearsighted person, can sense only a small region of the much larger DNA to which it is bound, surely not an entire DNA [molecule]. How can the enzyme manage to make the correct moves, such as to untie a knot rather than make the knot even more tangled? How could a nearsighted enzyme sense whether a particular move is desirable or undesirable for the final outcome? (Wang 2009).

And do we not know about the several hundred molecules that collaborate with surgical precision to remove parts of an RNA and splice together the remaining parts, typically preserving only a small fraction of the original molecule — all of which is accomplished *this way* rather than *that way* in order to produce the needed form of a protein under the current circumstances in a particular cell type (Chapter 8; Talbott 2024)? There are no mechanical linkages enforcing the outcome, and no instructions telling the diffusible molecules what sort of protein the larger context requires *here* and *now*. And yet, the molecular "surgeons" display all the expertise one could ask for. Do we have any idea how this expertise actually comes into play — or what part of our biological theorizing would remain if we only went as far as accepting the fact in front of our faces that it somehow *does* come into play?

Perhaps most importantly, do we not know of the technically overwhelming ways in which the whole cell brings all its intricately interwoven, almost infinitely complex regulatory resources to bear upon the expression genes — something I tried to give a slight hint of in Chapter 14 ("How Our Genes Come to Expression")?

We certainly have a right to worry about McClintock's use of words like "sensing," "knowledge," and "thoughtful," which so strongly suggest something like consciously directed human activity. But, however we care to think about organisms lacking our sort of conscious self-awareness and powers of reflective thought, the *effective knowledge* is somehow, undeniably there. Unfathomably, as far as our current thinking goes. Freely moving molecules are guided moment by moment — not by any *mechanism* contrived in the past, but with what we can only think of as a practical understanding of the detailed nature of the current task and, equally perplexing, a firm grasp of the needs of the present context. What can we make of this, and can we really afford to ignore it?

Think about it for a moment. *A superior wisdom vastly dwarfing any understanding we can consciously claim, is at work in all our bodily functioning.*

It's a wisdom through which the body, early on, launches each of innumerable cells upon one of hundreds of perfectly targeted, altogether different, multi-generation journeys, each of which eventuates in a differentiated state of a highly specific character. Each journey is a venture into the future requiring all the cell's resources, including its DNA, to achieve a whole-cell organization for which no roadmap or set of specifications is given in advance. And even if there were a map or set of explicit instructions, no one has any suggestion as to how molecules in the watery medium of a cell might be informed and guided by such instructions as they collide a million times every second with other molecules.

What we witness is not merely a set of complexly interwoven physiological processes impossible to encompass from moment to moment with our understanding minds. As an *effective* power, the wisdom of our bodies apparently acts (as you and I cannot in any conscious sense)[4] at the root of material manifestation just as physical laws do. And so this organic wisdom is able to entrain lawful material performances within the current of its own higher intentions and meanings.[5]

We might have asked long ago: Could such a wise and knowledgeable power, intrinsic and prerequisite to our material being, possibly *not* be enlisted in service of the future evolutionary state toward which an organism, as a member of an evolving species, was being "called"? And can we reasonably think that mutations or transformations of DNA, with which the immanent wisdom of our bodies is already so deeply engaged, are the one aspect of activity around DNA that this wise power is clueless about?

If you are skeptical at this point, what *do* you make of the bodily wisdom that accomplishes so much that is incomprehensible to you?

WHERE ARE WE NOW?

Does the organism's life have a bearing on evolution?

The powerful adaptive plasticity whereby organisms undergo concerted developmental change looks like exactly the sort of change that might translate, upon a wider stage, into the diverse organic transformations of evolution. But, oddly enough, the bare logic, or algorithm, of natural selection makes no reference to any specific potentials for organic transformation. On the other hand, we do discover such potentials playing out in the distinctive developmental trajectories leading from a single-celled zygote to osteoblast and endothelium, neuron and neutrophil. And we see them also when we watch the goliath beetle larva (or human embryo) metamorphosing into the adult form.

Only because we ignore the *living powers* required for such transformations do we subconsciously transfer our ineradicable sense of these powers to the working of a blind evolutionary algorithm — something we looked at in Chapter 16 ("Let's Not Begin with Natural Selection").

But the discussion of evolutionary issues and questions in the previous chapter and this one has so far been sketched on far too narrow a canvas. After all, it is not organisms individually that evolve, but populations or species or even larger groups. Furthermore, there is a very real sense in which we cannot even say that a collection of organisms evolves. The analogous truth would be this: we cannot say that it is a *collection* of cells that develops ("evolves") from a zygote to a human adult. That's not what we see. Starting with the zygote, and all along the trajectory, it is a whole, an undivided unity, that develops, and the cells come to be and gain their identity by being differentiated out of that unity. They are produced by the developing whole; they do not produce it.

There is no reason not to think similarly about the evolution of a population or species. What prevents us from doing so is our reluctance to recognize biological agency as the interior power of activity it is. But once we do recognize this — once we understand that the agency playing through a developing organism informs and governs perhaps trillions of cells with their relatively independent lives — we have no ground left for thinking it odd that something like this agency must play through a honey bee colony or school of fish or wolf pack or an entire species with countless individual members.

Just as individual cells participate in the life of a complex organism, so, too individual organisms participate in the life of a population, or species. In neither case is it always easy to distinguish what is individual from what is collective. And this suggests that the agency we recognize in individual organisms cannot be cleanly separated from the agency at work in the species — surely an idea the evolutionary theorist might run with.

But these remarks are only a kind of "advance warning" to brace you for some (I hope stimulating) intellectual turbulence ahead. Our task now is to keep our eyes open to the reality of organic transformation as we shift our focus from the development of individual organisms to the evolution of populations. We will begin to take up the issues in the next chapter.

NOTES

1. Gould 1976. By the time Gould completed his 2002 masterwork, *The Structure of Evolutionary Theory*, he would offer a richly nuanced qualification of these statements. But his fundamental belief in the creative role of natural selection — or, as he would say, its "efficacy" — remained.
2. Figure 17.1 credit: Goliath beetle larva: Ximonic, Simo Räsänen (https://commons.wikimedia.org/wiki/File:Mecynorrhina_torquata_ugandensis_-_Third_instar_larva.jpg), CC BY-SA 3.0. Goliath beetle adult: courtesy of Frantisek Bacovsky.
3. The organism's ability to transform its cells (that is, to transform itself) independently of genetic mutations during development becomes especially significant when we consider those evolutionary lineages where change seems to occur at an unexpected, almost preternaturally rapid pace. See, for example, the discussion of cichlid fish evolution in the lake region of East Africa (Chapter 19, "Development Writ Large"). But we would expect germline DNA, over generations, to be caught up in a species' self-transformation, just as are all other available resources. The main point is that DNA would not be the sole or leading factor in the change. It would, in its own way and like all the other parts, express the evolving whole, not govern it.
4. Actually, this is not true. We consciously move our own bodies, although we are not conscious of how it is done.
5. In this way our bodies show us in every detail of their activity why the supposed problem of physical determinism versus freedom is a falsely contrived problem whose solution stares us in the face every day. Every biological activity testifies magnificently to the fact that physically lawful interactions readily lend themselves to being caught up in larger, more fully meaningful performances. The laws of physics — which, we shouldn't forget, are ideas — are themselves expressions of one kind of meaning in the universe. And, like all meaning, they can serve the expression of higher meaning.

SOURCES

Arthur, Wallace (2004). *Biased Embryos and Evolution*. Cambridge UK: Cambridge University Press.

Bissell, Mina (2020). Asking the Question of Why (an interview with Mina Bissell). *Cell* vol. 181, no. 3 (April 30), pp. 503–6. https://www.cell.com/cell/fulltext/S0092-8674(20)30290-7

Gould, Stephen Jay (1976). This View of Life: Darwin's Untimely Burial, *Natural History* vol. 85, pp. 24–30.

_______ (2002). *The Structure of Evolutionary Theory*. Cambridge MA: Harvard University Press.

Kirschner, Marc W. and John C. Gerhart (2005). *The Plausibility of Life: Resolving Darwin's Dilemma*. New Haven CT: Yale University Press.

Lillie, Frank R. (1927). The Gene and the Ontogenetic Process, *Science* vol. 46, no. 1712 (October 21), pp. 361–68. https://www.jstor.org/stable/1651140

McClintock, Barbara (1983). The Significance of Responses of the Genome to Challenge, Nobel lecture (December 8). https://www.nobelprize.org/uploads/2018/06/mcclintock-lecture.pdf

Page, Scott L. and R. Scott Hawley (2003). Chromosome Choreography: The Meiotic Ballet, *Science* vol. 301 (August 8), pp. 785–89. doi:10.1126/science.1086605

Ryan, Frank (2011). *The Mystery of Metamorphosis: A Scientific Detective Story*. Foreword by Dorion Sagan and Lynn Margulis. White River Junction VT: Chelsea Green Publishing.

Talbott, Stephen L. (2024). How Do Biomolecules 'Know' What To Do? https://bwo.life/bk/ps/biomolecules.htm

Wahls, Wayne P. and Mari K. Davidson (2012). New Paradigms for Conserved, Multifactorial, *cis*-Acting Regulation of Meiotic Recombination, *Nucleic Acids Research* vol. 40, no. 20, pp. 9983–89. doi:10.1093/nar/gks761

CHAPTER 18

Teleology and Evolution

> It is somewhat odd that so many millions should have come to feel that the existence of, say, a poppy *as well as* a rose is a mystery deserving the closest attention, whereas the transformation of a minute poppy-seed into a full-blown poppy can be comfortably taken for granted. (Barfield 1971, p. 45)

EVERY ORGANISM IS continually dying in order to live. Breaking-down activities are prerequisites for building up. Complex molecules are synthesized, only to be degraded later, with their constituents recycled or excreted. In multicellular organisms such as vertebrates, many cells must die so that others may divide, proliferate, and differentiate. Many cancers reflect a failure to counterbalance proliferation with properly directed tearing-down processes.

You and I have distinct fingers and toes thanks to massive cell death during development. The early fetus' paddle-like hands give way to the more mature form as cells die and the spaces between our digits are "hollowed out." In general, our various organs are sculpted through cell death as well as cell growth and proliferation. During development the body produces far more neurons than the adult will possess, and an estimated ninety-five percent of the cell population of the immature thymus gland dies off by the time the mature gland is formed.[1] Even in our adult bodies, a million or so cells die each second.[2]

Despite all this life and death, I doubt anyone would be tempted to describe an animal's cells as figuratively "red in tooth and claw." Nor do I think anyone would appeal to "survival of the fittest" or natural selection as a fundamental principle governing what goes on during healthy development. *The life and death of cells appear to be governed, rather, by the form of the whole in whose development the cells are participating.*

But this has been a truth hard for biologists to assimilate, since it has no explanation in the usual causal sense. One way to register the problem is to ask yourself what you would think if I suggested that members of an evolving species thrive or die off in a manner *governed* by the evolutionary outcome toward which they are headed — that the pattern of thriving and dying off becomes what it is *because* of that outcome. It is not a thought any evolutionist is likely to tolerate.

But perhaps the occasional intrepid researcher will be moved to inquire: "Why not?" After all, we can also ask about the cells populating our bodies: do they thrive or die off in a manner governed, in some sense, by the forthcoming adult form? And here the answer appears to be a self-evident "yes."

Perhaps, when we can allow ourselves to reflect on what we see so clearly in individual development, we will find ourselves asking the "impossible" question about evolutionary trajectories: Does natural selection really drive evolution, or is it rather that the evolving form of a species or population drives what we think of as natural selection? Are some members of an evolving species — just as with the cells of an embryo's hands — bearers of the future, while other members, no longer able to contribute to the developing form of the species, die out?

What makes this idea seem outrageous is the requirement that inheritances, matings, interactions with predators, and various other factors in a population should somehow be coordinated and constrained along a path of directed change. Unthinkable? But the problem remains: Why — when we see a no less dramatic, life-and-death, future-oriented coordination and constraint occurring within the populations of cells in your and my developing bodies — do we not regard our own development as equally unthinkable?

Is it that we have simply learned to take for granted the coordination of cell births and deaths in the developing animal, since our extensive familiarity with animal development doesn't seem to leave us much choice in the matter? But apparently we do have a choice about whether to reckon in any profound way with the implications of this coordination for our thinking about biology in general, and our choice seems to be: "Let's just ignore it." And perhaps we are most assiduous in our ignoring when our thoughts turn to evolution.

So the question I am raising is this: once we accept the all-too-evident fact of an immanent coordinating agency at play in a population of cells pursuing a developmental trajectory, do we not have good reason to inquire whether an immanent coordinating agency is also at play in any population of organisms that is in fact pursuing an evolutionary trajectory?

Our approach to this question will undoubtedly be influenced by the degree to which we have taken seriously a general truth stressed throughout the first half of this book: agency and intention, wisdom and meaning, are given expression in all biological activity in a way that belies our expectations for collected bits of inanimate matter.

It will be part of my contention in forthcoming chapters that a coordinating power at work in evolving populations is as obviously apparent as the analogous power at work in developing organisms. It's not a conclusion based on radical new evidence, but rather one that depends only on a willingness to look at evolution with eyes that see, just as we do when observing a developing individual. We will do our best to look in this way. But first we need to deal with some of the prejudices blocking our way forward. That's what this chapter is about.

Are there obvious reasons to reject agency and teleology in evolution?

Every living activity we actually observe is purposive, or "teleological," or, as I have at times called it, "*telos*-realizing." It always has a contextual (holistic) dimension, and it always represents a further addition to a life story. We find ourselves watching, not necessarily a conscious planning (which is natural to humans), but rather the self-expression, or self-realization, of living beings. Physical events and causes are *coordinated* in the *interests* of a more or less centered *agency* that we recognize in cell, organism, colony, population, and species.

This coordination, these interests, this agency — they are already assumed, consciously or otherwise, by all biologists in the case of the individual organism's development and behavior. They are assumed, that is (as I have frequently been pointing out), insofar as one is doing biology, and not merely physics and chemistry. See Figure 18.1 for a well-known and meaningful performance in nature that would hardly raise an eyebrow among biologists.

Figure 18.1. A waggle-dancing bee (blurred image in center). The waggle dance is one of the ways honey bees communicate in a colony — in this case by providing information about the direction, distance, and richness of a food source.[3]

I tried to suggest in the opening section of this chapter that the agency and purposiveness so clearly manifest in the development of individual organisms is something we might also want to consider in relation to evolution. But, to most biologists, this is bound to seem a rather wild conjecture, and an impossible one at that. Let's listen to a few of the concerns that can so easily disturb our thinking about the role of agency — or, indeed, any sort of wisdom or intention (or, more broadly, interiority) — in evolution.

"How can you jump so casually from the hypothesized agency of a single, developing individual to that of vast, evolving populations?"

When we speak, not about physical processes as such, but rather about an underlying biological agency, intention, and purposiveness, then any radical distinction between an individual animal as a collection of molecules, cells, and tissues, on one hand, and entire populations as collections of animals, on the other, disappears. In neither case are we talking about the causal effect of discrete material things upon each other. Rather, we're trying to apprehend a governing unity — a principle of form, or what has been called a "formal cause" — in which those things are caught up.

The whole business of *telos*-directed biological activity as we observe it throughout all biology is to bring radically diverse physical processes — for example, those in the brain, heart, liver, intestines, and skin of a developing mammal — into a harmonious context, making a unified whole of them. As we have seen in Chapter 6 ("Context: Dare We Call It Holism?"), the unity of idea and meaning that makes for a context is not graspable in purely physical terms. Ideas and meanings are not physical entities, nor is their unity delimited by the boundaries we so easily imagine between physical things. Further, once we acknowledge the reality of purposive coordination and harmonization in one organic context involving many interrelated physical things, we have no reason not to look for it in other organic contexts involving different collections of physical things.

Of course, if one insists that all aspects of our understanding of life must be couched solely in terms of lawful physical interactions, then it will be impossible to accept anything I am saying here. But this insistence is exactly what I am now questioning. The biologist may want to quarrel with the present argument, but the quarrel cannot be furthered by endlessly repeating the very idea that is being disputed. It adds nothing to the discussion when one says, "The meanings you claim to be evident in all biological activity shouldn't be mentioned because we're bound by the rule that only physical interactions can be considered." That rule is the entire issue at hand. Those who want to argue that the meanings aren't really there should do so, or otherwise hold their peace. (Their argument, if they have one, is with the entire first half of this book.)

It may be that we no more understand the nature and origin of the observed powers of meaningful coordination in living organisms than we do the nature and origin of physical laws. Nor do we have any reason to assume that the powers of coordination are less fundamental than the physical laws. Neither the powers nor the laws are physical things, and what we *can* assume is that both come into play jointly, harmonizing with each other at the very origins of material manifestation. (We do not have matter first, and *then* the lawful ideas it obligingly "obeys.") If anything, an inherent power to choreograph physically lawful activity in a meaningful manner, however poorly understood, would seem more fundamental — and more fundamentally generative or creative — than the physical lawfulness of the processes being choreographed.[4]

Given our ignorance of the ultimate nature of things, the most immediate path forward when the teleological question arises in a particular context, is simply to observe everything we can about the adjustment of means toward the fulfillment of needs and interests. In this way we get to know agency at work, just as we also get to know physical laws at work.

But this much can be said already. Wherever we find *telos*-realizing entities somehow bound together in a larger unity, we see one example after another where the more comprehensive entity or context manifests in turn a teleological character of its own that somehow "irradiates" and harmonizes the teleology of its parts. (Compare the whole organism and its relatively independent cells and organs.) Whether it is all the molecules in a cell, or all the cells in an organism, or all the organisms in a coherent group (say, an insect colony or mammalian social group), we always find a weaving of lower-level narratives into the distinctive and harmonious intentional fabric of a larger story.

So we can hardly help asking the teleological question in an evolutionary context: When a species consisting of many *telos*-directed individuals moves along a coherent evolutionary trajectory, do we see the species displaying its own sort of *telos*-directed developmental potentials? We must be willing to look with open and honest eyes.

"You speak of a harmonization of physical elements in tune with meaning and purpose. But how can this occur without relevant causal connections between the elements? We can see such connections clearly in a developing organism, where all the parts are contiguous. But huge numbers of different organisms in different, interacting species, often scattered over a large geographic area, are a different matter."

Yes, a very different matter. And I know of nothing preventing us from hypothesizing that any purposive pattern of physical events requires that there be efficient causal connections between those events.

But this requirement for causal relations is hardly a problem, given that the members of evolving populations of organisms have no fewer or less relevant physical interactions than the aggregated cells in an individual body. Eating and being eaten are surely causal in the usual physical sense! And, of course, not only predator-prey relations, but also mating choices, territorial movements, various means of communication, lateral gene transfer mediated by microorganisms and viruses, and everything bearing on survival and death already figure importantly in conventional evolutionary theory.

Isn't the entire body of evolutionary theory today concerned with physical causation? Surely conventional theory is a *physical theory*, and its proponents

Figure 18.2. Animals gathered at a dry watering hole in a Namibean national park.[5]

believe they have all the causal interactions they need. The question about purpose, intention, and meaning is a question about the transformative *organization and coordination* of the physical processes already identified by evolutionary biologists. Or, to employ an older concept: it's a question about the *formal causes* at work ordering all the efficient (physical) ones.

Actually, the reality of a coordinating power weaving through and governing large, scattered populations of organisms is already put on display for us before we even think about evolution. It is displayed, for example, in instinctual behavior such as that of migrating monarch butterflies in eastern North America. Huge numbers of these gather from throughout a wide area, including parts of the United States and Canada, and may travel thousands of miles over multiple generations to a precise spot in Mexico — all this along aerial pathways they have never traveled before. Later this well-directed journey is reversed.

Or consider the sophisticated collective behavior of a wolf pack, an ant colony, or even the cells — bacterial and otherwise — of a biofilm. The latter has been termed a "city for microbes," and the complex, teleologically rich organization of a city is an apt picture of the life of a biofilm. In all these different sorts of collectives, the power of end-directed coordination, whatever we take it to be, seems to work across the relevant communities, all the way down to the molecules that actively participate in the bodily performance of the various organisms.

So I come back to my initial line of thought. Suppose, for the sake of argument, that an animal's mating choices, its preparation of inheritances for its offspring, and all the other relevant causal factors are guided, or end-directed, in a manner leading to coherent evolutionary change. My point right now is not (yet) that this supposition is easily confirmed. I am only saying that it poses no further problems for our physical understanding beyond those already posed by all the cellular inheritances and other interactions within the many proliferating and radically diverging cell lineages in a complex, developing organism. These, too, are guided toward the future in a manner leading to coherent and progressive change.

And yet it all happens in what we might call a causally "closed" manner. There are no gaps in the succession of physically lawful interactions where we say, "Here a miracle occurs." As I remarked above, the *telos*-direction of biological processes, like their physical lawfulness, appears to enter the picture at the very origins of material manifestation in organisms — not as an external imposition upon an essentially teleology-free material reality.

(Unfortunately, the "origin of material manifestation" is not a topic likely to come before the biologist's attention. But see Chapter 24, "Is the Inanimate World an Interior Reality?")

The relevant causal connections of an organism's development do not suggest that the individual cell must consciously "know" about the developmental trajectory in which it is caught up. We can assume that the same would be true of an individual organism caught up in an evolutionary trajectory. In neither case would this "not knowing" prevent the individual entity from lending itself to, and being interpenetrated by, the larger end-directed process in which it participates. And it's worth noting that neither the interpenetration nor the fact of a "larger process" is possible without an immaterial unifying idea, or formal cause, working across thing-like boundaries.

None of this is to deny that the particular principles of coordination in evolution must in some ways differ from those in individual development, as we will see shortly. But whatever the principles are, we will not discover them by looking at the laws of physics and chemistry. We will begin to grasp them only when we are able to read each particular context in terms of its own meanings, self-realizing powers, and directions of movement. We are already pretty good at this in the case of individual development. There is no reason not to try looking in an analogous way at evolving populations.

"It sounds as though your wonderful 'agency' can accomplish just about anything. But can you explain the presence or the nature of this agency?"

I pointed out above that we no more understand the nature and origin of the observed powers of meaningful, narrative coherence in living organisms than we do the nature and origin of physical laws in the inanimate realm. We don't doubt the laws because we see them so clearly at work, and observing the patterns of their working has been enough for us. Perhaps this disciplined observation will also be enough when it comes to the powers of organic intention and purposiveness, once we overcome the taboo that says we can't even look seriously at them.

We certainly know this much already: in every organism (that is, in all biology), diverse processes are coordinated toward a common end. This points toward a principle of interpenetration and unity (holism) that is universal in biology. The general rule is that we always find ourselves looking at wholes embedded within still larger wholes, and contexts overlapping other contexts. This is clearly evident when we consider the integrated unity of a physical body with all its cells, tissues, organs, and organ systems. It may take some effort, but we have to learn to think routinely in terms of this embeddedness of wholes and overlapping of contexts.

In Chapter 6 we heard how the botanist Agnes Arber described the relative character of organic wholes:

> The biological explanation of a phenomenon is the discovery of its own intrinsic place in a nexus of relations, extending indefinitely in all directions. To explain it is to see it simultaneously in its full individuality (as a whole in itself), and in its subordinate position (as one element in a larger whole).

From flocks, herds, and schools, to bee and ant colonies, to parasitic and symbiotic pairs, to more or less closely aggregated communities of cells, to the collective, highly differentiated, and elaborately integrated communities of cells in our own bodies — there are many different contexts of agency. We discover agency and intention wherever we find participants bound together in a larger, more or less focal community that unfolds its purposive activity along a continuous and well-directed pathway according to its own distinctive meanings.

The honey bee hive functions, in this sense, as a (relative) whole with its own agency. We have no difficulty recognizing this agency in the hive's pattern of coherently directed activity. The participants in the hive have no absolute discreteness or wholly independent identity. But neither do they lose all individual identity. It is a matter of one identity participating in a greater one.

If, as Arber suggests, biology presents us with interpenetrating wholes, then we should also expect to see interpenetrating agencies expressed in those wholes. The distinctive character of, say, a mammalian genus (or any other taxonomic group) is not silenced by, but rather informs, the character of each species within the group.

The question, "How does it all work?" calls only for closer observation of the detailed playing out of various general principles for which we already have ample evidence. (See, for example, Chapter 2, "The Organism's Story.") We may of course hope to come upon a deeper understanding of the nature and origin of biological agency. But I suspect that the demand for "something more" commonly stems from the errant feeling that if we can only identify more physical causes, they will spare us the need for formal causes and for the implied interiority.

"Isn't the idea of agency, when applied to organisms in general, a rather disastrous anthropomorphism?"

Anthropomorphism is indeed a supreme danger in biology. Think, for example, of all the human activity we rather blindly import into the organism when we analogize it to a machine. (See the section about the machine model of organisms, in Chapter 10.) Similarly, it would be highly misleading to think of biological agency in general as if it were like the directive activity of an individual human agent.

To begin with, human agency itself is not as neat and unambiguous as we may be inclined to suppose. A fully sovereign individual does not exist. Who among us can say that he is motivated solely by his own will? Who does not at times yield gladly to internalized and inspiring "voices" — for example, of parents, teachers, and mentors, or religious figures, or uplifting literature? And who does not also wrestle with lower, less worthy urges? What young child subjected to extreme abuse does not carry into adulthood the burden and unfreedom of a psychic complex expressing some of the disastrous ideational, affective, and volitional powers of his abusers? Or again, which of us is absolutely immune to the collective ecstasy, hysteria, or rage of a massive crowd "rooting for the home team" or submitting to the spell of a charismatic leader?

It is true that, when we speak of agency, we speak of capacities we ourselves routinely and, at times, consciously exercise. But we must also admit that our experience of our own agency is closely bounded on all sides by mystery. We do not fully understand where our thoughts and actions come from, or how our intentions move our bodies. It would be a mistake to clothe the mystery of biological agency in the imagined form of a grandly sovereign, all-knowing, perfectly harmonious human individual.

And if we cannot be entirely clear about the sources of agency in our own lives, we can hardly be dogmatic about the nature of the agency — or diverse agencies — at work in a single bee colony, a particular species of rodent, or the biosphere as a whole.[6]

But nothing prevents us from being good observers of living beings, which is also to be observers of the clear *manifestations* of biological agency in its different forms. In this way we become familiar with the complex and perhaps many-voiced character — the way of being — of particular organisms. We learn to know "from the inside" one species as distinct from another, and from ourselves. See, for example, the description of the sloth in Chapter 12.

If we humans are part of nature and the evolutionary process, why should we think that anything in ourselves is *absolutely* alien to other organisms. It has been pointed out often enough that we carry something of the animal within ourselves. This seems to suggest that all animals carry something of the human within themselves. Recognizing sameness as well as difference, and difference as well as sameness, seems fundamental to the scientific project. If the anthropomorphic projection of human traits onto other organisms can be a problem, so can the denial of human traits as "unnatural."

"But the simple fact is that evolution is not individual development. Don't you need to reckon with this fact?"

Yes. One obvious difference between development and evolution is that cycles of individual development are endlessly and reliably repeated before our eyes, so that no one can avoid at least unconsciously recognizing their teleological character. Time and again, amid all the inconstancies of life and environment, mouse zygotes develop into adult mice, just as newborn dogs and cats become full-grown.

Evolution, by contrast, encompasses the totality of life on earth, and occurs only once. No more than in reading a good novel can we predict, mid-way through the story, its later outcome, even if that outcome turns out to be the end toward which everything was tending.[7] This non-repeatability of evolution makes it all too easy, for those bent on doing so, to "forget" everything they know about the creative and end-directed character of all the life processes through which evolution occurs.

There are, of course, other distinctions between individual and evolutionary development. In the latter case we see (in those organisms reproducing sexually) a continual merging of separate hereditary lineages. There is also the fact of hybridization across species, genera, and even families. None of this occurs among the cells of a developing organism, even if cells have shown remarkable plasticity enabling, for example, the transformation of one cell type into another — up to and including the reversion of a differentiated cell into a stem cell — given the right contextual signals. And some evolutionary features figuring strongly in current theorizing — predator/prey relations, collective migrations, symbioses of various sorts, cultural inheritance, and lateral gene transfer — also serve to remind us that, while communities of organisms (think of the human microbiome) can be vitally important even for individual development, they become central in evolution.

We have no reason to assume that the play of purposiveness across all the cells of a complex, developing organism is exactly analogous to its play among the members of a species or population. Nor need we asume that the more or less fixed stages through which individual development passes give us a neat roadmap for the course of evolution.

We do, however, have at least one foundational principle: *nothing can become a fact of bodily evolution that was not first a fact of individual development.* The material substance of evolutionary transformation must first of all reveal itself within individual organisms. How these organisms subsequently merge (or fail to merge) their heritable features is another story.

The shortest path to confusion is circular

One last question, or objection, deserves a section of its own. It could be put this way: *Aren't you committing an egregious sin of omission by ignoring what everyone knows? Natural selection explains everything that needs explaining about the appearance of agency and purposiveness.* And it's true, I'm sure, that any reader with a conventional biological training will share this concern. Natural selection, so the story goes, "naturalizes," or explains away, the agency and purposiveness we observe in organisms. That is, explains it without appeal to any principles other than purely physical ones.

Biologists often think of purposiveness, or teleology, under the concept of function, as when they say that a trait is "for the sake of" this or that, or an organ exists "in order to" achieve a particular end. And so, as philosopher David Buller has summarized common usage, "the function of the heart is to pump blood, the function of the kidneys is to filter metabolic wastes from the blood, the function of the thymus is to manufacture lymphocytes, the function of cryptic coloration (as in chameleons) is to provide protection against predators."

But all this poses difficulties for a science that would honor its materialist commitments, since the concept of function, as Buller observes, "does not appear to be *wholly* explicable in terms of ordinary causation familiar from the physical sciences." Whereas kidneys may continually adjust their activities and their own structure *in order to* do the best possible job of filtering metabolic wastes from the blood, no physicist would say that falling objects adjust their activities and their own structure *in order to* reach, as best they can, the center of the earth. More generally, organisms may strive to live, but physical objects do not strive to maintain their own existence. Organisms, so it seems, have intentions of their own, whereas physical objects are simply moved from without according to universal law.

So the problem for biologists has been to explain, or explain away, their persistent and seemingly inescapable language of purpose, even if it is couched in terms of function. And the need is to do the explaining in a respectable, materialistic manner — that is, without acknowledging that organisms really are purposive beings in the sense of exercising, or being possessed by, an interior (immaterial) activity of a wise, meaning-infused and intentional sort. But this problem — so it might seem — has been fully solved in recent decades.

Buller, who was writing at the turn of the twenty-first century, was able to point to a "common core of agreement" representing "as great a consensus as has been achieved in philosophy" — an agreement that "the biological concept of function is to be analyzed in terms of the theory of evolution by natural selection." More particularly, "there is consensus that the theory of evolution by natural selection can provide an analysis of the teleological concept of function strictly in terms of processes involving only efficient causation" — the kind of purposeless causation physical scientists accept as applicable to the inanimate world[8] (Buller 1999).

So we no longer need to think of organisms as having genuine intentions, purposes, or *telos*-realizing drives issuing from their own interiors — no longer need to struggle with the problem of teleology, or end-directed activity. Teleology, we must believe, has been tamed, leaving biologists safe in a world without meanings, wisdom, purposiveness, intentions or other signs of living interiority.

To put the most common version of the idea very simply, organisms are said to possess teleological, or purposive, features *because those features are present by virtue of natural selection*. That is, they were selected for the very reason that they effectively serve the organism's crucial ends of survival and reproduction. And since natural selection is supposed to be a perfectly natural process — meaning that it is supposed to involve nothing "mystical" like *real* purpose, intention, or thought — we can know that the functionally effective traits given us by natural selection are straightforward exemplars of physical lawfulness and nothing else, whatever they might *look* like.

The solution Buller alludes to amounts to saying: (1) purposive traits arise through natural selection; and (2) because natural selection is defined as a matter of genes, mutations, and patterns of life and death — all in a straightforward, mechanistic fashion — we can say that the organism's evident purposiveness has been "naturalized." Nothing of real purposiveness, as opposed to an apparent purposiveness, remains.

The assumption seems to be that one needs only to invoke natural selection in order to "naturalize" and explain any feature of life. In this manner one could say, for example, that natural selection explains faster-than-light teleportation simply because it is a useful trait conducing to survival. But if (hypothetically), upon seeing teleportation actually happening, we invoked natural selection to explain it, wouldn't we still need to reconcile teleportation with our existing understanding of physical reality? By "solving" the problem with an appeal to natural selection, haven't we simply shifted the burden of explanation from the biologist to the physicist — while also showing how natural selection can be said to solve a problem without actually solving it? (On unrealistic expectations for explanation by natural selection, see Chapter 16, "Let's Not Begin with Natural Selection.")

Buller does not even attempt to perform the work of the physicist here. He does not enlighten us in even the vaguest terms about how traits expressing the physically "impossible" aspects of purposiveness might have come about through natural selection. Which is to say that he does not solve the puzzle we saw posed in Chapter 8 ("The Mystery of an Unexpected Coherence") regarding RNA splicing: How can we understand the intricate, complex, sequential operations of scores of molecules in a fluid milieu as they perform a kind of molecular surgery requiring a moment-by-moment exercise of intelligently directed intention allowing them to carry out this coherent surgery when they could just as well do a million other things? What they accomplish in a remarkably well-coordinated way is guided, not by gears, wires, levers, or incised channels of communication in silicon or any other material, but rather by the current and ever-changing needs, tasks, and functions of the local and more distant contexts.

What is happening with Buller here seems to me fairly obvious. He poses his initial problem in terms of reconciling a real interiority (purposiveness, intention, intelligence) with a materialistic conception of reality. But when he achieves his solution in terms of natural selection, he is no longer thinking of *real* interiority, but something more like a machine's *appearance* of interiority, an appearance that comes about thanks to the design activity of a human engineer. So he has radically switched the terms of his problem, and feels no need (as long as he manages to ignore the human engineer) to explain how real interiority can be reconciled with his materialistic conception of the world. He's just not dealing with the real problem of interiority any more.

The lacuna in conventional understanding here is truly astounding — or would seem so if biologists had not so routinely learned to hail natural selection as a kind of divine "Invisible Hand," accounting for whatever needs accounting for.

As for natural selection more generally, there are three serious difficulties I will note here:

(1) The problem of the "arrival of the fittest" remains

To say that natural selection preserves traits promoting the survival of organisms does nothing to explain how those traits might have arisen, or even (as we have just seen) whether they are compatible with materialist thought. This depends on what the preserved traits are and how they arose. The preservation of an already existing trait is an entirely different matter from its nature and origin. In our present example, claiming that teleological features or activities already existed at some time in the past and then were preserved by natural selection merely pushes the problem of "naturalizing" them — making them acceptable solely in physical and materialist terms — back to an earlier time, without solving it.

We heard about this in Chapter 16, where prominent figures in evolutionary biology over the past century and more complained that natural selection — even if it explains the survival of the fittest — cannot explain the arrival of the fittest. In conventional evolutionary thought the arrival of traits is mostly taken for granted, with natural selection then playing a role in their preservation and their spread throughout a population.

Let's put it this way. Yes, purposive features — *if* they could be had in a strictly physical world — would conduce to the survival of organisms, and therefore might be preserved. But the mere fact of preservation doesn't show us that the features *did* in fact arise in a strictly physical world — doesn't show us, in Buller's words, that they are fully "explicable in terms of ordinary [physical] causation," or are not expressions of a *real*, interior purposiveness.

Given the historical persistence of the complaint by leading biologists about natural selection and the arrival of the fittest, it is remarkable that the arguments today about how natural selection explains teleology generally proceed without so much as an acknowledgment of the problem.

(2) The explanation assumes what it is supposed to explain

It is important to realize that purposiveness is not just a particular, late-arriving trait, but runs through *all* biological activity. It is reflected in the coordinating principles that account for the integral, interwoven unity of the organism's life. The complexity theorist and philosopher of biology, Peter Corning — who appears to hold a conventional, materialist view of life — was nevertheless gesturing toward this purposive unity when he wrote that living systems "must actively seek to survive and reproduce over time, and this existential problem requires that they must also be goal directed in an immediate, proximate sense ... Every feature of a given organism can be viewed in terms of its relationship (for better or worse) to this fundamental, in-built, inescapable problem" (Corning 2019).

Rather than being just one more discrete trait that might have been neatly evolved at some particular point in evolution, the *telos*-realizing capacity of organisms reflects their fundamental nature. It is what "living" means. We are always looking at a live performance — a future-directed performance, improvised in the moment in the light of present conditions and ongoing needs — not a mere "rolling forward" of some blind physical mechanism.

Here we encounter a staggeringly obvious problem. You will recall from Chapter 16 that natural selection is supposed to occur when three conditions are met: there exists variation among organisms; particular variations are to a sufficient degree inherited by offspring; and there is a "struggle for survival" that tends to put the existing variants to a life-or-death test. But — and this is the crucial point — *all* the endlessly elaborate means for the production of variation, the assembly and transmission of inheritances, and the struggle for survival just *are* the well-regulated, end-directed activities whose teleological character biologists have tried to explain away. So the basic conditions enabling natural selection to occur could hardly be more thoroughly teleological.

In other words, the purposive performance of an organism is a pre-condition for anything that looks at all alive and capable of being caught up in evolutionary processes of trait selection. So the common form of the argument that natural selection explains the apparent purposiveness of all biological activity appears to assume the very thing it is supposed to explain. Purposiveness is built into the core presuppositions of natural selection itself, which therefore presents us with the problem instead of removing it. It would be truer to say that teleology explains natural selection than that selection explains teleology.[9]

Although this problem regarding the supposed explanation of teleology has been almost universally ignored among biologists, it has not been entirely overlooked. Georg Toepfer, a philosopher of biology at the Leibniz Center for Cultural Research in Berlin, has stated the matter with perfect directness:

> With the acceptance of evolutionary theory, one popular strategy for accommodating teleological reasoning was to explain it by reference to selection in the past: functions were reconstructed as "selected effects." But the theory of evolution obviously presupposes the existence of organisms as organized and regulated, i.e., functional systems. Therefore, evolutionary theory cannot provide the foundation for teleology[10] (Toepfer 2012).

(3) There are no stable mechanisms for selection to work on

As was mentioned in Chapter 16 ("Let's Not Begin with Natural Selection"), biologists conceive evolution by natural selection as tinkering with mechanisms that stably remain in place from generation to generation. These mechanisms can then be tinkered with further in the future. In this way, so it is thought, they can be improved, making it possible for complex and well-developed traits to be achieved over great spans of time.

The problem with this conception is that we never find the required sort of mechanisms in organisms, let alone stable mechanisms passed as such between generations so that they can be tinkered with over millions of years. This can perhaps be seen most clearly by looking at the molecular level, which virtually all contemporary biologists take as the ultimate basis for explanation.

Pick any substantive molecular process, such as RNA splicing (Chapter 8, "The Mystery of an Unexpected Coherence") or DNA damage repair and you see hundreds of molecules cooperating in a task fully as complex as any brain surgery. It all happens while the molecular "surgeons," lacking brains of their own, are moving in a watery medium with almost infinitely many physically possible interactions to "choose" from. There are no wires, gears, levers, or other mechanisms of control even remotely capable of simultaneously guiding the intricate, fluid interactions of all those interacting molecules and ensuring that the interactions occur, not haphazardly, but in exactly the right sequence — and no computer-like software able to coordinate and organize such mechanisms of control (Talbott 2024). We never see a command center issuing detailed instructions from which the interacting molecules might "read off" their moment-by-moment roles as each one of them collides with hundreds of thousands or millions of other molecules every second (Chapter 15, "Puzzles of the Microworld").

The entire picture of the evolutionary tinkerer acting upon preserved mechanisms in a cumulative fashion through natural selection is a blatant fabrication. No mechanisms are there to be tinkered with, and none are even imaginable. There is not, and never was, anything like a mechanism to perform RNA splicing or DNA damage repair. Whatever is going on needs to be explained, not as the after-effect of a mechanism assembled at some time in the distant past, but as a wisdom intelligently at work in the present moment while taking full account of the changing and never fully predictable circumstances.

The same problem of the working of a present wisdom inheres in virtually every molecular process in every organism. Whether a cell is dividing or food is being digested, crucially important molecules are embarked upon an elaborate, reasonable, and meaningful journey the overall details and pattern of which are underwritten by no physical mechanisms. If there have been efforts to make the conventional picture of tinkerable mechanisms even marginally more realistic, I am not aware of them.

WHAT, THEN, CAN WE MAKE of the theory of natural selection? It's a theory misleadingly focused on the idea of individual fitness rather than on the evolutionary outcome that is the final object of explanation — an outcome that could involve a crucial role for distinctly "unfit" organisms; a theory that has not even managed to define its pivotal parameter — the single organism's fitness — with any scientific clarity; a supposedly causal, evolutionary theory quite unable to understand the genesis of the "ubiquitous variation" in traits it takes for granted; a theory that, when it does allude to these traits, says no more than that they *somehow* correlate with randomly generated mutations, where the appeal to randomness not only is the very opposite of scientific explanation, but also seems to contradict the infinitely complex, wholly intentional, thoroughly qualitative, interwoven unity of being of a shark or kangaroo or a dog or cat; and a theory founded aggressively and restrictively on the genome — that is, on a single aspect of the cell rather than on the whole-cell life that so obviously governs the genome.

It appears, then, that there's not much compelling about the contemporary theory of evolution by natural selection. Without a clear definition of the problem it is meant to solve, it has veered off into various scientific research strategies that certainly have led to many profitable observations of organisms and their interactions, and certainly have produced a great deal of data about changes in relatively minor traits — changes that are often reversed later, as in the famous case of the coloring of peppered moths in England — all in the absence of anything we could call a fundamental theory of evolution. Perhaps it is no wonder that so many who would speak for science on the topic of evolution today, as we saw in Chapter 16, prefer to celebrate the "inevitable" logic of natural selection instead of demonstrating its explanatory contribution to our understanding of actual evolutionary transformations.

An aversion to meaning

The theory of natural selection gives us no argument explaining away the self-evident purposiveness of organisms. To the contrary, it confirms the theorist's largely unacknowledged recognition of this purposiveness. For we can make sense of natural selection only after we have thoroughly internalized, from childhood on, a vivid awareness of the lively agency, whether of cats and dogs, birds and squirrels, worms and fish, or of the animals in our laboratories. The scientist can take this agency for granted without having to mention or describe it, since everyone else also takes it for granted. It's what we observe every day.

This may be an extraordinarily naïve way to do science and philosophy, but, well, there it is. And so one speaks ever so casually of individual "development," or the "struggle for life," or the "production of variation," or "reproduction and inheritance" — all in order silently to import into theory the full range of the living powers that made biology a distinct science in the first place, but that few today are willing to acknowledge explicitly in their theorizing.

Several decades ago the British biologists Gerry Webster and Brian Goodwin had already noticed that "the organism as a real entity, existing in its own right, has virtually no place in contemporary biological theory" (Webster and Goodwin 1982). Goodwin later elaborated the point in his book, *How the Leopard Changed Its Spots*:

> A striking paradox that has emerged from Darwin's way of approaching biological questions is that organisms, which he took to be primary examples of living nature, have faded away to the point where they no longer exist as fundamental and irreducible units of life. Organisms have been replaced by genes and their products as the basic elements of biological reality (Goodwin 1994, p. vii).

The banishing of organisms from evolutionary theory was also an obscuring of biological purposiveness. It may even be that the banishing happened mainly *for the sake of* this obscuring. Yet who can doubt that, if we ever do take the purposive organism into account at anything like face value, the results could be of explosive significance for all of evolutionary theory?

It is difficult to pinpoint whatever lies behind the extraordinary animus the biological community as a whole holds, not only toward teleology, but indeed toward any meaningful dimension of life or the world. But the animus seems as deeply rooted as it could possibly be. Michael Ruse, who might be regarded as a dean of contemporary philosophers of biology, once briefly referred to an article by the highly respected chemist and philosopher, Michael Polanyi, in this manner:

> Polanyi speaks approvingly, almost lovingly, of "an integrative power ... which guides the growth of embryonic fragments to form the morphological features to which they embryologically belong."

And what was Ruse's response?

> One suspects, indeed fears, that for all their sweet reasonableness the Polanyis of this world are cryptovitalists at heart, with the consequent deep antipathy to seeing organisms as being as essentially physico-chemical as anything else ... Shades of entelechies here! (Ruse 1979)

Figure 18.3. Michael Polanyi. Having made many scientific contributions, Polanyi became a Fellow of the Royal Society. He was a Gifford lecturer, and author of the books, *Personal Knowledge: Towards a Post-Critical Philosophy* and *The Tacit Dimension*, among others.[11]

The assumption that the Polanyis of this world are antipathetic toward the idea that organisms are "as essentially physico-chemical as anything else" is a mere distraction from the real issue. No one needs to, or should, deny that organisms are perfectly reliable and unexceptionable in their physical and chemical nature. (Certainly Polanyi did not deny this.) By injecting his unfounded "suspicions" into his argument, Ruse simply abandons his responsibility as a philosopher to deal honestly with his antagonist's thought.

The real question has to do with the distinctive organizing ideas we find to be characteristic of organisms. After all, no one claims that the lawful ideas of the physicist are mystical just because laws are not physical things. They belong to the immaterial nature of inanimate phenomena. So why should we refuse to acknowledge the readily observable organizing ideas characteristic of animate phenomena? There is a burden of explanation here that Ruse seems not even to recognize, let alone to engage.

The real antipathy appears to be on Ruse's part. One wonders exactly what violation of observable truth he saw in Polanyi's reference to "an integrative power" that "guides" embryological growth. No biologist would dare deny that embryological development is *somehow* integrated and guided toward a mature state. And it is difficult to understand how any actual integrating and guiding could be less than the expression of an effective "power," however we might end up understanding that term. Just think how much less justification there is for all the conventional references to the "power," "force," and "guidance" of natural selection! (On that, see Chapter 16, "Let's Not Begin with Natural Selection.")

As for Ruse's shuddering at the term "entelechy" (sometimes rendered as "soul"), the scholar who is perhaps the foremost interpreter of Aristotle today translates the Greek *entelecheia* as "being-at-work-staying-itself" (Sachs 1995, p. 245). What better characterization of an organism and its distinctiveness relative to inanimate objects could there possibly be? Every biologist who uses the conventional term "homeostasis" (a system's maintenance of its own material stability) or, better, "homeorhesis" (a system's maintenance of its characteristic activity) is already saying something similar to "being-at-work-staying-itself." It's the way of being of any organism. The Aristotelian term is useful for reminding us that an organism is first of all an activity, and its activity is that of a centered agency possessing a remarkable coordinating and integrative power in the service of its own life and interests.

On our part, we will now do our best to begin reading the organism and its activity back into evolutionary theory. In doing so, we will ignore the strange taboo against acknowledging living powers and purposiveness wherever we see them at work.

WHERE ARE WE NOW?

Is teleology disallowed in the theory of evolution?

An animal's development from zygote to maturity is a classic picture of *telos*-realizing activity. Through its agency and purposiveness, an animal holds its disparate parts in an effective unity, making a single, ever more fully realized whole of them. This purposiveness extends "downward" from the whole so as to inform the parts, and "outward" from the inner (immaterial) intention toward the exterior (material and perceivable) expression. It is invisible to any strictly physical analysis of the interaction of one part with another.

Biologists in general have failed to take seriously the reality of the organism's agency, and have considered it unthinkable that something analogous to the agency playing through all the cells of an individual organism could play through all the members of an evolving population of organisms. The main lesson of this chapter is that we have no obvious grounds for making a radical distinction between the two cases.

A central point is that we no more understand the origin and nature of physical laws than we understand the origin and nature of biological agency. Nevertheless, we are quite able in both cases to observe how they work.

Moreover, the current unwillingness of biologists to reckon with the possibility that evolution gives us a coherent, *telos*-realizing narrative does not appear to be explained by the differences between individual development and evolution (which are very real), but rather by a refusal to take seriously the problem of active biological wisdom and agency in either case.

The uncomfortable truth is that biologists have yet to come to terms with the physically puzzling fact of purposive biological activity — which is to say, *all* biological activity. To suggest that evolution is *telos*-realizing is not to suggest some new kind of problem. It is merely to say: let's face up to the reality of teleological development and behavior that has already long been staring us in the face.

We also looked at three closely related problems with the general consensus among biologists that natural selection somehow explains (or explains away) biological agency and purposiveness:

- The *preservation* of purposive (functional) traits — or any traits at all — by natural selection neither explains their *origin* nor shows how they can be understood solely in terms of physical lawfulness.
- Selection itself is defined in terms of, and thoroughly depends on, the purposive lives of organisms. This purposiveness must come to intense expression in order to provide the basic pre-conditions for natural selection. These conditions are the production of variation; the assembly and transmission of an inheritance; and the struggle for survival. Since the entire logic of natural selection is rooted in a play of purposiveness, it cannot explain that purposiveness.
- Biologists conceive natural selection as tinkering with mechanisms that survive as such from generation to generation, so that they can be tinkered with further in order to achieve complex and well-developed traits. The problem with this conception is that there are no such enduring *mechanisms* to be progressively tinkered with. It is impossible to think of scores or hundreds of molecules cooperating in a watery medium to achieve an intricate task such as RNA splicing or DNA damage repair as if they constituted fixed, stable mechanisms that natural selection could tinker with over evolutionary time spans.

My aim in this chapter has been to clear away some of the major stumbling blocks biologists inevitably conjure up whenever they hear it said that evolution has a purposive, or teleological, character. Of course, there remains the question whether evolution does *in fact* show such a character. Does the evolution of species show the same kind of thought-imbued creativity we see at work in the development of individual organisms? Or, perhaps: can we intelligently imagine such an organic wisdom *not* being at work?

We will see that — just as with individual development — the question is answered as soon as it is asked. Once the metaphysical biases against the very idea of teleology are removed, all we need to do is look, and it's as if our eyes themselves are enough to give us our answer.

NOTES

1. Rich, Watson and Wyllie 1999. The authors go on to mention that, while researchers naturally tend to focus on cell survivors, "it is striking that, even with a sophisticated understanding of survival signals, we still know remarkably little of the reciprocal process by which, of the seven million germ cells present in the ovary of the mid-term human foetus, the vast majority is lost by the time of birth."

2. Green 2022:

    Every second, something on the order of one million cells die in our bodies. This is a good thing, because cell death is central to efficient homeostasis and adaptation to a changing environment. When, for some reason, it does not occur, the consequences can be catastrophic, manifesting as cancer, autoimmunity, or other maladies.

3. Figure credit: Courtesy of Dr. Christoph Grüter, leader of the Insect Behaviour and Ecology research group at the University of Bristol in the United Kingdom.
4. This choreography of physical processes occurring at the root of material manifestation is very different from the human engineer's arrangement of a machine's parts "from the outside." I discuss the machine model of organisms in Chapter 10 ("What Is the Problem of Form?").
5. Figure 18.2 credit: mejaguar (https://www.pickpik.com/africa-namibia-nature-dry-national-park-water-hole-11349), public domain.
6. Owen Barfield once wrote a kind of didactic fantasy novel (Barfield 1965, p. 163) in which the protagonist had conversations with a higher being modeled after the "maggid" of Jewish mystical tradition. The final words of that being — and of the novel — depict an "interwoveness" of a hierarchy of living and guiding agencies that may perhaps be suggestive in the present context:

    Twice, answered the gentle but inexorable voice, twice now you have called me "Master." But what you shall do shall be taught you not by me, neither by my masters. You may only receive it direct from the Master of my masters; who is also their humble servant, as each one of them also is mine; as you — if your "doing" should be only a writing — will strive to be your reader's, and as

    I am
    yours.

7. Actually, the same unpredictability is true of individual development. If we were watching a developmental sequence for the first time, we would not be able to predict its mature outcome based on what we saw during the early phases. And yet we might recognize retrospectively that this outcome was the end toward which everything was tending all along.
8. Here is a more complete statement from Buller:

    Consider how natural selection provides an explanation of why humans, for example, have hearts. The heart is a complex organ and all complex traits are the product of accumulated modifications to antecedently existing structures. These modifications to existing structures occur randomly as a result of genetic mutation or recombination. When they occur, there is variation in a population of organisms (if there wasn't already) with respect to some trait. If one of the variants of the trait provides its possessor(s) with an advantage in the competition for survival and reproduction, then that variant will become better represented in the population in subsequent generations. When this occurs, that variant of the trait has increased the relative fitness of its possessor(s) and there has been "selection for" that variant. That variant can then provide the basis for further modification. Thus, humans have hearts because hearts were the product of randomly generated modifications to preexisting structures that were preserved or maintained by natural selection due to their providing their possessors with a competitive edge. So natural selection explains the presence of a trait by explaining how it was preserved after being randomly generated.

9. From Georg Toepfer (see also the immediately following citation of Toepfer in the main text):

> Most biological objects do not even exist as definite entities apart from the teleological perspective. This is because biological systems are not given as definite amounts of matter or structures with a certain form. They instead persist as functionally integrated entities while their matter and form changes. The period of existence of an organism is not determined by the conservation of its matter or form, but by the preservation of the cycle of its activities ... Biologists can identify in every organism devices for protection, feeding, reproduction or parental care irrespective of their material realization. These functional categories play the most crucial role in biological analyses (Toepfer 2012).

10. There is also this from University of Toronto philosopher of biology, Denis Walsh. Natural selection, he says, occurs

> because individuals are capable of mounting adaptive responses to perturbations. This capacity to adapt allows individuals to survive in unpredictable environments and to reproduce with startling fidelity, despite the presence of mutations. It is adaptation which explains the distinctive features of natural selection in the organic realm and not the other way round. (Walsh 2000).

Therefore, he adds, "the programme of reductive teleology cannot be successfully carried out." Then there is the following succinctly stated criticism by the independent philosopher, James Barham:

> Selection theory does nothing to help us understand what it is about functions that makes it appropriate to speak [in a physically unsupported way] of their "advantage," "benefit," "utility," etc. for their bearers. Natural selection is like a conveyor belt which transmits a biological trait or function from one generation to the next ... But natural selection cannot explain how the capacity of biological functions for success or failure arose out of physics in the first place, for the simple reason that the selection process has no hand in constituting biological traits as functions (Barham 2000).

And, finally, in 1962 the philosopher Grace de Laguna wrote a paper in which she remarked that only when we regard the organism as already "end-directed" does it "make sense to speak of 'selection' at all" (de Laguna 1962).

Given my limited familiarity with the literature, I would not be surprised if there exist a few similar criticisms along the same line, at least among philosophers. But my own experience suggests that finding them amid all the conventional evolutionary thinking requires some serious digging.

11. Figure 18.3 credit: Store Norske Leksikon (https://snl.no/Michael_Polanyi), public domain.

SOURCES

Arber, Agnes (1985). *The Mind and the Eye: A Study of the Biologist's Standpoint*, with an introduction by P. R. Bell. Originally published in 1954. Cambridge UK: Cambridge University Press.

Barfield, Owen (1965). *Unancestral Voice*. Middletown CT: Wesleyan University Press.

________ (1971). *What Coleridge Thought*. Middletown CT: Wesleyan University Press.

Barham, James (2000). Biofunctional Realism and the Problem of Teleology, *Evolution and Cognition* vol. 6, no. 1.

Buller, David J. (1999). Natural Teleology, introduction to *Function, Selection, and Design*, edited by David J. Buller. Albany NY: SUNY Press, pp. 1–27.

Corning, Peter A. (2019). Teleonomy and the Proximate–Ultimate Distinction Revisited, *Biological Journal of the Linnean Society*, vol. 127, no. 4 (August), pp. 912–16. doi:10.1093/biolinnean/blz087

Goodwin, Brian (1994). *How the Leopard Changed Its Spots: The Evolution of Complexity*. New York: Charles Scribner's Sons.

Green, Douglas R. (2022). A Matter of Life and Death, *Cold Spring Harbor Perspectives in Biology* vol. 14, no. 1 (January 4). doi:10.1101/cshperspect.a041004

de Laguna, Grace A. (1962). The Role of Teleonomy in Evolution, *Philosophy of Science* vol. 29, no. 2 (April), pp. 117–31. doi:10.1086/287855

Rich, Tina, Christine J. Watson and Andrew Wyllie (1999). Apoptosis: The Germs of Death, *Nature Cell Biology* vol. 1 (July), pp. E69–71. doi:10.1038/11038

Ruse, Michael (1979). Philosophy of Biology Today: No Grounds for Complacency, *Philosophia* vol. 8, (October), pp. 785–96. doi:10.1007/BF02379065

Sachs, Joe (1995). *Aristotle's Physics: A Guided Study*. New Brunswick NJ: Rutgers University Press.

Talbott, Stephen (2024). How Do Biomolecules 'Know' What To Do? https://bwo.life/bk/ps/biomolecules.htm

Toepfer, Georg (2012). Teleology and Its Constitutive Role for Biology as the Science of Organized Systems in Nature, *Studies in History and Philosophy of Biological and Biomedical Sciences* vol. 43, pp. 113–19. doi:10.1016/j.shpsc.2011.05.010

Walsh, Denis (2000). Chasing Shadows: Natural Selection and Adaptation, *Studies in History and Philosophy of Science, Part C: Biological and Biomedical Sciences* vol. 31, no. 1 (March), pp. 135–53. doi:10.1016/S1369-8486(99)00041-2

Webster, Gerry and Brian C. Goodwin (1982). The Origin of Species: A Structuralist Approach, *Journal of Social and Biological Structures* vol. 5, pp. 15–47. doi:10.1016/S0140-1750(82)91390-2

CHAPTER 19

Development Writ Large

WE HAVE FOUND throughout the preceding chapters that all biological activity, including at the molecular level, is thoroughly and irreducibly purposive and end-directed. Successive states of an organism are stages in a life narrative — a drama of progressive self-realization and self-expression. Life stories are *meaningful* in the fullest sense of that word. Significant form comes into being; inner character finds exterior expression; processes emerge from implicit intentions; needs and interests trigger pursuit of means for their satisfaction; injury leads to an effort of healing; ends correspond to initiatives. Physical causes and effects are *organized*, or given their form, by the meanings of a particular way of life. Few biologists explicitly acknowledge this truth of a living narrative, but all biologists implicitly recognize it in their choice of descriptive language. (Extremely relevant here are the few paragraphs contrasting a dog and its corpse in Chapter 2, "The Organism's Story.")

This leads naturally to a conclusion of this book — a conclusion I will develop in this chapter — which is that *we already know more than enough to say that evolution is a purposive, or directive or* telos-*realizing, process*. I understand that you may have difficulty coming to terms with this conclusion. But, as I hope to show, it is really just a matter of admitting to ourselves what we in fact know quite well. After all, an at least implicit recognition of the directiveness of living activity, however repressed in intellectual consciousness, is the only thing that lends to the mass of biological description and theory any appearance of plausibility. ("This cell is dividing; that cell is replicating its genome; this other cell is engaged in a deliberate process of self-destruction; just about all the cells in that lioness' body are in one way or another supporting its effort of running at top speed in pursuit of an antelope." Can we avoid assuming, in one way or another, that the molecules in all these cells are fully informed by their context, so that they "know" what to do?) Organisms are beings whose lives give purposive expression to their own needs and interests, in accordance with their own character.

Living beings, in other words, are quite unlike inanimate things, which don't have needs and interests and don't carry out directive tasks. Whether consciously or unconsciously, every organism directs its actions toward the future. An unmet need leads to activity whereby the need can be satisfied. On our part, we find ourselves framing every organism's purposive actions within the time dimension

— this despite readily seeing that most organisms themselves have no *experience* of a future contrasted with the present. We can hardly believe that a bird building its nest is consciously anticipating its future chicks.

Nevertheless — and this is a crucial observation too easily ignored in evolutionary theorizing — there is a clear sense in which the objective character of all biological activity does exhibit what, from our own standpoint, we cannot help viewing as a future-oriented meaning and directionality. Every organism's life consists of initiatives temporally correlated with ends, ideas progressively being given form, needs and interests prompting exploration — even if the initiatives and ideas, the needs and interests, never become conscious for the animal itself.

The animal's purposive narratives are so "boringly" familiar that we cannot help taking them for granted. We assume their decisive role in every biological context we look at, and cannot "un-know" them even when we are theorizing from a position that ignores or denies them. And so we have the two paradoxical sides of biology today: an uneasy, theoretical disregard of what seems ungraspable or dangerously mysterious or unable to be captured in purely physical terms; and a carefree, unexamined taking-for-granted of the powers so obviously at work in those all-too-familiar mysteries — powers that unconsciously inform our scientific thinking and only by this means enable us to believe that our purely physical, cause-and-effect theorizing actually makes sense.

My aim in this chapter — an aim grounded in all the preceding chapters — is to facilitate the changed angle of vision that can enable readers to grant full recognition to what is already known.

Evolution as a transformation of developmental processes

Few developmental biologists will dispute that a wolf's development, proceeding from a fertilized egg cell through embryonic and fetal stages to the pup's birth, and then on through life to maturity, is highly directive. It is an improvisationally coordinated, inherently meaningful, and adaptive movement emerging out of the past while oriented toward the future. It is part of an ever more complete self-expression by the being we know as a wolf. We would never say of a geyser or meteor that it is, in this sense, moving toward fuller self-expression.

Yes, a wolf's development is a path full of unpredictable variation, never exactly repeated in different wolves. But this makes it all the more impressive that the entire trajectory remains persistently wolf-like despite all the adjustments to disturbances and despite all the adaptations to changing conditions — and also, despite the wolf's feeding day after day on the flesh of other animals, which it never fails to convert into its own flesh and way of being. The individual wolf, embedded within its physical and social environment, exhibits the organizing power of its species, and remains throughout its life capable of negotiating a wolf-like path through the uncertainties of its existence.

The three-week-old heart of the unborn wolf is dramatically different from that of the six-week-old heart, which in turn differs from the heart immediately following birth (at about nine weeks), and this again differs from the heart of the mature wolf. It is presumably uncontroversial to say that any biologist who studies wolf physiology and development (something you can be assured I have never done myself) will expect the processes leading from one stage to another to show all the features of organic activity.

Organic causation is inescapably holistic: a change in one part of the heart can happen only together with change in other parts, so that context-dependence is a theme in all physiological, morphological, and behavioral explanation. Change is never anything less than transformation of a whole. Everything proves thoroughly directive, plastic, and adaptive, with earlier features serving as an effective preparation for later ones.

"Effective preparation" here does not refer to a machine-like necessity by which earlier stages automatically determine later ones. Instead, it's rather as if, at each moment, reference were made to contextual *meanings*, so as to guide the current activity in light of those meanings — meanings that are continually being elaborated further.[1]

Actually, it's easy to recognize this process of ever fuller revelation when we consider our own lives. For example, when we are writing a scientific paper, the sentences we have already written do not determine the next sentence. Instead, we have to keep the overall flow of meaning in mind, and then make a creative effort — we might call it an *effort of origination* — in formulating the next sentence. Something has to come freely, without necessity or physical cause, but rather in tune with a governing meaning to which we are trying to remain faithful (even if we are still in the process of trying to grasp it).

This required activity of origination remains more or less conscious and recognizable (if we are paying attention) throughout the entire writing project. In the case of an embryo's transformation, the moment-by-moment, originative activity may be unconscious (on the part of the embryo), but it is nevertheless recognizable *by us* as outside observers. At no point can we predict the next stage based on physical necessity, but only with reference to the unfolding meaning.[2] So at the very least we can say that something is going on with the embryo that looks akin to what is going on with us when we try to write up what we have come to understand of its transformation.

As for the wolf: how can we possibly believe that there are ways to get from the embryonic heart to the mature heart via a purely mechanistic pathway *not* meaningful, originative, and directive in the sense of all development? Who can point to even a single biological example of a predetermined pathway of transformation, wholly written in the physical conditions preceding the transformation?

The heart of the evolving horse

We now switch to an evolutionary sequence, such as the classic textbook lineage of the horse and begin asking a question analogous to the one we have just asked about the wolf. How might we imagine that a heart, structured *that way* fifty million years ago in the fox-sized horse ancestor, *Hyracotherium*,[3] becomes this heart, structured *this way* in the Triple Crown winner, American Pharoah?

Can we realistically picture this evolutionary metamorphosis being achieved by processes *less* well coordinated toward a future state than the developmental transformations bridging the differences between, say, a two-month- and five-month-old horse fetus? Or *less* end-directed than we find all living activity to be?

Is there any basis whatever for us to assume that the requirements for changing the heart between two stages of an evolutionary lineage is *altogether different in character* from the requirements for changing the heart between two stages of a single organism's development? It is, after all, the heart itself that must proceed along a transformative pathway in both cases, and from its vantage point the nature of the task doesn't really change. If the heart's transformation requires all the directive, contextual, highly coordinated activity we observe during development, how could its evolutionary transformation require anything less than the same directive, contextual, highly coordinated activity?

We have no evident ground for hypothesizing that the transition between two evolutionary versions of the horse's heart could be largely a matter of accumulated accidents such as could be imposed, so to speak, "from outside" by a series of interiorless, non-directive, decontextualized mutations. By "from outside" I refer to mutations conceived as unexpressive of the kind of living, unifying, thought-infused power through which a cell or organ or whole organism holds together as the kind of thing it is. We know too much about the directed character of the development and functioning of all hearts — all biological entities — to make this proposal sensible.

Challenges of the heart

Imagine the heart of a horse ancestor (*Hyracotherium*). That heart must, in a coordinated way, have participated with the lungs, the brain and its complex signaling, the liver, the kidneys, and just about every other organ, as well as the vasculature of the circulatory system, and the metabolic and specialized chemistry in the blood-suffused bodily tissues — all in order to meet the "goals" of homeostasis. These include maintenance of oxygen levels in the blood and tissues within narrow bounds; maintenance of the levels of carbon dioxide and other gases within similarly narrow bounds; maintenance of blood sugar levels (every diabetic is acutely aware of the dangers of elevated or depressed blood sugar); maintenance of blood flow and supply according to the differing needs of various parts of the body; maintenance of blood temperature in the face of external temperature extremes; maintenance of blood pH (acidity or alkalinity); maintenance of blood pressure; excretion of toxins and waste products; and much more.

Each of these requirements is intertwined with all the others, so that the mere attempt to analyze everything that is simultaneously going on easily flummoxes the researcher or medical clinician. But the heart's "assigned function" is to participate in all these tasks with an "eye" on how they relate to the health and immediate needs of the body — all without the benefit of blood test readouts or other diagnostic tools. And all with no conscious knowledge of the highly specialized operations in which it is involved. Yet the "knowledge" is in some sense there, fully operative, and we find ourselves willingly or unwillingly admiring the incomprehensible "wisdom" of our heart's meaningful engagement with all our bodily systems.

So it must have been for the small *Hyracotherium*. Now consider the much larger and wholly transformed heart of American Pharoah. It, too, can only have differentiated out of the zygote via the kinds of developmental processes that every embryologist is familiar with. And it, too, in addition to (or as part of) its own development, must carry out all the well-aimed and expertly calculated functional responsibilities such as the homeostatic ones we alluded to above. It achieves this, not through a master control room such as we might find in a jet airliner's cockpit or a nuclear power reactor, but rather through immediate, "expert" participation in the whole body that it somehow "knows" well, without any of the training of a jet pilot or reactor operator.

So can we really think that the one heart might be transformed into the other except by means of the same play of wisdom and the same global coordination of "impossibly" intricate processes such as we see separately in the development and functioning of these two hearts? Dare we suddenly step back from all we know about the two hearts and try to imagine an intervening evolutionary process lacking the wisdom, the powers of purposive coordination, and the future-oriented performance without which neither of the hearts could exist? What justification can we find for abandoning the living picture in that way?

My point is that, once we have noticed the character of all biological activity, it's impossible to imagine a mechanistic, accident-ridden, non-directive, mutation-based, evolutionary transition from *Hyracotherium* to American Pharoah. We can, of course, try. But this immediately violates everything we know about actual organisms and hearts. An organism makes each of its parts and each of its traits an expression of the whole it belongs to. *The part comes into existence only as such an expression. It has no separate, de-contextualized, existence of its own.*

Not nearly enough attention and reflection has been given to the seemingly impassable gulf between a mere mutational rearrangement of genetic particles (as these are mechanistically conceived), on one hand, and, on the other hand, the functioning of a complex organ intimately expressive of a whole organism while also capable of purposively and expertly pursuing all the ends so efficiently carried out by the heart. Trying to bridge that gulf between discrete particles and the meaningful unity of an organism and all its parts looks rather like the frustrating effort of cognitive scientists to bridge the gulf between matter and mind. We might in fact wonder whether the two gulfs are one and the same.

Figure 19.1. An artist's conception of the fox-sized horse ancestor, *Hyracotherium*.[5]

If we take seriously the truth that a living transformation is always the re-forming of an integral whole in the sense that there are no separate parts pursuing an isolated existence, then the idea of a non-organic accident that merely acts physically upon a tissue or organ without participating in its living, transforming, purposive unity has no place in our picture of evolution.[4]

Figure 19.2. Triple Crown winner, American Pharoah.[6]

Evolutionists should not forget the directedness of biological activity

The processes of organic transformation, which we observe in endlessly different contexts, just are what they are. For centuries biologists have recognized the distinctive character of these processes as definitive of life. How is it that we forget the directedness of biological activity as soon as we turn to evolution? How is it that, when we imagine the matings, predation, hybridization (see below), and other activities yielding an evolutionary transformation of the mammalian heart, we forget the ways in which cells in disparate parts of our own developing bodies must live, die, and transform themselves in a manner governed, not by their immediate physical necessities, but rather by the emerging form of a body that does not yet exist — a body that belongs to a future from which no physical causes are able to act?

It is, after all, the whole nature of a developmental narrative to proceed directionally and seamlessly *from here to there*. It would require a powerful and unexpected set of arguments to show that nature, employing any conceivable set of historical processes, could effectively transform such a developmental narrative otherwise than by entering into and respecting the terms governing all such narratives. The need is to improvise as necessary while managing a frightfully complex, systematic, re-organizing, future-oriented activity that is the only basis for developmental transformation we have ever witnessed in organisms.

The racing champion, American Pharoah, is as remarkable an endpoint for the evolutionary trajectory from *Hyracotherium* as it is for the developmental trajectory from its own zygotic stage. If we can hardly help taking for granted the directive activity required for the latter sort of development, can we find any justification for overlooking the necessity for directive activity in the former sort?

Or again: If evolution employed fifty million years and countless generational inheritances and matings (and much else) to accomplish the transformation of the *Hyracotherium* heart into the heart of American Pharoah, is this likely to require *less* directive oversight than what is so visibly evident in the development of American Pharoah's maturely functioning heart from its embryonic stage?

Remember what was perhaps the main point of Chapter 18 ("Teleology and Evolution"): the decisive issue is whether we are willing to acknowledge explicitly the directed character that everyone implicitly recognizes in an individual animal's development. If we do acknowledge it, then the present argument is in part that nothing further needs to be assumed in order for us to recognize an analogous directive aspect in evolutionary change; we are watching a coherent process of organic transformation in both cases. If we do not acknowledge it, we are left with what would appear to be the hopeless mystery of the developmental process.

A thought experiment

Let's assume that horse-racing enthusiasts never stop breeding horses. We'll assume further that, having magically transported ourselves into the future, we are holding in our hands the exhaustive, generation-by-generation, molecular-level and phenotypic documentation for a thousand-year evolutionary lineage running from the 2015 Triple Crown winner, American Pharoah, to the greatest mega-champion of all time. Call him Chinese Ceasar if you wish.[7]

It is a safe bet that Chinese Ceasar differs significantly from American Pharoah. The specific differences will depend, among other things, on the qualities that breeders valued throughout those thousand years — running speed presumably being one of them. Due to the principle of holism, through which a change in one feature of an organism is linked to change in many others, it is hard to imagine what sort of horse we would be looking at a thousand years from now. But surely it would be a horse of a different color.

Surely also, this would be a case of *directed* evolution — "directed," not merely in the usual biological sense, but also in the sense of "intended by breeders." But *would we recognize this fact if we were unaware of the breeders' role*? That is, could we discover, solely from the horse lineage itself, the fact that it progressively realized certain ideas, or guiding principles, or intentions?

The question seems to me important. Strongly held opinion has it that actual evolutionary history shows no directive or progressive aspect — not, at least, in an intentional sense bearing much resemblance to the directionality imposed by breeders. But if the answer to our question is, "No, we wouldn't necessarily be able to recognize Chinese Ceasar as the result of directive evolution," might the reason be that all evolution is directive, and we have no contrasting experience of any other sort with which to compare our observations?

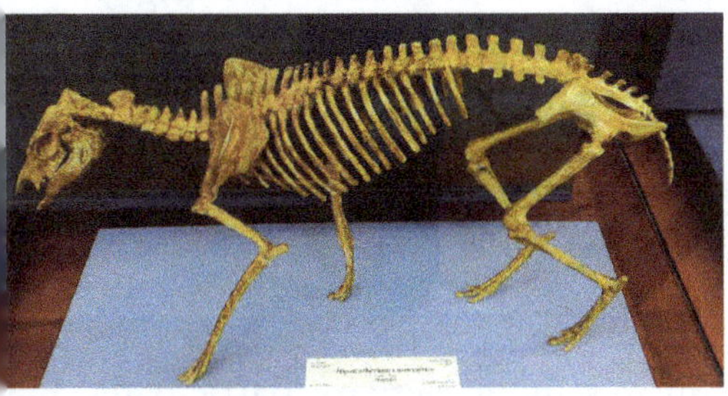

Figure 19.3. A museum skeleton of *Hyracotherium*.[8]

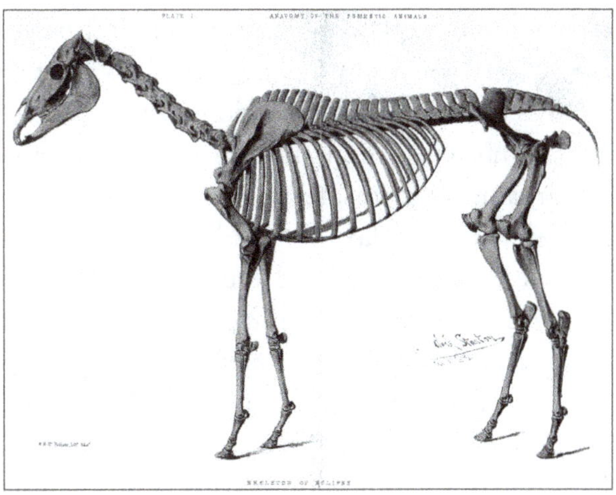

Figure 19.4. Skeleton of a modern horse. The two images are not shown at the same scale. The horse skeleton is, in reality, several times larger than the ancestral skeleton.[9]

We might also put the question this way: If, in fact, we have not yet found a way to distinguish the features of a directive evolutionary lineage from those of a supposedly non-directive one — then on what grounds has there arisen the consensus claim that normal evolution is not directive? Have we, all this time, been speaking of "non-directive" evolution while unconsciously assuming all the features of a directive evolution?

If I am not mistaken, then, here is a fair request we can make of evolutionary theorists. Show us how we might distinguish, at least in principle and in the metamorphosing organisms themselves, a non-directive evolutionary process from a directive one. Or, if you cannot do this, then explain the evidence upon which you have concluded that evolution is, in general, nondirective.

Experiment concluded?

As it happens, a year or so after I first wrote the preceding section, I discovered that philosopher Daniel Dennett had already pursued the same thought experiment — and had received an answer. He pictured aliens visiting earth and tampering with natural selection for a while, then departing. He asked: "Would their handiwork be detectable by any imaginable analysis by biologists today?"

Dennett did the sensible thing: he consulted some biologists. "All the biologists I have queried on this point have agreed with me that there are no sure marks of natural, as opposed to artificial, selection" (Dennett 1995, pp. 316–19).

This is a dramatic acknowledgment, although the real significance of it seems to have escaped Dennett. He was clearly thinking of intelligent design when writing this passage, and feared that ID advocates might seize on the idea that you can't disprove the intervention of an external Designer in evolutionary history. So he was quick to reply that, barring discovery of a feature positively *requiring* a Designer's intervention — a feature that natural selection without a Designer could not explain — there was no refutation of Darwinism to be had here.

If you want a measure of how thoroughly the organism has dropped out of sight in today's evolutionary theory, Dennett's account offers it. Apparently it did not even occur to him to ask, not about an intelligent Designer, but about organisms themselves, whose powers of directive development, physiology, and behavior, displayed right before our eyes, constitute their entire life story. The question, still ignored today, is how the organism's living activity participates, *out of its own purposive, cognitive, and intentional nature*, in the broader intentional coherence displayed so clearly in evolving populations.

To be as unambiguous as possible: the question here is not about an external designer, but about a purposiveness inherent in populations of organisms analogous to the purposiveness we see playing through the many more or less independent cells within a developing individual body. And the question is hardly out of bounds, given biologists' apparent agreement that "there are no sure marks of natural, as opposed to artificial, selection." One wonders how it is that the idea of meaningfully directive evolution has been so scorned if in fact all biological processes we can observe are irreducibly directive, and if we have no ready means for distinguishing a non-directive evolution from a directive evolution — or even from an *artificially* directive evolution.

I have been suggesting that there is a good reason why we cannot point to a distinction between non-directive and directive evolution. Lacking an ability to conceive clearly any sort of non-directive evolution, we have no basis for comparison. It is impossible even to imagine a coherent, organic transformation that is not in fact a coherent, organic transformation. That is, it is impossible to imagine an organic transformation not subject to the guiding principles or organizing ideas belonging to the form progressively being realized. The *coherence* of the transformation could lie only in such principles or ideas; that's just what we mean by "coherence," which would have no referent if there were only inert, quality-less, meaningless things. So the only thing biologists struggling toward a clear conception of evolution are ever in fact thinking of is directive evolution, whether they have acknowledged it to themselves or not.

The wisely purposive lives of organisms — their striving for life and survival, the intricate wonders of their capacity to reproduce, their masterful ability to gather and organize a unified, workable inheritance for their offspring — these "miracles" of directive activity (in terms of which, as we saw in Chapter 16, natural selection is defined) are so thoroughly imprinted upon our experience that not even an entrenched scientific materialism can dislodge them as implicit assumptions of our evolutionary theorizing.

So it is not that we have a choice between directive and non-directive evolution. The only biological transformation we ever see or can consistently imagine, whether developmental or evolutionary, is directive transformation. In order for us to realize that evolution is directive, it is enough for us to see that the group of once living and functioning animals known as *Hyracotherium* belonged to a complex evolutionary lineage that led to today's living and functioning *Equus*. We can understand biological transformation as such — for example, its holistic and organic character — only in a way more or less analogous to how we understand the transformation of an embryonic American Pharoah into the adult American Pharoah.

Every smallest (and largest) step and every degree of the change must first be realized in the development of an individual organism. But the contributing role of the relations between organisms may show very different emphases in development and evolution. Even in the case of sexually reproducing organisms, we are usually looking, in part, at individual development following upon a "vertical" inheritance associated with the merger of two conspecific lineages. But, when we try to trace the causal relations extending through the many generations of an evolutionary process, the situation may become much more complex.

We will now take a closer look at some of these relations.

It is, in the primary sense, populations that evolve, not individual organisms

The potential functional shapes of proteins are virtually infinite. So a major question in evolutionary studies has been, "How, amid this vast landscape of possibility, can more or less random mutations in DNA lead, in any reasonable amount of time, to the particular proteins useful for an organism's current adaptive needs?" This question has been a flashpoint for debate between intelligent design advocates and conventional biologists. The debate is, to say the least, perplexing. That's because the foundational assumption on both sides — that *natural* biological processes are inherently non-directive — is so dreadfully wrong.

The relevant fact is that *nothing* in an organism escapes being caught up in meaningful and directive processes. There is simply no available context for talk of "random" mutations. The processes of DNA maintenance, replication, damage repair, and mutational change are among the most fully characterized and the most intricately purposive and directive activities we have so far explored at the molecular level. DNA damage repair and the closely related incorporation of mutational change are, perhaps, orders of magnitude more complex than the spliceosomal activity we looked at in Chapter 8 ("The Mystery of an Unexpected Coherence"). It takes place in the same fluid environment as RNA splicing. And there is the same play of organizing *ideas* and ideal *reasons* for what goes on — living reasons of a sort that cannot be derived from concepts of physical lawfulness.

At this point — without ever addressing the decisive problem of the rational coherence of molecular activity in the cell — evolutionary theorists are quick to tell us that, although genetic mutations are in general nonrandom, they are nevertheless crucially random in one regard:

> Mutations are claimed to be random in respect to their effect on the fitness of the organism carrying them. That is, any given mutation is expected to occur with the same frequency under conditions in which this mutation confers an advantage on the organism carrying it, as under conditions in which this mutation confers no advantage or is deleterious (Graur 2008).

So then another debate arises: "Are mutations really random relative to their benefit for the organism, or are they 'directed'?" This is where the question of purposiveness or direction in evolution is thought to come to a sharp focus. The effort to prove or disprove the existence of "directed mutations" is often pursued as if it would tell us about the directiveness of evolution.

The question about mutations in the individual organism is certainly significant and worth pursuing. But here, too, the underlying assumption of most debate makes little sense. As I pointed out in the previous chapter, if we are talking about a *telos*-realizing evolutionary process, then the question is not about a mutation's benefit for the individual organism, but rather about its relation to whatever is being realized in the overall evolutionary process. We are not helped much in this by making assumptions about the relation between mutations and individual fitness. Rather, we must investigate how the individual organism is *caught up* in, and *participates* in, directive processes involving populations, species, and even larger groupings.

This is much the same as with individual development. We recognize the meaningful path of development, not merely by looking at what happens to an individual cell, but by picturing the coordinated activity of all the cells in the body. Any individual cell, or group of cells, may, as we saw in the introduction to Chapter 18 ("Teleology and Evolution"), be caught up in a coordinated dying-off process essential to the formation, say, of a particular organ. In this context, it is not, primarily, the welfare or fate of individual cells we are interested in, but the larger developmental transformation. Or: we are interested in the individual cell because of the way it participates in, and is informed by, that larger movement.

But the evolutionary parallel here requires some explanation.

Far from a simple, linear process

We know that individual development is marked by more or less dramatic periods of especially rapid, intense transformation. In our own development, profound changes occur around the time when the young child is taking its first steps and speaking its first words. Likewise with puberty and menopause. Then, too, there is the entire, nine-month period of human embryological development, from the zygote onward. This pre-natal phase is marked by vastly more physiological and morphological change than occurs throughout all the subsequent decades of life.

Perhaps even more dramatic are the millions of species — for example, many insects and amphibians — that undergo one or another kind of metamorphosis. A larva becomes a butterfly, a tadpole transforms into a frog. This reorganization can be both swift and virtually total. (See the description of insect metamorphosis in

Chapter 17, "Evolution Writ Small.") But such times of emphatic change typically occur between extended periods of relative stasis, or slower change.

That a similar pattern often, but not always, holds in evolution was argued in 1972 by paleontologists Niles Eldredge and Stephen Jay Gould, who called the pattern "punctuated equilibrium." Since then various forms of the idea have been broadly accepted, so that another prominent paleontologist, Robert Carroll, could write of vertebrate evolution that "instead of new families, orders, and classes evolving from one another over long periods of time, most had attained their most distinctive characteristics when they first appeared in the fossil record and have retained this basic pattern for the remainder of their duration" (Carroll 1997, p. 167).

It's not just the relative suddenness of change that matters in the present context. More significant is the remarkably nonlinear character of the processes by which major evolutionary innovations occur. My colleague, the whole-organism biologist Craig Holdrege (to whom I am deeply indebted for many of the insights in this section),[10] has drawn attention to one of the central lessons emerging from paleontological work: when something dramatically new arises in the fossil record, it is typically foreshadowed by fragmentary "premonitions" (not his word) in various taxonomic groups, some of which may then go extinct. There is no smooth, continuous, single line of development leading to the new form, which may arise not only rather suddenly, but also as a novel synthesis and transformation of the earlier, scattered, premonitory gestures.

Holdrege shows this very clearly in his book chapter on the frog ("Do Frogs Come from Tadpoles?").[11] After mentioning that no frog fossils have come to light from before the Jurassic period of the Mesozoic era, he notes that "the first frog fossils have virtually the same proportions and the same skeletal morphology as today's frogs." Earlier, there were indeed rare transitional forms possessing some frog features, especially features of the head. These were "a far cry from frogs, but if you know frog morphology well, you can see hints of what is to come." He goes on to say of the paleontological record that

> the hints or foreshadowing of what will come later are not manifest in only one type of fossil, but in several. Various elements of what appears later in the new group are manifest in earlier periods, but in different lineages. Evolutionary scientists often speak in this connection of "mosaic" evolution, since various characteristics appear in different arrangements in different organisms … Even when a trove of fossils is available, such as in the horse family (Equidae), it is not the case that they line up in a neat series. Rather, there is surprising diversity in the forms that predate modern horses (Holdrege 2021, p. 249).

In some of his other work Holdrege has pointed to the same reality in the human and pre-human fossil record. Using accurate models or professional drawings of the available skulls, done to scale, he asks students to arrange them in an order showing an apparent progressive movement toward the human form. It can be an informative (if frustrating) exercise, since no definitive sequence emerges. One skull may show a seemingly more "advanced" feature than the other skulls, while at the same time showing more "primitive" ones (Holdrege 2017).

All this resonates with other facts that have been in the news these past few years — news bearing on the most recent human evolution. We have heard a good deal about cross-breeding between humans, Neanderthals, and Denisovans, and also about the prevalence of variation within populations. The genomes of a major part of the present human race contain a significant proportion of Neanderthal and/or Denisovan DNA, and these elements are thought to play significant roles in human biology.[12]

Figure 19.5. A cichlid fish (*Pundamilia [Haplochromis] nyererei*), one of hundreds of cichlid species in the lake region of eastern Africa.[13]

Then, too, there is the broader fact that *hybridization* between species and genera — and even between families — is now linked to rapid evolutionary change. One impressive story was reported in the journal *Science*, in an article titled "Hybrids Spawned Lake Victoria's Rich Fish Diversity." Among cichlid fish in Africa's Lake Victoria, the rate and extraordinary extent of diversification has, we're told, "baffled biologists for decades." A mere 15,000 years ago there were only a few ancestral species, whereas today — as a result of a remarkable "adaptive radiation" — 500 or so species exist. Some of them "nibble plants; others feed on invertebrates; big ones feast on other fish; lake bottom lovers consume detritus." Varying in length from a few centimeters to about 30 centimeters, they "come in an array of shapes, colors, and patterns; and dwell in different parts of the lake."

The report goes on:

> Now, researchers have evidence that ancient dallying between species from two watersheds led to very genetically diverse hybrids that could adapt in many ways to a new life in this lake. Increasing evidence has shown that hybridization, once considered detrimental, can boost a species's evolutionary potential. Suspecting that might be the case in these fish, researchers sequenced hundreds of cichlid genomes, built family trees, and compared the genomes of fish throughout that part of

> Africa. They discovered that parts of cichlid genomes have been mixed and matched in different ways through time, with various descendants being repeatedly separated and reunited as lakes and rivers dry up and refill. These hybrids had extensive genetic diversity that fueled rapid speciation (Pennisi 2018).

This brings to mind the claim of Mae-Wan Ho and Peter Saunders in 1979 that the isolation of a population — isolation that is commonly thought to figure in evolutionary change — "considerably reduces the role of natural selection. In fact, one cannot but notice that the really conspicuous factor in all cases of rapid evolutionary change is the relaxation of natural selection. Could it be that the rapidity of changes observed is due as much to the lack of competition as to 'genetic revolution'? A review of the fossil history shows that the decisive evolutionary steps involving the sudden appearances of major classes of organisms invariably occurred in isolation, or when ecological niches were relatively empty ... that is, when competition was minimum."[14]

What strikes me in the ideas of hybridization and isolation is the fact that both point to a certain "open-endedness" and world of possibility lying before the populations about to evolve quickly. The open-endedness is genetic in the case of hybridization, where there are presumably many different pathways open to the organism for bringing its inherited genomes into harmony. And it is ecological in the case of populations entering a largely unoccupied niche — that is, without a lot of competition lying in wait along the possible paths of self-transformation.

As for the idea of a "relaxation of selection," I'm not sure there's more to notice here than the fact that natural selection never was much of an explanation of evolution. It gives us, rather, as I pointed out in Chapter 16 ("Let's Not Begin with Natural Selection"), a description of "the pattern of the natural history of life" (Langer), sketched in terms of a mere tabulation of all the extant (surviving) organisms. That is, it points us to whatever the patterns are that need explaining, and is not itself an explanation, if only because a list of survivors doesn't tell us what the survivors were *doing* transformatively along their evolutionary trajectories. Perhaps the survivors would have told us more if biologists had not been so single-mindedly focused on what *happens to them* in the way of physical (molecular) accidents.

Even more radical than hybridization has been the dramatic, *endosymbiotic* origin of different life forms at the cellular level. This has yielded some of the most decisive evolutionary transitions of all time. For example, the presence of chloroplasts (in plant cells), mitochondria (in animal cells), and perhaps a number of other cellular organelles — including possibly the eukaryotic cell nucleus — are now thought to have resulted from the merger of very different life forms. That is, a once free-living, single-celled organism becomes permanently internalized as a functioning part of a different (host) single-celled organism.

It took a long time for biologists to accept theories of endosymbiosis, which were first put forward more than a hundred years ago. This is hardly surprising because of the seemingly insuperable nature of the problem: once joined together, the two cells, with their entirely different life cycles, would have had to "figure out" how to proceed harmoniously through all the necessary and diverse functions of the new, united entity, including cell division. So it seemed that a successful merger of two very different organisms would have required an almost unthinkable and well-directed sort of "management" by both the host organism, and the internalized one. But the truth appears to be that, at critical moments in evolutionary history, such powers were indeed exercised.

Still further, we should not forget the broad fact of *horizontal gene transfer* — that is, the movement of genetic material laterally between different kinds of organisms rather than vertically through inheritance from biological parents. This movement is often mediated by bacteria or other microorganisms, and can involve the transmission of genes between widely differing organisms. This gene mixing is known to have occurred extensively, especially in simpler life forms. It, along with the other processes discussed immediately above, raises serious questions about the branching-tree model of diversifying life, based on vertical inheritance from parents to offspring, and also about the conventional idea of slow, linear, evolutionary change based on random mutations.

As if that were not enough, we have to reckon with the major role viruses have played in shaping many genomes, including those of mammals. For example, every human genome is thought to contain several times as much DNA of viral origin as the DNA of all the protein-coding genes combined.

Then again, there is the entire mass of microorganisms constituting the *microbiomes* of humans and other organisms. "The human body is a complex biological network comprising ten microbes for each human cell and 100 microbial genes for each unique human gene" (Smillie et al. 2011). It is easy to overlook that these microbes living in our gut and elsewhere belong to our bodies, and can be as functionally crucial for our lives as the cells we call "our own." Further, they are extremely dynamic and adaptive, freely exchanging DNA and other substances among themselves.

A sprawling narrative

So you get the picture. Traditional questions about "directed mutations," their effect upon the evolutionary "fitness" of individual organisms, and their spread through a single population via "normal" genetic inheritance — these have been rendered less relevant by our growing knowledge of actual evolutionary processes. We need to raise our sight to the larger collective sphere in which profound and relatively rapid evolutionary change can occur — the sphere where we can discover the kind of unexpected synthesis of diverse and scattered, "premonitory" features described above.

Within this larger sphere, one thing we can truthfully say about mutations (or the creation of genetic variation) is that they can be healthy for the species. They provide resilience in the face of changing environments. This is true regardless of any "fitness benefit" for the individual. And it is, of course, the species as a whole, not just the individual organism, that is evolving. But not *only* the species. There are (as we have seen) diverse interactions of various sorts among different groups of organisms, resulting in the movement of both genetic and non-genetic material between individuals, populations, species, and higher-level groups.

And so we arrive at an extraordinarily complex picture. A "strange dalliance," a few Neanderthal genes here and Denisovan genes there, the hidden and genetically seething world of microorganisms and symbionts constituting a vital part of the substance of higher organisms, the wholesale, lateral exchange of genetic resources among lower organisms (including those in our own microbiomes), the thriving of some lineages and the extinction of others that nevertheless carried for a time part of the essential "mosaic" of evolutionary potentials, and, finally, the relatively sudden convergence, or synthesis (evolutionary metamorphosis), of all those potentials in a new evolutionary configuration — well, if you want to ask about the directiveness of evolution, then all this, along with the overarching agency so clearly recognizable both in the outcome and in the only conceivable path of coordination for getting there, is the relevant stuff of your question. We are not looking at the isolated matter of a genetic mutation's fitness for an individual organism.

One thing is certain: we see no lack of room for a play of intentional, coordinating activity in evolution, just as we see a play of developmental intention through all the cells of, say, a mammal's body. And in both cases it is the *living results* of the activity, together with the necessarily coordinated, well-organized, harmonious nature of the processes for getting there — processes in which wholes, not isolated parts, must change — that tell us a directive and purposive activity has been going on.

Our current ability actually to *trace* this directive activity in evolution may be rather poor, if only because the fossil record tells us so little about the sprawling evolutionary interactions we know must have occurred. But we do know that the development of the individual horse, American Pharoah, required all the familiar, directive powers we have observed in developmental biology generally, all the intricate coordination, adaptation, and compensatory adjustment to disturbances, all the evident wisdom, thoughtfulness, and well-directed intention.

And we also know that much *more* than the wisdom of individual development was required for the evolutionary transformation of *Hyracotherium* into American Pharoah. For not only was it necessary for every ancestral animal in the relevant lineages to be capable of undergoing its own development, but so, too, the relations between mates and between predators and prey, together with all the other "complications" hinted at above, had to come under a directive, coordinating agency capable of realizing all the various metamorphoses of interacting lineages along the way.

However much we may not yet understand, we see the *fact* of this kind of directive evolutionary metamorphosis in the picture already given to us.

WHERE ARE WE NOW?

Evolution as a form of collective development

We have been led by all the preceding chapters to this present one, in which we have concluded that the question of the directiveness of evolution turns out to be almost trivially simple, with an unproblematic answer: evolutionary "development" must be directive in a manner roughly analogous to the development and life processes of an individual organism. In fact, evolution consists precisely *of* these processes, along with their directive coordination.

Their ignoring of the fundamental reality of directiveness in the life of organisms is a central reason why biologists have, for decades, denied all possibility of a coherent *telos*-realizing aspect of evolution. This emphatic denial has taken hold despite their admission that they wouldn't know how to distinguish a directive form of evolution from a non-directive one. I have suggested that the reason for this inability is that they (at least unconsciously) know too much about living beings and therefore cannot realistically imagine a non-directive form of evolution. It's just impossible.

But it's obviously not impossible for biologists to convince themselves that they believe in a non-directive evolution. They can do this only by unconsciously importing into the picture the highly coordinated, end-directed biological processes they see and read about and imagine every day. The organism's thorough-going directiveness is just too undeniable. It is fundamental to any comprehension of life. Trying to imagine evolution without it would be like trying to imagine an evolution of stones.

We have also seen in this chapter that the coordinating agency at work in evolution, while perhaps in some sense centered in individual organisms, must also play through complex interactions among many organisms and populations. We have noted a distinctly nonlinear aspect of much of evolution, where foreshadowings of changes to come ("glimpses of the future") can be found scattered through diverse lineages, leading, at certain critical points, to a more or less dramatic and sudden reconfiguration and synthesis of much that had gone before. This reconfiguration can involve hybridization, lateral gene transfer, and symbioses, among other things, in addition to the predatorial, mating, and migratory activities that have long figured centrally in evolutionary theorizing.

All this means that the relation between a mutation and the individual fitness of an organism is no more central to the origin of species than the "fitness" of an individual cell is central to the development of a complex organism's adult form. In fact, the death of many cells constitutes their positive contribution to the adult form. Similarly, the coordinated patterns of life and death within evolving populations can be recognized as essential to evolution.

In general, we have seen that the directive processes of evolution present us with no fundamental problems of purposiveness and agency that have not already been presented to us by the directive processes of development. Purposiveness and agency are definitive of biology, and their denial destroys biology as an independent science of life.

But while this chapter, building on the preceding ones, sets forth my core argument for acknowledging the essential directiveness of evolutionary processes, the discussion nevertheless remains incomplete. We have yet to look at the way whole organisms and whole-organism inheritances have been effectively negated or rendered invisible within current evolutionary theory. We take this up, along with questions about the role of genes, the supposed requirement for stable mutations, and the "disreputable" topic of holism, in our next chapters.

NOTES

1. This may remind us of the discussion of Paul Weiss' work in Chapter 6 ("Context: Dare We Call It Holism?"). Weiss shows how the relatively chaotic and unpredictable molecular activity within a cell is constrained at a higher level toward the order we can observe in the cell as a whole. I take Weiss' "order" to be closely akin to what I mean here by "meaning."

2. The nature of this "meaning" was illuminated from one angle in the example of leaf sequences" given in Chapter 12 ("Is a Qualitative Biology Possible?").

   Also, what I am saying about writing is basically true for all kinds of writing, not just writing about living processes. But it is best here not to raise too many different issues. Only consider this: writing, regardless of one's topic, is itself a living activity, and is wholly concerned with meaning. It makes no sense to radically or absolutely distinguish between such human activity and the performances of animals — and least of all to insert a natural/unnatural dividing line between the two.

3. We could also speak of *Eohippus* or any number of other horse ancestral groups often without clear relations to each other. But I will stick with *Hyracotherium* because I am proud of having learned to pronounce its name.

4. What we call "accidents" surely can happen. But for the organism's ongoing life, the decisive thing is how it responds or adapts to the accident. Accidents as such simply have nothing to do with the nature of life. A classic example of a significant accident was "Slijper's two-legged goat," who learned to walk upright on its two hind legs, with profound anatomical changes to its skeleton and musculature. Mary Jane West-Eberhard has prominently argued that this sort of plastic developmental response may have "played a role in the evolution of bipedal locomotion in vertebrates, including humans" (West-Eberhard 2005).

5. Figure 19.1 credit: Ricardalovesmonuments (https://commons.wikimedia.org/wiki/File:Natural_History_Museum,_London_Hyracotherium.jpg), CC BY-SA 4.0.

6. Figure 19.2 credit: Courtesy of Coolmore Corp. (https://coolmore.com/farms/america/stallions/american-pharoah).

7. Your spellchecker is missing the point. Try spellchecking the errant official name, American Pharoah.

8. Figure 19.3 credit: Ghedoghedo (https://commons.wikimedia.org/wiki/File:Hyracotherium_vasacciense.JPG), CC BY-SA 3.0.

9. Figure 19.4 credit: Wellcome Images (https://commons.wikimedia.org/wiki/File:Skeleton_of_Eclipse_(a_horse)._Wellcome_L0000443.jpg), CC BY-SA 4.0.

10. I also owe a good deal of my understanding of evolution in general to the writings and lectures of Holdrege, as well as to personal conversations with him. See especially his chapters on the giraffe and the frog in *Seeing the Animal Whole — And Why It Matters* (2021).

11. Holdrege 2021, pp. 213–56. Holdrege's answer to the question, "Do frogs come from tadpoles?" is, in a very important sense, "no." The appearance of the frog represents the achievement of something new, not the mere "rolling forward" of fully determinative "mechanisms" already present in the tadpole. Upon seeing a tadpole for the first time — and even after analyzing it to our heart's content — we could not predict the existence of the forthcoming frog, as if it were a physical necessity like the movement of a planet in accord with the law of gravity.

12. I speak of DNA not only because it is the focus of evolutionists today, but also because DNA is inert (dead) enough to be recovered from some fossils in the human evolutionary lineage. It would be another thing altogether to witness how the Neanderthals or Denisovans livingly incorporated DNA into their life processes — much as we today can witness the adaptation of a single human's DNA to the requirements of a pancreas or bone, a mouth or nose. But, of course, we have no means to look back into the evolutionary record in this living way.

13. Figure 19.5 credit: Kevin Bauman (https://commons.wikimedia.org/wiki/File:Pundamilia_(Haplochromis)_nyererei_male.jpg), CC BY 1.0.
14. Ho and Saunders went on to say:

> There is ample evidence that the relaxation of natural selection increases phenotypic variability. Darwin (1868) first noted that domestic animals and plants were much more variable than their wild counterparts. Similarly, laboratory stocks are phenotypically more variable than wild populations (Waddington, 1957). Ford & Ford (1930) showed that wild populations themselves are much more variable during the phase of rapid expansion than when they are saturated in number, suggesting that part of the uniformity exhibited by wild populations is simply due to "stabilizing selection" (Schmalhausen, 1947), or the selection for some restricted range (usually the mean) of a phenotype (Ho and Saunders 1979).

SOURCES

Carroll, Robert L. (1997). *Patterns and Processes of Vertebrate Evolution*. Cambridge UK: Cambridge University Press.

Dennett, Daniel C. (1995). *Darwin's Dangerous Idea: Evolution and the Meanings of Life*. New York: Simon and Schuster.

Graur, Dan (2008). Single-Base Mutation, Wiley Online Library, *eLS*. doi:10.1002/9780470015902.a0005093.pub2

Ho, Mae-Wan and Peter T. Saunders (1979). Beyond Neo-Darwinism: An Epigenetic Approach to Evolution, *Journal of Theoretical Biology* vol. 78, pp. 573–91. https://www.sciencedirect.com/science/article/abs/pii/0022519379901917

Holdrege, Craig (2017). *Diversity in Human Fossil History — A Teaching Unit on Hominid Evolution*. https://www.natureinstitute.org/book/craig-holdrege/diversity-in-human-fossil-history/about

________ (2021). *Seeing the Animal Whole — And Why It Matters*. Great Barrington MA: Lindisfarne.

Mayr, Ernst (1961a). Cause and Effect in Biology, *Science* vol. 134, no. 3489 (November 10), pp. 1501–6. Available at https://www.jstor.org/stable/1707986

Pennisi, Elizabeth (2018). Hybrids Spawned Lake Victoria's Rich Fish Diversity, *Science* vol. 361, no. 6402 (August 10), p. 539. https://science.sciencemag.org/content/361/6402/539

Smillie, Chris S., Mark B. Smith, Jonathan Friedman et al. (2011). Ecology Drives a Global Network of Gene Exchange Connecting the Human Microbiome, *Nature* vol. 480 (December 8), pp. 241–44. doi:10.1038/nature10571

West-Eberhard, Mary Jane (2005). Developmental Plasticity and the Origin of Species Differences, Chapter 5 in *Systematics and the Origin of Species: On Ernst Mayr's 100th Anniversary*. Washington, DC: The National Academies Press. https://doi.org/10.17226/11310

CHAPTER 20

Inheritance and the Whole Organism

IN 1923 Wilhelm Johannsen, the Danish plant physiologist and pioneering geneticist who had earlier given biologists the word "gene," expressed concern about the way genes were being conceived as neat, cleanly separable causal units. He made the following curious remark, which remains today as intriguing as ever, despite its never having had much effect on the direction of genetic research:

> Personally I believe in a great central "something" as yet not divisible into separate factors. The pomace–flies in Morgan's splendid experiments continue to be pomace–flies even if they lose all "good" genes necessary for a normal fly–life, or if they be possessed with all the "bad" genes, detrimental to the welfare of this little friend of the geneticists (Johannsen 1923, p. 137).

The pomace fly, of course, was the fruit fly (*Drosophila melanogaster*) that Thomas Hunt Morgan, in his Princeton University laboratory, was famously converting into a "model organism" for genetic studies. Thanks to procedures for mutating genes, controlling the mating of the flies, and tracing the inheritance of traits, this was the fateful period during which the word "genetic" was inflating wildly until it swallowed up (among other things) the meaning of "heritable." Oddly enough, the fact that whole cells — and not merely genes — pass as inheritances between generations was progressively losing its significance in the minds of biologists interested in inheritance and evolution.

This violent abstraction of genes away from the whole cell and whole organism was a crime against understanding from which biologists have yet to recover. The evidence of their own bodies should have established beyond doubt that genes do not single-handedly account for biologically and evolutionarily significant features of life. Cells of the human body possessing the same inherited genome differentiate as wholes into hundreds of unique cell types — hundreds of often dramatically different kinds of living things. The differentiation of each cell type along a progressive pathway of coherent, well-directed change extending across successive cell generations, enables the body to form the substance of bone and skin, liver and brain, lens and retina. (See Chapter 17, "Evolution Writ Small.")

Nevertheless, disinterest in this all too obvious and fundamental fact of life took over evolutionary biology as if the disinterest were somehow a prerequisite for the preservation of the discipline. Genes came to be seen as discrete and particulate entities, making them nicely definable and easily trackable, fit to be considered primary causes of the organism.

The consequence for the organism was that it lost its unity, becoming instead an aggregate of discrete parts and traits mapped to genes. Holistic, contextual understanding was severely downplayed. The organism's interior agency became alien to the biologist's manner of thinking — displaced by the informational gene, where the idea of information, with its inescapable connection to meaning, was conveniently conflated with material entities. With the aid of information one could import meaning into biology "under the table," thereby making biological description tolerable, while pretending that one's reference was really only to inherently meaningless matter. And so, as far as explicit theorizing went, the unifying play of organizing idea and intention through all biological activity (Chapter 6, "Context: Dare We Call It Holism?") could no longer be mentioned in decent circles.

On his part, Johannsen realized that the new genetic work, based as it was on the assumed existence of separate and independent causes of traits, left untouched what might easily be seen as the central problem of inheritance: the faithful reproduction of kind, or type — that is, the maintenance of the materially perplexing, integral unity that harmonizes all the particular traits and parts of an organism and expresses a species' characteristic way of being. While mutated genes might result in (typically pathological) *differences* in certain narrowly conceived traits, this sort of change never reached through to the fundamental identity ("that great central something") defining an organism as *this kind* of organism. Whatever the artificially induced and disfiguring mutational horrors, the pomace flies always remained pomace flies.

Figure 20.1. Bust of Wilhelm Johannsen at the Botanical Laboratory in Copenhagen.[1]

Johannsen's problem arises because we can hardly help recognizing the distinctive unity of a living being — a unity we cannot equate to any particular parts. Rather, the unified whole seems in some way *responsible* for its parts which, in turn, always appear to be *expressions* of the whole. We never see an organism being constructed or assembled from already-existing parts. In its development it works to bring them about — to differentiate them out of a prior and continuing unity.

Every organism is the power to do this work, and the power is not derivable from its material results. If some of its parts become deformed, the organism works out of its unity to compensate for the deformities as best it can, doing so according to the way of being of its own kind.

E. S. Russell picked up Johannsen's problem

Figure 20.2. E. S. Russell.[3]

But what sort of genetically investigated *differences* was Johannsen dismissing as disconnected from the problem of the whole? In his brilliant, and still decisively relevant[2] 1930 book, *The Interpretation of Development and Heredity*, the British marine biologist E. S. Russell took up Johannsen's concern. "When we say that a child shows a hereditary likeness to its father," Russell wrote, "we mean that it resembles its father more closely than it does the average of the population. *The likeness is observable in respect of those [rather incidental] individual characteristics that distinguish the father from the rest of the race*" (emphasis added).

Much the same can be said of the child's resemblance to its mother. It's also possible that there will be no particular resemblance to either parent. "But yet in all three cases the child would show the characteristics of its species and its race — it would be a human child, distinguishable as belonging to the same racial type as its parents." As Russell then noted, this general resemblance in type, whereby all members of a species *share* an entire manner of development and way of being, can hardly be understood by referring to the inheritance of this or that variation wherein a parent happens to *differ* — although not in its central identity or type — from most other members of the same species. But such incidental variations have been a main focus of geneticists' investigations for the past century.

In general, isolated "characters" — for example, the color of a pea or of an animal's eyes — are much more easily assessed and compared in two similar organisms than are the *characters* of two whole organisms of different types. The usual genetic breeding experiments that compare differences in isolated traits of closely related organisms can hardly be applied to the different natures and ways of being of an antelope and a bison — let alone an eagle and a pig — if only because the fact of infertility between fundamentally different types renders routine experimental inter-breeding impossible in such cases.[4]

And so biologists have long been forming their idea of heredity against the backdrop of carefully selected, inessential, experimentally accessible traits that scarcely touch the problems presented by every organism's essential unity and inalienable character. They have not been asking themselves, "How can we begin to think about the *organizing power* by which a mammal differentiates and maintains in its proper place the indivisible uniqueness of a whole liver cell (and every one of its billions or trillions of other cells) as long as it is needed?"

The distinction between a fundamental, shared nature, on one hand, and individual peculiarities incidentally distinguishing organisms sharing that nature, on the other, has practical implications for genetic research:

> The broad general resemblances of type give no hold for experimental or statistical treatment, and have accordingly on the whole been ignored. But it is this *general* hereditary resemblance which constitutes the main problem. [The gene theory] deals only with differences between closely allied forms, and with the modes of inheritance of these *differences*; it leaves the main problem quite untouched as to why, for example, from a pair of *Drosophila* only *Drosophila* arise. It takes for granted the inheritance of Johannsen's "great central something" — the general hereditary equipment of the species (Russell 1930, pp. 269–70).

Every organism is thoroughly holistic (contextual). Its entire business might be seen as the continual, total reorganization of its own part-relations, or causal interactions, in response to different environments — all in harmony with its own essential way of being. Given this organization, harmony, and unity of being through which the organism's central, governing character is expressed, it seems perfectly reasonable to surmise that this character could never be dissected or analyzed into a sum of causal relations between separate parts. Rather than being causal in the sense of "resulting from the impact of discrete thing on discrete thing," the organism's unity is an intimate interweaving, a participation of one part in the very being of another.

In sum: genetic analyses, in which we try to isolate the effects of specific genes, do not seem to be bringing us nearer to understanding the mystery of why pomace flies always remain pomace flies. The key issue here concerns the distinction between, first, individual features of an organism imagined as discrete and more or less separable parts (traits or "characters") somehow thought to be *caused* by particular genes; and, second, the integral unity whereby every cell and organism exists and functions as a whole, employing genes and all other resources according to its own, well-organized purposes.

The problem of the whole

The earlier parts of this book have provided the necessary basis for apprehending the wholeness of organisms. We will briefly review a few of the relevant understandings. In Chapter 6, for example, we looked at how, in contemporary biology, we hear over and over that one or another molecular process in an organism is "context-dependent" — albeit without anyone paying attention to what "context" means. What escapes notice is that the appeal to context is always an appeal to the governing *ideas* that make the context a coherent reality — make it what it is. Without a specific set of ideas, we don't know whether a game board is a checker board or a chess board, which provide very different contexts for play. In the same way, a quiescent cell and a dividing one give entirely different meanings to the "play" of the cell's constituent molecules.

I also noted in Chapter 6 how the reference to context-dependent processes suggests that there must be causes running from the larger context, or whole, to one or another local part. But the starting point of the "causal arrows" in such a case would have to be *everywhere*, which makes no sense in terms of our usual causal notions. If the arrow cannot be spatially located, it's because what we're trying to get at is not a link in a physical causal chain, but — consistent with the role of ideas in establishing a context — something more like a possible implication of a broad understanding. We're looking at the intelligible structure of an unfolding play of meaning, not the mechanistic structure of an array of point causes. How unfolding meanings will proceed into the future is never exactly predictable — no more than the precise outcomes of good stories are predictable. Stories generally *make sense*, and to that degree they are predictable. But they can also have an element of surprise. The perfectly predictable, almost by definition, can never be surprising in the same way.

If a cell manages to eject or degrade a certain toxin, it may be, for example, because in one way or another the toxin threatens the cell's health or viability. And if the cell initiates expression of a certain gene, it may be that the gene is associated with a regulatory product that will aid in the production of a protein needed for a journey of differentiation the cell has just now entered upon. Neither the threat in the one case nor the need in the other is a specific force or physical factor directly producing the cell's response. Physical things do not participate in stories where threats or needs can be felt.

In all living circumstances we will sooner or later discover a coherent set of physical connections "explaining" events. But if we look only at that level of explanation, we will never see the meaningful story that is going on. We will not see that this cell is dividing, or that cell is engaged in apoptosis ("committing suicide"), or this other cell is entering a process of differentiation. Local causal arrows — arrows specifically anchored at both ends — make no reference to the division of a cell, which is a fact of the whole.

That's why, as I mentioned in Chapter 2, death is not a recognizable event from the strictly physical point of view. The physical aspects of molecular interaction in an organism's tissues continue on uninterrupted after death. Of course, a purely physical, causal sequence is meaningful in its own terms. The question is whether we want to understand events fully, in all dimensions of their meaning.

Then, in Chapter 11, we saw that there is no explanation for the *form* of organisms that is not itself a consideration of form. Form can meaningfully be understood only in terms of form. This is because form is a principle of explanation *above* the physical, a principle pertaining to wholes. Just as the fact of death could never appear in a purely physical description, neither could the loss of form. The form of an organism belongs to the intelligible and holistic *idea* of the organism. Any supposed physical explanation of form is either itself a principle of form (that is, it is not really just physical), or else it doesn't connect with the form it is intended to explain.

And in Chapter 12 we were given three case studies illustrating what it might mean to have a *qualitative* science — which is to say, a science that is a kind of photographic negative of the science we currently have, which has rejected (or, at least, claimed to reject) all qualities. Qualities lead us upward toward the whole, since the qualities of a part become what they are only in light of their participation in the whole. We saw in that earlier chapter how the dominance of retractor muscles lends to the sloth a quality that gains its fuller meaning only in light of the animal's slowness, passivity, plant-like nature, and receptivity to its environment.

I have also commented more than once about the almost universal use by biologists of "organize" / "organizer" / "organization," pointing out that, without reference (implicit, but central) to organizing *ideas*, these words are void of content. Without governing principles of organization, there is no organizing going on. We call something "disorganized" when we cannot detect an ordering idea or meaning in the arrangement of its parts.

Idea, form, quality, organize — these words, so crucial for biology, all point to the interior, mind-like aspects of the organism. (We could also cite words like *intention*, *purpose*, *consciousness*, and *agency*, which are showing up ever more insistently around the periphery of biology today.) Such words testify to a reality that cannot be reduced to discrete, thing-like elements standing side by side and relating only externally.

Ideas in general, including those of form and organization, tend to interpenetrate and blend into each other, as is shown by the words of a human language, which are so strongly shaped to their context. This interpenetration of "idea-soaked" parts is essential to the unity of every organic whole. Much the same goes for qualities. It is well-known, for example, that our seeing of one color is strongly affected by the colors around it. The influences of colors are not rigidly delimited, which is why the colors of a painting can participate in, and reflect in themselves, the character of the whole work.

Actually, *all* interior contents, including feeling, intention, and will, seem strongly marked by this character of interpenetration. And this is a decisive fact when we are trying to come to terms with the unity and wholeness of organisms. A mere aggregation of physical *things* never makes for an organic, unified whole. If we are thinking of a whole, it is because, consciously or unconsciously, we have formed an idea of the character of a thing, and it is this idea that supports the meaning of wholeness. Or it might be that the qualities inherent in a thing and all its parts blend together (as in a well-executed painting) so as to give a powerful impression of a unified whole.

If it's true that the unity of an organism derives from its interior being, and if this unity is maintained across generations, then it doesn't make much sense to try to understand heredity solely in terms of the transmission of physical things. The power by which an organism holds itself together as a meaningful whole cannot derive from one of the parts held together. This helps us to understand why biologists can hardly help themselves when they make so much of the *informational* significance of DNA.

This is one of the ways they make their theoretical position tolerable by importing meaning "under the table." But, unfortunately, the reality of an interiority that could truly render the idea of information meaningful has never taken root in biological thought. And, in fact, it seems clear that biologists typically see in DNA only the molecule as physically and chemically understood, not the play of organizing ideas and meanings through which the molecule comes to exist as a functionally useful entity.

Biology is not yet ready to accept the reality of the organism's interiority, and therefore is not ready to take on the problem of wholeness and its implications for heredity and evolution. And that is where we must leave the matter, except for this final thought from Russell: the performance of a whole organism "can be [hereditarily] transmitted only by a whole, i.e., by the egg in its entirety, which already at the very beginning of development *is* the new individual" (Russell 1930, p. 283).

Wholes are *in fact* the only things ever inherited. Their importance for heredity is given right before our eyes. But we don't seem to know what to do with them.

To conclude the chapter I will look at one way in which holism has figured in current biological thought. This has to do with how the seeming intractability of the problem of wholeness from a conventional and thoroughly non-holistic viewpoint has been used in attacks against the very idea of holism in biology.

Is holism incompatible with evolution?

In 1978, and again in 1985, Harvard geneticist Richard Lewontin wrote that adaptive evolution is possible only if traits and trait variants are *quasi-independent*. That is, if evolution of a trait is to occur, it must be changeable (subject to mutation) in at least some ways whereby the mutation does not dramatically alter other traits. This is, he said, because the incidental (and random?) alteration of other traits is very likely to harm the organism.[5]

Think of it this way. The vast majority of mutations in organisms are thought to harm the organisms' fitness. If, then, organisms are so thoroughly holistic that any beneficial mutation in one trait will change (mutate) many other traits, the mutation's beneficial effect on fitness, so the thought goes, will likely be overshadowed by the negative effects of all those other changes. And so evolution in the direction of greater overall fitness would require a highly improbable number of secondary, beneficial mutations in order to counteract the deleterious "side effects" of the original beneficial mutation. But if traits and their variants are quasi-independent — not incurably holistic — they can, at least some of the time, undergo beneficial mutation without carrying in their train countervailing and damagingly large side effects. These changes could then become part of the evolutionary record.

The relation of parts to whole in biology is a subtle and difficult matter, mainly because the issue is generally treated in a materialistic way, which ignores the dynamic essence of the matter. I will return to this briefly in the next section. But my immediate concern has to do with how Lewontin's "quasi-independent" criterion has been picked up by others in order to make jabs against the idea of holism. The Australian philosopher Kim Sterelny, for example, has written that "It is hard to change developmental sequences if the development of any characteristic is linked to the development of many characteristics. For a change is likely to ramify, having many effects on the developed phenotype, and some of these are nearly certain to be deleterious":

> Thus, to the extent that development is holistic, the more complex the organism, and the more it has been elaborated over evolutionary time, the less significant further change there can be in that lineage. The point that adaptive change would be impossible if development

> were holistic has been made before. Lewontin, for example, has pointed out that such change requires traits to be "quasi-independent" ... (Sterelny 2001).

Figure 20.3. Richard C. Lewontin.[6]

But there is something strange here. The argument starts by assuming that, in holistic organisms, the effects of a trait change are likely to conflict with each other and be unhealthy. In other words, the assumption is that organisms *cannot* function (with respect to the assembly of inheritances) integrally, coherently, and holistically. But if this is the starting assumption, then there is only brute assertion and no argument at all. The argument, such as it is, becomes possible only *because* of this assumption that organisms cannot really adapt in a holistic manner. And it overlooks the interior wisdom through which all the cell lineages in humans and other multi-cellular species are, amid unfathomable complexity, orchestrated into a harmonious and intricately differentiated whole where accidents, injuries, and unexpected circumstances are commonly overcome.

So we haven't heard much of a case against holism — especially given how often it is admitted that we don't presently have the tools (let alone the inclination) even to begin investigating the possiblities of holistic inheritance and evolution. As Lewontin himself put it:

> If a change in a trait as a solution to one problem changes the organism's relation to other problems of the environment, it becomes impossible to carry out the analysis part by part, and we are left in the hopeless position of seeing the whole organism as being adapted to the whole environment (Lewontin 1978).

However hopeless we may feel, the truth of our observations remains. If an organism's life and development is in fact holistic, why should we suddenly lose sight of this holism as soon as we turn our attention to its management of traits, genetic mutations, and inheritance, or its participation in a species-wide pattern of evolutionary change?

Why, for example, should we ignore the fact of an organism's future-oriented, holistic capacities when it comes to the preparation of a coherent inheritance for its offspring. Least of all should we ignore this when we consider an organism invested (as described in Chapter 19), with the directive intentions at work in an ongoing, complex, multi-lineage evolutionary process. And why should we lose sight of the developing organism's remarkable capacity to integrate and reconcile as far as possible its various physical resources — or, for that matter, the even more stunning capacity of two gametes to organize their separate lineage inheritances into a single, viable zygote with new possibilities of life?

When Lewontin spoke of quasi-independence, saying a trait must be changeable (subject to mutation) in at least some ways that do not dramatically alter other traits, he was apparently accepting the particulate view of inheritance and the random view of mutations.[7] He therefore overlooked the possibility that an organism caught up in the evolving destiny of its kind might, by virtue of that very fact, be capable of *coordinating* the elements of its hereditary bequest to the next generation — and doing so, as we saw in that earlier chapter, by participating in the winding, "mosaic," perhaps unexpected pathway leading indirectly yet coherently from the past of its own kind to the future. But, insightful as he was in so many other regards, Lewontin did not seem to consider this possibility even to be on the table, despite his familiarity with the highly complex, coordinated, and directive aspects of individual development.

It seems that the very idea of holism is so alien to biologists that the attempt to think it is aborted before it gets very far. This is all the more odd given that some of those discouraged by the idea of holism in general are also (and with justification) enamored of the inescapably holistic idea of *phenotypic plasticity* — the individual organism's ability to alter itself in order to adapt to a particular environment. If organisms are phenotypically plastic, then their different internal systems — for example, those involved in bone growth, muscle growth, and nerve growth — must be tightly integrated, so that they can respond adaptively and mutually to changes in each other. "Phenotypic plasticity," we read in one enthusiastic author, "pre-adapts lineages to evolutionary change, by connecting the development of distinct organ systems":

> Limb development requires simultaneous and co-ordinated development in other organs and tissue systems: cartilage, muscle tissue and attachment points, innervation of soft tissues; circulatory connections to tissues and bone marrow. If bone structure or muscle mass is plastic, responding to signals from the environment, co-ordinated systems must be plastic too, responding to signals from the systems developing with them ... This same sensitivity of integration to the contingencies of development will make functional integration possible in the face of genetically-caused changes in crucial organ systems.

The author of these remarks (Sterelny 2009) happens also to be the author of the comment above about the problem holism presents for evolutionary change. It's as though, when one's attention turns to evolution, one is obligated to begin thinking of change as if it were brought about, not by the plastic and adaptive character and agency of the organism and its kind, but by random disturbances to a mere aggregate of particulate genes that somehow (in their separation and relative isolation) map to and determine the organism's phenotype.

And, yes, it is then very hard to imagine a set of scattershot changes that would, in harmony, alter the intricately interwoven, holistic way of being of an organism. But once we have acknowledged that an organism's holistic nature includes the power actively and adaptively to coordinate its physical resources, why should we so quickly forget this, especially when, in evolutionary theory, we are actually addressing the issue of holism?

I have not said anything about the degree of "quasi-independence" some organismal traits might have. I may indeed be inclined to start with the thought that organisms are far from being machines; they are not assemblages of independent, pre-fabricated parts. But if organisms consist of parts — cells and organs — that are *relative* wholes in their own right, then we would expect to see not only a principle of profound interpenetration among parts, but also manifestations of partial independence. This is worth a further look.

Holism and the independence and integration of parts

The organism's holistic integration implies neither that further evolution is impossible nor that it is equally possible in all directions or for all species. Every type has its own future potentials, which are not the same as those of any other type. We do not look for signs that pigs will develop wings. Or, to take a different example: for all we know (and I am not proposing this) *physically* evident evolution may no longer be occurring in humans — or not occurring nearly as much as in previous evolutionary eras. It might be argued, after all, that in humans a major evolutionary transition is placing the power to direct evolution into our own hands. And this looks more like an evolution of consciousness and culture than a further bodily evolution.[8]

As for "quasi-independent" traits and holism, I think Samuel Taylor Coleridge, writing during the first half of the nineteenth century, put the question into the right perspective:

> The living power will be most intense in that individual which, as a whole, has the greatest number of integral parts presupposed in it; when moreover, these integral parts, together with a proportional increase of

> their interdependence, as *parts*, have themselves most the character of wholes in the sphere occupied by them (Coleridge 1848).

Or, re-phrased: Life will be fullest in the individual that most fully integrates the greatest number of interdependent parts; and when those parts are themselves most like independent wholes.

Perhaps we can begin to glimpse the unity underlying these apparently contrary principles when we realize how, in human society, ever stronger and more centered selves are required if we want those selves to contribute ever more strongly and selflessly to the good of the larger society. Society becomes more complex and healthier to the degree the many movements toward a strengthened independent identity and toward interdependence are mutually reinforced.

Or think of your heart or brain. These wonderfully "perfected" organs, while possessing the strongest possible identity and wholeness in their own right, are — as an expression and extension of their wholeness — bound together with everything else that goes on in the body. No part of our bodies can be separated from the circulatory and nervous systems, just as the heart and the brain cannot meaningfully function in isolation from everything else in our bodies.

In other words, the potential for holism and the potential for a (relatively) independent perfection of parts are two sides of the same coin. An overall, deeper holism depends on a greater independence and perfection of parts in their own right, and a greater independence and perfection of parts depends on a deeper holism. The two principles, despite their contrary natures, are complementary in such a way that each exists only by grace of the other. This principle of *polarity* might almost be considered definitive of the organism. For example, every organism lives by distinguishing itself from its environment — and also lives only by virtue of what it takes into itself from its environment.

Coleridge's remark derived, I believe, from a straightforward observation of living beings and required no evolutionary theorizing. He was, of course, writing before Darwin's *Origin*. And he was willing to look at whole organisms as they actually presented themselves. As it happens, there is nothing in evolution that contradicts this profound holism of organic life. Holism, far from making evolutionary change more difficult, is what makes whole-organism transformation, and therefore evolution, possible.

At the same time, the "hopeless" situation Lewontin imagined, where we must see the "whole organism as being adapted to the whole environment" is not so hopeless after all. Our analytical skills, whereby we mentally dissect a whole organism into its parts, are not useless, since the parts of an organism manifest a certain analyzable independence.

But what really leaves us in a hopeless position is analysis alone without a complementary movement of thought. For then we find ourselves unable to reconstitute the parts of our analysis in an integral whole. We can see only parts side by side, interacting like inanimate objects pushed and pulled by external forces. And it will be impossible to see the organism otherwise until we come to terms with the fact that organisms have an interior dimension. (See the discussion of *idea*, *form*, *quality*, and *organize* above.)

An organism is able to act coherently as a whole because it is in fact a profoundly integral whole. But within that whole, each part is able, in its relative independence, to give its own intelligent and discriminating expression to the whole in which it participates so intimately. Yes, we have to learn to look with new eyes in order to see the integral unity of the organism. But, in Chapter 12 ("Is a Qualitative Biology Possible?"), we have at least glimpsed intimations of what that might be like.

WHERE ARE WE NOW?

When the organism was seen whole

During the first half of the twentieth century a considerable number of biologists, among whom E. S. Russell was a leading figure, sought to articulate a biology that kept the whole organism in view. We could, perhaps, call theirs a "common-sense view" since, as I argue throughout this book, all biologists even today reveal in their direct, observational language that they see, at least unconsciously, the truth of the agential organism — its story-telling, directive, *telos*-realizing life — in a perfectly practical sense. (See Chapter 2, "The Organism's Story.")

A key point emphasized here is that inheritance is never anything other than a qualitative, whole-cell inheritance; we always find ourselves watching the uninterrupted life of whole, integrally organized, living entities. It happened, however, that the seductive possibility of tracking and statistically analyzing the passage of genes from one generation to the next opened the way for the kind of logically clear, mathematized results that felt to most biologists "more like science" than did the difficult effort of acquainting themselves with the less clear-cut, qualitative character of whole cells and whole organisms.

And yet, as Russell pointed out, this narrowed the biologist's view down to the observation of some of the genetic causal factors playing into more or less minor differences between closely allied organisms, such as parents and their offspring. (Geneticists also learned to produce monstrosities by grossly interfering with normal development, but these didn't have a whole lot to teach us about the evolutionary potentials of viable organisms.) On top of this, geneticists blithely ignored the multicellular organism's dramatic capacity to orchestrate the "evolution" (differentiation) of numerous cell lineages that are, in their own terms, as phenotypically distinct as distantly related species.

We have also seen here how the organism's wholeness, or merely the thought of that wholeness, has tended to repel some biologists, who have responded with the idea that holism would make evolution impossible. Or, at least, holism is nearly impossible to work with scientifically. We heard how the argument in this direction can invoke the difficulties of holism while refusing to consider the ways in which those difficulties are overcome if we consistently keep in view the organism's holistic character. The problems arise only when we forget what we know about that character.

In the next chapter (which can usefully be read in close conjunction with this one) I will try to pinpoint the decisive inclinations underlying the "genetic distraction" that has so powerfully wrenched the past century's evolutionary biology away from any reckoning with the actual life of whole organisms.

NOTES

1. Figure 20.1 credit: Daderot (https://commons.wikimedia.org/wiki/File:Wilhelm_Johannsen_-_Botanisk_Laboratorium_K%C3%B8benhavn_-_DSC08099.JPG), CC0 1.0.
2. On the relevance of Russell's work today, see "Heredity, Development and Evolution: The Unmodern Synthesis of E. S. Russell" by Maurizio Esposito (2013). For a view of Russell along with W. E. Ritter, Kurt Goldstein, Agne310, and J. H. Woodger, see "A Reflection on Biological Thought: Whatever Happened to the Organism?" by Robin W. Bruce (2014).
3. Figure 20.2 credit: From Ramster 2003.
4. Hybridization does in fact sometimes occur between distinctly different species (within limits, but way more often than most biologists believed not long ago) and, as I mentioned in Chapter 19 ("Development Writ Large"), this can contribute to rather dramatic evolutionary change. But such hybridization is likely to generate massive genetic and cellular reorganization, far too extensive and global to allow for conventional genetic approaches. So one is still facing the unsolved "problem of the whole" — the problem that genetic analyses were designed to steer clear of by focusing on particular genes causing particular trait differences under well-defined conditions.
5. See Lewontin 1978 and Levins and Lewontin 1985, pp. 79–80. Lewontin actually spoke of two requirements for adaptive evolution. In addition to the quasi-independence of traits and their variants, he also cited the need for *continuity*: "small changes in a characteristic must result in only small changes in ecological relations; a very slight change in fin shape [of a fish] cannot cause a dramatic change in sexual recognition or make the organism suddenly attractive to new predators" (Lewontin 1978).

6. Figure 20.3 credit: Museum of Comparative Zoology, Harvard University; copyright President and Fellows of Harvard College, CC BY-NC 2.5.
7. On the supposed randomness of mutations, see the discussion in the concluding section of Chapter 17 ("Evolution Writ Small").

   Of course, apparently random events may figure in a scientific theory. But when the whole point of the theory is to explain evolutionary change, the assignment of that change to random mutations doesn't yet give us an explanation we can reasonably call "scientific." It's basically a way of saying, "We have no scientific explanation." For the thing we want to explain isn't random at all. What we really want is an understanding of how we can characterize the origin or transformation of a trait that is present in the only way normal traits can be present — as part of the wondrous and tightly organized unity of a living being. The relation between such traits and the genes in our cells — or between such traits and the overall organization of our cells — does not look simple, and has scarcely been approached in the modern era of biology.
8. German philosopher, Dieter Wandschneider, has commented that "In a world in which sickness can effectively be cured, clinics and spas are at people's disposal, artificial limbs are applied, and replacement organs are implanted, the biological principle of survival has been 'unhinged.' And that means, too, that natural evolution has come to an end":

   > One could object that the human species changes biologically even today — for example, in muscle structure, susceptibility to sickness, and life span. That cannot be denied. But these changes are manifestations of the "self-domestication" of man and thus consequences of civilization, which as such are not the results of *natural* selection. On the contrary, they are expressions of an evolution that is now taking place under completely different conditions, namely those of *cultural evolution* (Wandschneider 2005, pp. 204–5).

SOURCES

Brady, Ronald H. (1979). Natural Selection and the Criteria by Which a Theory Is Judged, *Systematic Biology* vol. 28, pp. 600–21. Available at https://natureinstitute.org/ronald-h-brady/natural-selection

________ (1982). Dogma and Doubt, *Biological Journal of the Linnean Society* vol. 17, no. 1 (February), pp. 79–96. Available at https://natureinstitute.org/ronald-h-brady/dogma-and-doubt

Bruce, Robin W. (2014). A Reflection on Biological Thought: Whatever Happened to the Organism?, *Biological Journal of the Linnean Society* vol. 112 (June), pp. 354–65. doi:10.1111/bij.12188

Coleridge, Samuel Taylor (posthumous/1848). *Formation of a More Comprehensive Theory of Life*, edited by Seth B. Watson. London: John Churchill. Facsimile edition from Ann Arbor MI: UMI Books on Demand.

Esposito, Maurizio (2013). Heredity, Development and Evolution: The Unmodern Synthesis of E. S. Russell, *Theory in Biosciences* vol. 132, no. 3 (September), pp. 165–80. doi:10.1007/s12064-013-0177-4

Johannsen, Wilhelm (1923). Some Remarks about Units in Heredity, *Hereditas* vol. 4, pp. 133–41.

Levins, Richard and Richard Lewontin (1985). *The Dialectical Biologist*. Cambridge MA: Harvard University Press.

Lewontin, Richard C. (1978). Adaptation, *Scientific American* vol. 239, no. 3 (September), pp. 212–30.

Oyama, Susan, Paul E. Griffiths and Russell D. Gray, editors (2001). *Cycles of Contingency: Developmental Systems and Evolution*. Cambridge MA: MIT Press.

Ramster, John, Niels Daan, Chris Frid et al. (2003). Sixty volumes of the Journal du Conseil / ICES Journal of Marine Science, *ICES Journal of Marine Science* vol. 60, no. 6, pp. 1169–71. https://academic.oup.com/icesjms/article/60/6/1169/651056

Russell, E. S. (1930). *The Interpretation of Development and Heredity: A Study in Biological Method*. Reprinted in 1972. Freeport NY: Books for Libraries Press.

Sterelny, Kim (2001). Niche Construction, Developmental Systems, and the Extended Replicator, in Oyama, Griffiths and Gray 2001, pp. 333–49.

_______ (2009). Novelty, Plasticity and Niche Construction: The Influence of Phenotypic Variation on Evolution, *Mapping the Future of Biology*, pp. 93–110. doi:10.1007/978-1-4020-9636-5_7

Wandschneider, Dieter (2005). On the Problem of Direction and Goal in Biological Evolution, in *Darwinism and Philosophy*, edited by Vittorio Hösle and Christian Illies, pp. 196–215. Notre Dame IN: University of Notre Dame Press.

CHAPTER 21

Inheritance, Genetics, and the Particulate View of Life

THIS CHAPTER about the gene-centered ("genocentric") view of evolving organisms shouldn't have needed to be written. Today genocentrism has been challenged from so many different sides, by so many leading biologists, and in such an ever more insistent manner, that it might easily seem a waste of time to raise the usual issues afresh here. So I won't.[1]

But there are caveats. One is that, despite the criticism, the idea of the masterful, controlling gene remains as strongly entrenched as ever in the minds of most biologists. This is especially true of evolutionary theorists, for whom the word "genetic" has long been synonymous with "heritable." In other words, for purposes of evolutionary theory genes substitute for the entire, one-celled living being that in fact passes between generations — and also for the whole organisms capable of producing such heritable cells. This means that, as participants in an inheritance-based evolutionary lineage, organisms themselves scarcely exist for the theorist.

The century-long habit of genocentrism is seemingly resistant to all criticism. As three Duke University biologists summarized the matter in 2017, "Everyone understands" that the idea of a definitive *gene* for this or that feature of an organism "is a distortion of the biological facts, yet, as a profession, we have yet to rid ourselves of this crutch" (Gawne et al. 2018).

Much of the criticism of genocentrism has arisen from the field of evolutionary-developmental biology ("evo-devo"). Yet even here, according to a leader in that discipline, "increasing gene centrism characterizes the field today":

> This reductionist attitude continues to be upheld, even though overwhelming evidence points to the fact that it is not gene expression and regulation that singularly define body structures but the systemic processes of interaction between genes, cells, and tissues as well as the physics and physiologies of the involved entities and their interactions with numerous factors of the environment (Müller 2019).

A second caveat, even more discouraging, is that, for the most part, the critics themselves leave the door wide open for the persistance of genetic reductionism.

This is because few if any well-positioned, reputable biologists are willing, at the risk of reputation and career, to speak out against the reigning materialism of their profession.[2]

It is this dogma that supports machine models in biology and that requires there to be physical "controlling factors" determining the lives of organisms. Whether these factors be genes or something else doesn't much change the distorting effect upon biological understanding. The qualitative *way of being* of every organism, the immaterial *organizing ideas* shaping their lives, and the purposive *coordination* and *direction* of physically causal processes in order to satisfy the organism's *needs* and *interests* — these foundations for biological understanding are swept away before the religious fervor supporting the restrictions upon thought imposed by the materialist faith.

Most immediately, however, this dogma invites an almost worshipful regard for the all-determining, informational gene — a machine-like gene intelligently designed and engineered from outside by the "creative forces" of evolution. In this way the theorist employs the gene as a lifeless stand-in for the present and effective wisdom that moment by moment lives and expresses itself in every animal's inner, qualitative, perceptual experience (however primitive) and in its meaningful response to that experience.

In what follows I will review the genocentrism of the received evolutionary theory, and then explore some fundamental problems with genocentrism — problems the present critics of genocentrism cannot allow themselves to recognize at risk of violating the materialist taboo. This exploration will continue into the next chapter, where we will look at the contrast between the evolutionarily stable ("potentially immortal") genetic particles that Richard Dawkins celebrates so vehemently, and the evolutionarily stable yet *dynamically transformative* whole organism.

DNA as the essential substance of evolution

As we saw in Chapter 16 ("Let's Not Begin with Natural Selection"), evolution is said to be inevitable once three conditions are met: (1) There must be trait *variation* among individuals in a breeding population; (2) This variation must to some degree be *inherited*, so that offspring generally resemble their parents more than they resemble others; and (3) Individuals possessing different variants of a trait must, at least in some cases, exhibit *differential fitness* (or differential survival) — that is, they must produce, on average, different numbers of offspring, whether immediate offspring or later descendents. This is often referred to as *survival of the fittest*, or the principle of *competition*.

We also saw in that earlier chapter that these three conditions — insofar as they are abstracted from the life of the organism and its agency — become a hollow formula that tells us nothing about why a given species in a particular location evolves in one direction, while another species evolves in a different direction. That is, until we reckon with what organisms *do* as a function of their entire way of being, the so-called "core logic" (Gould), or "algorithm" (Dennett), of evolution can tell us little about their evolution or about the diversity that may arise from them — or even about whether they will evolve at all. And when we do reckon with the active life of organisms, it is this reckoning itself, not some core logic of inheritance and selection, that elucidates the evolutionary trajectory of a species.

The chief excuse for ignoring what organisms do has been found in a distorted picture of genes and DNA. This picture gives to the core principles of natural selection described above what little biological content they have:

- Variation: All or nearly all the variation that matters for evolution (so the theory goes) is ultimately accounted for by mutations in the genetic sequence.
- Inheritance: So far as it bears on evolution, inheritance equates completely or nearly completely to the replication and transmission of genetic sequences.
- Differential fitness: The organism's differential fitness, so far as it matters for evolution, is regarded primarily as the result of traits that in turn result from instructions carried by genetic sequences.

So genes are the one unquestioned material foundation and efficient cause upon which, from the organism's side, the modern edifice of evolutionary theory has been erected. As defined in a classic introductory text, the process of evolution "includes all mechanisms of genetic change that occur in organisms through time..." (Hartl 1988, p. 143). Evolution, in this view, looks very much like a matter of genes and their fate — and not much more.

Once one adopts this view wholeheartedly enough, it becomes all but impregnable. If genes are First Causes of the organism, then even the gene-independent lipids and sugars (for example) that play such fundamental roles in the cell can be said to have their fates interwoven with the proteins that genes "code for," and therefore to be under the "control" of genes. And it's true that, because of the integral unity of the cell, proteins are involved in just about everything that goes on, including lipid metabolism and the formation of the lipid-based cell membrane.

But, as we saw in Chapter 4 ("The Sensitive, Dynamic Cell"), that same integal unity of the cell makes it possible to claim (as some have) that the cell membrane or the cytoskeleton is the true "master controller" of the cell. But recognizing this integral unity, on one hand, and claiming for any element of this unity the role of master controller or First Cause, on the other, are entirely different things.

It's the latter sort of claim on behalf of genes that (as biologists Tobias Uller and Heikki Helanterä have pointed out) may lead one to ignore a beaver's agency and activity in dam building as an evolutionary cause and instead claim that dam building "is itself an adaptation" brought about by gene mutations in the past. And so "the beaver's impact on its environment, which both maintains the adaptive value of its phenotype and may bias further evolution" counts for nothing (Uller and Helanterä 2019).

In this way the organism's agency — *an actual power of purposive, end-directed activity* — drops out of the picture of evolutionary causation, replaced by inert genetic "particles." And yet, genetic "effects" themselves reveal (by being dependent on) a whole-cell physiological organization and power of activity — a power never brought into mainstream evolutionary theory. Do you find anything strange in this picture?

The aggressive claims of population geneticists

It was preeminently the population geneticists who framed the twentieth-century "Modern Synthesis" as the definitive formulation of evolutionary theory. It is likewise the population geneticists who still today speak most forcefully about the primacy of genes. And it is the population geneticists (regarding themselves as working at the pinnacle of evolutionary theory) who continue to dominate the evolutionary field today.

Their life's work has always centered on calculations relating to the transmission of genes between parents and offspring, and the changing distribution of genes within populations. The more advanced, technically impressive results of this work often take the form of sophisticated equations that tend to be more or less opaque to working biologists outside the field of population genetics.

Here are two examples of comments from population geneticists:

Michael Lynch, who holds the Distinguished Professorship of Evolution, Population Genetics and Genomics at Indiana University and was formerly president of the Genetics Society of America, has remarked that "the litmus test for any evolutionary hypothesis must be its consistency with fundamental population-genetic principles." He freely admits that organisms themselves, as "phenotypic products," result from "more than a change in gene frequencies." But the crucial conclusion remains, if only by brute assertion: "If we are concerned with the process of evolutionary change, then evolution is indeed a change in genotype [gene] frequencies" (Lynch 2007a; Lynch 2007b, p. 371).

And then there is Dan Graur, author of a textbook on *Molecular and Genome Evolution*, who proceeds in the same confidently dogmatic spirit (to which he adds his own unique brand of arrogance, in which — judging by his presentation of himself in social media — he apparently revels):

> Evolutionary biology is a mature science. It is a coherent discipline with a handful of logical principles, each of which repeatedly withstood rigorous empirical and observational testing. Evolution is not difficult to define. If one ignores the obfuscations of the creationists, the casuistry of the philosophers, and the ruminations of the "sophisticates," evolution turns out to be merely the process of change in allele [gene] frequencies over time. The only mandatory attribute of the evolutionary process is a temporal change in allele frequencies (Graur 2015).

This is truly amazing — a stunning contraction of human understanding among the would-be elder statesmen of evolutionary theory, smugly satisfied that they do indeed sit at the pinnacle of their discipline. The organism's entire way of being along with its needs, interests, and agency have been reduced, for purposes of evolutionary theory, to one material part — DNA. The underlying drive appears to be the reduction of mindedness to mindlessness, *logos* to meaninglessness, *telos* to chance — all reflecting a horror of interiority.

What seems to these two population geneticists so obviously and incontrovertibly true is nothing but the extremely one-sided, absolutely genocentric, twentieth-century theorizing about evolution, which yielded a body of impressive technical knowledge about gene flows in populations — knowledge to which they have wedded themselves while wearing blinders. The main problem, disastrous in any science, lies in their inability and unwillingness to step outside their particular training and try to see their discipline from outside, as it might appear to critics. This requires, not an aggressive assertion that the fundamental principles of their theory cannot be questioned, but rather an ability to become questioners themselves.

The habit of ignoring organisms is so stubbornly entrenched among population geneticists that their ability to recognize crucial evolutionary issues seems scarcely to exist. Think, for example, of the problem of the origin of innovation — that is, the problem of the "arrival of the fittest," or the arrival of new, living performances (traits) for natural selection to act upon. (We looked at this in

Chapter 16.) Armin Moczek, a prominent evolutionary developmental biologist at the University of Indiana, remarked that "fields such as population-genetics have long stopped asking the question how evolution innovates, not because it is not a foundational question in evolutionary biology, but because population genetics lacks the ability to even frame the question" (Moczek_2022).

They have stopped asking in part because the fact that some genes can make a more or less stable difference in some existing traits (Chapter 22) has convinced them that they need not consider all the other, less experimentally and mathematically tractable features of an organism's life that also make a difference. Nor need they inquire into the principles of organization through which cellular and organismal identity are stably maintained, and through which alone a trait can come into existence as an integral and viable aspect of the larger whole. And they are least of all inclined to consider whether the organism's most dramatically demonstrated capacity — its capacity for directed developmental change and metamorphosis consistent with its own way of being — might be relevant to evolution.

Who defines what counts as an evolutionary process?

The population geneticists, in the manner of those adhering to many an aging and rickety scientific viewpoint, have constructed for themselves an institutional and intellectual fortress whose final crumbling must, as the rather brutal saying has it, await the dying out of its last, well-known defenders.

In the meantime, one thing giving a sense of impregnability to the fortress is the fact that evolution has been aggressively *defined* in terms of genes, as we heard above. One is then bound to argue that whatever is not adequately gene-like cannot be important for evolution.

This argument by definition is why population geneticists routinely dismiss epigenetic factors (Chapters 7 and 14) as irrelevant to evolution: such factors, they tell us, often don't have the long-term, transgenerational stability usually ascribed to genes, and therefore can't contribute much to evolution.[3]

In other words, "Since epigenetics doesn't give us the kind of genetic stability we want to see as the essence of a mindless and strictly mechanistic sort of evolution, we refuse to consider what evolutionary potentials it *does* give us." Those who think this way can scarcely imagine that epigenetics presents us with a revealing expression of the highly adaptive processes of continual, directive change we discover in every sort of whole-organism activity — for example, in the many differentiating cell lineages of our own developing bodies. These lineages result from the changing organization of whole cells, which includes their changing ways of employing their genes.

Such transformative processes — which one might think would be the first things looked for by evolutionists — are ruled out of evolutionary theory on the grounds (now known to be false) that nothing occurring in development affects the *genetic* inheritance of the next generation. But even if that were true, so what? This constricted focus on genetics overlooks the holistic nature of cellular inheritance revealed during cell differentiation. This inheritance is not only dramatic, but undeniably a *whole-cell* phenomenon. If biologists dismiss the significance of developmental processes for evolution, it is only because they have planted their flag and staked their claim in advance: "We're not interested in the potentials of the whole cells that contribute an inheritance to the next generation, but only in the genes those cells contain. So we're simply going to ignore the actual performances of real cells."

As with so many perverse doctrines, there is a dim and distant reflection of the truth in the gene's-eye view of evolution, although it is a truth lost on Lynch, Graur, and their kin. DNA is indeed caught up in, and informed by, the character of the whole organism, including its adaptive character. Therefore we can in one way or another expect to find the whole organism reflected in DNA. Such is the case with all the other major aspects of any organism, as we have already seen in our discussion of cell membranes and the cytoskeleton (Chapter 4, "The Sensitive, Dynamic Cell").

Getting to the bottom of things?

One of the most common strategies for honoring the materialist taboo in all sciences is to describe a microscopic level of supposedly meaningless and inherently inert, mindless things, or particles, possessing fixed, well-defined natures with causal powers. (The fact that causal powers are *powers*, not material things, is conveniently ignored.) Then one claims that whatever really counts in the explanation of phenomena derives from various effective combinations of these particles.

In Richard Dawkins' biological theorizing, the particles at issue are genetic elements with a wonderfully computational nature. "Digitalness," he has said in what must have been one of his rashest statements, "is probably a necessary precondition for Darwinism itself to work" (Dawkins 2006, p. 163). "What is truly revolutionary about molecular biology," he wrote, "is that it has become digital." We know that genes "are long strings of pure digital information ... The machine code of the genes is uncannily computerlike. Apart from differences in jargon, the pages of a molecular-biology journal might be interchanged with those of a computer-engineering journal" (Dawkins 1995, pp. 17–19).

The meaning of "digital" can be illustrated by the game of basketball. The game is so designed that the making of a basket is always definite and unquestioned.

The ball either goes through the hoop, or it does not. A team either earns points for the basket or it does not. How graceful or awkward, skillful or random the shot may have been has no bearing on the matter. A player's approach to the basket can be ugly as sin, but if the ball ends up going through the hoop, the points are counted. There's a clean, yes-or-no, "ones-and-zeros" aspect to the proceedings. The number of points earned is always exact and countable. There are no half or three-quarter points, but only whole numbers. Everything is precise, and there is no ambiguity.

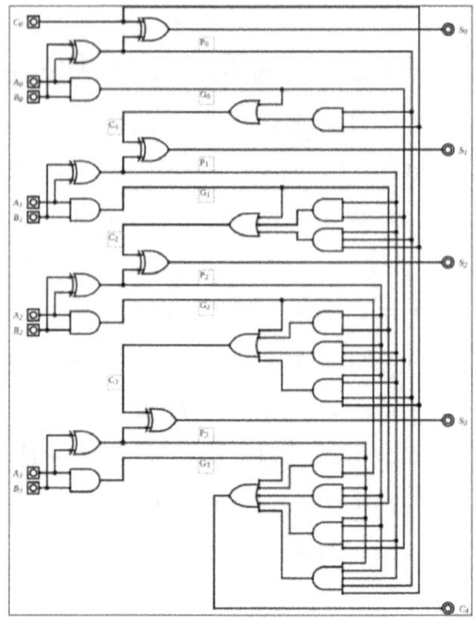

Figure 21.1. Schematic diagram of a computational (digital) logic circuit. The circuit is a simple one. There might be millions of such circuits on the central processing chip (CPU) of a computer.[4]

Dawkins' claim accordingly is that, when it comes to the identity and role of genetic particles, "there are no half measures and no intermediates or compromises" — certainly nothing we might recognize as qualitative or as a power of subtle self-transformation. "Our particles of inheritance … don't blend, but remain discrete and separate as they shuffle and reshuffle their way down the generations" (Dawkins 2006, pp. 159–63).

So this is what Dawkins assumes to be the essential character of our genetic material, which he situates at the causal root of every organism: it consists of discrete and separate causal elements that do not interpenetrate, neither are they subject to half measures, intermediates, or compromises. His reference to genes as strings of "pure digital information" makes it clear how paradoxically abstract his thinking has become — a common feature of materialist thought, which tends to be not nearly material or "enfleshed" enough. His "causal" factors come very close to being nothing more than elements of computer logic.[5]

The convergence in Dawkins' mind of (1) causal force and (2) compelling, computational logic — a symptomatic conflation running through much biology today — is ironic in light of his materialist commitments. If in fact the pure conceptuality of logic finds revelatory application in science, it would only reveal a world causally governed or organized according to thought, even if, in the case of thought reduced to logic, it is highly constricted and vacuous thought. By contrast, the very real mindfulness we encounter in our concrete and engaging interactions with the world, and above all with organisms, is far more profound — and far more replete with meaningful content — than a mere play of logical form.

It hardly needs saying that Dawkins' genetic informational bits are, by definition, incapable of participating in living wholes. Because their identity is conceived as almost eternally fixed, unblending, and quantifiably (digitally) specifiable, they cannot lend themselves to being qualitatively transformed or reimagined in harmony with the fluid, organizing ideas and intentions at work in every organism. It is the mutual interpenetration ("blending") of features and organizing ideas in the organism that makes its unity possible, and since Dawkins wants none of this, the unity is hidden to him.

But Dawkins gets his digital, non-blending particles of inheritance only by defining them into existence. Reality gives us a very different story.

How the image of "particles" has distorted the biologist's imagination

Digital, unblending genetic elements of pure, computational *logic* or *information*: where is one even to begin a critique of these impossible notions, upon which so much biology and evolutionary theory is erected? Where in the living organism do we find the slightest justification for them? Can Dawkins show us even one DNA sequence that functions in a strictly digital fashion?

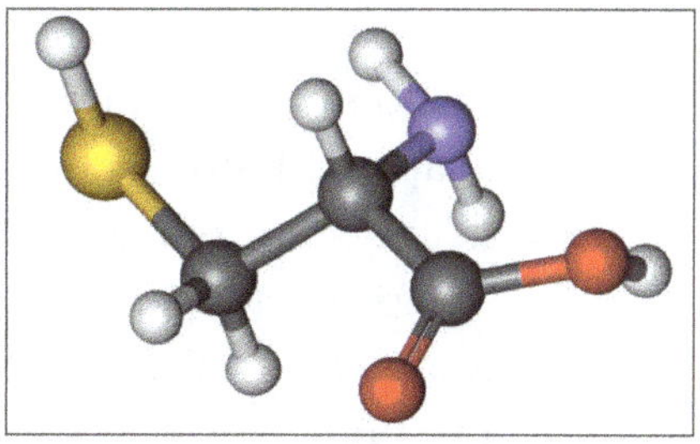

Figure 21.2. A conventional ball-and-stick model of a cysteine molecule.[6]

We might start thinking about this at a fairly remote distance from Dawkins' immediate genetic meanings by looking at an illustration offered by twentieth-century cell biologist, Paul Weiss. He reminded us of the commonplace pictures and models of molecules, represented as conglomerates of colored, spherical "billiard balls," each standing for an atom, and each seeming to be a stable, self-contained unit. (An example — not Weiss' — is shown in Figure 21.2.) He then contrasted that with the image in Figure 21.3, which shows how the constituent atoms of a molecule interpenetrate ("blend" into) each other as continuous fields or a complex system of mutually shaped forces. These "blur the former sharpness of the outer boundary of the molecule and let it melt into the surrounding molecular domains" (Weiss 1971a, p. 9).

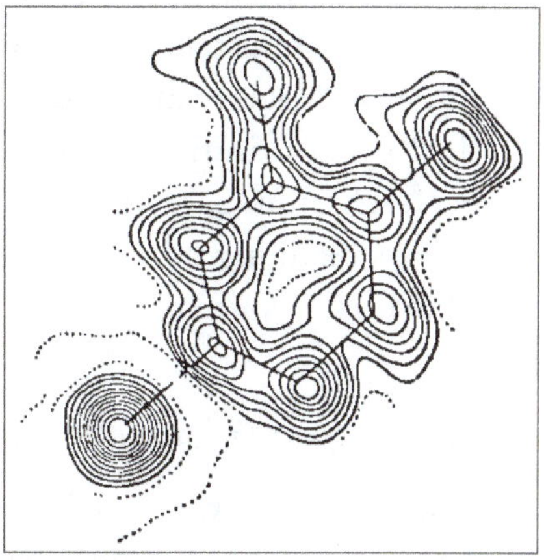

Figure 21.3. Schematic, two-dimensional representation of a molecule, where concentric "circles" are intended to map electron distributions of individual atoms, and straight lines artificially represent bonds between the atoms. NOTE: this drawing is not intended to represent the same molecule shown in the figure above.[7]

Weiss also remarked of figures such as this that "one is reminded of the contour maps of mountain ranges. Domains of particles are no more truly isolated than are mountain peaks" (Weiss 1971b, p. 235). Moreover, the parts of such an interactive system are rather "like islands," so that they must be "conceived as interconnected, though not so solidly as by a bed of rock, but loosely by the all-pervading mesh of forces and interactions" (Weiss 1971a, p. 11).

This may remind us of the discovery (mentioned in Chapter 5) of how water interacts with DNA. Lifting one paragraph from that earlier chapter:

> Early efforts to develop a computer simulation of a DNA molecule failed; the molecule (in the simulation) almost immediately broke up. But when [the researchers] included water molecules in the simulation, it proved successful. "Subsequent simulations of DNA in water have revealed that water molecules are able to interact with nearly every part of DNA's double helix, including the base pairs that constitute the genetic code."

Needless to say, this interaction of DNA with water alters the entire landscape (or seascape) of DNA, very much in the manner of Figure 21.3 above. None of it suggests much of a defense "at the bottom" for the idea of independent, digital, unblending genetic particles. But this point will become more explicit and more directly aimed at genes as we move along.

The fundamental reality of the matter was already glimpsed in the nineteenth century by the great experimental physicist, Michael Faraday. He recorded his prescient, "final brooding impression" that "particles are only centres of force" and that, for instance, water does not consist of atomic particles "side by side," but rather of "spheres of power mutually penetrated." And in notes for a talk given to the Royal Institution, he suggested that "matter is not merely mutually penetrable, but each atom extends, so to say, throughout the whole of the solar system, yet always retaining its own centre of force" (quoted in Barfield 1971, pp. 201n11 and 244n17).

Facing up to the gestural reality of DNA

What, then, do we see when we take even a cursory look at DNA? Exactly what Faraday and Weiss would have expected.

We have already heard about the positive and negative supercoiling of the double helix (Chapter 3, "What Brings Our Genome Alive?"). Both reflect a dramatic re-organization of forces along the DNA molecule — forces from which neither genes nor single nucleotide bases ("letters") of DNA can claim to be immune. These forces are not respecters of the boundaries between genes, and their re-shaping effects are known to help determine the functional role of genes within the organism.

But supercoiling is only one of many ways the genome is continually restructured. The DNA molecule, while inherently rather stiff and inert, lends itself, under cellular influences, to endless, plastic, structural change, both subtle and not so subtle, under the forceful pressure of its cellular environment.

Consider, for example, the many protein transcription factors that come to bear upon gene expression. In the simplistic thought of an earlier day, they were assumed to be just about the sole factors implementing the rigid "control logic of the genetic program." Moreover, they were said to bind DNA in a manner strictly determined by the abstract sequence of DNA "digital letters." The determining role of the abstract sequence was then supposed to justify the geneticists' belief that genes were the real agents in control of their own expression. But this conclusion sounds farcical today.

Transcription factors are now well known to engage in a gestural dance with the matrix of physical forces constituting the reality of the DNA sequence. They contribute their own forces to the infinite variations in the way genetic sequences "blend" together with each other and with innumerable regulatory molecules.

Actually, even within the terms of the earlier view, there were all sorts of unanswered questions about transcription factors. For example, what determined when and how often a particular factor was bound to the supposedly controlling DNA sequences? Why did it bind more to some of those sequences and less to others? And how was its activity coordinated with that of the many other transcription factors participating in the complex task at hand, which might involve the need for balanced expression of hundreds of genes?

Beyond this, however, the old concept of transcription factor binding (or not binding) to DNA is now recognized as hopelessly one-dimensional. The reality of DNA is difficult to think of as anything other than a play of ongoing *gestural form* orchestrated by the cell as a whole. You don't need to know the complex details of the terminology in order to get a sense for what it means when an article in the journal *Nature* reports how transcription factors interact with "shape features of the DNA sequences, such as minor-groove width, roll, propeller twist and helical twist" (Burgess 2015). These ever-changing shape features are subject to influences arising from the limitless cloud of regulative molecules (including water molecules, as we heard above, as well as the transcription factors themselves) that more or less transiently swirl around a cell's DNA. None of the analog shape features looks very much like Dawkins' concept of digital or computer-like elements of a genetic program.

The vast majority of our DNA is tightly and forcibly bent so as to wrap around millions of nucleosome core particles in the cell nucleus. The nucleosome, as we saw in Chapter 14 ("How Our Genes Come to Expression"), is perhaps the central integrator of signals bearing on gene expression and coming from all corners of the cell.

The wrapping of DNA around these protein core particles is probably never exactly the same in the case of any two of the thirty million or so nucleosomes in our genome, nor ever the same at any two different times in the case of a single nucleosome. The bending and shape changes in the DNA require a great variety of force interactions between DNA and the core particle, all in the presence of numerous associated molecules. So we hear (as before, in the case of transcription factors) that the functionally critical spacing and location of nucleosomes along a stretch of DNA can vary, depending on transient DNA features influenced by the cell "such as propeller twist, opening, electrostatic potential, minor groove width, rise, stagger, helix twist, and shear and roll … and buckle" (Kurup 2019). Everything is subject to dynamic variation.

The gestural form we are looking at in the case of DNA and related molecules appears to have no limit in its extent or its significance for gene expression. It cannot even be said that the double helix is always even close to being a standard double helix, or a double helix at all:

> Alternative conformations (including left-handed DNA, three-stranded triplex DNA, four-armed cruciforms, slipped-strand DNA with two three-armed junctions, four-stranded G-quadruplex structures and stable, unpaired helical regions) can exist in the context of chromosomes. Rather than being a static helix, DNA possesses dynamic flexibility and variability, as evidenced by helix regions that can be curved, straight or flexible. Differences result from variations in base stacking and twist angles inherent in different DNA sequences. DNA supercoiling[8] [induced, for example, by transcriptional enzymes], particularly unconstrained supercoiling, plays a major part in the dynamic flexibility and topological contortions of the DNA double helix (Sinden 2013).

Everything we have been hearing about is fundamentally qualitative and gestural rather than digital, and it counts not only toward whether a "basket is scored" (a gene is expressed), but also toward the "size of the score" — for example, will there be a large, rapid burst of expression, or a low-level, steady-state expression? Crucially, there is also the question of the *nature* of the score: for example, which of several functional variants of a protein will be synthesized — that is, what sort of protein is the DNA sequence effectively "coding" for? So the qualitative, gestural features mentioned above bear hugely on the practical meaning of a gene for the organism. They help to define what any given gene is at a given moment.

DNA, when caught up within the whole cell, is a phenomenon of movement brought about by its context. Of the endless variety of its movements, two are referred to as "DNA breathing." One of these (which we heard about in Chapter 3, "What Brings Our Genome Alive?") involves the rhythmic loosening of parts of the double helix from the nucleosome core particles around which they are wrapped. This has great influence — rhythmic influence — on the accessibility of portions of the DNA to gene-regulatory proteins. The other sort of breathing consists of local, transient separation of the two strands of the DNA double helix, which also affects the accessibility of the DNA.

Furthermore, the foregoing represents only a minuscule introduction to all the ways gene expression turns out to be a fluid, non-digital, and non-machinelike expression of dynamic, gestural form. I have not even mentioned what is widely considered to be the most prominent way DNA "letters," or nucleotide bases, are modified and the play of forces re-sculpted — namely, by the attachment of methyl groups to nucleotide bases in a process called "DNA methylation." This is the most common of a fair number of modifications to DNA, affecting many millions of nucleotide bases in our genomes. By this means the letters become *different letters*. This is one of the ways the cell makes a lie of the supposed constancy of the "digital DNA code."

Neither have I mentioned the large range of factors affecting the structure of RNA, an essential molecular carrier of the "genetic code" (and a mediator between DNA and actual protein production) whose liveliness of functional form is, if anything, even more obvious than what we see in DNA. So once a stretch of DNA is transcribed into RNA, we are again looking at a vast range of potential transformations of the "coding" sequence.

And then there is the actual translation into protein at the hands of what is called a "ribosome." Here, for a third time, a huge array of molecules somehow "aware" of the cell's contextual state and needs, come to bear on the proceedings.

And, perhaps most important of all, I have not cited the massive research effort today dealing with the form and movement of chromosomes — for example, the critical looping movements that bring genes and regulatory elements of DNA into functional proximity within the three-dimensional space of the nucleus. We are looking here at a gestural performance that many investigators can hardly resist referring to as a "dance" or an elaborate "choreography." We encountered some of this in Chapter 3 ("What Brings Our Genome Alive?"). How that dance occurs critically shapes how genes will be expressed. Genes are as far from being discrete, well-defined, independent causes as two ballet dancers engaged in a *pas de deux*.

Finally, all the foregoing can give you at least some slight sense for how the cells in our bodies — all working from a single inherited genome — can nevertheless pursue such diverse pathways of differentiation, from pancreas to blood to brain to muscle to retina to bone (Chapter 17 "Evolution Writ Small"). It seems just obvious that the range of uses to which the cell can apply its genes is all but unbounded. The cell can gesture or dance in countless different ways with whatever genome it has been dealt.

A gene that, within the full contextual life of the organism, can be interpreted as a self-contained and determinate cause or a bearer of strict, univocal, digital logic does not exist.

To view from the bottom or view from the top?

Evolutionary theorists, so it appears, have great difficulty recognizing as significant either the stable and highly distinctive character of the whole cell and whole organism, or the remarkable interior (mind-like) capacities through which that character is consistently expressed and sustained amid all the transformations of individual development. And so they find it easy to discount everything living. They discount, that is, the future-directed powers of self-realization, adaptation, and whole-cell (whole-organism) reproduction — the very powers that hold the most obvious relevance for inheritance and evolution.

Organisms as such simply don't show up through the death-shroud that is the particulate view of life. As for the particles themselves — the supposedly unblending, unexpressive, qualitatively inert genes and nucleotide bases ("letters" of the genetic code) — they are, in reality, illusions. The upshot of the foregoing discussion is that the discrete, non-blending, genetic particles that Dawkins, for example, is so quick to idealize as controllers of the organism's evolution, do not actually exist as real, material entities. They exist only as logical or digital constructs fancifully projected upon the living cell by a materialistic mindset that can't seem to face actual materiality with its insistent expression of interior being.

The situation can also be summarized by saying that evolutionary biologists are currently blind to organisms as beings *organized from within*. To speak of interiority is anathema to them, and therefore any genuine recognition of *organizing ideas* is also anathema. Which is too bad, because either organizing ideas in the material world have real consequences, or else the terms biologists freely use to invoke such ideas subconsciously — "organizers," "organization," "self-organization" — are symptoms of shamefully vacuous theorizing.

Think of it: take away the organizing *ideas*, and what would be left of any sort of "organizing" or "organization"?[9] "Capable of constituting a kind of *meaningful* order or coherence" is just what the word "organizing" means. When we think of any organizing principle, we are thinking of a principle, not a thing. Every science necessarily seeks immaterial (ideational) *principles of order*.

Consider the much-criticized but still endemic idea that there are *genes for* particular traits. If we believe that genes possess, in their own right, the essential, organizing or directing power to realize traits — traits that are in fact qualitative, non-discrete, interpenetrating, and expressive of a specific (species-related) "way of being" — then we are ascribing to genes a living power to *organize* almost unthinkably complex physiological processes requiring a kind of moment-by-moment active judgment governing a virtually infinite number of molecular interactions in a fluid, continually changing context. (I focused especially on such processes in Chapter 8, "The Mystery of an Unexpected Coherence.")

Genes surely do *participate* in such a power, but it is a power of activity belonging to the whole organism and is not properly attributed to any collection of material elements, such as genes, that are caught up in it.

Those who want to adhere to materialist principles gain nothing by contradicting them. They gain nothing, that is, by transferring the interior, organizing ideas of the cell or organism to genes. If genes really possessed their own "informational" powers for sensing their wider context; if they really could encode messages tuned to moment-by-moment changes under infinitely varying circumstances; if they really had a way to direct the interactions among countless billions of molecules in a fluid medium, enabling those molecules to carry out indescribably intricate operations such as RNA splicing, and if in general they really were able to *inform* and *organize* the life of the entire cell[10] ... well, once we have accepted this impressive play of wisdom through the genome, what reason would remain for denying it to the vividly expressive cell or the whole organism, where we actually observe it?

If "context matters," as so many biologists are now telling us, it can only be because it really does substantively matter. It makes a difference to what happens. The context exercises, in other words, its own, over-arching sort of causal power. It's not a kind of power that can be transferred to materialistically conceived particles — particles abstracted from cell and organism ("the context") as if, by themselves, they carried a decisive causal logic independent of the living matrix in which they find their real existence.

WHERE ARE WE NOW?

Looking beyond particulate inheritance

Biology today is governed by a taboo: *The biologist must never acknowledge stepping outside the materialist framework — or at least must remain unaware of doing so*. She must never grant that animals have an interior — that every animal is an integral, unified whole possessing, or possessed by, an active, wise agency.

The biologist's materialist commitments are impossible to keep. They would render the organism, as a living being, invisible to scientific investigation. This is why biologists can hardly avoid preserving the organism's agency by covertly transferring it to special molecules (DNA). But at the same time — in order to keep an illusion of observing the taboo — they speak of these wonderfully effective, "informational" molecules as if they were "things" uninformed and ungoverned by the agency of the whole.

This double-talk, which would have DNA possessing the creative powers of life while at the same time consisting of "mere chemical stuff," is somehow easier to stomach in the case of molecules than in the case of whole organisms. Molecules, being non-phenomenal (invisible to sense perception), lend themselves more obligingly to the projection of our mechanistic/animistic fantasies.

It is population genetics, above all else, that has converted evolutionary theory into a theory about genes rather than organisms. As we heard from one population geneticist, "The only mandatory attribute of the evolutionary process is a temporal change in allele [gene] frequencies."

There is little beyond quantifiable (digital or logical) entities in this picture — nothing material, plastic, and expressive, nothing qualitative, nothing through which the interior life of organisms can shine. We cannot connect anything in the particulate gene to our own conscious awareness, or to animal sentience, or to perceptual experience and cognition, or to the organizing ideas underlying animal form and behavior. If particulate genes account for these aspects of life, no one has a clue how it could be. Yet these genes are routinely posited as the evolutionary basis for understanding *all* life.

We have also found more than ample reason to question the notion of genes as fixed, inert, particulate, unchanging entities passing down through the generations unaffected by their bodily hosts. There are, in reality, countless ways, both subtle and obvious, in which the whole cell lends to DNA its present and ever-changing meaning for the life of the organism. This is, in the first place, how cells manage to turn their DNA toward the end of cell differentiation, resulting in the hundreds of cell types in our bodies, from muscle, blood, and bone to brain and liver.

In the next chapter we will consider Richard Dawkins' defense of the idea that the heritable basis of evolution must consist of "potentially immortal" elements passed down through the generations without undergoing change. We will see that this apostle of evolution offers us a principle of stability and duration, but nothing like a principle of evolution (change or transformation).

NOTES

1. For the record, currently relevant criticisms of genocentrism go all the way back a century. There is, for example, the brilliant work by marine biologist E. S. Russell, especially his 1930 book, *The Interpretation of Development and Heredity: A Study in Biological Method*, which I discussed in the previous chapter. In the modern era, one could start with *Exploding the Gene Myth* by Harvard professor of biology Ruth Hubbard and Nobel Prize recipient Elijah Wald, or the essay, "Unraveling the DNA Myth," by cell biologist Barry Commoner.

   The physicist, biologist, and philosopher of science, Evelyn Fox Keller, has illuminated genocentrism from many sides, including in her book, *The Century of the Gene*, published in 2000, and her chapter on "Genes as Difference Makers" in 2013. The one-time molecular biologist and now philosopher of science, Lenny Moss, wrote an incisive and influential critique titled *What Genes Can't Do* in 2003. And his book chapter, "Darwinism, Dualism, and Biological Agency" (2005), has perhaps never been exceeded for the succinctness and penetrating depth of its take-down of the controlling gene.

   There are many other worthy commentaries on genocentrism, of which one (Gawne, McKenna, and Nijhout 2018) is cited in the main text below. See also Holdrege 1996, Rose 1998, Moczek 2012, Noble 2013, Walsh 2015, Noble 2018, and any number of other books and journal articles published over the past couple of decades.

   Special mention also goes to University of Chicago microbiologist James Shapiro's book, *Evolution: A View from the 21st Century* (second edition, 2022). Evolution can only occur if there is useful variation, or potentials for variation, in the evolving organisms. And Shapiro provides overwhelming evidence that, whatever whole-organism features may count as viable heritable variation, organisms certainly have the genetic aspect very well covered. That is, they possess a sophisticated and wide-ranging ability to revise their own DNA — and they put it to use in a huge variety of ways. One could prefer that Shapiro not rely so heavily on computer and program metaphors, but nevertheless he makes abundantly clear the organism's effective exercise of a well-directed agency with respect to its DNA.

2. Disappointingly, those who continue giving support to genetic reductionism include proponents of the "extended evolutionary synthesis" and the "third way of evolution." As near as I can tell, these movements remain as thoroughly materialist in their fundamental assumptions as the evolutionary mainstream. It truly does appear that any questioning of the materialist creed in biology is likely to spell the immediate dimming, if not the end, of an otherwise promising career. The persistence of such metaphysical dogma in science ought to be ringing alarm bells on all sides.

3. For a discussion of the question of stability, see Chapter 22, "A Curiously Absolute Demand for Stable Variation."
4. Figure 21.1 credit: Trex43 (https://commons.wikimedia.org/wiki/File:Four_bit_adder_with_carry_lookahead.svg), CC0 1.0.
5. Here is a slightly different angle on the present paragraph in the main text and the two following paragraphs:

   We might say that, with his digital information, Dawkins relies on formal principles, or formal causation. But he doesn't realize this because his "form," having become one dimensional and logical/mathematical, shorn of full-bodied expression and meaning, has been reduced toward the vanishing point. This severely abstracting tendency is the usual end result of a strong materialistic bias. The real essence of that bias is a flight from meaning into mindlessness. This necessarily becomes a flight from the material world, because this world, if received in its vital, perceptual ("sensual") reality, is always found to be an expression of meaning. Hence the preference for abstract thought. Compare Dawkins' seeming abhorrence of the stuff of bodies:

   > There is no spirit-driven life force, no throbbing, heaving, pullulating, protoplasmic, mystic jelly. Life is just bytes and bytes and bytes of digital information.

   See also Chapter 24, "Is the Inanimate World an Interior Reality?"
6. Figure 21.2 credit: MarinaVladivostok (https://commons.wikimedia.org/wiki/File:Cysteine_ball-and-stick.png), CC0 1.0.
7. Figure 21.3 credit: From Weiss 1971a.
8. I offer a very brief explanation of supercoiling in Chapter 3 ("What Brings Our Genome Alive?").
9. To speak of organizing ideas at work in an animal's life is not to imagine the animal *thinking them*. As the facts of instinct can remind us, an organism may be more *possessed by* its intelligence than in willful or conscious possession of it.
10. Everything becomes nonsensical if we overlook or deny the inner power of the whole when in fact we have quietly transferred it to a part while pretending not to believe in it. Already in 1930 E. S. Russell recognized the consequences of this transfer:

    > The germ-plasm, even in its modern genic form, is [thought to be] something which itself remains unaltered while acting as the cause of visible change in the organism. Aristotle would have recognized in this almost mystical conception something strangely like his "soul"! (Russell 1930, pp. 267–68).

    I have no wish to belittle the idea of the soul. But anyone who believes in it should believe in it — not transfer it "under the table" to particular material particles, genetic or otherwise.

    More recently the philosopher of biology Jason Scott Robert remarked on the "*animistic* (and otherwise problematic) idea of a genetic programme" (Robert 2004, p. 37). One of the most obvious ways DNA is treated as if it were by itself an animated, living being lies in the common, yet false, conviction that it exercises not only the powers of a human programmer to maintain and modify a single inherited "program" for the highly divergent purposes of hundreds of cell types (and trillions of cell contexts), but also the power to reproduce itself. In reality both the replication of DNA and its adaptation to the needs of different cell types are extraordinarily complex activities of whole living cells and organisms — activities of the sort we looked at throughout the preceding chapters of this book.

SOURCES

Barfield, Owen (1971). *What Coleridge Thought*. Middletown CT: Wesleyan University Press.

Burgess, Darren J. (2015). Shaping Up Transcription Factor Binding, *Nature Reviews Genetics* vol. 16, no. 259. doi:10.1038/nrg3944

Commoner, Barry (2002). Unraveling the DNA Myth: The Spurious Foundation of Genetic Engineering, *Harper's Magazine* (February), pp. 39–47.

Dawkins, Richard (1995). *River Out of Eden: A Darwinian View of Life*. New York: BasicBooks.

________ (2006). *The Blind Watchmaker*, third edition. New York: W. W. Norton. First edition published in 1986.

Gawne, Richard, Kenneth Z. McKenna, and H. Frederik Nijhout (2018). Unmodern Synthesis: Developmental Hierarchies and the Origin of Phenotypes, *Bioessays* vol. 40, no. 1, article 1600265. doi:10.1002/bies.201600265

Graur, Dan (2015). Introduction to Molecular and Genome Evolution. https://judgestarling.tumblr.com/post/129931383826/introduction-to-molecular-and-genome-evolution

Hartl, Daniel L. (1988). *A Primer of Population Genetics*, 2nd edition. Sunderland MA: Sinauer Associates.

Holdrege, Craig (1996). *Genetics and the Manipulation of Life: The Forgotten Factor of Context*. Hudson NY: Lindisfarne.

Hubbard, Ruth and Elijah Wald (1993). *Exploding the Gene Myth: How Genetic Information is Produced and Manipulated by Scientists, Physicians, Employers, Insurance Companies, Educators, and Law Enforcers*. Beacon Press.

Keller, Evelyn Fox (2000). *The Century of the Gene*. Cambridge MA: Harvard University Press.

________ (2013). Genes as Difference Makers, in *Genetic Explanations: Sense and Nonsense*, edited by Sheldon Krimsky and Jeremy Gruber. Cambridge MA: Harvard University Press, pp. 34–42.

Kurup, Jiji T., Ion J. Campeanu and Benjamin L. Kidder (2019). Contribution of H3K4 Demethylase KDM5B to Nucleosome Organization in Embryonic Stem Cells Revealed by Micrococcal Nuclease Sequencing, *Epigenetics & Chromatin* vol. 12, no. 20. doi:10.1186/s13072-019-0266-9

Lynch, Michael (2007a). The Frailty of Adaptive Hypotheses for the Origins of Organismal Complexity, *PNAS* vol. 104, suppl. 1 (May 15), pp. 8597–604. doi:10.1073/pnas.0702207104

_______ (2007b). *The Origins of Genome Architecture*. Sunderland MA: Sinauer Associates.

Moczek, Armin P. (2012). The Nature of Nurture and the Future of Evodevo: Toward a Theory of Developmental Evolution, *Integrative and Comparative Biology* vol. 52, no. 1, pp. 108–19. doi:10.1093/icb/ics048

_______ (2022). When the End Modifies Its Means: The Origins of Novelty and the Evolution of Innovation, *Biological Journal of the Linnean Society* (July 6). https://doi.org/10.1093/biolinnean/blac061

Moss, Lenny (2003). *What Genes Can't Do*. Cambridge MA: MIT Press.

_______ (2005). Darwinism, Dualism, and Biological Agency, in *Darwinism and Philosophy*, edited by Vittorio Hösle and Christian Illies. Notre Dame IN: University of Notre Dame Press, pp. 349–63.

Müller, Gerd B. (2019). An Evolutionary Biology for the 21st Century, in *Perspectives on Evolutionary and Developmental Biology*, edited by Giuseppe Fusco. Padova, Italy: Padova University Press, pp. 29–39.

Noble, Denis (2013). Physiology Is Rocking the Foundations of Evolutionary Biology, *Experimental Physiology* vol. 98.8, pp. 1235–43. doi:10.1113/expphysiol.2012.071134

_______ (2018). *Dance to the Tune of Life: Biological Relativity*. Cambridge UK: Cambridge University Press.

Robert, Jason Scott (2004). *Embryology, Epigenesis, and Evolution: Taking Development Seriously*. Cambridge UK: Cambridge University Press.

Rose, Steven (1998). *Lifelines: Life Beyond the Gene*. Oxford: Oxford University Press.

Russell, E. S. (1930). *The Interpretation of Development and Heredity: A Study in Biological Method*. Reprinted in 1972. Freeport NY: Books for Libraries Press.

Sinden, Richard R. (2013). The Helix Turns at 60: Writhing Free in Chromosomes, *Nature Structural and Molecular Biology* vol. 20, no. 3 (March), pp. 251–53. doi:10.1038/nsmb.2533

Uller, Tobias and Heikki Helanterä (2019). Niche Construction and Conceptual Change in Evolutionary Biology, *British Journal for the Philosophy of Science*, pp. 351–75 (published online August 30, 2017). https://doi.org/10.1093/bjps/axx050

Walsh, Denis (2015). *Organisms, Agency, and Evolution*. Cambridge: Cambridge University Press.

Weiss, Paul A. (1971a). The Basic Concept of Hierarchic Systems, in *Hierarchically Organized Systems in Theory and Practice*, edited by Paul A. Weiss, pp. 1–43. New York: Hafner.

_______ (1971b). One Plus One Does Not Equal Two, in *Within the Gates of Science and Beyond: Science in Its Cultural Commitments*, by Paul Weiss, pp. 213–61. New York: Hafner. Reprinted from *The Neurosciences: A Study Program*, edited by Gardner C. Quarton, Theodore Melnechuk and Francis O. Schmitt. Rockefeller University Press, 1967.

CHAPTER 22

A Curiously Absolute Demand for Stable Variation

THERE COULD HARDLY BE a more frequently stated requirement for natural selection than this: *any beneficial variation occurring in an organism, if it is to be evolutionarily relevant, must be faithfully replicated, stable, heritable, and long-lasting down through the generations*. The idea is that, if a given variation is likely to pass away after a generation or two, or if it quickly suffers further change, then the normally long and slow process (by selection) of spreading the variation throughout a population will not have enough time to reach completion. Patrick Bateson was giving voice to a universal consensus when he wrote, "For the Darwinian evolutionary mechanism to work, something must be inherited with fidelity" (Bateson 2017, p. 77).

But we might want to ask: If stable, unchanging heritable elements are a *prerequisite* for evolutionary change, what about the change itself? What is our principle, not of stability or changelessness, but of transformation — the coherent transformation of an evolutionary lineage from *that* kind of a viable, living, whole organism to *this* one? And is our principle consistent with everything we know about organic processes of identity and change in general?

After all, we do have almost an over-abundance of examples of organic change. They are given to us in every complex organism's developmental life cycle, which illustrates continuous and often intense transformation — including the kind of radical "melt-down" of old structures and subsequent "re-creation from scratch" we saw in the metamorphosis of the goliath beetle (Box 17.1).

This near-total digestion of the old form, followed by the emergence of a dramatically different form leads us to a second, equally important question: Is it primarily certain material *products* of an organism's activity that must be stably maintained along a path of transformation? Or is it rather the contextual (holistic) *capacity for an integrative, unifying activity* that must in the first place be preserved — an activity out of which not only are particular structures precipitated, but also the coherent life and character of a species is sustained. Whatever this sustaining activity may be, and wherever we can imagine it occurring, would this not also be the activity and the place where transformation becomes possible? After all, every organism is already a process of continual change, without which it is dead.

When the Goliath beetle larva is overtaken by the seemingly chaotic "catastrophe" through which it will gloriously re-emerge in adult form, what is the organizing power, and what are the organizing ideas, through which this all-encompassing transformation of materials occurs? And how are the organizing ideas and power passed from one generation to the next?

When we talk only about the inheritance of discrete *products* of activity, we have already shown a willingness to ignore the more fundamental problem of the origin of viable new traits, which require much more than some new bits of matter. Even if we are talking only about the development of the color patch (speculum) on a duck's wing feathers (Figure 11.2), we still need to embrace in thought a huge range of dispersed but precisely coordinated molecular interactions that are possible only as well-directed expressions of an integrated and living whole that functions as an unfathomably wise power of activity.

The obstacle to a proper reckoning with change and inheritance lies in the focus on *things* (whether isolated or in the aggregate), which can never themselves be understood as agents of change. This remains true when the things are genes whose mathematically calculable spread through a population is naïvely thought of as equivalent to the spread of discrete traits, and when this in turn is taken to be evolution. It is the demand for this sort of sterile calculability that leads to a one-sided emphasis on stable variation (long-lived gene mutations) rather than on the potent activity of self-transformation that organisms put on such obvious display. Here, then, is the decisive point: the "thing" that remains stable (is "replicated") across generations is not really a *thing* at all, but a process and an organizational unity whose dramatic formative potentials are shown in the wide array of stably achieved cell types in our bodies, each of which faithfully reappears "endlessly" from human generation to generation.

Unfortunately, the philosophical bias of today's biologists prevents them from even beginning to take seriously an organism's dynamic *capacities*, as opposed to the *things* that precipitate out of that dynamism. The only kind of stability they can imagine is the stability of such things, not the ever-reliable, interior, form-giving powers through which these thing come about and are maintained.

It would be interesting to put this question to any evolutionist: Can you tell us how much of the difference between a chimpanzee and a human is attributable to different genes, and how much is attributable to different principles (ideas) of cellular organization? And, of the part you attribute to genes, how much of that is dependent on the aspects of cellular organization affecting how those genes are used?

Meanwhile, it is well to recognize the mystery we are up against in talking about these things. The differentiating cell acts as though it somehow "knows" where it is along the larger path of transformation. It "knows" how to use its inheritance from its parent cell not only to venture upon its own variation from that parent, but also to provide a distinctive inheritance for use by its daughter cells as material for still further variation. In this way each cell participates in a coordinated movement toward an ultimate "goal" of differentiation it can neither "see" nor consciously plan for. The cell participates, that is, in the intention or directiveness of its larger context, just as its constituent molecules participate in its own directiveness.

Obviously, we are up against great mystery when we take note of all this. But we will never make progress against the mystery if we cannot allow ourselves to take note of what is in front of our eyes.

A good place to begin getting a further grip on the issues may be the popularly effective case Richard Dawkins has made for the all-but-immortal, thing-like stablity of the causal elements of evolution.

Richard Dawkins, genes, and the biologist's "ultimate particles"

Dawkins has been articulating his genocentric view of Darwinian evolution for some fifty years, evoking, at the extremes, both passionate support and vitriolic criticism. Despite the ebb and flow of controversy, however, one oddity remains constant: the decisive failure of his view rarely or never comes into clear focus, presumably because it is a defect found in virtually all conventional (and nearly all unconventional) thinking about evolution.

Figure 22.1. Richard Dawkins[1]

In order for a genetic variation to be useful, Dawkins says over and over, it must be "potentially eternally heritable." "I'm not wedded to DNA," he assures us, but "I am wedded to this operational criterion that alterations in it go on forever potentially" (Dawkins 2009).

What he means is that, in order to be evolutionarily useful, variations must be *selected for* — perhaps not eternally, but at least for a long time. The ones that are harmful are selected *against*, and therefore tend to pass out of existence. But the truly beneficial adaptations can be selected and selected again, generation after generation, without any in-principle limitation. They are in this sense "potentially eternally heritable," which can only be the case if they are extremely stable.

The transgenerational longevity (stability) of genes is why Dawkins favors them, rather than whole organisms, as the true reproducers, or replicators, upon which natural selection works. "Bodies don't get passed down the generations; genes do" (Dawkins 2006b, p. 79). Just about all the details of one's body can change from one generation to the next. Bodies are, compared to genes, "like clouds in the sky or dust-storms in the desert. They are temporary aggregations or federations. They are not stable through evolutionary time" (Dawkins 2006a, p. 34).

For Dawkins, then, the body is a non-repeatable collection of material bits. Its identity lies in those variously aggregated bits, not in the character or the principles of organization at work bringing them to expression. But it is difficult to see how a conglomeration of material bits, without any integrating and unifying principle, amounts to much of an identity at all. He apparently offers no principle of identity for organisms as particular sorts of living beings able to maintain their own character. And so he is quite right in saying that an organism rendered essentially invisible in this way — discounted as a kind of non-entity — can hardly be a significant evolutionary cause. "An individual organism is not [an evolutionarily relevant] replicator, because alterations in it are not passed on to subsequent generations" (Dawkins 1982). While an organism as a whole may be "the all important instrument of replicator preservation: it is *not* that which is preserved."[2]

It would have been a different matter if Dawkins had asked himself about the holistic capacity of an organism to remain true to type, or to adjust itself adaptively to circumstances in accord with its given character, or — in all its processes of cell differentiation, including those of its germline — to move directively toward a complex future form that does not yet exist and is not dictated by its earlier states. It is just a fact that all these capacities *are* passed between generations; they all manifest, with stable character, in offspring as well as in parents. Who today has even bothered to look for evolutionarily significant variations in this stable power of directive movement, which is the one sort of power that could transform a species? But this is to get ahead of ourselves.

No one, incidentally — neither Dawkins nor any other biologist — is saying that the organism's phenotype is irrelevant to evolution. Their claim is that the real relevance has to do with the fact that certain genes have contributed to this phenotype and therefore to the survival capabilities of the organism and its offspring. This in turn influences which genes will be passed down the line and survive in the larger population. It is, in this picture, the changes (mainly the beneficial variations) in genes that most directly explain and map to adaptive evolutionary change.

But beyond the question of the organism's survival, in which they themselves have a say, genes are said by Dawkins to "live" independently of the particulars of an organism's life. They follow their own stable arc down through the generations, remaining just what they are except for the rare beneficial mutation that is indefinitely preserved. They constitute a nearly eternal "river of information." This river "passes through bodies and affects them, but it is not affected by them on its way through" (Dawkins 1995, p. 4).

Implied in all this — and very important for Dawkins — is the idea that genes can be conceived in something like a particulate fashion. "I insist on an atomistic view of [genes]," he wrote in *The Extended Phenotype* (Dawkins 2008, p. 113). And elsewhere he has elaborated: "What I have done is to define a gene as a unit which, to a high degree, *approaches* the ideal of indivisible particulateness. A gene is not indivisible, but it is seldom divided. It is either definitely present or definitely absent in the body of any given individual. A gene travels intact from grandparent to grandchild, passing straight through the intermediate generation without being merged with other genes."[3]

Development versus evolution

Dawkins is well aware that much of the criticism he has received comes from those studying the development of organisms. These observers find it very hard to recognize his genes in the developmental processes they investigate. It is, in many developmental contexts, impossible to assign genes long-lasting, discrete, well-identified causal roles, and also impossible to view genes as passing through these contexts unchanged in their functional significance for the developing and evolving organism.

In offering repeated responses to such criticism, Dawkins has made it clear that he considers the intricate choreography of development — in which many non-genetic factors figure prominently — to be irrelevant for evolution. But he emphatically rejects the charge that the gene-centered view denies "proper respect to the Great Nexus of complex causal factors interacting in development" (Dawkins 2008, p. 99). "I yield to no one," he told an Oxford debate audience, "in my admiration of the complexity of feedback loops, of the details — the immensely complicated details — whereby genes actually do influence phenotypes. There's absolutely no suggestion that it's irrevocably deterministic, there's absolutely no suggestion that it's simple" (Dawkins 2009).

And yet he fears that too many people get carried away by the intricacies of development. It is true, he grants, that it is precisely through development that we see how an organism grows and adapts toward maturity through complex and holistic processes. But these all too easily distract us from the decisive role of genes in evolution — a mistake he derisively equates to the lament, "Dear oh dear, development is a terribly complicated nexus, isn't it?" (Dawkins 2004).

He himself prefers "frankly facing up to the fundamental genetic nature" of Darwinian selection (Dawkins 2008, p. 28). Development may be a "complicated nexus," but evolution is merely a matter of pristine bits or bytes in an informational DNA sequence.

The fact just is, he says, that we learn nothing of interest to evolutionary theorists by looking at the dynamic interrelations — the forming and dissolving, spaghetti-like causal arrows — through which DNA is fitted to its proper place among all the cellular activities. "There may be backwards arrows in all sorts of other senses but, in the sense that specifically matters for Darwinian evolution, the causal arrow of biological development from genotype to phenotype really is a one-way arrow" (Dawkins 2004).

The key to all this lies in two features of genes, as Dawkins sees them: (1) They can, through the occasional gene mutation, produce evolutionarily relevant variation in organisms. And (2) this genetic variation is characteristically stable; through the replication of DNA it can be passed down the generations potentially forever. This means that natural selection can, over as much time as necessary, change the distribution of genes in an evolving population. Very little other than genetic change, so the argument goes, yields discrete variation with such stability and lasting power and therefore little beside genetic change can contribute to evolution.

So, however vague and non-determinate genes may be for development, they are decisive for evolution. That's why, for the evolutionist, "the complexity of development itself is an obscurantist red herring" (Dawkins 2004). Yes, he says,

> development is terribly complicated, and we don't yet understand much about how phenotypes are generated. But *that* they are generated, and *that* genes contribute significantly to their variation are incontrovertible facts, and those facts are all we need in order to make neo-Darwinism coherent (Dawkins 2008, p. 22).

And again:

> It doesn't matter how complicated the developmental support structure, nor how utterly dependent DNA may be upon it, the central question remains: which elements … of development have the property that *variations* in them are replicated, with the type of fidelity that potentially carries them through an indefinitely large number of evolutionary generations? (Dawkins 2004).

So it's not just that genes qualify as drivers of evolution, but also that the other players in development do not. When we shift our attention from development to evolution, "the special status of genetic factors rather than non-genetic factors is deserved for one reason only: genetic factors replicate themselves, blemishes and all, but non-genetic factors do not" (Dawkins 2008, pp. 98–99). And, in perhaps his most succinct summary:

> The quality of hi-fi variation … is a precious, rare, onerous, difficult talent, possessed by genes and computer viruses and a few other things — but *genuinely* few … In order for anything to evolve by natural selection, there has to be variation in something that is both potentially long lasting and causally powerful, so that there emerges a difference, on the evolutionary time-scale, between the state of the world if one variant survives compared with the state of the world if an alternative variant survives. If neither variant survives more than a couple of generations anyway, we are not talking evolution at all (Dawkins 2004).

Genes, according to Dawkins, survive this rare and onerous test.

A short critique of Dawkins' view

Dawkins is admirably forthright about his desire for a purely genocentric explanation of evolution, and therefore also about his need to put evolution and development into different boxes. But it doesn't work. One thing both he and his critics could probably agree on is that development shows genes actually carrying out their biological roles. If we want to know what genes are and what they mean for the organism, then all the abstract talk in the world about "pristine bits and bytes" and "rivers of information" cannot supplant what we actually observe about genes in living contexts.

And this is where the problems begin. If Dawkins really is willing to concede the reality of the "Great Nexus of complex causal factors interacting in development" — if, that is, he recognizes the holism implied by the fact that the "causal arrows" of development run in all directions as guided by the unity of the larger context — then how can he assign a determinative causal role to genes entirely independent of that context? How can he conclude, based on the non-evolutionary, non-transformational criterion of stability ("immortality") rather than any demonstrated capacity for consistent, directive change, that genes alone are the decisive basis for evolutionary transformation? Has he shown, counter-intuitively, that the organism's employment of its genes in the service of radically different cellular phenotypes is irrelevant to the evolutionary role of genes? How is this even conceivable?

Dawkins has his principle of stability, but (as I have been pointing out) this is not yet a principle of change or evolution. The fact of the organism's dynamic unity, its capacity to bring about coherent change, has never seemed to register with him. This fact is already implicit in his talk of a one-way causal arrow from genotype to phenotype, which necessarily assumes an adequate *organizing wisdom* capable of managing, for example, the movement of the Goliath beetle from larva through chaos to adult. But he needs to reckon with this reality in his thinking, which he never seems to have done.

But Dawkins' lapse of thinking here is even worse. His conviction that, "in the sense that specifically matters for Darwinian evolution, the causal arrow of biological development from genotype to phenotype really is a one-way arrow" is exactly wrong. Not only is the causal arrow always a work of whole organisms, but even within a single organism's body such as that of a mammal, the causal arrow is spun around a dial whereby it is made to run from a single genotype to just about any and every imaginable cellular phenotype. So it is not really *an arrow* at all. The sense of the "arrow" that matters for evolution is the sense in which the organism shows that it is in charge and that it can harmoniously aim the arrow in many directions at once — all consistent with the unified way of being of the kind of organism it is.

The overriding causal relations in an organism run from the whole to the part. There is no way to sketch such an arrow; it certainly cannot be sketched in the way we often (erroneously) imagine a single gravitational causal arrow running from the sun to the earth. We may, for analytical purposes, separate that particular arrow from the one that runs in the reverse direction (and from the ones involving all the other bodies in the solar system), but in fact we know of no such separable causes. Our knowledge of matter and its lawfulness does not humor us in this.

The simple fact is that an organism changes the meaning of its genes seemingly at will. How could this potent and always adaptive redirection of its genetic resources toward hundreds of distinct cellular endpoints be irrelevant to the organism's provision of an inheritance for its offspring — an inheritance that can be materially expressed only in the form of one of those masterfully managed, future-facing, whole-cell, context-governed achievements, the gamete?

Dawkins' evolutionarily significant bits of material substance not only lack any agency of their own for directing cells along a pathway of differentiation, but they are, according to his own testimony, even incapable of participating in the organism's agency. They are, unlike any real biological entities, passive and inert, impervious to influence from the organisms ("vehicles") that carry them along. Contrary to his professed respect for the "Great Nexus of complex causal factors interacting in development" he sees genes as distinguished by their stasis and longevity, and by the fact that they have *some* relevance to the organism's phenotype, however vaguely conceived. Accepting this shroud of vagueness without pausing to consider the whole-organism reality behind the shroud is crucial to Dawkins' thought.

The vagueness is necessary because it is impossible to imagine concretely how his bits of genetic material account for the actual powers of organic life, and therefore impossible to imagine concretely how these bits pull off any genuine evolutionary transformation of the organism. The idea seems to be: if we can only shove the problem back into the mists of geological time, maybe we can *vaguely* picture everything happening under the aegis of the all-powerful, god-like logic of natural selection acting with designing intelligence and unexplained power upon beings possessing no living, organizational, and transformational powers of their own (Chapter 16, "Let's Not Begin with Natural Selection").

Time, it seems — if only we are given large enough quantities of it — can magically paper over the explanatory void at the heart of our theorizing, just as when natural selection is said to "solve" the problem of the organism's purposive, intentional, end-directed activity, despite such activity widely being seen as "impossible" within the accepted terms of physics and at the same time being a *prerequisite* for natural selection. (See "Natural selection: The shortest path to confusion is circular," in Chapter 18.)

In sum: if, as Dawkins acknowledges, genes accomplish their effects only in the context of tortuous interactions with innumerable constituents of the cell, and if a different organization of these interactions is required for each of the drastically different cellular "creatures" our bodies produce — red blood cell, bone cell, lung cell, liver cell — then how do we avoid the conclusion that an evolution of gene roles in a particular lineage of organisms must require a coherently organized evolution of the entire context of interactions?

The evasion of this conclusion leaves one aghast. Every one of the trillions of cells in our bodies, from the brain to the soles of our feet, has its own organization and character. That's what makes us what we are. Our existence is unthinkable without this organizing power that gives every cell a distinctive life and place of its own. And Dawkins is telling us that this power is irrelevant to evolution? — that the unique *organizational powers* associated with the cells passing between generations have somehow been emptied of their usual sort of significance? — that the only heritable factors for us to consider are the pitifully incapable and inert *things* we know as DNA molecules, which cannot even replicate themselves?[4]

Saying that development is irrelevant to evolution is a way of saying that organisms don't exist as far as evolutionary theory is concerned. Or, rather, that organisms exist only under servitude, as bearers of DNA. And, of course, Dawkins has found rhetorically effective ways to make exactly this point. He has presumably realized all along that taking development seriously would torpedo his gene's-eye view of evolution. My suspicion is that he simply never closely engages with the problems of development because he is uninterested in them and they are alien to his entire point of view, based as it is on the masterful, almost magical power of genes as purely physical First Causes of every organism.

"Immortal" genetic variation

Is genetic variation "potentially immortal" (Dawkins 2008, p. 83), stably holding to its own identity and remaining mostly unchanged by the organisms hosting it down through the generations?[5] Does Dawkins even have the principle of stability he thinks he has?

We might begin with the fact that all life on earth is commonly thought to be the flourishing, down through all generations and all species, of the life of a single original cell of unknown origin. We can reasonably assume that the successors of that original cell have retained many elements of the essential character — the characteristic organization — of cellular life.

What, then, is the means by which the distinctive character of cellular life has been maintained in these "potentially immortal" cells? What keeps cells alive and as the kind of things they are, more or less forever? I mentioned above that principles of process and organization — powers of activity — are decisive. But Dawkins seems to have no interest in such principles and powers.

But while cells are potentially immortal in Dawkins' sense, genes certainly are not. In fact, their causal involvement in the cell's work does not even remain constant in a single cell over its lifetime. This is because local activity within the cell — the context that gives genes their meaning and makes the difference between a neuron and muscle cell — is always being coordinated and repurposed according to the changing needs and interests of the whole cell and whole organism, and the cell's genes are thus "captured" by the cell's purposes.

Dawkins gets his river of fixed, unchanging, selectable "particles" of inheritance only through an act of intellectual violence — only by mentally wrenching certain molecular parts of DNA (in particular, the nucleotide bases constituting the genetic "code") from their meaningful, dynamic chromosomal and cellular contexts. He abstracts them from the stream of life in which alone they gain their function moment by moment. He wants dead, material things for theoretical manipulation. You might say that he abstracts the genetic "letters" from the cellular "sentences" or "conversations" in which they find their life and meaning.

Only such an act of abstraction can give him what he needs: heritable particles to which he can apply the same name (even if not the same meaning) from one context to the next — particles that can be tracked within both individual organisms and breeding populations. Then, because the name of any particular genetic "letter" remains the same whether it is functionally the "same" letter or not, he can pretend he is always talking about precisely the same, unchanging thing.

And yet we know very well that, in the living and meaningful sense, genes never remain unaltered for long. The "same" genes can have radically different causal implications in different contexts within a single cell or organism, and also in different kinds of cell or organism. Even when we think a gene has analogous functions in different organisms, it can turn out that the functions have unrecognizably different physiological realizations. This reflects the very different ways of being from one kind of organism to another — ways of being in which genes are caught up rather than being the decisive orchestrating factors.

As one of endless evolutionary examples: the PAX gene is found in both fruit flies and mammals, and has been thought of as a "master control gene" for the formation of eyes. But not only is its activity now known to be interwoven and interdependent with that of countless other genes and their regulating factors and functions; the compound eyes of a fruit fly are altogether different organs compared to the eyes of a mammal. If we try to picture a gene, *as a single, supposedly well-defined causal unit*, independent from the stream of life in which it finds itself and yet at the same time fitting itself into the completely different physiological contexts of fly and mammal, the picturing does not go well.

Throughout the first half of this book we have seen over and over how DNA is caught up in, and given its functional identity by, its context. And in Chapter 21 ("Inheritance, Genetics, and the Particulate View of Life"), I illustrated some of the countless ways in which the DNA content of genes is itself twisted, untwisted, bent, distorted in various ways, chemically modified, moved around in the nucleus, converted to nonstandard double-helical forms or even non-double-helical forms, and otherwise driven by the cell into conditions that transform its genetic role and identity.

So the truth counter to Dawkins' claims is that genes participate in, and are informed by, the whole cell and organism. Whether we're talking about a goliath beetle (before and after metamorphosis) or a human being moving from embryo to adulthood, genes simply are not capable of determining the form of an organism. It would be truer to say that the organism determines the meaningful form of genes and the chromosomes on which they reside.

Genes, then, are not "potentially immortal" — or even potentially "lifetime-lasting" within individual organisms — because the whole organism defines and redefines their functional nature and identity as it proceeds through the cycle of its life.

To put the matter in these terms, however, would doubtless provoke Dawkins' skepticism. For he has very publicly worried that "a kind of 'holistier than thou' self-righteousness has become fashionable," and further, "There are times when holistic preaching becomes an easy substitute for thought" (Dawkins 2008).

This may be true. But it is also true that, when it comes to understanding the life of organisms, holism is just about the most disastrous thing to lose sight of.

The embodied organism is not like a cloud

Recall Dawkins' claim that bodies don't exhibit the constancy required for natural selection and evolution. They are, we heard him say, "temporary aggregations" like clouds or dust storms in the desert. Changes in bodies (as when someone devotes much of his life to weight-lifting) are not inherited. This truth is obvious enough. But the fantasy of the whole organism's transitory existence is as wildly off the mark as Dawkins' invention of the "almost eternally" durable genetic particle immune to its host organism's influences.

The well-directed development of a whole organism no more represents change without stable identity than does cell differentiation. If a differentiating cell "knows" very well what kind of cell it is and where it is going, all the more a developing organism follows a reliably defined path of development and self-expression leading to its own fullest realization. Crucially, this path is never *precisely* defined or *materially* determined (which seems to be Dawkins' main concern). For example, adjustments to unpredictable environmental disturbances during development may occur at any time. But the adjustments are *in the service of* the organism's fullest expression of its own nature.

The identity here is that of a *dynamic* unity. But a dynamic unity capable of maintaining itself while responding in its own way to its context is a far more profound center of identity, power, and life than a supposedly static particle disengaged from its surroundings. The capacity of an organism to maintain its identity wholly escapes Dawkins' world of thought because it cannot be understood materially. Rather, it reveals principles that can be principles of true understanding only because they are formal causes standing above and governing the material aspects of biological phenomena.

Never, except in the imagination of someone preferring inert particles to living beings, could the organism have been likened to a cloud or dust storm. Its unity and stability lies in its giving material expression to a specific *kind* or *species* (Chapter 20, "Inheritance and the Whole Organism"). Every complex organism is an almost miraculous, stable, reigning unity governing many differentiating cell lineages. The reliability and constancy of this sort of identity is so evident that for millennia nearly everyone erroneously assumed that all species were eternally fixed in their essential nature.

In order to correct this view today, we do not need to throw out the constancy everyone can observe, but only to bring it alive by rendering it more dynamic, adaptive, and transformative in line with modern evolutionary insights. An organism's identity is crucially reflected in its ability to adapt to a changing environment in a manner consistent with its own species-character, and to transform itself and its gametes according to the future (evolutionary) potentials of its kind. We see such dynamic identity and adaptability even in our own biographies. Our immediate experience as selves gives no support to the idea that growth, development, and change imply the lack of a coherent, guiding identity.

To observe life, then, is always to watch an interplay of change within continuity, plasticity within constancy. Both aspects of the living interplay run through all biology. Perhaps our most convenient access to them comes through the study of development, as when we watch a whole organism "coming to itself," or watch a differentiating cellular lineage progressing toward its fully realized type.

Here the principle of change is easy enough to verify — and it no more relies on the absolute constancy of the material products of change than does the metamorphosis of the goliath beetle larva into a mature beetle (Box 17.1). I described in Chapter 17 how hundreds of cell lineages in our bodies "evolve" (are differentiated), not by remaining mostly unchanged, but rather by compounding change upon change down through the cellular generations. The result is a profound, qualitative transformation of whole cells, explainable only as a power of activity, not as a determination by previous material constituents.

It is difficult to doubt that this compounding of change upon change is owing to an orchestrating power that works throughout the entire cell — a power not at all one-sidedly determined by genes, their mutations, or any other material constituent of the cell. We are watching a whole cell radically re-organize or "re-think" itself.

If many developmental changes in a cell lineage are not stable and heritable over any large number of cellular generations, it is because *they had better not be*. After all, the lineage is *on the way to somewhere*, proceeding directionally along a pathway of integral, holistic transformation. This suggests how differently we will have to look at evolutionary processes once we are willing to acknowledge that they centrally involve *coherent change* — coherent expression — against a background of the constant flow of physical substances. What we're *not* looking for are random, atomistic mutations in otherwise dead (if preternaturally durable) little bits of matter.

As I have already suggested, Dawkins has no theory of evolution, no theory of change. He has only a theory of genes — a false theory of independently existing, unchanging particles being carried down the river of time and passing inertly through generation after generation of organisms. How these particles magically account for living traits — how they *could*, even in principle, account for such traits — is not part of the story Dawkins has cared to share with us. All we know is that he holds, vaguely, to the conventional view that genes must *somehow* explain organisms, their traits, and their evolution.

But there is no need to avert our eyes from the "dangerous fluidity" of a picture involving continuous change and transformation. For the other aspect of the process of cell differentiation lies in the fact that the differentiating cellular lineage is remarkably sure of its identity: it is *this* lineage, and is powerfully "insistent," in tune with its context, about reaching *its own* mature character. And, having reached its maturity, it is capable of stably maintaining it as long as necessary, while never losing a degree of adaptability.[6]

So here in the organism's development and cellular differentiation, we see compelling identities involving, not the stasis of inert particles, but rather a marriage of constancy and change, selfhood and transformation — a reality it would be very strange for any student of evolution to lose sight of.[7] What is constant is not a mere physical product, but an active way of being. As I have put it a number of times (drawing on a phrasing of the Aristotle translator, Joe Sachs), an organism is "continually at work remaining itself." We could say much the same of a cell. Physical entities — aggregations of particles, if we insist on thinking that way — are a precipitated residue of such activity, not a cause of it.

A counter-picture to the gene's-eye view of evolution

The central problem that Dawkins avoids reckoning with is this: everything we have noticed about the enduring unity, purposiveness, and transformational powers of the cell and organism applies, not merely to development, but also directly to evolution. That's because *what passes hereditarily between generations is never less than a whole cell, and this whole cell is never less than a whole organism.* And, as we have seen, both the cell and the organism exhibit an enduring, purposive identity possessing not only a dynamic constancy, but also a profound transformative capacity — exactly the combination an evolutionary theorist must look for.

Anyone who wants to claim that genes, rather than whole living cells (organisms) are the basis for evolution needs to explain why we must ignore everything we know about cells — about how they proceed so effectively along evolutionary (developmental) trajectories and how they intricately, flexibly, and authoritatively enlist their genes along with all their other resources in achieving their ends.

Further, why should we assume that the totipotent zygote (capable of engendering all the cell lineages of the adult organism) brings nothing of its self-transformative, re-organizing powers *across the generational gap* — nothing, that is, of evolutionary relevance? There is, of course, really no gap at all, but only continuous life. And the activity of whole-cell transformation is fully as insistent on the parental side — in the unique differentiation of germ cells and the seemingly against-all-odds merging of distinct germ cells with distinct genomes in a single zygote — as it is on the offspring side.

We know a great deal about the powers of reliable change possessed by cells and organisms, and we also know about a gene's complete inability to represent on its own an organism's expressive traits. If it happens (as it does) that an organism's abilities as a living, developing, striving-to-survive "vehicle" for its genetic cargo require all its vast array of transformational powers, and if we see it employing those powers with almost infinite intricacy and adaptational effectiveness in all its cell lineages, why should we imagine these powers going dead or inert at those especially intense moments when one generation is actively preparing for a successor generation.

One can only assume that Dawkins has proven blind to the only agents of biological transformation we know because, as a materialist and reductionist, he simply cannot tolerate the idea of biological agency, despite what he witnesses in every organism he has ever seen. He *must* overlook active and purposive organic wholes by conceiving organisms as built, bottom-up, from collections of inert particles and mindless processes. But, as I have pointed out repeatedly in this book, such a conception is never possible to hold consistently, and all observation-based biological description immediately controverts it. (See, for example, Chapter 2, "The Organism's Story.")

The organism's dynamic, transformative capacities are so clearly underwritten by everyday perception that they need not even be mentioned in order to be casually, if also silently, taken for granted by evolutionists. This gives their theorizing a sheen of plausibility. But when the failure to mention these transformative capacities turns into an active "conspiracy of silence," so that our theories of evolution must ignore the obvious, then something has gone badly wrong.

WHERE ARE WE NOW?

Sterile, immortal particles, or living cells?

It would hardly be surprising to suggest that a great deal of one's evolutionary theorizing depends on whether one approaches the topic with the mindset of a materialist and reductionist or, to the contrary, ascribes to every cell and every organism a wise agency and purposiveness expressed in all its living activity. But before tackling that issue directly, one can always begin by looking at what we know of cells.

This chapter is grounded in a seemingly incontrovertible fact observed in all complex, multicellular organisms: their cells proceed through dramatic processes of differentiation reflecting stable organizing ideas in which the entire cell is caught up and through which it undergoes profound transformation. Cells can, to all appearances, become as different from each other as any two organisms with dramatically different genomes. But the differences between cell lineages in a single organism cannot, in any fundamental sense, be attributed to genetic differences, because all the lineages derive from a single, inherited genome.

I have argued that Richard Dawkins, with his stable, "almost immortal" genes has a principle of stability and continuity, but not a principle of evolutionary change. What change he does have is a mere redistribution of inherently meaningless particles within a species or population, achieved by the agonizingly slow process of random mutation and natural selection. He acknowledges that we know almost nothing about the relation between these particulate rearrangements and the development of an organism's actual traits. And he offers nothing to suggest how, even in vaguest principle, meaningless and immortally inert particles could possibly contribute to the traits expressing the integral, qualitative, and coherent meanings of a life.

But, as mentioned immediately above, we do see such a principle of meaningful change illustrated in cell differentiation within a complex, multicellular organism. Here organizational change is not preserved from cell generation to cell generation, but is repeatedly compounded (changed again) — and *must* be compounded because the cell lineage is *going somewhere*. It is proceeding toward its mature form. We have every reason to look for a similar principle of change-on-top-of-change within an evolutionary lineage that is going somewhere — proceeding toward the coherent, living form of a transformed organism.

The question this poses for evolution is straightforward — and is puzzlingly ignored on all sides: Why should this transformational capacity of whole cells be ignored as a source of evolutionary change — ignored, that is, when we shift our view away from genes toward the cell-organism that in undisputed fact bears the full inheritance passing directly between parents and offspring? It's true that the question is a difficult one because the kind of dramatic, qualitative, whole-cell transformation we encounter so often in developmental studies cannot be reduced to mathematically analyzable changes in gene sequences and gene frequencies. But why wouldn't we expect *any* explanation for the evolutionary transformation of an organism to be challengingly complex? The complexity of holistic processes is certainly no reason to turn away from their investigation when we are seeking to understand evolution.

There is one other thing we should notice. Dawkins' disinterest in a meaningful principle of change, as opposed to a meaningless principle of stability, is one reflection of a fact we have registered throughout our discussion of evolutionary theory: the organism as such has dropped out of the theory. If Dawkins has no principle of change, it is because his theory has been purged of the life of organisms. In fact, he himself stresses this point when he describes organisms as mere robotic "vehicles" for genes. We might wonder why his genes take on such transcendent importance in his mind when there is no *life* for them to participate in.

NOTES

1. Figure 22.1 credit: Alex Folkes/Fishnik Photography (https://www.flickr.com/photos/libdems/3940859791/in/photostream/), CC BY-ND 2.0.
2. Dawkins 2008, p. 114. Dawkins is quoting himself from an earlier paper in *Zeitschrift für Tierpsychologie*.
3. Dawkins 2006a, pp. 33–4. Along these same lines, Dawkins has written that the "digitalness [of genes] is probably a necessary precondition for Darwinism itself to work" (Dawkins 2006b, p. 163). I discuss this remark along with the non-digital character of genes in Chapter 21, "Inheritance, Genetics, and the Particulate View of Life."

   I would add here that my criticism of Dawkins' "atomistic" genetic particles is in no way intended to deny the fact that genes, like much else in the organism, possess their own *specificity*. Just as an enzyme has an affinity for a particular substrate or substrates, so also a gene is associated with its own suite of protein variants. The cell flexibly employs the gene as a resource for the generation of one or another of those variants. The cell, of course, requires many other resources for its various tasks, and the decisive, repeated modification of a protein over its lifetime is achieved *after* the gene-resource has been brought into play for the initial synthesis of the protein.
4. It is false when Dawkins says that genes "replicate themselves." It is perfectly well known that they are utterly helpless to replicate themselves — and also to perform error correction on themselves. It is the cell that both replicates them and carries out DNA

damage repair. (He surely does understand this, despite his not wanting to focus on it.) On the power of proteins to manage DNA and reassemble shattered genomes, see Chapter 8.

On a different note: I can't help wondering whether one among many valid ways to think of genes is as relatively dead, crystallized, anti-life structures (representatives of the past) that provide the *resistance* that the organism needs in order to assert its creative, organizing forces and bring alive its own story.

5. More specifically, it is genetic variation *in the germline* that Dawkins says is potentially immortal.
6. It would be good not to forget what we learned in Chapter 6 about cells changing their "identity" in the hair follicle niche, depending on circumstances. We have to understand the cell's identity in a broad enough way to recognize this possibility of metamorphosis. It illustrates how cells are subordinate to, and receive their identity from, the larger context and the organism as a whole.
7. Change and continuity: every organic whole embodies — lives by means of — a harmonization of these contrasting principles. But these are exactly the principles that any theory of evolution must somehow hold together. It's obvious enough that you can't have evolution without change. But so, too, without continuity there is only the arbitrary substitution of some elements of a mere aggregate for others, with nothing that lends significance to the result. If the change is to be non-arbitrary or coherent, there must be a persistent character attributable to the whole. Without an underlying continuity no enduring, nameable entity or being exists of which we can meaningfully say, "Yes, *this* is evolving." There is instead just "one damned thing after another."

   And we should add this: if, as in the physical organism, *all* the material elements are subject to change, then the underlying continuity must be immaterial (Barfield 1965, p. 96). But then, that is how we have understood the organism all throughout this book — as a bodily exterior "shone through" by an interior. Or: as an interior partly condensed into a translucent exterior.

SOURCES

Barfield, Owen (1965). *Unancestral Voice*. Middletown CT: Wesleyan University Press.

Bateson, Patrick (2017). *Behaviour, Development and Evolution*. Cambridge UK: Open Book Publishers.

Dawkins, Richard (1982). Replicators and Vehicles in *Current Problems in Sociobiology*, edited by King's College Sociobiology Group, pp. 45–64. Cambridge: Cambridge University Press.

_______ (1995). *River Out of Eden: A Darwinian View of Life*. New York: BasicBooks.

_______ (2004). Extended Phenotype — But Not *Too* Extended. A Reply to Laland, Turner and Jablonka, *Biology and Philosophy* vol. 19, pp. 377–96.

_______ (2006a). *The Selfish Gene*. Oxford: Oxford University Press. Originally published in 1976.

_______ (2006b). *The Blind Watchmaker*, third edition. New York: W. W. Norton. First edition published in 1986.

_______ (2008). *The Extended Phenotype*, revised edition (1999), reprinted with corrections (2008). Oxford: Oxford University Press. Original edition published in 1982.

_______ (2009). Comments during Homage to Darwin debate, Balliol College, Oxford University (May 8), featuring Richard Dawkins, Lynn Margulis, Steve Bell, and Martin Brasier. The event was chaired by Oxford's Denis Noble. Links to recordings of all three parts of the event can be found at http://www.voicesfromoxfordorg/?s=homage+to+darwin

Sachs, Joe (1995). *Aristotle's Physics: A Guided Study*. New Brunswick NJ: Rutgers University Press.

CHAPTER 23

The Evolution of Consciousness

"THE ECONOMIC AND SOCIAL STRUCTURE of Switzerland," wrote Owen Barfield in his major work on the evolution of consciousness, is owing in part to the tourist industry, which in turn depends upon the fact that "the *mountains* which twentieth-century man sees are not the mountains which eighteenth-century man saw" (Barfield 1965, pp. 145–46).

Barfield is gesturing toward an evolution of consciousness that, as he saw it, implied an evolution of phenomena. Taken literally (as indeed he intended it), his claim easily baffles even serious attempts to fathom it. Most observers will conclude that the claim is either outrageous or else trivially expresses the idea that, while our *experience* of the mountains has changed, the mountains themselves have certainly not. But the question of the relation between reality on one hand and our experience of reality on the other is subtle and potentially disorienting for us today. It is also inseparable from the troubled, several-hundred-year quest for an empirical (experience-based) science of the objective world. Does "experience-based" mean science is inescapably subjective, or, rather paradoxically, that our experience bears the marks of a mind-independent (experience-independent) reality? Or is there another alternative, as I try to sketch in Chapter 24 ("Is the Inanimate World an Interior Reality?").

In this chapter we will use Barfield's remark as little more than a stimulus to begin looking at the evolution of consciousness. By the time we are done, however, the question about the relation between human experience and the world we so readily think of as wholly mind-independent may have gained a more intriguing aspect. But we will forego until the next chapter any effort to throw light on the underlying issues. These have to do with the way we are situated in the world through our cognitive experience.

We now begin by looking at a few aspects of the evolution of language.[1]

There was a primordial unity of inner and outer meaning

According to the evolutionary story that most of us have forcibly absorbed through our education, humankind somehow raised itself above the beastly, mindless, material substrate of its origin so as to achieve, step by step, the mystifying wonders of language and poetry, music and art, politics and science, and all the other sublimations contributing to high culture. The sea of meaning within which we now swim — without which we would have nothing we could recognize as human life — *somehow* bubbled up from *somewhere*, if only as an illusion of the human mind, and cast a kind of spell over the bedrock meaninglessness of brute matter.

"Somehow," I say, since the meaning at issue (whether illusory or not) and the question how it could have emerged from an eternal silence of Unmeaning is so great an enigma that we can scarcely even articulate the question coherently.

What is not enigmatic — and is clearly available to investigation — is the fact that when we look further and further back through history, we see an ever richer language, not an increasingly material and "de-meaned" language reflecting our supposedly brutish origins. As the nineteenth-century English poet Percy Bysshe Shelley observed, "In the infancy of society every author is necessarily a poet, because language itself is poetry" (Shelley 1840).

We do not, that is, discover ancient literature to be impoverished relative to modern literature. It is more like the reverse of this: we still debate today whether, for example, the Homeric epics — composed orally before the development of writing in ancient Greece — have ever been surpassed for psychological depth, dramatic power, poetic subtlety, and human interest.

We will take the philologist and historian, Owen Barfield, as our primary guide, first, to the evolution of language, and then to the evolution of consciousness more generally. Barfield devoted a long life spanning the entire twentieth century to the study of these two topics, and about the former he wrote:

> The farther back language as a whole is traced, the more poetical and animated do its sources appear, until it seems at last to dissolve into a kind of mist of myth. The beneficence or malignance — what may be called the soul-qualities — of natural phenomena, such as clouds or plants or animals, make a more vivid impression at this time than their outer shapes and appearances. Words themselves are felt to be alive and to exert a magical influence (Barfield 1967, pp. 87–88).

The "enchanted" landscape of ancient consciousness, as Barfield sketches it for example in *Poetic Diction*, could not have been one of conscious invention, unrestrained metaphor, or causal speculation. The earliest historical evidence shows us that humans were not yet possessed of the sort of selves, or the resources of language, conducing to such invention and hypothesis. They simply observed nature as it was given to them. Their meanings did not arise from anything like modern reflection or theorizing, but were encountered directly, as if spoken by the earth itself.

This truth has been disguised from us by what Barfield referred to as "logomorphism" — the projection of modern thought processes onto "that luckless dustbin" of the primitive mind. "The remoter ancestors of Homer, we are given to understand, observing that it was darker in winter than in summer, immediately decided that there must be some 'cause' for this 'phenomenon,' and had no difficulty in tossing off the 'theory' of, say, Demeter and Persephone, to account for it" (Barfield 1973, pp. 74, 90).

But we are given no evidence that the mythic mind had any concern with such explanations, if only because the conditions for them did not yet exist. Our modern ideas of cause and effect lay far in the future. The ancient fact of the matter was more like this: "In the myth of Demeter the ideas of waking and sleeping, of summer and winter, of life and death, of mortality and immortality are all lost in one pervasive meaning" (Barfield 1973, pp. 90–91).

Think for a moment about what we mean today by "explaining the world." Such explanation requires two distinct awarenesses: that of something "out there" posing a puzzle for us, and an understanding "in here" that clarifies the puzzle. But our ancestors did not possess these separate awarenesses. Unlike us, they were not in a position to dualize the world into outer material fact and interior explanatory idea. They lacked the requisite psychological distance from the world, and therefore did not experience the otherness of "things" as we do. The mythically enchanted landscape was, for them, an unanalyzed interfusion of outer and inner, of sense perceptions and soul content.

For example, the story of the Greek sun-god "Helios" could hardly have originated as an animistic effort to account for a material sun, given that neither the history of language nor what we can surmise of mythic consciousness affords any evidence that a purely material sun (for example, as a ball of gases) had yet been conceived. The sun's glory, its light and warmth, were directly and non-reflectively experienced as ensouled realities.

We still find remnants of such indivisible meaning in later eras, as when we read in the *New Testament*,

> Truly, truly I say to you, unless one is born of water and the spirit, he cannot enter into the kingdom of God ... The wind blows where it wishes and you hear the sound of it, but do not know where it comes from and where it is going; so is everyone who is born of the spirit. (John 3:5–8)[2]

Translators into English have been forced to use two different words, "spirit" and "wind" (in other texts "breath" is required) where the original Greek has a single word, pneuma. "We must, therefore, imagine a time," Barfield noted, "when [Latin] *spiritus* or [Greek] *pneuma*, or older words from which these had descended, meant neither *breath*, nor *wind*, nor *spirit*, nor yet all three of these things, but when they simply had *their own old peculiar meaning*, which has since, in the course of the evolution of consciousness, crystallized into the three meanings specified"[3] (Barfield 1973, pp. 79–81).

"Nor yet all three of these things" — not the addition of one distinct meaning to another, but a single unity encompassing wind, breath, and spirit. The distinct meanings had not yet arisen, and so were not available to be added together. Our current dualism of "inner" and "outer" was not yet part of human experience. It is hard for us to appreciate this at a time when our language forces a dichotomous choice between the terms of outward, sensible reference and those drawn from our interior life.

We will take one further example, this one drawn from Barfield's *History in English Words:*

> As far back as we can trace them, the Sanskrit word "dyaus," the Greek "zeus" (accusative "dia"), and the Teutonic "tiu" were all used in contexts where we should use the word *sky*; but the same words were also used to mean *God*, the Supreme Being, the Father of all the other gods ... If we are to judge from language, we must assume that when our earliest ancestors looked up to the blue vault they felt that they saw not merely a place, whether heavenly or earthly, but the bodily vesture, as it were, of a living Being (Barfield 1967, pp. 88–89).

Summing up the historical picture, the nineteenth-century American transcendentalist, Ralph Waldo Emerson, wrote in his 1836 book, *Nature*: "As we go back in history, language becomes more picturesque, until its infancy, when it is all poetry; or all spiritual facts are represented by natural symbols." And again: "It is not words only that are emblematic; it is things which are emblematic"[5] (Emerson 1836, pp. 33, 37).

Figure 23.1. Egyptian sky goddess, Nut, and earth god, Geb, held apart by Shu, a god associated with air and wind.[4]

So the direction of the evolution of language and meaning is, so far as we can discern from the historical record, the opposite of an "ascent from brute materiality." Before humans could speak in their individuated voices, or could even conceive of devising theories about nature, the natural world spoke to and through them — meaningfully and poetically. The rhythm and meter we find, for example, in the epic Homeric hexameters with their "thundering epithets" were, Barfield wrote, relics of a time "when men were conscious, not merely in their heads, but in the beating of their hearts and the pulsing of their blood — when thinking was not merely *of* Nature, but was Nature herself" (Barfield 1973, pp. 146–47).

Looking back upon myths such as that of Demeter and Persephone (where *you and I* are likely to see metaphor or simile) Francis Bacon pointed out the error of this view: "Neither are these only similitudes, as men of narrow observation may conceive them to be, but the same foot-steps of nature, treading or printing upon several subjects or matters."[6] And regarding these "footsteps of nature," Barfield adds:

> Men do not *invent* those mysterious relations between separate external objects, and between objects and feelings or ideas ... The language of primitive men reports [these footsteps] as direct perceptual experience. The speaker has observed a unity [for example, of sky and God, wind and breath], and is not therefore himself conscious of *relation*. But we, in the development of consciousness, have lost the power to see this [unity] as one (Barfield 1973, pp. 86–87).

What words can teach us about the evolution of consciousness

There is one province of reality, one domain of the material world, where we humans have gained a knowledge unexcelled in its sophistication, its fine detail, and its almost infinite nuance of meaning. It is a domain that, perhaps more than any other, shapes our lives and influences our happiness day in and day out. And knowledge of events within this domain comes naturally: nearly all humans achieve a level of expertise dwarfing the scientific researcher's mastery of material phenomena within this or that field of inquiry.

The phenomena I am referring to are those coming to expression in the human face. I have specifically in mind, not the power of producing those expressions, but rather of objectively reading them. For, of course, we do read them objectively. Our lives and society would be impossible if we could not navigate the universe of facial gestures with a largely shared understanding. *This* means *that*. The face illustrates how, in physical features, we are dealing with meaning borne upon a material dynamic of force and substance, but not explicable as if the meaning arose from, or were *caused by*, that dynamic. We naturally think of the cause as operating in some sense interior to its outer manifestation. There is an *interior* expressing giving rise to an outer expression.

And what we have seen in the preceding section is that the face of nature herself presented our ancestors with a countenance whose inner significances were inseparable from what we today would consider its outer manifestation. Natural phenomena constituted a living language, rather as, still for us today, the sense-perceptible human face can at times scarcely be distinguished from its expressive eloquence — from the meaning it communicates.

The history of language gives us ample evidence pointing back to the kind of inner/outer unity we are presented with in the Greek *pneuma*. Barfield shows how we can see this especially in two broad classes of words:

Nearly all those words now bearing immaterial meaning in the form of high abstraction, or else referring to our interior life, were once inseparable from sensible experience.

Emerson was not the first to recognize this truth when he wrote in 1836:

> Every word which is used to express a moral or intellectual fact, if traced to its root, is found to be borrowed from material appearance. *Right* means *straight*; *wrong* means *twisted*. *Spirit* primarily means *wind*; *transgression*, the *crossing of a line*; *supercilious* the *raising of the eyebrows* ... *thought* and *emotion* are words borrowed from sensible things, and now appropriated to spiritual nature (Emerson 1836, chapter 4).

The idea is not that the interior or psychic aspect was lacking in the perception of ancient folk, but rather that it was bound together inseparably with the outer, material meaning. And, as Barfield reminded us, this truth extends far beyond words like *spirit, thought*, and *emotion*:

> To what, precisely, does each one of them refer — the tens of thousands of abstract nouns which daily fill the columns of our newspapers, the debating chambers of our legislatures, the consulting rooms of our psychiatrists? *Progress, tendency, culture, democracy, liberality, inhibition, motivation, responsibility* — there was a time when each of them, either itself or its progenitor in another tongue, was a vehicle referring to the concrete world of sensuous experience with a tenor [immaterial meaning] of some sort peeping, or breathing, or bursting through (Barfield 1977, p. 38).

Moreover, as Barfield stresses, high-sounding scientific terms "are not miraculously exempt" from the general rule. A great part of the explanatory apparatus of science consists of largely abstract and dematerialized words such as *stimulus, cause, effect, reference, control, repress, information, code*, and *program,* all of which can be shown to have been once inseparable from an "outer clothing." Only with time did the abstract or inner meanings become detached from sense perception. By abstracting away from that clothing we gained the powers of thought necessary for our current science[7] (Barfield 1973, p. 134).

The other group of words, now referring to material, sense-perceptible phenomena, once also connoted sentience or inwardness.

We have already seen how ancient words for "sky" also meant "divine being." The very words by which we today designate the materiality of the world are sufficient to make the point. "Matter" likely traces back to Latin *mater,* "mother." And "physical" derives from Greek *phyein*, "grow." So the Greek *ta physika* — "natural things" or "things of external nature" — was rooted in living activity. Of course, as we also noted earlier, words by which we now refer to purely physical bodies such as "sun," "Venus," "Mars," "Jupiter," and "Saturn" can be traced back to the names of various deities.

So words now having a purely immaterial significance once also referred to sensible phenomena, and words now purely sensible or material in reference once also referred to interior experience. Taken together, these two groups of words testify to the primeval experience of nature as a material/immaterial, outer/inner unity before the dualization of this unity in the modern sense was even conceivable.

But none of this is to say we should look to etymology for current meanings. Will anyone claim today that when we say someone is "wrong," we really mean he is bent like a stick, or that to "conceive" something is to grasp it physically? The dualization of the world *has* occurred, and one result is that we now enjoy a vast panoply of meanings and a diverse range of distinctions formerly unavailable. Nevertheless, the history of meaning raises its own questions.

How could the unitary meanings of our ancestors have possessed their primordial, immaterial aspects if the associated, sense-based images (a bent stick, the hand's grasp) were not inherently expressive of an immaterial reality?[8] If the indissoluble unity of sensible image and immaterial meaning were arbitrarily invented by early speakers and were not inherent in the phenomena themselves — if things were not, as we heard from Emerson, essentially *emblematic*, but were instead subject to any speaker's arbitrary, metaphoric invention — how would others have picked up on the speaker's invented, immaterial meanings? Indeed, how could the very possibility of immaterial meanings ever have come about, if the original reality out of which humans emerged was (what we think of today as) solely physical?

The cognitive experience of the ancients was *given* by nature. Its inner, expressive content was not added by a reflective or theorizing perceiver, but was already experienced in perception. Things meant something *on their face*. Our ancestors were, you might say, participant-observers entranced by an ensouled drama staged within their own consciousness by the world's phenomena.[9]

What the historical record shows is that those ancestors recognized, in whatever was expressed through natural phenomena, a speaking agency akin to themselves. "Whether it is called 'mana,'" wrote Barfield, "or by the names of many gods and demons, or God the Father, or the spirit world, it is of the same nature as the perceiving self, inasmuch as it is not mechanical or accidental, but psychic and voluntary" (Barfield 1965, p. 42).

Today our evolutionary trajectory has brought us to a vastly different place — a place where we are routinely taught to think disparagingly of the ancients as astonishingly naïve. But whatever our thoughts and meanings may be, we ought to acknowledge with some humility that they are available to us only because the world first mimed them, so to speak, thereby enabling them to light up in human minds "naïve" enough to read the face of nature in a way that few of us today can.

At the same time, we will need to acknowledge that, so far as the historical record testifies, our evolutionary trajectory has not accorded with the usual assumptions. There is no evidence that we slowly ascended from a crude life of material unmeaning to a humanly and artificially contrived realm of meaning, value, culture, and spirituality. Our life today, with its materialistic convictions and experience of a meaningless world, has required a long descent from the living, ensouled landscape upon which our ancestors were nurtured.

Our evolutionary heritage, culminating in Cartesian dualism, has taught us to insist upon a radical separation of the inner and outer dimensions of our experience, which once formed so compelling a unity. And then, under the further influence of materialist thought, we have learned to regard the inner dimension as "merely subjective" or somehow less than fully real.

But perhaps, instead of projecting our current mental processes upon the "woefully subjective and ignorant" ancients, we might want to consider how our own history may have cut us off from an ancient wisdom, finally concreting in our deepest, most unyielding, and largely unconscious (and therefore unfree) habits of thought and experience. Through such reflection, perhaps we would gain the freedom within ourselves to inquire in all seriousness whether we today are the ones who lack ready access to much of the world's reality.

All this suggests how advisable it might be for us to take a closer look at the evolution of consciousness through which our own thinking has gained (and become limited by) its current character.

Europeans and their changing landscape

In his book on *The Changing Nature of Man*, the Dutch historical psychologist Jan Hendrik van den Berg described the dawning among Europeans of something like our modern "sense of nature." This emerging sense, he claimed, can be recognized in Jean-Jacques Rousseau's *Confessions*, where he describes a trip through the Alps in 1728. It came to full expression in his novel, *Julie, or the New Heloise* (1761). With surprising rapidity for such a fundamental historical change, the new appreciation of nature took Europe by storm:

> Like an epidemic the new sensation spread through Europe. Every one wished to see what Rousseau had seen, to experience the same ecstasy. Everybody visited Switzerland and climbed the Alps. This had not happened before Rousseau. It was then that the Alps became a tourist attraction. Previously they had been an obstacle .... Even in 1750, Henault, a poet and a friend of Voltaire's, crossed the Jura and the Alps without the least enthusiasm, merely observing, "There is always a creek at my side and rocks above my head, which seem about to fall in the creek or upon me." These words would nowadays disqualify him as a poet (van den Berg 1961, p. 233).

If there was an "epidemic" of sightseeing, it was not caused by Rousseau's published descriptions. Rather, his descriptions were themselves an early symptom of the epidemic.

Figure 23.2. A photo of the Alps taken from above the town of Flims in the Imboden Region in the Swiss canton of Graubünden.[10]

Before commenting on Rousseau, van den Berg had mentioned Da Vinci's *Mona Lisa*. People came from far and wide to see this painting because it was, as van den Berg put it, "the face of later generations," the revelation of a new way to live. Mona Lisa was smiling over the delicious and unexpected discovery of an interior secret, a hidden subjectivity, powerful enough to remake the world. The sudden flowering of the Renaissance, the childlike fervor of the Scientific Revolution, the compelling urge that sent Magellan and the other great voyagers steadfastly beyond the edges of the world, where sea monsters once dwelt — all testified to a humanity *waking up* from its medieval enchantment. We stretched, blinked, rubbed our eyes, looked out upon a fresh world we were seeing for the first time. And, in that moment, we became aware of the one who was inside, looking.

A subject becomes a subject by virtue of its ability to stand apart — to stand within itself — and make whatever is now "out there" an object for itself. A new subjectivity is necessarily married to a new objectivity. So it was not only Mona Lisa's smile that became famous, but also the landscape behind her. We must see her smile and its interior significance against that external backdrop. Van den Berg saw it as

> the first landscape painted as a landscape, just because it was a landscape. A pure landscape, not just a backdrop for human actions: nature, nature as the middle ages did not know it, an exterior nature closed within itself and self-sufficient, an exterior from which the human element has, in principle, been removed entirely (van den Berg 1961, p. 231).

Figure 23.3. Leonardo da Vinci's *Mona Lisa*[11]

Van den Berg proceeds to quote Rilke: "This landscape is not … the judgment of a man on things at rest; it is nature coming into being, the world coming into existence, unknown to man as the jungle of an unknown island. It had been necessary to see the landscape in this way, far and strange, remote … It had to be almost hostile in its exalted indifference, if, with its objects, it was to give a new meaning to our existence"[12] (van den Berg, p. 231).

So, what was going on with the changing relation between Europeans and their landscape? Were people just "talked into" seeing the Alps differently, or was a deeper, underlying change at work? Were our forebears several centuries ago becoming situated in their environment in a fundamentally new way? Did Da Vinci, foresighted as he was in so many ways, catch a first, premonitory glimpse of nature detaching herself from the human being — a strange sight at first? And did Rousseau testify to a historical transition toward a more comfortable, aesthetic *appreciation* of this new reality?

History of ideas – or an evolution of consciousness

A familiar task for any philosopher or historian would be to trace the impact, say, of Aristotle's or Descartes' or Darwin's thought upon subsequent thinkers. We think of it as a history of ideas. But what if there are changes of consciousness that run mostly along subterranean channels of which we have no immediate awareness? After all, we might well wonder how we got from the undivided, inner/outer (neither subjective nor objective) consciousness inherited from the age of myth to our own detached-observer subjectivity today, where we find ourselves confronted by "mindless natural objects."

A fundamental premise of Barfield's work was that there is a crucial distinction to be made between the history of ideas and the evolution of consciousness: "A history of thought, as such, amounts to a dialectical or syllogistic process, the thoughts of one age arising discursively out of, challenging, and modifying the thoughts and discoveries of the previous one" (Barfield 1965, p. 67). This is, for example, the way the history of philosophy is normally taught.

On the other hand, any method for approaching the evolution of consciousness must be quite different. What matters is not so much *what* people are thinking as *how* they are thinking, and how they are connected, in the greatest depths of their being, to what is happening in the world, both material and immaterial. Intellectual thoughts or theories about this or that are less relevant to the evolution of consciousness than the unconsidered habits of thought and the qualities of experience determining what one *can* think.

We need to notice, in particular, qualities of *meaning*. To focus on "propositional content," as we think of it today, is to make the ancients into objects of ridicule by assuming that they were engaged in something like our own detached, self-aware habits of intellectual debate. We mistake their immediate *perceptions* for our own philosophically loaded *thoughts*, and so we discover in the ancients only confusion.

It was to evolutionary studies that Barfield continually returned as he illustrated, in a series of works spanning several decades, how the meanings of words "are flashing, iridescent shapes like flames — ever-flickering vestiges of the slowly evolving consciousness beneath them" (Barfield 1973, p. 75). He tried to show that the processes of evolution, while not determining the particular ideas of a given era, do circumscribe the kinds of things one can conceive and mean.[13]

As an example, the historian Herbert Butterfield describes how the Aristotelian worldview gave way during the sixteenth and seventeenth centuries:

> Through changes in the habitual use of words, certain things in the natural philosophy of Aristotle had now acquired a coarsened meaning or were actually misunderstood. It may not be easy to say why such a thing should have happened, but men unconsciously betray the fact that a certain Aristotelian thesis simply has no meaning for them any longer — they just cannot think of the stars and heavenly bodies as things without weight even when the book tells them to do so. Francis Bacon seems unable to say anything except that it is obvious that these heavenly bodies have weight, like any other kind of matter which we meet in our experience.

Butterfield adds that there was, during this period, "an intellectual transition which involves somewhere or other a change in men's feeling for matter" (Butterfield 1957, pp. 130–31). Sometimes, as other historians have noticed, certain ways of thinking just seem to be "in the air." For whatever reason, their time has come.

Barfield suggests that even the history of ideas, when looked at closely, can reveal "forces at work beneath the threshold of argument." Using the linguist's term "Aryan" in something like the modern sense of "Indo-European," he notes that "the comparatively sudden appearance, after millennia of static civilizations of the oriental type, of the people or the impulse which eventually flowered in the cultures of the Aryan nations can hardly have been due to the impact of notion on notion. And the same is true of the abrupt emergence at a certain point in history of vociferously speculative thought among the Greeks."

And still more remarkable, he says, is the "historically unfathered impulse of the Jewish nation to set about eliminating participation." By "participation" (see following section) he refers to the ancient sense of a numinous presence in nature that was akin to the human interior. And so,

> Suddenly, and as it were without warning, we are confronted by a fierce and warlike nation, for whom it is a paramount moral obligation to refrain from the participatory heathen cults by which they were surrounded on all sides; for whom moreover precisely that moral obligation is conceived as the very foundation of the race, the very marrow of its being. We owe to the Jews the pejorative significance in the word *idol*. The representative images, the totemic *eidola*, which

> ritually focused the participation of the surrounding Gentile nations, are either condemned by their prophets as evil or denied as unrealities ... (Barfield 1965, pp. 67–68).

It is good to realize how, even in studying relatively recent history (or the cultural realities of our own day), we are always looking at more than a discursive play of ideas. We may indeed be focused on the history of ideas, but there is always a deeper current to be aware of. In a moment we will glance at earlier stages in the evolution of consciousness. But, first, we will draw from Barfield one further example illustrating how even the usual historical narratives can suggest something about an underlying evolution of consciousness.

Speaking of the introspective psychology that yielded the theories of Freud and Jung, Barfield refers to the "startling phenomenon" whereby "a literal-minded generation developed a sympathetic response to the psycho-analytical gnosis of dream-imagery, and accepted the (one would have thought) fantastic idea of an immaterial realm of 'the unconscious'." Who, he wonders, could possibly have foreseen this, say, in the year of the Great Exhibition (1851)?

The question is meant to voice our inevitable puzzlement if we look merely at an intellectual history, tracing the impact of idea upon idea. For, in intellectual terms, the second half of the nineteenth century represented the zenith of that literal-mindedness exemplified by a science recognizing, at bottom, only the mindless and deterministic interaction of atom with atom. Whatever sort of change yielded the possibility of psychoanalysis was not the mere product of the discursive play of extant ideas, but rather of the changing (evolving) structure of human experience.

And so, leaving aside the question of the long-term significance of psychoanalysis, Barfield remarks that "for the historian of consciousness the most significant thing will always be the way it 'caught on'; the number of its technical terms — and still more the characters out of Greek mythology — which had become household words even before the death of its founder. Pan, it seems, has not only not retired from business; he has not only gone indoors; he has hardly shut the door, before we begin to hear him moving about inside"[14] (Barfield 1965, pp. 133–34).

The long historical arc of the evolution of consciousness

Barfield devoted much of his life to tracing the evolution of consciousness, so far as the historical record — and especially the record of language use in western civilization — can reveal it. He schematized this evolution in the form of a 'U,' where the left leg marks a long descent from mythic "original participation" to detachment, following which we might hope for a (still to be achieved) re-ascent toward what he called "final participation."

The detachment, which western civilization has been experiencing with particular intensity for the past several hundred years, involves disconnection from a material world that now seems wholly "out there" and independent of the human interior. But the important flip side of this independence is the birth of the self-possessed and more or less free human individual. It is this individual who, without giving up her independence, can enter into "final participation" by reconnecting with the world's interior through love and consciously directed cognition and activity.

Original participation was a kind of unfree or instinctive inner union with the world — a union we saw reflected in the words of ancient languages. Our ancestors were relatively un*self*-aware, yet conscious of an intimate, interior connection to what lives in the world. Or perhaps we should say: they themselves simply *lived* in and through this interior connection. Their experience was collective rather than highly individualized or private.

Crucially (as I already noted above in the discussion of the history of language), theirs was a time when the meaning of things was directly given to the human being from the world — when, as Barfield observed, "thinking [was] at the same time perceiving." Without a detached and individuated self-consciousness to question it, experience just was what it was. There could have been no philosophers or scientific researchers.

To have our thoughts given to us directly along with our perceptions (rather than our having the responsibility to enliven perceived contents at least in part with thinking we experience as "our own") would have been a condition we can scarcely imagine today. There was, in the most remote era we can glimpse through the literate, historical record, "a picture-thinking, a figurative, or imaginative consciousness, which we can only grasp today by true analogy with the imagery of our poets, and, to some extent, with our own dreams" (Barfield 1973, pp. 206–7).

We have come a long way from that original participation. (Only a very late stage of the journey was indicated in the changing relation to the landscape briefly discussed above.) But the freedom we have thereby gained is not unproblematic. Disconnection from the world in which we live is a kind of death. It amounts to being severed from the sources of our own life, as reflected in the widespread feeling that we live in a meaningless world. The freedom of detachment easily loses all content — we can find no more reason to do *this* than to do *that* — and therefore freedom can become mere emptiness. The question is whether we can employ our freedom and independence in order to reconnect with the spiritual sources of our lives.

It is important to realize the change in directionality here. In the unfree state of original participation we were gaining from the world a language that could eventually serve for our *selves* — we were, you might say, *being spoken* into being as individuals, thereby gaining the potential to become the modern selves that we are. To move forward now, however, requires us freely to participate in the creative processes by which the world itself first served us: we must play our own part in *speaking the world* into the coming phases of its existence. This is the reversal of direction — the movement from the left leg of the evolutionary 'U' upward into the right leg.

We are not lacking stimulus for pursuing this movement. Our current era of detachment presents us with a picture of centrifugal forces threatening society with disintegration. Former institutions, traditional values, and blood ties become ever weaker factors in holding societies together, leaving many with a kind of vertigo suggesting that everything is falling apart. This in turn may produce a backlash in the form of various "fundamentalist" defenses of an "old order" that no longer holds promise of helping us along our way to the future.

Perhaps we see signs of that future in the way our present situation has called forth not only burgeoning volunteer activities from free individuals, but also flourishing non-governmental organizations across the political spectrum and a growing sense of individual and social responsibility — responsibility for local and global environments (including social and political environments), for the protection of all forms of life, and for the fruitful direction of evolution itself. The change I spoke of a moment ago — between *being spoken* on one hand and *speaking* on the other — marks nothing less than a critical transformation of the very agency of evolution, which is inevitably falling more and more to our own conscious choices.

An ideal degree of detachment – before and after

If Mona Lisa hinted at a new, more private and individuated sense of the human interior, and if, correlative to this detachment of the human being, there was a dawning sense of a landscape that was "pulling away," gaining its own independent existence so that it could begin to be noticed as such, then we might wonder where this change was coming from, and where it might be going — or where we, in our current state, might help to direct it.

Here is one way to contextualize these particular changes. If, along the way from original participation toward near-total disconnection from the world, there is a certain ideal aesthetic distance, a point of maximum fascination, a mutual interplay of subject and object wherein humans and their world resonate in the most exquisite tension, then, Barfield maintained, it was the Romantics (1770–1870) who lived closest to that condition. It was the point where humans had become sufficiently detached from the world to notice and appreciate the independent life of "things," but not so detached that they lost all consciousness of their inner connection to them. Their separation from the world only allowed them to savor all the more their resonance with it as relatively independent beings.

This was the state being entered by those who, as we heard above, first rushed out to see the mountain vistas and to revel in what became known as "picturesque"[15] scenes of nature. The distancing process, however, was not arrested or reversed by the Romantics, so that van den Berg is correct in observing how "the estrangement of things, which brought Romanticism to ecstasy, belongs, for the most part, to the past." We are no longer close enough to the world even to feel the conscious fascination of our estrangement. Today,

> Many of the people who, on their traditional trip to the Alps, ecstatically gaze at the snow on the mountain tops and at the azure of the transparent distance, do so out of a sense of duty. They are only imitating Rousseau; they are simulating an emotion which they do not actually feel. It is simply not permissible to sigh at the vision of the great views and to wonder, for everyone to hear, whether it was really worth the trouble. And yet the question would be fully justified; all one has to do is see the sweating and sunburned crowd, after it has streamed out of the train or the bus, plunge with resignation into the recommended beauty of the landscape to know that for a great many the trouble is greater than the enjoyment (van den Berg 1961, p. 233).

Which one of us doesn't feel at least some symptoms of this detachment from nature? But perhaps, in order to contextualize a little more fully the changes running from the Renaissance of Da Vinci through the Romanticism of Coleridge and Goethe to the alienation of our own day, it will be useful to add a picture from the period immediately preceding the Renaissance. Here is Barfield trying, in just a few words, to give an impression of the qualities of medieval consciousness — a consciousness still possessing more than a few echoes of original participation:

> If it is daytime, we see the air filled with light proceeding from a living sun, rather as our own flesh is filled with blood proceeding from a living heart. If it is night-time, we do not merely see a plain, homogeneous vault pricked with separate points of light, but a regional, qualitative sky, from which first of all the different sections of the great zodiacal belt, and secondly the planets and the moon (each of which is embedded in its own revolving crystal sphere) are raying down their

> complex influences upon the earth, its metals, its plants, its animals and its men and women, including ourselves ... Our own health and temperament are joined by invisible threads to these heavenly bodies we are looking at ...
>
> We turn our eyes on the sea — and at once we are aware that we are looking at one of the four elements, of which all things on earth are composed, including our own bodies. We take it for granted that these elements have invisible constituents, for, as to that part of them which is incorporated in our own bodies, we experience them inwardly as the "four humors" which go to make up our temperament. (Today we still catch the lingering echo of this participation, when Shakespeare makes Mark Antony say of Brutus:
>
> *... The elements*
> *So mixed in him, that Nature might stand up*
> *And say to all the world, This was a man.*)
>
> ... A stone falls to the ground — we see it seeking the center of the earth, moved by something much more like desire than what we today call gravity ... (Barfield 1965, pp. 76–77).

From myth to literacy: the coming into focus of the human individual

The earliest "histories" of which we can catch a glimpse were not centered on human events. Indeed, the idea of distinctively *human* events can hardly have been available. The accounts were more like spiritual and cosmic histories. Humans — their gaze riveted by fascinating goings-on in what we today might denigrate as "supernatural" realms, but which they experienced (pre-reflectively) simply as *nature* — did not narrate their own histories. Rather, as is still echoed in Hesiod's *Theogony* long after the primary age of myth, they told stories of the genesis of gods and nature spirits. Only with time would history become more human-centered and prosaic.

We saw earlier in this chapter how the most ancient historical evidences and the linguistic shards that remain of mythical experience suggest a language, a manner of consciousness, and an experienced world, very different from our own. There existed a unity of the "inner" and "outer" dimensions of experience that has almost wholly disappeared from our modern engagement with the world, strongly polarized as the latter is between self and other, subject and object. What we know today as the "material world" was then alive or "enchanted," and humans were aware that in the world they met powers akin to, even if other than, their own power of directed activity. So, too, what we know today as the "inner world" was then embodied, inseparable from sense-perceptible expression.

The trajectory from the age of original participation to the present has been a long one, and — apart from some hints (derived from language) about the mythic consciousness — we have looked only at a few relatively recent "snapshots." It is time to get some sense for the larger picture.

At the edge of literacy and beyond

Between the age of myth and the medieval era, there lay the long period beginning (at different times in different places) with the earliest forms of literacy. In Greece, the Homeric epics, first in oral form and then in writing, straddled the beginnings of literacy. In those epics we find "meaning still suffused with myth, and nature all alive in the thinking of man":

> The gods are never far below the surface of Homer's language — hence its unearthly sublimity. They are the springs of action and stand in place of what we think of as personal qualities. Agamemnon is warned of Zeus in a dream, Telemachus, instead of "plucking up courage," meets the goddess Athene and walks with her into the midst of the hostile suitors, and the whole earth buds into blossom, as Zeus is mingled with Hera on the nuptial couch ... And these august beings, speaking now from the mouths of the characters, and again passing and repassing invisible among them, dissolve into a sort of *largior aether* [greater, or transcendent, atmosphere], which the Homeric heroes breathe all day; so that we, too, breathe it in the language they speak — in their ροδοδακτυλοσ ηωσ ["rosy-fingered dawn"], their ιερον ημαρ ["sacred day"], in the sinewy strength of those thundering epithets which, for all their conventionality, never fail to impart life and warmth to the lines (Barfield 1973, pp. 93–94).

Following Homer something like a miracle occurred within Greek culture. In his widely used textbook, *The Story of Art*, the eminent art historian, E. H. Gombrich, refers to the "Great Awakening" that took place in Greece from the seventh through the fifth centuries BCE (Gombrich 1989, chapter 3). Painters and sculptors began to do more than follow the rather schematic rules of representation handed down through the centuries, but also observed for themselves, and tried to be faithful to their observations.

The older style is shown in Figure 23.4. It is hard for us to appreciate the strange forms given to the human figures on this vase — forms in what has

Figure 23.4. Top portion of a large Greek funerary vase ("krater") with a prothesis scene — presentation of the deceased's body on a table. The vase originated around 750–735 BCE. Note the geometric decorative patterns on the vase, and the "geometrically" repeated human figures.[16]

become known as the "geometric style." Surely, we might well think, Greek eyes were as capable of physically registering the actual form of the human body as ours are. But apparently — and so various authorities have argued — Greeks before and during the Homeric era experienced their bodies rather differently from us. The classicist Bruno Snell described how the art gives the impression that "the physical body of man was comprehended, not as a unit but as an aggregate" (Snell 1960, p. 6). Or, as Mark Vernon puts it, the Greeks of this period must have experienced their bodies "rather as baggy gatherings of spirited factions," not as well-integrated entities.

Vernon is a theological scholar and psychotherapist who has traced the evolution of consciousness down through both Greek and Jewish cultures. Describing the geometrically styled human figures as having "bull-like thighs, wasp-like waists, barrel-like chests, pin-like heads," he goes on to say that the "locus of aliveness wasn't set within a person's frame and physique ... Instead, their identity came from the outside in, with different limbs and organs attuned to external divine influences. The inner life of the cosmos was their inner life." Further, he says,

> They had little or no notion of the isolated individual ... and little sense of a unified self who was or could hope to be in charge. To be alive, to be functioning, was implicitly tied up with being porous to society, spirits, gods. [On the vases] the people appear to move as one, as if swaying in a field of consciousness like as many wheat ears blown by the wind" (Vernon 2019, pp. 47–49).

But by degrees with the beginning of the Great Awakening, the figures begin to gain individual and personal traits (Figures 23.5 and 23.6), while at the same time the artist takes up a personal point of view, and perspectival foreshortening starts to come into play. Correlative with this, Athenian democracy took form, reflecting an individualizing mindset. In the case of the fifth-century sculptor, Pheidias,

> His figures weren't generic presences with blank eyes. They looked at you. They conveyed a sense of alertness and interiority ... His works were immediately recognized as spell-binding, displaying a dignity and beauty that called forth an interiority from within the viewer ...

Figure 23.5. Greek (Attic) Panathenaic prize amphora. The painter employed the black-figure technique to depict runners in a race. From about 530 BCE.[17]

Figure 23.6. Ancient Greek bronze statue of a youth, from the Antikythera shipwreck. The statue dates to circa 340–330 BCE.[18]

> His work was instrumental in showing a clear image of the integrated person, thereby spreading a sense of it in others ... The best sculptures could now show the interactions of individuals. They left behind the collective swaying of the masses (Vernon 2019, pp. 50–51).

In 1953 Snell published an influential book called *The Discovery of the Mind*. It contained discussion not only of the intimate relation between the Homeric heroes and the speaking of the gods, but also traced in the tragedians (Aeschylus, Sophocles, and Euripides) a movement from the centrality of fate to that of the individual conscience. And he described how "the early Greek lyrists had awoken to the fact that man has a *soul*; they were the first to discover certain features in the feelings of men which distinguished those feelings sharply from the functions of the physical organs." Further,

> For the first time it was noticed that these feelings do not represent the intercession of a deity or some other similar reaction, but that they are a very personal matter, something that each individual experiences in his own peculiar fashion, and that originates from no other source but his own person. Further they had found out that different men may be united with one another through their feelings, that a number of separate people may harbour the same emotions, memories, or opinions. And finally they discovered that a feeling may be divided against itself, distraught with an internal tension; and this led to the notion that the soul has intensity, and a dimension of its own, viz. depth (Snell 1960, p. 301).

The Great Awakening was a time when the individual human thinking activity was vigorously detaching itself from perception and gaining a sense of its own free powers. Barfield, referring to the work of the Greek philosophers, particularly Aristotle, wrote:

> Struggling to fit herself as into a glove, to the processes of cause and effect observed in physical phenomena, the mind became suddenly conscious of her own shape. She was astonished and delighted. She had discovered *logic* (Barfield 1967, p. 109).

Turning toward the Jewish scriptures and culture, to which Vernon brings his theological training: he recalls, for example, how the Jewish prophets urged the people toward monotheism — toward acknowledgment of a God whose name was I AM. This was inseparable from an awakening of the individual worshipper. Only someone who was becoming an individual in his own right could "perceive the singularity of the divine nature." "A felt sense of 'I am,' even if transient, is a prerequisite for feeling the inner power of the divine I AM." As Barfield summarized it, the locus of participation was narrowed down to the divine name, which Jews could hardly speak without invoking their own inwardness (Barfield 1965, p. 155). Thereafter,

> Nature can be experienced as [the] speaking of God rather than itself being divine, enchanted and haunted, and God can speak through creation but not be held within creation ... Henceforth, monotheistic knowledge of God would be inextricably tied up with self-knowledge, and introspection would become a key spiritual task (Vernon 2019, pp. 36–41).

Under Hezekiah (the king of Judah who reigned from the later eighth into the early seventh centuries BCE), a general literacy was encouraged for the first time. The individual worshipper could now read the sacred texts for herself and ruminate over them internally. Old idols and sacred groves were banished, and clan tombs were replaced with burial sites for single families or individuals. All these developments, Vernon points out, were associated with a transition from collective religious ritual and experience to the importance of the emerging life of the ethically responsible and self-aware individual.

But the path from original participation through detachment and then toward reconnection of our now-independent consciousness with the spirit in the world that gave birth to us is not easy. The vigorous philosophical speculations and disciplines of the Greeks would eventually be frozen into the mathematical, rule-bound, one-sidedly cause-and-effect mindset of modern science — a kind of wooden materialism from which we have yet to find any decisive exit.

So, too, the growing Jewish awareness of individual moral responsibility would eventually (during the centuries leading up to the destruction of the temple in Jerusalem in 70 CE) be paralyzed under the weight of external laws and a prescriptive literalism that left little room for creative individual insight into the moral potentials inhering in every situation. Still today, in various religious fundamentalisms, we find the same tendency.

The evolution of consciousness, like all evolution, is not a straight line

As the preceding two paragraphs indicate, the evolution of consciousness has been far from a smooth, linear progression. As civilizations have risen and fallen, so too there have been periods of rapid advance toward new forms of consciousness, attempted returns to a more "paradisal" consciousness, and various sideways movements.

Consider, for example, the development of the art of visual perspective in the Renaissance. This reflected and powerfully contributed further to the development of individual points of view, while also supporting an individual and quantitative experience of space where previously there had been something more like a collective space of meaning:

> Before the scientific revolution the world was more like a garment men wore about them than a stage on which they moved. In such a world the convention of perspective was unnecessary. To such a world other conventions of visual reproduction, such as the nimbus and the halo, were as appropriate as to ours they are not. It was as if the observers were themselves *in* the picture (Barfield 1965, pp. 94–95).

And yet, dramatic and important as the late-medieval and Renaissance discovery and embrace of perspective proved to be, it was not altogether new. This is why it has been referred to as a "rediscovery" (White 1972). There was in antiquity — in Greek and Roman culture — a genuine anticipation, in theory and practice, of linear perspective. It was lost in subsequent centuries, but when the time was right, was rediscovered and flourished during the Renaissance in a way that took permanent hold and changed everything.[19]

Similarly, we find during the Hellenistic era that managing one's own subjectivity, or soul life, became a central problem addressed by Epicurean and Stoic philosophers. But here again the movement into this particular sort of self-awareness and concern for "care of the soul," did not lead directly to the dramatic emergence of the modern individual that we have witnessed since the Renaissance and Scientific Revolution.

WE BEGAN THIS CHAPTER with Barfield's comments about how the mountains of Switzerland we see today are not the same mountains our ancestors saw. Of course, in today's environment the nearly universal assumption will be that Barfield was not really talking about the mountains themselves, but "only" about how people see and experience the mountains. And we did learn above how different this experience became during the transition from medieval to modern culture.

Actually, however, Barfield really *was* talking about the mountains themselves, whose reality he did not believe could be radically or dualistically distinguished from our experience of them. His refusal to treat the distinction between reality and experience as fundamental reflects a long-running conviction within science that our knowledge of the world must be empirically based — rooted in experience. Our knowledge of the world is always a union of perception and thought that occurs at the same time in ourselves and in the world, much as the mathematics of physical law, which we discover in the world, can be found also in the mind of the knower of the world.

I did say at the outset that I was not about to attempt an explanation of Barfield's deeper meaning in this chapter. And I will hold to that. But I do hope that the preceding notes on the evolution of consciousness may at least make his mysterious suggestion about the Swiss mountains more intriguing — and may also fortify the reader for the perhaps unexpected challenges to contemporary thinking in Chapter 24. That chapter offers, among other things, a possible approach to Barfield's meaning.

WHERE ARE WE NOW?

We've come a long way.

Here is some of the ground we have covered in this chapter:

- The ancients, who were incapable of anything like our own theoretical and causal speculation, directly perceived a world that seemed to possess a powerful interior aspect. They recognized what lived in the world as akin to what lived in their own interiors. And they had not yet come to distinguish sharply between interior and exterior, or between immaterial and material.
- Our ancestors' thoughts were at first *perceived* more than actively *thought*. Their meanings and language were given in their immediate perceptions

of the world around them. Eventually, an independent inner being and independent powers of speech arose as a further, inward development of what had initially been the world's "speaking."

- Historically, there has been a change in directionality. We humans who were, in a sense, first spoken into being by the world, now find ourselves bearing a responsibility to speak this world's future into being — if only, to begin with, by accepting a responsibility to avoid destroying it.
- Our lately achieved independence from the world as self-aware individuals has given us the freedom to think and imagine the world with our own thoughts, even if in a highly distorted way. We are free to err. We are free to "forget" humanity's origin and past, if only by ignoring the study of it. We can, if we wish, retreat into a comfortable materialism requiring no burden of responsibility on our part.

The chapter as a whole concerns *human* consciousness, but the picture certainly suggests that *all* organisms make their way through a larger, meaning-soaked surround that comprises the givenness of their lives and the givenness of the world. And it is this same meaning that, by contracting into a bright focus in human minds, has engendered our consciousness and self-awareness. In this common, if diverse, interior aspect lies the unity of life on earth.

Our discussion of the evolution of consciousness does not suggest that it makes any sense to imagine an *origin* of consciousness. More particularly, it is not clear how the idea of a "first" meaning arising from bedrock meaninglessness can make sense. We cannot grasp any meaning except against a contextual background full of already existing meaning. Make an experiment: take any single word (or invent one) and try to understand or define it other than in the terms of many other words. You will find that any specific meaning can shine forth only in the light of a meaning-soaked universe.

The background of meaning is simply a given of our lives as children of what we might call a *logos*-world. We cannot even legitimately imagine an origin for meaning, because the only contents available to our thought-world are meaningful contents. An imagined leap from unmeaning to meaning can occur only via circular reasoning, whereby elements of meaning are (if only unconsciously) assumed from the beginning.

In short, there can be no meaninglessness in the *known* universe — in a universe that submits itself to human perception and understanding. For a more explicit treatment of these matters, see Chapter 24.

We today have learned to view just about everything through an evolutionary lens. The benefits to understanding have been many. The oddity is that these benefits have scarcely been extended to a knowledge of the evolution of consciousness — an evolution that includes the changing cognitive relation between the perceiver and what he perceives. There is a penalty to be paid for this: we lose the ability to understand the very different qualities of consciousness characteristic of earlier eras, and therefore we become trapped in modernity — in our own "moment" of evolution. And this at a time when we need to begin learning to carry responsibility, not just for one moment, but for the entire future course of evolution.

NOTES

1. The next two sections are adapted from Talbott 2018.
2. The translation is from the New American Standard Bible.
3. Barfield also tells us that "such a purely material content as 'wind,' on the one hand, and on the other, such a purely abstract content as 'the principle of life within man or animal' are both late *arrivals* in human consciousness. Their abstractness and their simplicity are alike evidence of long ages of intellectual evolution. So far from the psychic meaning of [Latin] spiritus' having arisen because someone had the abstract idea, 'principle of life ...' and wanted a word for it, the abstract idea, 'principle of life' is itself a *product* of the old concrete *meaning* 'spiritus,' which contained within itself the germs of both later significations" (Barfield 1973, pp. 80–81).
4. Figure 23.1 credit: source unknown.
5. Actually, words were inseparable from things. For the ancients, a word and its reference were not distinct things. This begins to make sense when one realizes (as we will see more clearly below) that the human being did not yet have a private or subjective interior where he could become aware of words as his own property set over against an objective world wholly other than himself.
6. From Bacon's *Advancement of Learning*, II.v.3., quoted (and translated) in Barfield 1973, p. 86.
7. It's worth reflecting on the fact that we have gained our material understanding of the world only with the aid of a massive linguistic superstructure consisting of words with immaterial meaning. In order to gain our material meanings, we have had to purge the world of its interior aspects, but those aspects have "hung around" as a dominant (and often distorting) language of abstraction. The distortions occur because we are inclined to treat many of our abstractions as if they were material (perceivable) things, which they are not. (Just consider the word, "particle," as we have it from particle physics. The effort to conceive particles *as if* they were perceivable things rather than high abstractions has caused no end of trouble for physicists.)
8. For a treatment of this and related questions, see Barfield's essay, "The Meaning of 'Literal'" in *The Rediscovery of Meaning and Other Essays*, pp. 32–43. Perhaps equally valuable is his essay on "The Nature of Meaning."

9. Barfield would say we must also come to terms with the reverse truth: the phenomena are themselves an evolving, ensouled drama staged as the "outer" world by conscious beings. That is, consciousness and the phenomena (whose objective nature is to occur within consciousness) are correlative. But this radical notion would take us far beyond the current exposition. For some related discussion, see Chapter 24.
10. Figure 23.2 credit: Zacharie Grossen (https://commons.wikimedia.org/wiki/File:Albula_Alps.jpg), CC BY-SA 4.0.
11. Figure 23.3 credit: public domain photograph of the painting in the Louvre (https://en.wikipedia.org/wiki/File:Mona_Lisa,_by_Leonardo_da_Vinci,_from_C2RMF_retouched.jpg), public domain. The image has been digitally lightened to counter darkening that has resulted from aging.
12. The foregoing paragraphs are drawn directly from my chapter, "Mona Lisa's Smile" (Chapter 21) in Talbott 1995.
13. Barfield, a philologist whose approach to the evolution of consciousness was primarily through the study of words, wrote that the evolution of consciousness requires us "to penetrate into the very texture and activity of thought, rather than to collate conclusions. It is concerned, semantically, with the way in which words are used rather than with the product of discourse. Expressed in terms of logic, its business is more with the proposition than with the syllogism and more with the term than with the proposition" (Barfield 1965, pp. 67, 90).
14. Somewhat tangential to, yet resonant with, Barfield's point, van den Berg describes one aspect of the process of human individuation over the past few centuries: "James Joyce used as much space to describe the internal adventures of less than a day than Rousseau used to relate the story of half a life. The inner self, which in Rousseau's time was a simple, soberly filled, airy space, has become ever more crowded. Permanent residents have even been admitted; at first, only the parents, who could not stand being outside any longer, required shelter, finally it was the entire ancestry. As a result the space was divided, partitions were raised, and curtains appeared where in earlier days a free view was possible. The inner self grew into a complicated apartment building. The psychologists of our century, scouts of these inner rooms, could not finish describing all the things their astonished eyes saw. It did not take them long to surpass Joyce, and their work became endless in principle. The exploration of one apartment appeared to disturb another; and if the exploration moved to the next place, the first one required attention. Something fell down or a threat was uttered; there was always something. The inner life was like a haunted house. But what else could it be? It contained everything. Everything extraneous had been put into it. The entire history of mankind had to be the history of the individual. Everything that had previously belonged to everybody, everything that had been collective property and had existed in the world in which everyone lived, had to be contained by the individual. it could not be expected that things would be quiet in the inner self" (van den Berg 1961, p. 232).
15. The word "picturesque," which is recorded as first appearing in 1703 and became widely used in the Romantic era, testifies to the ideal aesthetic distance Barfield refers to. On one hand, it suggests detachment, inasmuch as the world can now be *looked at* as an independent object by the observer, like a picture hanging on the wall. But, on the other hand, a picture or painting was itself appreciated as a production of the human spirit.

16. Figure 23.4 credit: Mary and Jon Hirschfeld Workshop (https://commons.wikimedia.org/wiki/File:Hirschfeld_Workshop_Terracotta_Krater_ca_750-735_BCE_by_alkalisoaps_on_flickr.png), CC BY 2.0.
17. Figure 23.5 credit: Metropolitan Museum of Art (https://en.wikipedia.org/wiki/File:Terracotta_Panathenaic_prize_amphora_MET_DP245713.jpg), CC0 1.0.
18. Figure 23.6 credit: From the National Archaeological Museum of Greece in Athens (https://commons.m.wikimedia.org/wiki/File:Ancient_Greece_Bronze_Statue_of_a_Youth,_from_Antikythera_Shipwreck,_c._340-330_BC.jpg), CC0 1.0.
19. This nonlinear character of the evolution of consciousness may remind the reader of our discussion of "mosaic evolution" in Chapter 19 ("Development Writ Large"), where we heard this (drawing on the work of Craig Holdrege): When something dramatically new arises in the fossil record, it is typically foreshadowed by fragmentary "premonitions" in various taxonomic groups, some of which may then go extinct. There is no smooth, continuous, single line of development leading to the new form, which may arise not only rather suddenly, but also as a novel synthesis and transformation of the earlier, scattered, premonitory gestures.

SOURCES

Barfield, Owen (1963). *Worlds Apart: A Dialogue of the 1960s*. Middletown CT: Wesleyan University Press.

_______ (1965). *Saving the Appearances*. New York: Harcourt, Brace and World. Originally published in 1957.

_______ (1967). *History in English Words*. Foreword to revised, 1967 edition by W. H. Auden. Great Barrington MA: Lindisfarne. Originally published in 1926.

_______ (1973). *Poetic Diction: A Study in Meaning*. Middletown CT: Wesleyan University Press. Originally published in 1928.

_______.(1977). *The Rediscovery of Meaning, and Other Essays*. Middletown CT: Wesleyan University Press.

_______ (1981). The Nature of Meaning, *Seven* vol. 2, pp. 32–43.

Butterfield, Herbert (1957). *The Origins of Modern Science 1300–1800*, revised edition. New York: Free Press.

Emerson, Ralph Waldo (1836). *Nature*. Boston: James Munroe and Company. https://archive.org/details/naturemunroe00emerrich

Gombrich, E. H. (1989). *The Story of Art*, fifteenth edition. Oxford: Phaidon.

Shelley, Percy Bysshe (1840). *A Defence of Poetry, in Essays, Letters from Abroad, Translations and Fragments*, by Percy Bysshe Shelley, edited by Mary Wollstonecraft Shelley. London: Edward Moxon. *A Defence of Poetry* was originally written in 1821 and published posthumously. Text is available at https://www.gutenberg.org/ebooks/5428

Snell, Bruno (1960). *The Discovery of the Mind*, translated by T. G. Rosenmeyer. New York: Harper and Row. Originally published in 1953.

Talbott, Stephen (1995). *The Future Does Not Compute: Transcending the Machines in Our Midst.* Sebastopol CA: O'Reilly and Associates. Available online at https://netfuture.org/fdnc/

________ (2018). A Physicist, a Philologist, and the Meaning of Life: Do We Have a Home in the Vast Cosmos? https://bwo.life/org/comm/ar/2018/meaning_33.htm

van den Berg, J. H. (1961). *The Changing Nature of Man: Introduction to a Historical Psychology*. New York: Dell Publishing.

Vernon, Mark (2019). *A Secret History of Christianity: Jesus, the Last Inkling, and the Evolution of Consciousness*. Arlesford, Hampshire, UK: John Hunt Publishing.

White, John (1972). *The Birth and Rebirth of Pictorial Space*. New York: Harper and Row.

Part 3:
BEYOND BIOLOGY

CHAPTER 24

Is the Inanimate World an Interior Reality?

IN THE EARLIER PARTS of this book, while learning about organisms and their evolution, we found it necessary to use terms such as *agency, purposive, intentional, end-directed*, and *telos-realizing*. We saw that every animal's life is the spinning of a wise and thoughtful *narrative* (however unconscious), more like a striving or a pursuit of interests or a satisfaction of needs than a mindless chain of physical causes and effects. Things happen for a *reason*, where "reason" encompasses all the potential significances we may encounter, for example, in a dog or cat's unfolding life. The *meaning* of an activity rather than its "mechanistic" causal interactions — its purposive or otherwise meaningful organization and direction rather than its strictly physical lawfulness — is our most reliably predictive guide to what will happen next. The entire bulk of a massive animal can be brought into motion, not by an external impact of great force, but rather by a significant image in its field of sight. We found ourselves recognizing how the living organism acts by giving continual, directed expression to its own *interior* way of being, where the idea of an interior dimension of nature is quite foreign to our usual scientific conception of a material world that can be understood without reference to sentience, will, intention, reason, or mind-like qualities.

("Interior" might be taken as a synonym of "experiential." And, of course — which is perhaps easy to forget in our day — experience has long been considered crucially definitive of science. Science is routinely referred to as "empirical," which just means "experience-based.")

But how far dare we push this notion of an *interior*? The question may arise, perhaps not forcefully but at least at the edge of our minds, when we consider that, throughout almost all human history, our ancestors believed they lived in a cosmos alive with spirits of every sort and with storytelling.

As we saw in Chapter 23 ("The Evolution of Consciousness"), this was not their theoretical understanding of the world, since they had not yet become modern individuals capable of looking out at, and theorizing about, a world wholly external to themselves. Rather, it was how the world immediately and intimately presented itself to their unquestioning experience. So now we are asking: What, if anything, was the truth of that experience? How could a non-theoretical, impossible-to-question experience of an interior dimension even have arisen from the ground of an essentially impersonal, psychically inert, mindless world in the first place? And is there any way at all in which we still today need to recognize such a dimension of the inanimate world?

And here we need to distinguish between two complementary aspects of "interior." A simple illustration can make the point.

A novel may be genuinely expressive of the author's interior being, and, as an interior expression, it can be apprehended and brought alive only within our own interiors. When we do bring it alive in this way, we gain intimate access to the author's mind. But we would not want to say that this makes the pages of the book, or even the meanings expressed on the pages, into minds or selves or agents capable of their own action. So, while we can speak of the novel — the story — as an objective reality existing crucially within an interior dimension (which is the only "place" where we could possibly encounter it), we must not treat it as if it were itself a living being or agent.

In the same way, we would not want to regard our own thoughts as agents, or to confuse them with our own thinking agency. All of which is to say that we can make a useful distinction between the interior and meaningful contents or products of mind, on one hand, and the generative activity of mind on the other.[1]

It appears, then, that when we speak of the interior dimension, we can refer either to the creative powers, or activity, of living agents, or else to the meaningful products of that activity. We shouldn't confuse the two — shouldn't confuse act and product, spiritual creating and spiritual creation — even if both belong to the interior realm. We needn't think that the products of living agency are themselves possessed of the agential powers by which they came about.

In this chapter, with its focus on the inanimate world, I will sometimes be speaking of the products or contents of the interior dimension, rather than the interior activities of agents. However, it will be quite otherwise when the focus is on the processes of human cognition. Here we may have to keep in mind both the cognitive activity of the perceiver and the inanimate (but interior!) objects of perception. I hope this will become clearer as we move along.

Having recognized these complementary aspects of the interior dimension, we can ask again, "What, if anything, was the truth of the unquestioned and direct experience of the ancients? Is there any way in which we still today need to recognize the interior nature of the inanimate world?"

I'm afraid we may not be very close to disentangling all the issues we stumble into when we begin enquiring about the relation between our own or other organisms' interiors and the inanimate world in which we find ourselves. But stumble around we must if we want to make even the vaguest possible approach to the question.

I will begin by expanding the very brief discussion of perception in Chapter 13, "All Science Must Be Rooted in Experience." Then, with open minds, we will see how much further we can go.

How the world lends itself to our knowing

Stand anywhere in nature and observe the scene. It can be a mountain or meadow, sea or sky, lake or desert — or a city street. Then ask yourself: what would remain of the scene if you were to remove every sensible (sense-perceptible) quality from your surroundings? The question has to do with the character of the world we know through experience and routinely take as real, from the luxuriant Amazon rainforest to the barren surface of the moon. Wherever you and I manage to get to, what would exist for us if there were no perceptible qualities? Does any *material thing* in the known cosmos present itself other than through qualities?

It is not a difficult question. Would that tree be there for us as a material object if there were no color of the leaves, no felt hardness of the trunk, no color and texture of the bark, no whispering of the breeze among the leaves, no smell of sap, wood, or flower, no possibility of song from birds flitting among the branches? Do we see, hear, touch, smell, or otherwise sense anything in the world apart from its qualities? Could we speak of a thing's form, substance, or even its existence if it did not present a qualitative, sense-perceptible face to us?

The hardest part of all this talk about qualities for most people lies in their feeling that the solid, objective reality of things is being questioned. But to point to the qualitative nature of the sensed world need not be to question its reality, or its felt solidity (which is one of its qualities), or its objective existence beyond the privacy of any single person's interior. It can, in fact, be just the opposite. We can say, with common sense, that the solidity we all feel is the real thing. Real solidity — the crushing weight of a boulder, the solidity we are given in experience and can collectively attest to when pursuing an experience-based science — is always and only *felt solidity*.[2] The sensed hardness of things is no less a perceptible quality than the taste, color, or sound of things. If we did not encounter that hardness, so that we passed right through things as if they were not there, then this would be another aspect of the material world's not existing for us if it were shorn of all qualities.

What we *cannot* say is that qualities belong to a mind-independent world or that they can be contents of anything other than the experience of a living being — contents of what we might call, rather inelegantly, "interiority." Qualities seem to be made of the very "stuff" of our interiors, and yet they are a requirement for the manifestation — and, as far as we can know, the very existence — of the material realm.

So we come back to the perfectly straightforward question: "Does anything exist materially, available to an empirical (experience-based) science, except as a presentation of qualities?" Would we have quantities to play with if there were no qualities from which to abstract them? And would we know what our mathematical formulae were about — what they meant — if we could not restore to our thinking the qualitative contexts from which those formulae were abstracted? Numbers alone do not give us a material world.

I think the conclusion you will come to is inescapable: whatever knowledge of the world we manage to gain is rooted in qualitative appearances, and the world would lose its reality for us — it would no longer be there as a content of experience or a subject for scientific investigation — were its qualities to vanish.

Given the more or less determined yet never fulfilled resolve among scientists from Galileo onward to have a science without qualities, it would seem that the integrity of science as a respectable knowledge enterprise rather than an empty pretense hangs on our answer to the question, "Would anything be left to investigate if we could be true to our ideals and remove all qualities from our science?" If the answer is as clearly "No" as I think it is, then we must learn to

Figure 24.1. A natural scene.[3]

integrate the world's qualitative aspects into a truly experience-based science.[4] (On the potentials for a qualitative science, see Chapter 12, "Is a Qualitative Biology Possible?")

We know the world through thinking as well as sensing

There are two primary portals for our experiential knowledge of the world: first our senses, and then the thinking that conceptually orders the contents of the otherwise inchoate sense reports, bringing them to meaningful and coherent appearance. If we could not perceive qualities through our senses, as I suggested in the previous section, we would not have a world. But it is equally true that without a conceptual ordering of whatever it is we receive through the senses alone, we again would have no world.

The thinking I am referring to here is not merely our theorizing *about* the world of objects. It is also the thinking that constitutes this or that thing as an object in the first place — a grain of sand or a cloud or a mountain. The common assumption that our perception gives us "things" directly and mindlessly, about which we then think and form theories is an untruth widely recognized by those who study cognition. We have no "things" at all except through an activity of thinking. In the case of familiar objects, this thinking typically becomes automatic and unconscious and, as such, may have informed our perception of those objects since childhood. But, with proper attention to perception, it is rather easy to catch this thoughtful, formative activity of thinking "in the act" so as to become aware of it.[5]

And so the general truth is this: if we are truly to recognize anything — a *this* of a particular sort as opposed to a *that* — we must be able to form *some* conception of what we are beholding. Which is to say: we must grasp the ideas that inform and are inherent in what we are beholding. The phenomenon can present itself to us as a given reality only so far as its real and inherent thought-content becomes at the same time *our* thought-content. To see a soaring hawk while having no idea of organism, bird, wing, flight, raptor, eyesight, predation, rodent, air, gravity, matter, and so on, would not be to *see a hawk*.

We would not recognize a tree if, in looking up toward a cluster of green leaves, we had no ideas to tell us that the relation of the leaves to branch, trunk, and roots is very different from their relation to the visually adjacent patch of sky-blue color. We would in general fail to recognize the tree at all if we had no idea of the thought-relations constituting a tree as what it is.

To stare in absolute, thoughtless incomprehension at the scene around us would be to stare at a meaningless blur — or not even that, since, in our utter thoughtlessness, we would lack even the concept of a "blur." Things come to meaningful appearance only by virtue of their distinct and interwoven meanings; we recognize them by means of the ideas lending them specific form and significance, through which we can identify them as being the kind of things they are. ("Oh, *that's* what I'm seeing!")

In only slightly different words: we could have no idea of things that, in their own nature, were entirely non-ideational. The traditionalist metaphysician, René Guénon, expressed the correspondence between thing and idea this way: "If the idea, to the extent that it is true and adequate, shares in the nature of the thing, it is because, conversely, the thing itself also shares in the nature of the idea" (quoted in Burckhardt 1987, p. 14n).

One way to approach the inherent idea of a thing is to realize the intimate relation between idea and form. We observe the form of a thing only by apprehending it in thought. The form of a rose or skyscraper reflects the thought through which the thing has become what it is — through which it has gained its specific, internal relationships and meaningful appearance. So one way to grasp the inseparability of sense and thought is to see how impossible it is even to imagine any particular material thing that is not already an expression of significant form. We never encounter a material substance that is not a manifestation of specific, intelligible form — or that is somehow separable from its own form.

Similarly, our laws of physics are *ideas*, mathematical or otherwise, that we find inherent in the material world. Typically, they come to expression as the form of the dynamic relations between things in space.

The spirit of our age makes it easy for us to overlook the obvious: if we, with our human thinking, can *make sense of the world*, it is because the world itself is in the business of *making sense*. The fact that thoughts are not merely the private property of individuals, but also come to manifestation within the world around us, remains virtually unapproachable for most of us today.[6]

I don't suppose there could be a more startling disconnect than when knowledge seekers aim to *articulate a conceptual understanding of a world they consider inherently meaningless*. A conceptual articulation, after all, is nothing other than the working out of a pattern of interwoven meanings. A truly meaningless world would offer no purchase for this effort.

My repetition in this section has been intentional, because the truth so easily escapes us. Let this be the sum of the matter:

> *Anything whose objective and true nature we can apprehend, in part, only through revealing thought or language, including scientific description, can hardly be said to possess a nature independent of mind, thought, language, or meaning.*

Finally, nothing in the foregoing is meant to suggest that the world has distinct "parts," the sensible and the thoughtful. We can no more imagine a sensible thing without thought than we can imagine substance without form. We can, of course, distinguish between the two aspects. But as soon as we ask "what is the sense content as such, apart from thinking?" we have a problem. To say anything at all about what the sense content is in itself — this would already be to characterize it with thought, so we would no longer be talking about a sensible content apart from thought.

I don't think there is any way around this, nor need there be. The world is a unity. It resists every rigid dualism. But surely we can say — as a matter of distinction rather than dualism — that whatever meets our senses must be inherently bound up with thinking, much as every substance is inherently bound up with form.

Facing up to the interior dimension

We have seen that the only world we could ever know is known interiorly, through qualitative sense perception and thinking. That's just what this world is, as far as we can judge: a "marriage of sense and thought" (Edelglass et al. 1997). We might surely ask: "If that is how the world presents itself to our understanding, and if our understanding is as genuine as we all find ourselves willingly or unwillingly taking for granted, might this not tell us something about the nature of things?" Of course, our knowing of the world requires other interior capacities as well as sense and thought, such as those of imagination and will. The appreciation of qualities such as color also seems to require an activity of feeling. But the main point at the moment is the rather obvious one that all our knowing calls upon interior capacities — powers of inner activity that presuppose "consciousness" in the broadest sense of that term (which includes the "subconscious," and even much of what we mean by "the unconscious").

This idea — that the only world we are ever given is an interweaving of, among other things, sense and thought — will be taken by those of a so-called "post-modern" bent as proving that we cannot talk about a "real" world, which is (they will say) hopelessly obscured behind all the relativizing subjective and cultural aspects of human existence. But this is actually a quintessentially modern approach born of Cartesian dualism (see next section) and doctrinaire materialism. Overlooked is the fact that we might also respond in an opposite way — not by denigrating the world's interiority as a "merely human construct," but instead by celebrating this interiority and recognizing ourselves as true natives of the world that has brought us forth, natives naturally equipped with the cognitive means for experiencing the terms of our own existence.

In other words, we can take the foregoing discussion of the role of sense and thought in human cognition as telling us, not only about our interior selves, but also about the objective character of the world in which we live — *because this world also possesses an interior, and our own interiors, historically derived from the world* (as we saw in the previous chapter), *put us in the deepest possible connection with the universe around us.*

It certainly stands to reason that whatever is required for understanding the world tells us something about the nature of the world being understood. It's either that or our understanding is a total illusion. If we can apprehend the world only through a marriage of sense and thought, and cannot even conceive any other way of apprehending the world, and if we all, in our practical, day-to-day lives, act as if the manifest world is the real thing — a world with which we routinely, materially, and consequentially engage in the immediate terms of our experience ...

But let me interrupt myself right here and emphasize the dishonesty of (1) behaving in accord with the practical conviction of the world's accessible reality throughout our daily lives while (2) at the same time intellectually professing radical skepticism about whether the world is actually there in the form in which it appears. At the very least, let the person taking this position point to *something* fake or unreal about the appearances (but preferably only after reading the considerations about human experience in Chapter 13, "All Science Must Be Rooted in Experience"). The simple fact is that we all find the world as we actually experience it to be perfectly natural and consistent, so that, hour by hour, it provides us with the effective reality principle and understanding of our lives.[7]

So, then, continuing where I left off: ... the most straightforward and consistent conclusion in the absence of contrary evidence is that the world itself, *in its own nature*, is just what it appears to be. It consists of the full and inexhaustible range of its appearances. Its true being lies in these appearances and takes form in the full-fleshed terms of our conscious experience. Or again: it is in the nature of the world to manifest interiorly. The interior experience can be our own or that of any creature capable — if only in the slightest degree, and whether with self-awareness or not — of bringing to manifestation within itself some experiential potential of the cosmos.

In other words, we should take seriously our own experience. To say that the world is essentially an appearance to consciousness — something we *experience* — is not to say it is insubstantial or a wisp of mere individual subjectivity. If you think this, you are forgetting your own experience, shared with others. To recognize that the world is a world of appearances, a world of experienced qualities, is only to say that it *really does* have, for example, the solidity we all encounter in experience — a solidity we can't help taking at face value in practical life. This is the real thing, an actually *experienced* solidity.

Admittedly, it is (for us today) a radical idea: qualitative and thought-informed, the world comes to its own characteristic expression — achieves its own reality, or fullest existence — as a manifestation within what we might call the *interior dimension*.[8]

There are many ways to speak of this interior dimension, none of which rings quite true in our culture. To say, as I have above, that the world consists of "appearances to consciousness" may be true enough, but the idea of an "appearance" has a falsely anemic and insubstantial feel for most people today. It *should* be taken as referring to the full, undiminished reality of the perceptible world as muscularly given in actual experience. The tree of our experience is an appearance, but it is an appearance of the sort we might crack our skulls against if we make a wrong move while skiing. We don't lose that solidity simply by recognizing that we would know nothing of it if it did not present itself as a sensible content of experience.

One proposal for how we might think of the material world in relation to its interior aspect comes, it's been said, from the philosopher, Gottfried Wilhelm Leibniz, who is supposed to have referred to materiality as condensed or coagulated spirit (*coagulum spiritus*).[9] We might want to think of this in terms of an analogy with ice forming in water as the temperature drops.

In any case, the main point of this chapter is indeed simple, and does not require us to range far afield in abstruse philosophical territory. The point is only that we cannot separate the concept of matter from that of mind, or interiority, or spirit. The idea that our perception of the world gives us a mind-independent reality is a strange importation into modern thought with no evident support and everything against it. The world, so far as we could ever know it, manifests itself within an interior space. We cannot even imagine it otherwise, given that the space of imagination is itself interior. Since nothing in our experience of the world gives us fundamental reason to distrust that experience (Chapter 13, "All Science Must Be Rooted in Experience"), and since we all find it impossible to avoid taking our experience as reality, it seems reasonable at least to test out in our thinking the hypothesis that what our experience gives us upon the stage of consciousness is the foundational substance and matrix of reality.

And perhaps we can add here, as I believe Owen Barfield has somewhere said, "There is only one interior." I will come back to this in the section below on language.

In this book I have mentioned quantum physics only rather trivially and in passing. It's not just that I can hardly claim to understand it, but also because it seems to refer primarily to a realm of potential, whereas I have been speaking throughout of manifest reality. However, it may be that quantum physics has something to say about the relation between the potential and the manifest. And in this regard, I can hardly help taking note of the comment by George Musser, a contributing editor of *Scientific American* and a well-regarded writer on quantum physics. When talking about the human mind as the "one place" where quantum potential becomes reality, he writes:

> The waves that ripple through the quantum ocean break on the shores of our conscious selves (Musser 2023, p. 107).

I do not know to what degree Musser accurately represents the convictions of physicists, and I rather doubt he would accept the framework of my discussion here. But his statement certainly resonates with the idea that mind is where the world comes to manifestation, and without mind there would be neither potential nor manifestation. (It is hard to see how one can speak of potential where there is no possibility of manifestation.)

I realize that all this way of speaking is problematic in the extreme for contemporary thinkers. But I hope in the course of this chapter to provide enough context to open our minds just a crack, so as to let in the light from some unexpected possibilities we might allow ourselves to explore.

Meanwhile, perhaps we can momentarily reflect on an observation by the respected French mathematician and physicist, Henri Poincaré, who once wrote: "A reality completely independent of the mind which conceives it, sees or feels it is an impossibility" (Poincaré 1913, Introduction).

The Cartesian diversion – is there a way to bypass it?

But the conclusion that the world in its fullest reality occurs within an interior dimension — that no world we could ever know exists independently of the union of sense and thought — collides with a centuries-long mental habit that tells us we look out upon a world of mindless objects utterly other than, and unlike, our cognizing selves — objects wholly alien to our own interior being.

The common suggestion, then, is that we have two different worlds: the *subjective* world of appearances — appearances not only mediated by, but also unknowably transformed by, our nervous systems — and a world of *real things* somehow hidden behind the terms of our experience. From this point of view, untrustworthy appearances are all we have, at least in any direct sense. Objective reality, on the other hand, is — well, it is presumably *out there somewhere*.

This secondary dualism of appearance and reality is descended from the primary "Cartesian dualism" of mind and matter. During the first half of the

Figure 24.2. René Descartes (1596–1650).[10]

1600s, the French philosopher René Descartes distinguished between "extended stuff" and "thinking stuff" — and did so as if they were separable and disconnected substances having little or nothing in common. Having echoed down through the last several centuries, dualistic thinking has crystallized especially in what we think of as the mind/body problem and, more generally, the mental/physical dichotomy.

Many scientists and scholars today disavow "Cartesian dualism," yet nearly all live intellectually by means of it. There is a very real sense in which Descartes' cleaving stroke through the heart of reality has been almost universally accepted — perhaps most of all among materialist-minded biologists. That is, they seem to have felt they must accept the stroke as a kind of *fait accompli* and then try to live with the violence thereby done to the unity and harmony of the world. They merely choose: which half of this improbably fractured whole shall they accept and which half reject? And so the "material" they embrace is dualistic material, bequeathed to them by the Cartesian sundering of mind from matter. Likewise, the mind they reject is dualistic mind.

Materialists they may be, but their materialism is defined by the dualism that has been drilled into our habits of thought and perception. Instead of going back before Descartes and searching for a different, non-dualistic way forward, they have accepted the original, dualistic fractionation of a living, unified reality, and been content merely to carry a torch for just one of its mutually estranged aspects.

It's not that the problem has gone completely unrecognized. John Searle, Professor of the Philosophy of Mind and Language at Berkeley, has suggested that materialism today "inadvertently accepts the categories and the vocabulary of dualism." It accepts, he says, terms such as "mental" and "physical," "material" and "immaterial," "mind" and "body" just as they have been handed down through the dualistic tradition. Searle, himself a materialist, went so far as to suggest that the deepest motivation for materialism in general "is simply a terror of consciousness" (Searle 1992, pp. 54–55).

So, then, the legacy of dualism has been extremely difficult to shake off, even as an endless procession of scholars have denounced it. Perhaps the primary symptom of the legacy is the seemingly immovable conviction that we face a mind-independent world. (Is this where Searle's "terror of consciousness" comes to expression most easily? If only we can convince ourselves that we live in a mind-independent world, then perhaps we will be spared unpleasant intimations of intelligences other than our own.)

Given the contradiction between belief in a mind-independent world on one hand, and the inescapability of our own minds on the other, we have done our best to get along with two apparently disconnected (dualistic) vocabularies — an objective one for the mindless world and a subjective one for our own minds.

Can we recover the unity of the world?

Instead of a "terror of consciousness," Searle could just as well have cited a "terror of interiority." He also could have said, "We're all materialists now" — because we are. It's built into our experience: we look out at a world that seems to have absolutely nothing to do with our own minds. But this experience is founded on contradiction — fortunately, a contradiction we can recognize and try to get around. True, the recognition may have little power to change our immediate experience. But recognizing and correcting the contradiction in thought may be an important step toward eventually healing the breach between ourselves and the world.

We all know that we are the ones perceiving and experiencing the world. But, at the same time, we experience the world as if it were *out there* independent of our own minds. This is the contradiction: *we seem unable to avoid regarding the world as if it were alien to the interior experience wherein we regard it.*[11]

The resolution of the contradiction requires recognizing that the judgment, "I behold things *out there*," is a judgment we make from *within* our experience. That is, our ability to *experience* things "out there" is an objective feature of the world's interiority. It reflects the fact that the world's phenomena are not my personal, private, merely subjective possession, but are *out there* to be shared with other sentient beings. But they clearly are not *out there* in an absolute and mind-independent sense. They belong to an objective interiority wherein we humans can make an appropriate distinction between our private subjectivity and the publicly shared world.

Further, we can recognize what has led us to distort "out there" to the point where it seems to mean "absolutely mind-independent." The fault lies with the Cartesian legacy, and with the entire drift of modern thought, and perhaps also with Searle's "terror of consciousness." As a result, we have become convinced, first, that our own interiors are shut up within our heads, and second, that the world itself altogether lacks an interior. So we feel in our bones that any world at all, if we are to share it with others, must reside mind-independently *out there*, so that we can all encounter it, so to speak, "from outside." This contrasts with our actual experience, where everything is encountered within an objective marriage of sense and thought in which we collectively participate with our own interiors.

There are other contradictions we can observe in ourselves on the way to freeing ourselves from implicit Cartesianism and the appearance/reality dualism. For example, our faith in the powers of an experience-based (empirical) science conflicts with the widespread conviction that we live in a world of mere appearances whose relation to reality is unknown. If the conviction were correct, how could we have a trusted science of the real world? But we find ourselves with every reason to believe that such a science is possible.

There is also the fact that the mindless-world assumption has given rise to a long-running perplexity, which is commonly framed as the epistemological question, "How can our minds 'in here' apprehend mindless substance 'out there'?" But this unsupported, dualistic framing of the question is proposed *before* one looks at the actual process of knowing, and *before* one has any ground for judging as mindless whatever is "out there." So the dualistic stance is arbitrarily imposed on the epistemological analysis in advance by our implicit Cartesian dualism, defining (and distorting) the entire shape of the philosophical problem.[12]

I MENTIONED A MOMENT AGO the possibility of going back before Descartes and finding a different way forward. That way forward has already been suggested in the foregoing. Instead of a dualism of incommensurate mind and matter, we can acknowledge the actual process of our knowing, with its marriage of sense and thought, both of which occur on the stage of consciousness. The world thus presented to us is unriven by the Cartesian stroke.

Our own experience testifies that there is nothing dualistically problematic about this intimate union of sense and thought. The perceived world shows itself to be a realm of appearances, or experienceable contents, existing in harmonious unity.

The objective world consists, so far as we could ever know, of knowable stuff (appearances), and we are given no positive reason to doubt that its knowability upon the stage of consciousness is perfectly natural, reflecting the nature of things. We ourselves, along with our neural structure and everything else involved in our understanding, are engendered by this world and we are, unsurprisingly, expressions of its character. As beneficiaries of its creative potentials, we are naturally constituted so as to participate meaningfully in our surroundings according to their real nature.

Our cognition does not give us representations of the world

We are not quite done with our focus on the Cartesian legacy and the way it blocks our awareness of the world's interiority. That's because the appearance/reality dualism and the unbridgeable fissure between mind and world have almost forced upon us the conviction that our perception gives us, not the world itself, but a *representation* of it. And this conviction in turn binds us all the more strongly to the dualism from which it arose.

A representation, by definition, is not the real thing. A map of the city is not the city; a photograph of a tree-covered hillside is not the hillside; a small-scale model of a village is not the village. We cannot walk among the trees in a photograph, birds do not make their nests in the branches, and we cannot carve our initials in the bark. If there were total fidelity between the representation and the thing itself, we would not call it a "representation"; it would be the actual thing. And actual things are what we are given in perception.

The proper response to those claiming a gap between appearance and reality might be: "Show us anything in our perception that hints at the existence of a second world beyond the perceivable one — a *real* world contrasting with appearances." A perceived tree appears *itself to be* the tree. We can climb in it, picnic in its shade, and (if it is a sugar maple) tap it for making maple syrup. So also the stream I sometimes sit alongside. If I pick up a small stone and toss it into the water, I perceive both the object and my own arm in picking up the stone and throwing it, and I likewise perceive the trajectory of the stone in relation to earthly gravity, the wind, and the energy at work in my muscles. I can be sure that, exactly as observed — and exactly where observed — the stone and all the other elements of the scene, from my arm to the water, are fully "respecting" the laws of nature. That is, these elements are lawful *in their own immediate, experiential terms* — without my needing to refer to some hidden, mind-independent non-qualitative, non-experienceable (non-empirical) reality behind, or in any way different from, the appearances.[13]

So the world I perceive, while it shows up within my experience and manifests itself upon the stage of consciousness, gives no sign of actually being inside my head, whether literally, or as a reduced representation, or as an illusion, nor any sign of somehow referring to an unknown substratum lying outside all possible experience. Rather, perceived objects testify with overwhelming force to their occurrence, *in their full-bodied presence and reality*, right where and as they are given in qualitative, thoughtful experience — experience that we consistently and objectively enter into along with other sentient beings.

So our perception gives us, not a representation of the world, but the world itself. This is a profound truth we have scarcely begun to reckon with. And the reckoning isn't easy. Perhaps the biggest obstacle lies in the widespread but insupportable conviction that our visual cognition is somehow analogous to the photographs (or videos) that a camera mounted on a robot might produce. The damage inflicted by this analogy upon our perceptual sensitivity can hardly be over-estimated. We may appreciate this more fully when we reflect on our camera-habituated age — an age when snapping a photograph of a significant event or beautiful sight often seems more important than noticing the reality of what it is we are photographing.

And so, sticking to the visual point of view: we need to grasp the difference between our looking externally (from a certain "distance") at a photographic representation of the world, and conjuring the things themselves, in all their reality, within our experience. It's difficult to distinguish between these alternatives until we recognize that "conjuring the *things themselves*," as opposed to looking at representations of them, must mean participating in the creative act of calling them into being, which means realizing them or bringing them to their fullest possible manifestation as interior contents — that is, as "appearances."

The idea that our cognition is a participation in creation is so huge and powerful that, I fear, it tends to stun us into a blank stare. If we were to attend to the idea, we would need to picture ourselves, not *looking* at things, but rather participating in a creative act, *much larger than ourselves*, wherein, by means of our perception of the world, we are continually cooperating in imagining or speaking things into being, which is to say, into their very real and substantive appearance all around us. We would not think our eyes were giving us a picture of things we must interpretively map to some other reality, such as a sub-microscopic, "particulate" one. Instead, we would think of our eyes, together with our other senses and our thinking, as invested with the very same power through which all things have come into being, thereby enabling us to walk and live our lives among them.

This is a thought we need to consider further.

We cognize the world by participating in its creation

There can be no overstating how dramatic and unexpected, for us today, is the view hinted at above. It is one thing to imagine that our eyes are little camera-like devices producing an image that someone, somewhere, somehow, manages to view and interpret as a representation of a mind-independent world. But it is quite another to recognize that, through our eyes and other senses together with our thinking, the world itself takes up its existence *according to its own nature and in the only way it can — as part of lived experience within an interior dimension that we, too, inhabit.*

During the first third of the nineteenth century Coleridge had to have come to terms with the difference between reality and a representation of it when he suggested that our power of perceiving and knowing the natural world is an analog within our own minds of the very same creative activity through which the world comes to exist and is sustained. Or, as he put it in his own unforgettable words:

> The primary Imagination I hold to be the living Power and prime Agent of all human Perception, and as a repetition in the finite mind of the eternal act of creation in the infinite I AM[14] (Coleridge 1906, Chapter 13).

Figure 24.3. Samuel Taylor Coleridge (1772–1834).[15]

In other words, so far as we truly and imaginatively perceive the world, we do not merely encounter it from outside. With our cognitional faculties, we stand within it as co-creators, so that the known world is always coextensive with the reach of our informed imaginations. We bring the expressive "words" of creation alive by making them the expressions of our own minds. After all — as I have been suggesting above — it is not that we "snap a picture" of an independently existing world. We have the very world itself through our cognitional activity — and we have it in a kind of "God's-eye" or creator's view rather than a camera view. We know it from inside its own way of being — which is inseparable from our own way of being — rather than as a representation projected upon a screen.

How much we have had to pay for the anemic belief that our senses give us mere picture-like representations of an alien world! But everything changes when we realize that, just as a boulder on a mountainside is fully and seamlessly embedded in the surrounding world of wind, water, light, and gravity, so, too, our own cognition and expressive powers embed us as knowing participants within a reality of universal expressiveness, and do not confront us with a mere representation of it.

This is not a strange view. It is easy to notice that everything we make into a content of our own experience requires a re-enacting of something like the interior activity that first produced that content. This re-enacting is, for example, the way one human being experiences the content of another's mind. If we attend a lecture (and are paying attention), we follow along by bringing the speaker's thought-content alive as the content of our own minds. So far as we do this faithfully, we live within the same thought-world as the speaker, not a copy of it.[16]

But something like this must also be true of the qualities and thought that constitute the interior dimension of the world as a whole. Here, too, our possibility of seeing and understanding depends on our ability to re-enliven the one world's interior by participating directly in it through the activity of our own interior — in particular, our sensing and thinking.

Coleridge's remark about human perception can help us keep in mind just how radical all this is. If we, in bringing the contents of the world alive within our own experience, must participate in the creative activity through which these contents are originated and sustained, and if this does not mean creating some kind of representation, but rather being active in the one world's ever-evolving manifestation of itself — well, then, this places us, as "mere" perceivers, in a position of high responsibility indeed.

The world as a form of speech

Human language gives us our most immediately accessible picture of the marriage of sense and thought. The outer, sense-perceptible sounds of speech are shone through by an inner meaning that we apprehend in thought. Only when we embrace, and are embraced by, the meaning in its own (and our own) interior realm do we have the phenomenon of language at all. And the point of all I have said earlier in this chapter is that this union of sense and thought, so easily recognizable in human speech, reflects, however dimly, the general character of the world into which we were born.

The suggestion, then, is that the world has the character of language. It is meaningful expression. Or, in more ancient terminology, it is the Logos on display. The whole universe, in its essential nature, is a continual coming into being — which is also to say, a continual speaking or expression or unfolding of meaning — and we are children of this meaning, and the responsible heirs of it. This proposal hardly seems more of a "reach" than one that says a universe that just "happens" to be scientifically accessible and explicable somehow came about from a meaningless "nowhere" of which we have no knowledge — and which we cannot even conceive, since we can only conceive that which is conceivable, or possesses meaning.

Numerous creation stories from around the globe have pictured the genesis of the world and all its creatures as occurring through the spoken word (or song). As we saw in the chapter on "The Evolution of Consciousness," this is how the ancients experienced the world — as thoughtful expression — and the experience was largely lost only in relatively recent history.

Language, then, is not a mere tool we somehow invented. Our minds and our speech precipitated out of language — a language of nature in itself too profound for (merely human) words. We were spoken into being so that we might eventually learn to speak for ourselves, however crudely. All along the way, the meanings inherent in the world nurtured us toward this end.

It would be a useful exercise to trace how, in so many naïve discussions of the supposed origin of language — that is, in discussions about how language is thought somehow to have arisen in creatures initially lacking any form of it — we find a hidden assumption that language already existed before its supposed origin.

For example, a grunt or a finger-pointing or an "excited" state of jumping up and down would typically be assumed (quite rightly) to have some initial, unaccounted-for *meaning*, rather than being merely part of a chain of physical causes and effects. So such actions are, from the very beginning, taken to be significant *gestures*, and therefore are already being imagined as language.

This is fine as long as we realize what we are doing. The grunting and finger-pointing are not the means whereby the non-meaningful becomes meaningful, or non-language becomes language, but stages upon the path by which language comes to ever greater clarity and focus in human consciousness. Human history does not record our moving from no language to language, but rather our learning to possess language rather than be unfreely possessed by it (as we might imagine many animals to be).

This is why Barfield once remarked that to ask about the origin of language "is like asking for the origin of origin." Language just *is* the origin of things, all of which had to be spoken into being. We ourselves had first to be spoken in the deepest and most meaningful language before we could begin internalizing that creative speech and making it our own.

A similar understanding shines through remarks by the German philosopher and linguist, Wilhelm von Humboldt, a contemporary of Coleridge:

> It is my overwhelming conviction that language must be viewed as having been placed in man: For as a product of his reason in the clarity of consciousness it is not explicable. It does not help to grant thousands upon thousands of years for the purpose of its invention ... For man to truly understand even a single word, not as a mere physical outburst, but as sound articulating a concept, language must already exist as a whole within him. There is nothing isolated in language, each of its elements only appears as part of a whole. As natural as it may seem to assume that languages develop, if they were also thus to be invented, this could only happen all at once. Man is only man through language; in order to invent language he would have to have already been man.[17]

The interwoven unity and indivisibility of language ultimately extends to all languages, human or otherwise, and even to the entire cosmos as "the book of nature." Just as we heard it said that "there is only one interior," so, too, language is One, and so also is Logos, and so also is the world that allows itself to be brought to light only through language. It is from this all-encompassing matrix of meaning that we, like all other organisms in one degree or another, emerged as meaning-bearers in a world of meaning.

But what about the billions of galaxies?

At the end of any discussion such as that above, a chilling thought will occur to many who were until then interested. They will say to themselves: "The vastness of the universe is so far beyond the customary dimensions of human experience that we can hardly accept any suggestion that humans participate in the creative process. Even if we were to credit this thought with respect to familiar earthly realities, it would become vanishingly insignificant relative to the universe as a whole."

The pre-eminent physicist, Richard Feynman, summarized the issue with almost poetic succinctness when he dismissed the idea that the universe as a whole might bear any sort of meaningful relation to the story of human life. "The stage," he said, "is too big for the drama" (quoted in Gleick 1992, p. 372).

But Feynman, with his intelligence, should have been self-critical enough to realize that he was doing no more than insisting that the human drama be reduced to the familiar terms of materialism. The alienation to which his remark points is the alienation of supposedly mindless matter from human life. The vast dimensions may intensify that alienation, but what is being intensified is the sense of otherness and indifference associated with the materialist stance. If, by contrast,

Figure 24.4. Portion of an image from the James Webb Space Telescope.[18]

we experienced the material universe as the glory and expression of an interior in which we share, then the vastness would only intensify the glory.

Instead of reducing the human drama to the mindlessness of his conception of matter, Feynman might have asked himself instead whether the universe's material spaciousness needed to be re-considered in light of its manifestation as appearance — and ultimately (in earth evolution) as *human appearance*. Why simply assume that the universe's *being known* (achieving manifestation) within human consciousness is not a significant development in the history of the cosmos? Wasn't Feynman's dismissal of the idea of a cosmic drama simply a re-assertion of his initial, materialistic (and dualistic) assumption about the disconnect between humans and the matter of which the universe is composed?

It's worth asking ourselves, to begin with: suppose we were each raised under a ten-foot ceiling, so that we never saw a sky reaching without limit above us. Would we ever have had any vivid notion of the transcendent? (Try imagining this the next time you leave a closed-in room and stand under a broadly visible sky.) Yet, the notion of the transcendent has been of decisive importance throughout human history. In fact, the earliest histories of which we have any record, as well as the stories echoing down to us from the primary age of myth, did not concern earthly events so much as the activities of divine, celestial beings — beings who were the centers of human interest. Perhaps this vast and ever-expanding celestial perspective not only elevated human aspirations, and not only (in some respects, anyway) raised the level of human culture, but also reflected truths we have long since forgotten.

But the main thing Feynman failed to take into account was the evidence of our demonstrable means of knowing — the evidence that material phenomena, wherever in the universe we encounter them, always present themselves as a union of sense and thought within an interior dimension. They *must* present their meaning and essence interiorly if we are to believe that our most trusted experience — including the science in which we place so much faith — gives us genuine understanding of the world.

In other words, we directly know (by paying attention to our own means of understanding) that the universe as a whole manifests itself to us within a cosmic-scale interiority. And Feynman apparently never asked himself whether this interior sort of manifestation could have originated anywhere other than from a commensurate interior power of creative imagination.

If, for this creative power, to imagine something is also to realize it as an objective appearance in which all beings can share through their conjoined interiors, what would "far away" mean? How many milliseconds would it take for that creative imagination to leap from one side of its interior space to the other?

How long does it take *us* to encompass in thought the most distant galaxies? If they, too, are phenomena — appearances to consciousness — how much of their glory first achieves anything like full reality in the imaginative perception of humans? Can we really say that this *being known* is not as important to their reality as *knowing them* is to ours?

Of course, in our present state we can hardly address the questions we have now brought ourselves up against. But, oddly enough, very many have been willing and eager to pre-judge these questions, whether with Feynman's succinctness or physicist Steven Weinberg's blunt but self-contradictory remark that "The more the universe seems comprehensible, the more it also seems pointless" (Weinberg 1984, pp. 143–4). To find the universe comprehensible is hardly a pointless exercise for human beings whose inner lives are a continual upward striving to understand ourselves and the world that has nurtured us.

Given our lives emerging from this universe as unbounded creative possibility, we have every reason to ask: could this same universe have presented itself to us as a neatly bounded appearance? If the naked eye, or now our telescopes, were ever to reach an "edge" of the universe, it would be as if we had reached an edge of our own consciousness. It would be rather as if we really were living under a ten-foot ceiling, instead of the limitless depths of blue sky and starry heavens.

We were not made for such limits. And neither was the universe in which our ancestors discovered the meanings of their (and our) lives.

> Ah, not to be cut off,
> not through the slightest partition
> shut out from the law of the stars.
> The inner — what is it?
> if not an intensified sky,
> flown through by birds and deep
> with the winds of homecoming.
> – Rainer Maria Rilke[19]

WHERE ARE WE NOW?

An interior world hiding in plain sight

We began this chapter by looking at how we would have no experience of a material world if it were not for our perception of sensible qualities together with the thinking through which we order those qualities and thereby grasp something of the nature of what we encounter in the world. The material world as we have it, then, is a "marriage of sense and thought," and it presents itself to us upon the stage of consciousness where our perception and thinking take form — or, we might say, it exists for us in the terms of the *interior dimension* of our existence.

That is how we know the material world, assuming we do know it. If we really don't know it, then we have nothing to talk about and could just as well keep our mouths shut. But if we do know it, as everyone seems to assume in practice, then the most straightforward and indeed necessary assumption seems to be that the world presents its true character when it comes to manifestation as an appearance to consciousness.

This should not be taken as a reduction of the world to some sort of wispy, airy-fairy notion of human subjectivity. After all, this train of thought begins with the reality of human experience in all its full-bodied presence and solidity. *That* is what we should mean by "appearance," since that is in fact the nature of the appearances; it is how they present themselves to consciousness.

The greatest obstacle to our receiving this truth lies in our dualistic Cartesian heritage, which lives on in the almost universal conviction (at least within western culture) that we look out at a mindless world. It also lives in the appearance/reality distinction, and in the idea that our perception gives us, not things themselves, but distorted representations of them. Nothing in our experience supports this view, which in fact is a judgment we make from *within* our experience, showing how much implicit confidence we unwittingly place in this experience.

I pointed out in the middle of the chapter how errant is the camera model of (visual) cognition. If, in fact, our cognition conjures up all around us the very body of the world — the "things themselves" that make up the world — then it seems that this cognition is actually a participation in the creative activity through which the world gains its powers of appearance, which may also be our participation in the creative activity through which things come to be in the deepest sense of their actual (as opposed to potential) presentation of themselves.

Moreover, our development as language users may testify to the depth of our participation in the world's manifestation of itself. For language is a pre-eminent example of the marriage of sense and thought, and many ancient traditions hold that the world was spoken (or sung) into existence.

Lastly, I have pointed out that the world's existing within an interior dimension can also counter the self-doubt by which so many question the significance of human life against the backdrop of the vastness of the universe. In the interior dimension our own inner being directly participates in that of the world. Physical size is not a relevant parameter when we are considering consciousness or an *appearing* world.

NOTES

1. This distinction might prove useful in contemporary discussions of pan-psychism. I have not seriously delved into the literature of pan-psychism, so (as far as I know) the distinction may already have been made. In any case, recognizing the interior character of the world may imply an origin of the world in some sort of interior agency, but does not imply that everything in the world is, to one degree or another, an agent. Likewise, the fact that material phenomena originate within consciousness, and may be said to occur within consciousness (as I argue below), does not imply that they themselves are "conscious."

2. There is a false way of dealing with the quality of solidity. How many times have we heard that the solidity of this or that object is an illusion, because it is "mostly empty space"? The irony is that we are denying the quality of solidity in the object where it actually occurs by transferring it to an invisible realm where it doesn't occur. That is, the solidity of the object is presumed to be disproved by our appeal to "particles" with vast tracts of empty space between them. The vastness of that empty space is demonstrated only by reference to the minuscule volume and great dispersion of the particles. These, despite the testimony of physics and despite their existence as purely theoretical, non-perceived constructs, are taken to be tiny, solid things. This false picture of the particles' contrasting solidity is the only thing that gives rhetorical force to the idea of "empty space." (I discuss this kind of thinking in Chapter 13, "All Science Must Be Rooted in Experience.")

   Far better to accept felt solidity as the quality it is where we can actually feel it — which is everywhere in the world around us — instead of transferring it to a notional realm of theoretical constructs where we cannot feel it or coherently speak of it. This willingess to stick with the experiential ideal of science involves no disruption of our scientific understanding — and it may save us from a great deal of confusion.

3. Figure 24.1 credit: Sunrise222se (https://commons.wikimedia.org/wiki/File:Wonderful_Nature_Beauty.jpg), CC BY-SA 4.0.

4. "But the science we already have works — nearly miraculously!" This is emphatically true. It works because working is just about the sole intent of the methods of those sciences whose working impresses us so much. But technological savvy — making things that work — is a very different matter from a fundamental understanding of the character of the world we live in. Finding ways to manipulate the world successfully is not at all the same as understanding what sort of things we are manipulating and how we might relate to them beyond our capacity for manipulation.

In many situations mere trial and error is sufficient for successful manipulation. Often sufficient, too, are scientific models that are known to falsify reality in one way or another. John Dalton's theory of the indivisible, indestructible atom and Niels Bohr's theory of the "solar-system" atom both served to further the manipulative powers of science, and both found crucial application in the experimental domains from which they were derived. But neither of them would possess any respectability if seriously put forward as the best summation of our understanding today.

Notice also that, with our manipulative powers, we are always addressing in one way or another the qualitatively given world — so we are *not* being true to the professed ideals of a quality-free science. The very idea of such a science is a gross absurdity deserving no respect at all. We can't have a science of a world that isn't there for us. Nor can we have a science without a world from which we can abstract our preferred quantities. And we can't have a quantitative science without a world we can go back to in order to fit our quantitative formulae to it.

5. See in particular the section, "How do things around us become what they are?" in Chapter 13 ("All Science Must Be Rooted in Experience"). If anyone should remain skeptical of the role of thinking in the constitution of things as whatever they are, I would strongly suggest reading Chapter 4 ("Intentionality") by philosopher Ronald Brady in the online, freely accessible book, *Being on Earth: Practice In Tending the Appearances* (Maier et al. 2006).

6. The philologist and historian of consciousness, Owen Barfield, in the second lecture of his little book, *Speaker's Meaning*, pointed out that, up until the Scientific Revolution, the conviction that ideas were the private property of individuals would have been fully as unapproachable as is, for us today, the conviction that ideas belong to the objective world. The "common sense" of every age can be remarkably difficult to come to terms with, or even to recognize as such. So we tend to be trapped within our own cultural era. The best escape from the trap is to become literate about how earlier eras differed from our own. And that literacy is not achieved by spinning naïve tales about our triumphs over the childish ignorance of our forebears. See Chapter 23, ("The Evolution of Consciousness").

7. Coleridge says of such an understanding, "If it be said, that this is idealism, let it be remembered that it is only so far idealism, as it is at the same time, and on that very account, the truest and most binding realism." He goes on:

> For wherein does the realism of mankind properly consist? In the assertion that there exists a something without them, what, or how, or where they know not, which occasions the objects of their perception? Oh no! … The realism common to all mankind is far elder and lies infinitely deeper than this hypothetical explanation of the origin of our perceptions, an explanation skimmed from the mere surface of mechanical philosophy. It is the table itself, which the man of common sense believes himself to see, not the phantom of a table, from which he may argumentatively deduce the reality of a table, which he does not see. If to destroy the reality of all, that we actually behold, be idealism, what can be more egregiously so, than the system of modern metaphysics, which banishes us to a land of shadows, surrounds us with apparitions, and distinguishes truth from illusion only by the majority of those who dream the same dream? (Coleridge 1906, Chapter XII).

8. It is certainly true that a person who is blind or deaf or who has had traumatic encounters in nature might have experiences of the world differing from those of someone whose senses are functioning "normally." There is in general a huge range of experiential potentials among different persons. Mozart would have ("normally") experienced the world of sound and music to a depth I cannot imagine, just as Picasso would have experienced the world of visual form in ways incomprehensible to me. I do not have a bat's sonar-like sense, nor an insect's infrared sense. The world lends its potentials of experience to all creatures according to their capacity. But we all find ourselves living side-by-side in *one world* — a consistent and shared world with diverse yet harmonious potentials of experience.

9. In the late eighteenth and early nineteenth centuries, the Romantic philosopher, Samuel Taylor Coleridge, used this phrase, (which was later picked up by the philologist, Owen Barfield) and attributed it to Leibniz. I am not a student of Leibniz's work (and I don't, by the way, know Latin), nor have I been able to identify the source of the phrase, *coagulum spiritus*. For a constructive use of the phrase, see Barfield's essay, "Matter, Imagination, and Spirit" in Barfield 1977.

   Since writing the above, I have learned from Peter Cheyne, author of *Coleridge's Contemplative Philosophy* (Cheyne 2020), that he has investigated Coleridge's use of the phrase. The results of the investigation are presented in Chapter 15, "Bloody Speck: How S. T. Coleridge Turned the Embryological *Punctum Saliens* into a Metaphysical Principle," in the book, *Matter and Life in Coleridge, Schelling, and Other Dynamical Idealist*s.

   It happens (Cheyne relates) that — as one of Coleridge's earliest editors, his daughter Sara Coleridge, found — Coleridge's use of *coagulum spiritus* could possibly derive from Frans (François) Hemsterhuis, whom he read and who is cited in Friedrich Schelling's *System of Transcendental Idealism* (1800) with which Coleridge was familiar. Schelling cites Hemsterhuis's concept of "congealed spirit" side by side with a reference to Leibniz. Cheyne adds that this concept goes back earlier to Henry More, whom Hemsterhuis read, as did Coleridge. More's introducing "coagulated spirit" in connection with "monades," a central concept in Leibniz, could also account for Coleridge's misattribution. For details see Cheyne's book.

10. Figure 24.2 credit: The Free Media Repository (https://commons.wikimedia.org/wiki/File:091717-34-Descartes-Philosophy.jpg), CC BY-SA 2.5.

11. Coleridge points to this contradiction in Chapter XII of his Biographia Literaria.

12. The philosopher Ronald Brady, in a posthumous treatise titled "How We Make Sense of the World" (Brady 2016), succinctly summarized today's Cartesian epistemological stance and its alternative this way:

    ♦ "If the question is: 'how can we know the world?' or 'how does the act of cognition take place?' we cannot begin with the very 'knowledge' that this investigation should justify, or we investigate no more than the logical implications of our presuppositions. Epistemology … cannot begin from any positive knowledge of the world, but must suspend all such 'knowing' in order to examine the act of knowing itself … if we do begin from such 'knowledge' our epistemology will necessarily validate present sciences, and deny the possibility of any other form of science." In other words, if we are undertaking a fundamental epistemological investigation, we cannot begin by presupposing the Cartesian diremption of mind from matter.

♦ "Most modern approaches, for example, take their starting-point from the apparent distinction between the thinking subject and the world external to that subject, and thus formulate epistemology after a Cartesian or Neo-Kantian framework. In this formulation … the basic question of epistemology becomes: 'what is the relation of thinking to being?' or 'what is the relation of subjective consciousness to external or objective reality?' These questions arise from the assumed separation of the two — that is, thinking attempts to know the world of objective reality, which world is itself totally independent of thinking. In such a formulation, however, we [assume that we] already know something of that world (such as its difference from thinking), and the problem is created by what we know — that is, the distance between the thinking and its object."

♦ "Since we cannot take the results of previous cognition for granted when we attempt to grasp cognition itself, another formulation of the problem is necessary. If we simply propose that knowledge is immanent in human consciousness (if it is not, then we are not speaking about anything), the basic question of epistemology could be simply: How? What is the act of knowing? Thus we face toward our own act of cognition, and the investigation turns on the *self-observation of thinking*."

13. We are free to theorize in terms of non-experienceable, theoretical constructs. But we typically do so by at least implicitly making models out of them, *as if* they were experienceable things (such as the "particles" of particle physics). And such models — because they are based on theoretical constructs abstracted from appearances and falsely conceived as if they were themselves actual appearances (phenomena) — always turn out in one way or another to be false to reality. (See Chapter 13, "All Science Must Be Rooted in Experience"). They also vex us to no end, as in quantum physics.

    There is no reason we should not investigate the appearances in all directions available to us, without limit. We can, for example, use instruments to explore the structure of forces at a level beneath the possibility of actual sight or touch. But the physics of the past century has taught us very well that we run into crippling trouble when we try to clothe unsensed theoretical constructs with sensible qualities, as we typically do when we talk about "particles" and then all too naturally assume that these must be more or less like solid things capable of traveling from point A to point B through space (or through narrow slits) in the manner of sense-perceptible things.

    If the world is by nature an interiorly experienced world (as I have been urging in this chapter), then we betray reality when we talk about non-appearing things as if they were phenomenal.

14. Along the same line, Coleridge also said that the productive power of becoming which we discover in or above the finished products of nature is a power we can call "Nature," or "Agency." And this Agency at work in nature, he claimed, is akin to the "intelligence, which is in the human mind above nature" (Coleridge 1969, pp. 497–98).

15. Figure 24.3 credit: Wikimedia (https://commons.wikimedia.org/wiki/File:Samuel_Taylor_Coleridge.jpg), public domain.

16. Regarding our attention to a lecture: it is also well known that we tend to mimic the lecturer's physical speech subliminally within our own vocal apparatus. As for copies of thoughts, it is well to realize that conceptual elements are not material structures, and we cannot create multiple copies of them. What would be the "thing" we are copying? If we are paying attention to our own thinking and not hypothesizing theoretical brain states or whatever, we can hardly help realizing that, no matter how many times we return to the same concept, we are not multiplying copies of it, and the same is true when different people take up the same concept. We may accompany a concept with varying mental imagery, but the images are no more the concept than our various pictures of "straight lines" are the concept of a straight line. All instances of the concept, as pure concept, are the *same* instance; they are numerically one, not many. Through our thinking we share, as it were, in "one spirit." It is a useful exercise to think of a pure concept (the straight line will do) while asking yourself, "How might this concept, *as a concept*, not as a mental picture, be multiplied?" It is difficult to imagine even what this might mean — or, at least, it is, so long as one stands within the actual experience of thinking, and not in some materialized image of it.
17. (Humboldt 1963, pp. 2–3). The translation from German is by Norman Skillen: https://journals.ucc.ie/index.php/scenario/article/view/scenario-16-1-10

    Speaking of consciousness rather than language, but with a meaning complementary to Humboldt's, William James had this to say:

    > The demand for continuity has, over large tracts of science, proved itself to possess true prophetic power. We ought therefore ourselves sincerely try every possible mode of conceiving the dawn of consciousness so that it may *not* appear equivalent to the irruption into the universe of a new nature, non-existent until then (James 1890, p. 148).

    Why should we call consciousness and thought "unnatural" as first principles for the understanding of the world? Are they more unnatural than atoms and molecules that suddenly appear from nowhere? Why not begin with consciousness, since in any case we cannot *conceive* of anything that is not an expression of articulate consciousness? Maybe this reflects the nature of reality.
18. Figure 24.4 credit: ESA/Webb, NASA & CSA, A. Martel (https://esawebb.org/images/potm2301b/), CC BY-SA 4.0.
19. Slight alteration of a translation by Stephen Mitchell of the poem, "Ah, not to be cut off."

SOURCES

Barfield, Owen (1965). *Saving the Appearances*. New York: Harcourt, Brace and World. Originally published in 1957.

________ (1967). *Speaker's Meaning*. Middletown CT: Wesleyan University Press.

________ (1973). *Poetic Diction: A Study in Meaning*. Middletown CT: Wesleyan University Press. Originally published in 1928.

________ (1977). *The Rediscovery of Meaning, and Other Essays*. Middletown CT: Wesleyan University Press.

Brady, Ronald H. (2016; posthumous). How We Make Sense of the World: A Study in Rudolf Steiner's Epistemological Work. https://natureinstitute.org/ronald-h-brady/how-we-make-sense-of-the-world

Burckhardt, Titus (1987). *Mirror of the Intellect: Essays on Traditional Science & Sacred Art,* translated and edited by William Stoddart. Albany NY: State University of New York Press.

Cheyne, Peter (2020). *Coleridge's Contemplative Philosophy.* Oxford UK: Oxford University Press.

_______ (2025), editor. *Matter and Life in Coleridge, Schelling, and Other Dynamical Idealists.* New York: Springer.

Coleridge, Samuel Taylor (1906). *Biographia Literaria.* London: J. M. Dent & Sons. Originally published in 1817. Available at https://archive.org/details/cu31924102776196/

_______ (1969). *The Friend* vol. 1, edited by Barbara E. Rooke. Princeton: Princeton University Press (Bollingen Series LXXV). London: Routledge and Kegan Paul. Originally published in multi-volume editions of 1812 and 1818.

Edelglass, Stephen, Georg Maier, Hans Gebert and John Davy (1997). *The Marriage of Sense and Thought: Imaginative Participation in Science.* Hudson NY: Lindisfarne.

Gleick, James (1992). *Genius: The Life and Science of Richard Feynman.* New York: Random House.

Humboldt, Wilhelm von (1963). Über das vergleichende Sprachstudium in Beziehung auf die verschiedenen Epochen der Sprachentwicklung, in A. Flitzer and K. Giel, editors, *Werke in fünf Bänden*, Band 3, *Schriften zur Sprachphilosophie*, pp. 1–25. Darmstadt: Wissenschaftliche Buchgesellschaft.

James, William (1890). *The Principles of Psychology* vol. 1. New York: Henry Holt. Available at https://archive.org/details/theprinciplesofp01jameuoft/

Maier, Georg, Ronald Brady and Stephen Edelglass (2006). *Being on Earth: Practice In Tending the Appearances.* Freely available online version: https://natureinstitute.org/book/being-on-earth. Hardcopy book is available from Logos Verlag Berlin (2008): http://www.logos-verlag.de/cgi-bin/engbuchmid?isbn=1887&lng=eng&id=

Musser, George (2023). *Putting Ourselves Back in the Equation: Why Physicists Are Studying Human Consciousness and AI to Unravel the Mysteries of the Universe.* New York: Farrar, Straus and Giroux.

Poincaré, Henri (1913). *The Value of Science*, translated by George Bruce Halsted. https://ebooks.adelaide.edu.au/p/poincare/henri/value-of-science/

Searle, John R. (1992). *The Rediscovery of Mind.* Cambridge MA: MIT Press.

Weinberg, Steven (1984). *The First Three Minutes.* New York: Basic Books. Originally published in 1977.

INDEX OF NAMES

A

Abbott, Larry, 148
Adami, Christoph, 279–280
Alberch, Pere, 292
Alberti, Simon, 75n7
Alberts, Bruce, 269
Albrecht-Buehler, Guenter, 82, 262–267, 269–271, 273n6. *See also* Subject Index
Allis, C. D., 99
Altucci, Lucia, 99
Anway, Matthew, 97
Arber, Agnes, 78, 88, 319
Aristotle, 13, 164, 180, 183, 390n10, 423, 432. *See also* Subject Index
Arnhart, Larry, 19
Arp, Robert, 19, 32n1
Arthur, Wallace, 302

B

Babu, M. Madan, 65
Bacon, Francis, 416, 423
Ball, Philip, 43, 71–73
Ballweg, Stephanie, 60
Barfield, Owen, iii–iv, 312, 333n6, 410, 412–419, 420–429, 432–434, 436n3, 437n9, 437n13–15, 449–450, 458, 464n6, 465n9. *See also* Subject Index
Barham, James, 334
Bateson, Patrick, 393
Bateson, William, 291
Beatty, John, 295
Behar, Marcelo, 126
Berger, Shelley, 123, 251
Bissell, Mina, 251, 303
Blackmore, Susan, 279, 281, 286
Blobel, Gerd, 128
Bobola, Nicoletta, 222
Boehme, Jakob, 59
Bohr, Niels, 271, 464n2
Brady, Ronald, 16, 166, 171–174, 177–180, 196–198n1, 7–9, 465n12. *See also* Subject Index
Brangwynne, Clifford, 65–68, 70
Brenner, Sydney, 124
Brown, Robert, 237, 264–265
Bruce, Robin, 370n2
Buller, David, 322–325, 333n8
Burckhardt, Titus, 446
Burgess, Darren, 383
Burnett-Stuart, George, 274n7
Butterfield, Herbert, 423

C

Cairns, Bradley, 248
Callier, Viviane, 274n10
Campeanu, Ion, 47
Carmo-Fonseca, Maria, 105
Carroll, Robert, 348
Carroll, Sean, 150, 155–160, 162. *See also* Subject Index
Carter, Jimmy, 81
Cervantes, Marlene, 96
Chaires, Jonathan, 221
Chen, Weijun, 107
Cheyne, Peter, 465n9
Churchland, Paul, 210–211
Clancy, Suzanne, 117n1
Coleridge, Samuel Taylor, 59n9, 150n3, 367–368, 373, 427, 456–458, fig24.3, 464n7, 465n9,11, 466n14,15. *See also* Subject Index
Commoner, Barry, 389
Conaway, Joan Weliky, 254
Conaway, Ronald, 254
Cope, Edward Drinker, 291
Copernicus, Nicolaus, 205, fig13.2
Corning, Peter, 325
Cosgrove, Michael, 241

Coulombe, Pierre, 58n4
Coyne, Jerry, 288–289
Crews, David, 97
Crick, Francis, 45

D

da Vinci, Leonardo, 421–422, 437n11
Daan, Niels, 372
Dalton, John, 464n2
Darwin Charles, 153, 277–283, fig16.1, 287, 292, 356n14. *See also* Subject Index
Davidson, Mari, 304
Dawkins, Richard, 64–65, 69, 153, 279–280, 374, 379–384, 390n5, 395–409, fig22.1, 410n3–5. *See also* Subject Index
de Almeida, Sérgio, 105
de Laguna, Grace, 334n10, 335
de Vries, Hugo, 290, fig16.2
Dekker, Job, 42, 112–113
Deng, Wulan, 128
Dennett, Daniel, 278–281, 344, 375. *See also* Subject Index
Descartes, René, 422, 451, fig24.2
Doi, Masao, 123
Dolgin, Elie, 68
Dolinoy, Dana, 98
Dunker, A. Keith, 66

E

Edelglass, Stephen, 216n2, 217, 447
Einstein, Albert, 261
Eisenberg, Eli, 254n14
Eldredge, Niles, 277, 348
Emerson, Ralph Waldo, 61, 415, 417–418. *See also* Subject Index
Esposito, Maurizio, 370n2

F

Fabre, Henri, 1
Faraday, Michael, 382
Farinella-Ferruza, N., 135
Ferris, Timothy, 209–210
Feynman, Richard, 260, 459–461
Finn, Elizabeth, 128
Fletcher, Daniel, 52
Fogarty, Elizabeth, 232
Ford, Brian, 138, 356n14
Franzmann, Titus, 75n7
Frauenfelder, Hans, 75n8
Frye, Michaela, 254n15
Fu, Xiang-Dong, 130
Fuchs, Gilad, 244, 245
Fukushima, Hiroto, 127
Furlong, Eileen, 128
Furst, Branko, 75n1
Fyodorov, Dmitry, 255n19

G

Galileo, 260, 444
Gan, Qiong, 246
Ganser, Laura, 64
Gatlin, Lila 3, 280
Gawne, Richard, 373, 389n1
Gehring, Neils 105
Gerhart, John, 302
Gerstein, Mark, 71–72
Gilbert, David, 127, 136
Gilbert, Nick, 250
Goethe, 8–9, 171–173, 197–198n9, 427. *See also* Subject Index
Goldstein, Kurt, 85 370n2
Gombrich, E. H., 429
Göndör, Anita, 128
Goodwin, Brian, 329
Gould, Stephen Jay, 279, 289, 298–299, 310n1, 348, 375. *See also* Subject Index
Grandjean, Valérie, 102
Grant, Barry, 64
Graur, Dan, 346, 377, 379
Graveley, Brenton, 106 121
Gray, Asa, 153
Gregory, R. L., 204–205
Grimson, Andrew, 232
Grobstein, Clifford, 74n1
Grosschedl, Rudolf, 128
Grüter, Christoph, 333n3
Gsponer, Jörg, 65
Guénon, René, 446

H

Haldane, J. B. S., 19, 291
Haldane, John Scott, 147, 85
Halfmann, Randal, 69
Harayama, Takeshi, 59

Harold, Franklin, 53, 58
Hartl, Daniel, 375
Hawley, R. Scott, 304
Helanterä, Heikki, 376
Hemsterhuis, Frans (François), 465n9
Hepp, Paula, 127
Hill, Meredith, 235
Hilser, Vincent, 65
Hinde, Elizabeth, 247
Hines, William, 251
Ho, Mae-Wan, 350, 356n13
Hoffmann, Alexander, 126
Holdrege, Craig, 62, 116, 144, 167–171, 181–182, 196n4, 273n6, 348–349, 355n10,11. *See also* Subject Index
Holdrege, Henrike, iv
Homer, 413–416, 429–431
Hubbard, Ruth, 389n1
Hubstenberger, Arnaud, 69,
Humboldt, Wilhelm von, 458–459, 467n17
Hume, David, 197n9
Huxley, Julian, 298
Huxley, Thomas, 161–162
Hyman, Anthony, 66–67, 274n10

J

Jacob, François, 32n4, 295n4
James, William, 467n17
Johannsen, Wilhelm, 357–360, fig20.1
Jonas, Hans, 29
Joshi, Sachindra, 242
Joyce, James, 437
Jukes, Thomas, 292

K

Kaiser, Jocelyn, 128
Kant, Immanuel, 132n1, 196–198n8,9, 466n12
Keller, Evelyn Fox, 389n1
Keyak, J. H., 273
Kidder, Benjamin, 47n8
Kim, Seyun, 58n4
King, Jack Lester, 292
Kirschner, Marc, 302
Klerk, Eleonora de, 226, 250
Kouzine, Fedor, 40
Krisko, Anita, 110–111
Kristjánsdóttir, Katla, 232
Kumar, Vivek, 123
Kurup, Jiji, 47n8, 384
Kwak, Hojoong, table14.1
Kwok, Roberta, 58n3

L

Laguna, Grace de, 334n10
Lai, William, 247
Landecker, Hannah, 58n1, 63
Langer, Susanne, 32n1, 74n1, 295n5, 350
Laplace, Marquis de (Pierre-Simon), 90n4
Lavelle, Christophe, 40, 46n6
Ledford, Heidi, 95, 119
Leibniz, Gottfried Wilhelm, 449, 465n9
Leonardo da Vinci, 421–422, 437n11
Levanon, Erez, 254n14
Levin, Michael, 135, 136, 140–149, 151n5. *See also* Subject Index
Levins, Richard, 370n5
Levitt, Michael, 71–72
Levy, Emmanuel, 125
Lewin, Roger, 124
Lewontin, Richard, 11, 35, 80–81, 135, 364–366, fig20.3, 368, 370n5. *See also* Subject Index
Lieberman-Aiden, Erez, 46n1, n2
Lillie, Frank, 85, 93–94, 100, 303
Lis, John, table14.1
Lockley, Martin, 192–194
Lu, Xin, 120
Luco, Reini, 105
Lührmann, Reinhard, 106
Lyell, Charles, 280, 291
Lynch, Michael, 376–377, 379

M

Magnani, Luca, 119
Maier, Georg, iv, 91n7
Makowski, Matthew, 246
Malik, Sohail, 254n9
Margulis, Lynn, 292
Martinez-Campa, Carlos, 240
Mayer, Bruce, 126
Mayr, Ernst, 278, 281
McClintock, Barbara, 306–307
McDougall, William, 23–24

McKenna, Kenneth, 389
Merabet, Samir, 222
Michieletto, Davide, 250
Miller, Courtney, 237
Misteli, Tom, 42, 63, 105, 128
Moczek, Armin, 378
Moore, Melissa, 107
Moretti, Rocco, 221
Morgan, Thomas Hunt, 357
Moss, Lenny, 34, 56, 59n8, 389
Moujaber, O., 58n4
Müller, Gerd, 373
Mullins, R. Dyche, 52
Musselman, Catherine, 243
Musser, George, 450
Mustoe, Anthony, 233

N

Nicholson, Edward Max, 27
Nilsen, Timothy, 106, 254n15

O

O'Neill, Bill, 279
Ohlsson, Rolf, 128
Olin, Ada, 237–238
Olin, Donald, 237–238
Oren, Moshe, 244–245
Oyama, Susan, 288

P

Page, Scott, 304
Pai, Vaibhav, 141
Pandolfi, Pier Paolo, 235
Papasaikas, Panagiotis, 105–107
Pearson, Hele, 63
Pennisi, Elizabeth, 350
Perrin, Jean Baptiste, 265, 273
Persinger, Michael, 267
Pezzulo, Giovanni, 148, 151n5
Piaceri, Irene, 127
Plankar, M., 58
Poincaré, Henri, 450
Polanyi, Michael, 329–330, fig18.3
Popper, Karl, 19
Portmann, Adolf, 153–154, 192, 292
Poulter, M., 98
Pugh, Franklin, 247
Punnett, Reginald, 291

R

Radanović, Toni, 55
Radman, Miroslav, 110–111
Rao, Suhas, 39
Rassoulzadegan, Minoo, 97
Relton, Caroline, 128
Rhine, J. B., 23
Rich, Tina, 332n1
Riegner, Mark, 192, 199n16
Riezman, Howard, 59
Rilke, Rainer Maria, 421, 461
Rinn, John, 39, 235
Robert, Jason Scott, 390n10
Roeder, Robert, 254
Rohs, Remo, 241
Roignant, Jean-Yves, 105
Rothman, Stephen, 65
Rousseau, Jean-Jacques, 420–422, 427, 437n14
Ruse, Michael, 329–330
Russell, E. S., 12–13, 25–28, 83, 85, 98, 136, 147, 359–360, fig20-2, 363, 369, 389n1, 390n10. *See also* Subject Index
Ryan, Frank, 301

S

Sachs, Joe, 13, 184, 330, 406
Sapp, Jan, 50
Sassone-Corsi, Paolo, 96, 123
Saunders, Peter, 350, 356n14
Schad, Wolfgang, 61–62, 181–195, 198n11, 12, 199n17. *See also* Subject Index
Schneider, Robert, 128
Seamon, David, 182
Searle, John, 451–452
Separovich, Ryan, 245
Severin, Philip, 236
Shakespeare, William, 428
Shapiro, James, 306, 389n1
Shelley, Percy Bysshe, 413
Sherrington, Charles Scott, 25, 85, 138–139
Shin, Yongdae 68
Sinden, Richard, 384
Singer, Robert, 112
Skinner, Michael, 97
Slattery, Matthew, 222
Smillie, Chris, 351

Snell, Bruno, 430–431
Sober, Elliott, 277, 281–283, 292. *See also* Subject Index
Stadhouders, Ralph, 128
Steensel, Bas van, 128
Steiner, Rudolf, 198n11
Sterelny, Kim, 364–365, 367
Stochaj, U., 58n4
Stoltzfus, Arlin, 291
Strahl, B. D., 99
Stunnenberg, Hendrik, 99
Sundvall, Maria, 105
Sweatt, J. David, 237
Szewczak, Lara, 70

T

't Hoen, Peter A. C., 226, 250
Takahashi, Joseph, 123
Talbott, Stephen, 34, 45, 65
Tata, Purushothama Rao, 79
Tay, Yvonne, 235
Technau, Ulrich, 95
Thompson, D'Arcy Wentworth, 85
Toepfer, Georg, 32n1, 326, 334n9
Tompa, Pete, 70
Tran, Nham, 235
Travis, John, 112–113
Trivers, Robert, 19
Tsankov, Alexander, 99
Turner, Bryan, 239–240
Turoverov, Konstantin, 64–65, 75n3

U

Uller, Tobias, 376

V

Valcárcel, Juan, 105–107
van den Berg, Jan Hendrik, 420–421, 427, 437n14. *See also* Subject Index
Vandenberg, Laura, 141
Varela, Francisco, 30
Vernon, Mark, 430–432
von Senden, M., 216n1
Vries, Hugo de, 290, fig16.2

W

Waddington, C. H., 137, 356
Wahl, Markus, 106
Wahls, Wayne, 304
Wald, Elijah, 389n1
Walsh, Denis 293, 295n7, 334n10
Wandschneider, Dieter, 371n8
Wang, James, 44–45, 307
Watson, E. L., 1–2
Watson, Christine, 332n1
Watson, James, 45
Weaver, Ian, 98
Weber, Andreas, 30
Webster, Gerry, 329
Wei, Li 237
Weinberg, Steven, 461
Weiner, Assaf, 127
Weismann, August, 278, 280
Weiss, Paul, 29, 55, 81–83, 89, 112, 115, 131, 136, 147, 149, 163, 283, 354, 381–382. *See also* Subject Index
West-Eberhard, Mary Jane, 355
White, John, 433
Willcockson, M. A., 255n19
Willyard, Cassandra, 46n5
Woelfle, Mark, 248
Wyllie, Andrew, 332n1

Y

Yang, Xiaoyong, 123
Yusufova, Nevin, 255

Z

Zhou, Yaoqi, 64
Zhu, Lian, 65, 70
Zimmer, Carl, 39

INDEX OF SUBJECTS

A

accidents, 339, 350, 365
actin, 52
agency. *See also* intention; purposiveness; teleology
  anthropomorphism in attributions of, 320–321
  biological activity and, 313
  biological theory and, 10–12, 147, 201, 314, 319, 358
  "blind" forms of, 287
  blindsight surrounding, 147, 299
  causation and, 315
  consciousness and, 442
  coordinating form of, 83, 313, 353–354
  definition of, 11–12
  development and, 352–353
  evolutionary theory and, 314–322, 426
  explanatory challenge of, 318–320
  form and, 147
  genes and, 34
  genocentrism and, 376
  germline and, 304
  holism and, 11–12, 78, 83, 149
  individual and population forms of, 314–316
  intention and, 11–12, 28–29
  interiority and, 309
  language and, 288–289
  misplacement of, 292–294
  mystery surrounding, 320
  narrative and, 26–29
  natural selection and, 287–290, 292–294, 298–299, 322–328
  organisms and, 10–11, 21–23, 26, 30, 147, 149, 304–305, 320–321, 358, 376, 400
  physical interactions contrasted with, 11–12
  rejection of, 314–322

agency (*continued*)
  self–transformation and, 304–305
  sources of, 320–321
  stable variation and, 400–401
  teleology and, 313, 314–322
Albrecht–Buehler, Guenter
  Brownian motion, 265
  cell membrane, 269
  cell size, 263
  electrical forces, 267
  gravity, 273n6
  holism, 82, 262–264
  "In Defense of 'Nonmolecular' Cell Biology", 262
  liquid computer possibility, 270–271
  polymerization, 268–269
  viscous drag, 264
Alps, 420–422
ambiguous figures, 205–207
American Pharoah (race horse), 339–340, 341, 342–343, 352–353
animals at watering hole, 317
anthropomorphism, 3–4, 25, 53, 201, 320–321
appearance
  ambiguous figures and, 205–206
  causation and, 179–180
  empirical science and, 206–214
  form and, 179–180
  inanimate world and, 448, 450–453
  interiority and, 158, 192, 324
  knowledge and, 444, 448, 460–461
  machines and, 324
  meaning and, 159, 446
  microworld puzzles and, 271
  reality contrasted with, 448–454
archetypes, 12, 287
Aristotelianism, 183, 330, 423

Aristotle
- *entelecheia*, 13, 330
- formal causation, 63, 80
- influence and reception of, 422–423
- integral unity of the organism, 13
- logic's discovery, 432
- motion types, 183–184

arrival of the fittest, 290–291, 298, 325, 377

atoms. *See also* molecules
- biological theory and, 12, 269
- chemical energies and, 226
- definition of, 12
- empirical science and, 209–211
- experience and understanding of, 12
- explanatory role of, 34
- organisms constituted by, 22
- perception and, 210, 239
- qualitative biology and, 166
- reality of, 12, 210, 382
- visualizations of, 209–210, 239, 270–271, 381

B

Barfield, Owen
- Aristotle, 432
- correlation between consciousness and phenomena, 437n9
- detachment, 424–425
- enchantment of the world, 413
- European sense of nature, 420–422, 434
- evolution of consciousness, 412–413, 422–424
- fantasy novel written by, 333n6
- on the footsteps of nature, 416
- history of ideas distinguished from consciousness, 422–424
- ideal aesthetic distance from nature, 426, 437n15
- ideas as private property, 464n6
- interiority, 417, 449–450
- language's evolution, 413–419, 424–425
- logomorphism, 414
- long historical arc of the evolution of consciousness, 424–425
- mana, 419
- medieval consciousness, 427–428
- methodology of, 437n13

Barfield (*continued*)
- origin of language, 458
- original participation, 424–425
- perception identified with thinking, 425
- *pneuma* as wind, breath, and spirit, 414–415, 417, 436n3
- psychoanalysis, 424
- scientific terms and the evolution of language, 417–418

basal cells, 79

"because," meaning of, 2

bees, 309, 314, 319–320

behavior, 7–9, 14, 19–20, 22–24, 81–82, 144–145, 149, 317, 388

"bioinformatics and bioengineering" conference (2008) (Harvard Medical School), 66

biological activity
- agency and, 313
- context-dependence of, 131
- end-directedness of, 7, 13, 227, 283, 305, 315, 337, 342
- intention and, 7, 358
- material coordination of, 50
- meaning and, 7, 22, 89, 313, 337
- mechanism and, 150, 340
- metabolic basis of, 249
- molecules and, 136
- movement as substance of, 63–64
- narrative and, 79
- natural selection and, 326
- purposiveness of, 16, 24–25, 325–326, 331, 336
- wisdom and, 313

biological clocks, 122–124

biological functions, 22, 59n8, 63, 69, 88, 111, 120, 322–323, 334n10

biological theory. *See also* evolutionary theory; molecular biology; qualitative biology
- agency and, 10–12, 147, 201, 314, 319, 358
- aims of current book on, 3, 7–8, 17–18
- as-if thinking in, 19, 23
- atoms and, 12, 269
- blindsight within, 10–11, 23, 181, 216n1
- causation and, 9, 26, 43, 87, 119
- complex organisms prioritized in, 17

biological theory (*continued*)
  context-dependence and, 9, 43, 45, 84, 86–87, 139, 361, 387
  conventional approach in, 139, 182
  Darwinian functionalism in, 193
  definition of key terms for, 11–16
  development and, 4–5, 22
  directiveness and, 12–13, 305, 336
  dogma of, 116, 143
  evidence in, 17–18
  form and, 137–139, 144, 153–154, 181–195
  function and, 322–323
  genes and, 3, 34, 98–99, 357–358, 375
  genocentrism and, 111, 376–378, 386–387
  genomes and, 96
  holism and, 9–10, 78, 81–82, 84–89, 319, 366
  homeostasis in, 330
  human beings and, 201–202
  inheritance and, 359–360
  integral unity of the organism and, 13
  intention and, 13–14, 201
  interests and, 201
  interiority overlooked in, 7–8, 23, 87, 88, 362–363
  key terms in, 11–16
  knowledge and, 10–11
  language of life denied in, 23, 119
  materialism assumed in, 14–15, 88, 293, 374, 388
  meaning as taboo in, 15, 20, 87, 328–331
  mechanism and, 43
  messaging and, 125
  molecules and, 12
  mysticism as threat to, 116
  narrative and, 4–6, 10–11
  natural selection in, 3, 8, 288–289, 293–295, 322, 324
  organization and, 362
  particles' distorting effect on, 381–382, 395–399
  perception and, 23
  physical interactions and, 158
  physics and chemistry contrasted with, 130
biological theory (*continued*)
  plausibility in light of current scientific assumptions of, 23
  purposiveness and, 15, 19–20, 25–26, 201
  qualities and, 362
  self-transformation and, 305
  systems biology, 43
  task-nature of problems underlying, 10
  teleology, role of, 15, 19–21, 30, 315
  unity and, 319
  value as taboo in, 20
  wisdom and, 253n5, 306
biomolecules, 63–64, 71–72
bison, 189–191, 359
bisphenol A, 97
blindsight
  agency and, 147, 299
  biological theory as subject to, 10–11, 23, 181, 216n1
  causation and, 119
  counteracting, 17
  definition of, 10
  meaning and, 23
  messaging and, 125
  metaphoric language use as form of, 125
  overview of, 10–11
  purposiveness and, 19
Blueprint (Plomin), 34
Brady, Ronald
  causation, 177
  final causes not justified, 178
  "Form and Cause in Goethe's Morphology", 171–172
  Goethe's morphological writings, 171–172
  "How We Make Sense of the World" by, 465n12
  on Kant, 196–197n8–9
  leaf sequences, 172–177, 180
  logical necessity, 179
  movement, 174–177
  undergraduate experience, 166
bronze statue of a youth, 430, 431
Brownian movement, 264–265
buttercup plant (*Ranunculus acris*), 172–173, 177

C

carnivores, 187–191
Carroll, Sean
  causation, 158
  critiques of, 158–160, 162
  development, 157–159
  DNA, 156
  evolutionary developmental biology, 155
  form, 157–159
  genetic switches, 155–157, 159–160
  invisible artist, 162
  mechanism, 158–160
  modules, 156
  qualitative nature of form, 159
  simple rules of animal form, 157–158
Cartesian dualism, 419, 447, 450–451, 453
causation
  abuses of language and, 121–122
  agency and, 315
  appearance and, 179–180
  arrow of, 130
  biological clocks and, 122–124
  biological form of, 88, 122, 132, 140, 144–145
  biological theory and, 9, 26, 43, 87, 119, 129–132
  blindsight and, 119, 125
  Caenorhabditis elegans roundworm and, 124–128
  causal confusion, 223
  cells and, 99–100
  challenge of identifying biological forms of, 119
  chicken-or-egg paradox and, 122
  collaborative decision-making and, 126–128
  complexity and, 223
  context-dependence and, 87–88, 338
  cross talk and, 125–127
  development and, 124, 318, 338
  DNA and, 94
  efficient causes, 316–317, 375
  epigenetics and, 99–100, 119
  evolutionary theory and, 316–317
  final causes, 16, 178
  form and, 142–143, 180–181
  form of organisms and, 158
causation (*continued*)
  formal causes, 163, 180, 315, 317–318, 320, 404
  functions and, 322
  fundamental role in biology of, 129–130
  future research directions for, 130–132
  future-oriented forms of, 142–143
  gene expression and, 127–128, 136, 141–142, 222–223, 248–251
  genes and, 51, 357–358, 403
  goal-directed activity and, 126
  hearts and, 338
  holism and, 9–10, 126–127, 338
  immaterial causes, 47n10, 88
  individuals as "coming from" and "passing to", 177–179
  leaf sequence and, 177–181
  master controllers and, 120–122
  meaning and, 88, 124, 125–128, 201
  mechanism and, 126
  messaging and, 126–127
  movement and, 177, 179–181
  moving beyond definitive forms of, 130–132
  narrative and, 88, 201
  natural selection and, 325
  organisms and, 125, 146, 223
  overview of, 119, 130–132
  problem of, 129–130, 142
  proteins and, 126, 222–223
  purposiveness and, 26, 223
  qualitative biology and, 177, 179–181
  reversibility of, 130
  RNA splicing and, 108–109
  stable variation and, 400, 403
  teleology and, 312–313, 315–318
  transcription factors and, 222–223
  unambiguous forms of, 119
  webs of, 222–223
cell membrane
  definition of, 54–55
  dynamic nature of, 55–56
  genes and, 56
  molecules of, 55
  movement and, 55–56
  proteins and, 375–376
  vesicle fusion with, 55
  visualization of, 54, 55

cells. *See also* cell membrane; cytoplasm; cytoskeleton; nucleus; ribosomes; stem cells
behavior of, 7
causation and, 99–100
chemical energies of, 266
coherence of molecular activity of, 346
complexity of, 45–46
context-dependence and, 84, 100
de-differentiated form of, 79
descriptive challenges posed by, 113
development and, 302–305
differentiation of, 302–303
directiveness and, 395
DNA held in intimate and instructive embrace in, 223–225
epigenetics and, 93–94
executive control of, 50–51, 56, 70
genetic control of, 56
gesturing and, 39
heritability and, 357
holism and, 79–83
identity of, 55, 79, 82–83, 89
information and, 56–57
interiority and, 6–7
knowledge and, 306–307
meiosis of, 303–304
mitochondria of, 350
mitosis of, 38, 43–44, 129
movement and migration of, 41, 51–52, 55–56, 62–63, 81
niche of, 79–81
organelles of, 50–51, 54–56, 66, 70, 82, 136, 350
overview of, 50–51, 57
parts as reflecting whole of, 53, 57
phase transitions in, 69
purposiveness and, 24–25
self–transformation and, 302–303
senses and limbs attributable to, 51–54, 306
shape of, 268–269
"skin" of, 54–56
specialization of, 79
teleology and, 312, 314–322
topoisomerases used by, 41
water and, 55, 84, 112, 308
Century of the Gene, The (Keller), 389n1
ceRNAs (competing endogenous RNAs), 235, 255n16
chaffinch, 27
Changing Nature of Man, The (van den Berg), 420
chemical energies, 188, 226, 266
chemical transformation, 96, 154, 236, 260
chicken-or-egg paradox, 122, 157
chromatin
complexity of, 243
definition of, 36, 223–224
as full substance of chromosomes, 223–224
functions of, 37, 224
gene expression and, 223–225
histone core particles constituting, 36
information-rich nature of, 225
proteins in, 223–224
torsion and, 40
visualization of, 239
chromosomes, 37–45, 223–226, 236–239. *See also* nucleosome
abstract code unable to capture performance of, 43
chromatin as full substance of, 223–224
coherence of movement in, 40
contemporary research on, 39
dance of, 42
dynamic nature of, 38
fractal globules formed in regions of, 37–38, 43
gene expression and, 39–42
lack of absolute rules for interaction within regions of, 39
loop domains in, 37–39
as manifestation of life, 41
mechanism and, 40–43
misleading visualization of, 38
movement and, 38, 40, 42
spatial organization of, 41
topoisomerases and, 41, 43–45
torsion and, 40
visualization of organization of, 37
cichlid fish, 349–350
coagulum spiritus, 449, 465n9
cognition, 88, 213, 279, 425, 442, 445, 448, 454–457, 462
coherence, unexpected. *See* unexpected coherence

Coleridge, Samuel Taylor
  becoming of nature, 466n14
  coagulum spiritus, 465n9
  image of, 456
  holism, 367–368
  idealism, 464n7
  machine and organism distinction, 150n3
  opposing forces, 59n9
  perception as primary imagination, 456–457
  on the productive power of becoming, 466n14
  representation, 456
  Romanticism of, 427
Coleridge's Contemplative Philosophy (Cheyne), 465n9
Confessions (Rousseau), 420
congealed spirit concept, 465n9
consciousness
  agency and, 442
  definition of, 12
  detachment and, 424–428, 432–433
  empirical science and, 412
  European sense of nature and, 420–422
  evolution of, 422–426, 433–434
  expression and, 416–417
  Greek understanding of, 430–434
  history of ideas and, 422–424
  human form of, 12, 428–435, 458, 460
  impairment and, 465n8
  interiority and, 15, 416–417, 423–424
  Jewish understanding of, 433
  language and, 413–419
  long historical arc of the evolution of, 424–426
  meaning and, 28, 412–419, 422–423, 425
  medieval form of, 427–428
  monotheism and, 432
  from myth to literacy in, 428–434
  non-linear evolution of, 433–434, 437n19
  original participation and, 424–427, 432–433
  overview of, 412, 434–436
  participation and, 423–424
  perception and, 15, 422, 425, 432–434
  purposiveness and, 28
  terminology for, 12
  translation and, 414–415
consciousness (*continued*)
  unconscious and, 424
  unity of inner and outer meaning in, 412–416
context–dependence
  biological theory and, 9, 43, 45, 84, 86–87, 139, 361, 387
  causation and, 87–88, 338
  cells and, 84, 100
  genes and, 100
  genocentrism and, 387
  holism and, 9–10, 84–86, 338
  ideas and, 86
  inheritance and, 361
  intention and, 86
  leaf sequence and, 179
  meaning and, 89
  molecular biology and, 84
  molecules and, 84
  narrative and, 86
  organisms and, 89–90, 99–100, 361
  qualitative biology and, 179
controlling genes, 155, 373, 389n1
"Controlling the Controllers" (Separovich et al.), 245
coordinating power, 154, 313, 317, 352–353. *See also* agency; purposiveness; teleology
cows, 168, 170, 187–188, 191
Critique of Judgment (Kant), 196n8
cytoplasm
  compartments in, 66–67
  enzymes and, 67
  extracellular influence on, 67
  functions of, 50, 53, 66–67
  holism and, 82
  movement and, 55
  organization of, 66
  phase separation and, 68–69
  proteins and, 67
  RNA and, 53, 67, 104, 220, 249
  translation and, 220
  well–structured droplets in, 67–69
cytoskeleton, 51–54
  as master regulator 50, 53–54
  meaning and, 54
  nucleus and, 53
  perception and, 51–54

D

Darwin, Charles. *See also* evolutionary theory; natural selection
- the eye, 153
- natural selection, 153, 277, 281–282, 287, 292
- photograph of, 277
- variation, 356n13

Darwinian functionalism, 193

Darwinism, 298, 344, 379, 398

Dawkins, Richard
- agency, 400–401
- bodies, 396, 404
- causation, 390n5, 395, 400
- coercion among animals, 279–280
- critique of, 64–65, 69, 399–407
- development, 397–399
- digital nature of genes, 69, 379–381, 409n3
- on DNA's structure, 64–65
- *Extended Phenotype* by, 397
- genocentrism of, 395–399
- holism, 399, 406–409
- immortality of genetic variation, 402–404
- independent life of genes, 396–397
- *Origin of Species*, 153, 291, 368
- photograph of, 395
- principle of change, 405–406
- on proteins as digital technology, 64
- stable variation, 395–407
- ultimate particles of biology, 374, 380–381, 395–399

death, 5, 22–23, 105, 312–313, 354, 362, 425

Deinococcus radiodurans (bacteria), 109–112, 306

Dennett, Daniel
- algorithm of natural selection, 279–281, 344, 375
- artificial vs natural selection, 344
- requirements of evolution, 278

Descartes, René. *See* Cartesian dualism

detachment, 425–428, 432

development
- agency and, 352–353
- artificial selection and, 343–346
- biological theory and, 4–5, 22
- bodies and, 61–63
- causation and, 124, 318, 338

development (*continued*)
- cells and, 302–305
- collective form of, 353–354
- coordinating activity and, 352
- directiveness and, 337, 342–343, 352–353
- effort of origination and, 338–339
- endosymbiosis and, 350–351
- epigenetics and, 93–95, 99–101
- evolution's relation to, 337–342, 353–354, 397–399
- form and, 136–137, 145–146
- form of organisms and, 155–163
- genes and, 34, 155–156, 226–227, 283–285
- genetic mutation and, 346–347, 351–352
- hearts and, 62, 339–341
- holism and, 78, 81, 341
- hybridization and, 349–350
- inheritance and, 364–366, 379
- intention and, 344
- knowledge and, 306–307, 340, 352
- as metamorphosis, 160
- movement and, 61–62
- narrative and, 162
- natural selection and, 292–293, 295, 350
- organisms and, 4–5, 22, 24, 344–345
- overview of, 336–337, 352–353
- perception and, 211–212
- predictability and, 148–150
- punctuated equilibrium and, 347–351
- purposiveness and, 344–345
- qualitative biology and, 180–181, 187–188, 191, 193
- RNA and, 234
- self–transformation and, 302–305
- stable variation and, 397–399
- teleology and, 16, 312–314, 316, 318, 321–322, 347
- thought experiment for, 343–344
- transformation of the processes of, 337–342

dinosaurs, 191–194

directiveness. *See also* purposiveness; teleology
- biological theory and, 12–13, 305, 336
- cells and, 395
- definition of, 12–13
- development and, 337, 342–343, 352–353
- DNA and, 346

directiveness (*continued*)
  end-directedness, 7, 13, 20, 30, 87, 201, 227, 272, 283, 305, 315, 337, 342
  evolutionary theory and, 342–345, 352–354
  genetic mutation and, 346–347, 351–352
  hearts and, 339
  materialism and, 345
  meaning and, 28
  natural selection and, 281–282
  organisms and, 346, 353
  self-transformation and, 305
  spliceosome and, 106
  tracing of, 352–353
Directiveness of Organic Activities, The (Russell), 13
Discovery of the Mind, The (Snell), 431
DNA. *See also* chromatin; chromosomes; nucleosomes; RNA; transcription; transcription factors
  as Book of Life, 34
  breathing of, 42, 385
  causation and, 94
  cells and, 223–225
  classical view of, 250
  computational understanding of, 157
  computer simulations of, 72
  coordination of splicing activity by, 108–109
  crystalline order attributed to, 64
  damage to, 109
  as dead molecule, 35
  definition of, 36–37
  description of, 35–40
  digital conception of, 64, 69, 222, 236, 241, 381–385
  directiveness and, 346
  discovery of, 45–46, 219
  double helix structure of, 36, 40–42, 44, 71–73, 106, 109, 219, 223–226, 229–231, 236–245, 248, 382–385
  as essential substance of evolution, 374–379
  form and, 221–222
  gene expression role of, 219–225
  genocentrism and, 374–379, 382–386
  gestural reality of, 382–386
  gesturing and, 382–386
DNA (*continued*)
  inert misconception of, 221–222
  information concept applied to, 56, 255, 363
  interiority and, 8
  loops of, 38–39
  mechanism and, 62
  methylation of, 236–237, 241, 243, 245, 251, 385
  movement and, 42, 63, 385
  personal identity and, 42–43
  proper functioning of, 35
  proteins and, 224, 237–242
  purposiveness and, 346
  repair of, 8, 68, 121, 327
  replication of, 38, 44–45, 149, 242, 244
  sequence of, 42–43
  structure and elements of, 35–40, 42
  terminology for, 36–37
  unexpected coherence and, 108–109
  water's interaction with, 382
  wisdom and, 307–308
  wrapping of around histones, 383–384
droplets of water, 263, 268
dualism, Cartesian, 419, 447, 450–451, 453

E

E. coli, 109
efficient causes, 316–317, 375
electrical forces, 141, 143, 150n2, 266–267
embryos. *See* development
Emerson, Ralph Waldo
  abstract language, 417
  evolution of language, 415, 418
  method of nature, 61
empirical science
  ambiguous figures and, 205–207, 212
  appearance and, 206–214
  aspects of human experience and, 202–203
  atoms and, 209–211
  bat sense and, 211–212
  chaos and, 203–208
  closing thoughts on, 212–214
  collective illusions and, 213
  conceivability and, 202
  deception and, 208–214
  definition of, 412

empirical science (*continued*)
desirability of, 207–208
Earth and sun relationship and, 208–209
error and, 208–214
identity of objects and, 205–207
knowledge and, 202–203, 212–213, 215
mathematics and, 202–203
meaning and, 202
organ transplantation and, 204
overview of, 201–202, 215
participation in thinking and, 205–207
perception and, 202–214
relations of thought and, 203–208
source of error and, 208–214
end as more constant than means of attaining it, 23–26, 83, 103, 136
end–directedness, 7, 13, 20, 30, 87, 201, 227, 272, 283, 305, 315, 337, 342
Endless Forms Most Beautiful, The (Carroll), 150, 155
endosymbiosis, 350–351
enzymes and substrate binding, 71, 269
entelecheia, 13, 330
Environmental and Architectural Phenomenology (Seamon), 182
enzymes, 35, 40–44, 67, 245, 269, 306–307, 384
epigenetics
causation and, 99–100, 119
cells and, 93–94
as code , 99
definition of, 95–96
development and, 93–95, 99–101
environmental impact on, 97–98
explanatory problem addressed by, 94–95
famine research and, 98
gene expression and, 99–100, 233–237
genetics' relation to, 100, 233–237
grasping at explanatory straws through, 98–100
human beings and, 98
inheritance and, 97
meaning and, 96
molecular biology and, 94
overview of, 93–94, 100–101
re–writing of transcription factor circuits attributed to, 99
RNA splicing and, 104
epigenetics (*continued*)
self–transformation and, 303–304
unexpected coherence and, 104
unexpected discoveries surrounding, 96–98
usefulness of term of, 95–96
Europeans' changing landscape of, 420–423
evo–devo (evolutionary developmental biology), 292–293, 373
evolution of consciousness, 412–439. *See also* Barfield
Evolution (Shapiro), 306, 389n1
evolutionary theory. *See also* biological functions; biological theory; inheritance; natural selection
agency and, 314–322, 426
anthropocentrism and, 3–4
archetypes and, 12
causation and, 316–317
conditions required for, 277–279, 374–375
consciousness and, 412–413, 422–426, 433–434
conventional approach to, 193–194, 316–317
coordinating power and, 317–318, 352–353
defining processes in, 378–379
development's relation to, 337–342, 353–354, 397–399
directiveness and, 342–345, 352–354
DNA as essential substance of, 374–379
endosymbiosis and, 350–351
explanatory challenges of, 2–3
fitness in, 182, 194, 278, 283, 294, 305, 328, 346–347, 351–354, 364, 374–375
as form of collective development, 353–354
genes as synonymous with heritability in, 373, 379, 402, 406–407
genomes and, 283
holism and, 364–369
hybridization and, 349–350
interiority and, 6–8
language and, 413–419, 424–425
mindlessness in, 30, 377
Modern Synthesis in, 376
narrative and, 4–6, 30

evolutionary theory (*continued*)
 new perspectives required for, 194
 organisms in, 3, 308–309
 populations as primary subject of evolution in, 321–322, 346–353
 purposiveness and, 4, 19–21
 qualitative biology and, 181–195
 randomness and, 30
 responsibility for explaining patterns by supporters of, 191–195
 survival of the fittest and, 277–279, 286, 290, 311, 325, 374
 teleology and, 314–322, 331–332, 347
 terminology of, 2
 third way of, 389n2
 as tinkerer, 295n4
 as transformation of developmental processes, 337–342
 use of "because" in, 2
executive control, 50–51, 56, 70
expression of genes. *See* gene expression
Extended Phenotype, The (Dawkins), 397

F

filaments of cytoskeletion, 52–53
final causes, 16, 178
fish, 15, 69–70, 260, 309, 328, 349
fitness, 182, 194, 278, 283, 294, 305, 328, 346–347, 351–354, 364, 374–375
fluidity of bodies. *See also* water
 bodies as process, 29–30
 in development, 61–66
 mechanism and, 62
 molecular machines and, 62
 movement and, 61–63
 overview of, 61–64, 73–74
 regulation of flow and phases and, 69–70
 stable variation and, 404
 unexpected phases of life and, 66–70
 water as mediator and, 71–73
 water composition of bodies, 62
 well–structured droplets and, 67–69
form. *See also* form of organisms
 agency and, 147
 appearance and, 179–180
 biological theory and, 137–139, 144, 181–195
 causation and, 142–143, 180–181
form (*continued*)
 coordination of events and, 140
 development and, 136–137, 145–146
 DNA and, 221–222
 dynamic forms, 180–181
 end–directedness and, 145–146
 explanatory challenge of, 137–140, 362
 gene expression and, 136
 holism and, 135–136, 142–143
 intention and, 146
 leaf sequence and, 177–181
 Levin and, 140–149
 mechanism and, 145–146
 molecules and, 137–140, 142
 movement and, 171–181
 organ bud transplantation and, 135–137
 overview of, 135–137, 149–150
physics and, 163
 problem of, 137–140
 qualitative biology and, 171–181
 RNA splicing and, 139–140
 setpoints and, 145
 technological application and, 144
 top–down methodological approach and, 143, 146
 transcription and, 221–222
 unquestioned model of, 148–149
 wound healing and, 139
Form and Causality in Early Development (Dalcq), 74n1
"Form and Cause in Goethe's Morphology" (Brady), 171–172
form of organisms
 biological theory and, 153–154
 causation and, 158
 design and, 146
 development and, 155–163
 earlier forms and, 157–160
 endless transformations and, 160–163
 explanatory challenge of, 153–154
 gene expression and, 156–157
 genes and, 155
 invisible artist and, 161–162
 as irreducible, 163
 meaning and, 158
 mechanism and, 158–159
 molecular biology and, 182
 overview of, 153–154, 163

form of organisms (*continued*)
perfection and, 153–154
physical explanation of, 155–163
as primary, 163
threefoldness and, 181–195
formal causes, 163, 180, 315, 317–318, 320, 404
FOX-P1 (protein), 121
fruit flies, 63, 156, 158, 357, 403
functionalism, Darwinian, 193
functions, biological, 22, 59n8, 63, 69, 88, 111, 120, 322–323, 334n10

G

galaxies, 459–461
gene expression
bioelectric fields and, 150n2
causation and, 127–128, 136, 141–142, 222–223, 248–251
chromatin and, 223–225
chromosomes and, 39–42
complexity and, 223
complications surrounding research on, 220–225
cytoskeleton and, 52–53
definition of, 219
DNA's role in, 219–225
end-directedness and, 232–233
epigenetics and, 99–100, 233–237
form and, 136
form of organisms and, 156–157
genocentrism and, 383
histone variants and, 240
initiation of, 225–229
as matter of sculptural performance, 232
Mediator molecule in, 227–228
molecular biology and, 219–220
molecular "village" required for, 221–225, 228
movement and, 38, 41, 232
narrative and, 227
nucleosomes and, 237–248, 383–384
overview of, 219, 250–253
phase transitions and, 250
physical interactions and, 239
post-translation modifications (PTMs) and, 249
gene expression (*continued*)
pre-initiation complex (PIC) and, 226–229
process of, 225–229
proteins and, 222–223, 228–229
requirements for, 41
ribosomes and, 249
RNA's role in, 53, 219–231, 234, 249–250
sculptural performance of, 232
specialized filament regulation of, 52–53
steps in, 219–220
switches in, 156–157
terminology of, 219–220, 232
topics not discussed, 248–250
transcription factors and, 219–222, 383–385
translation and, 53, 220
variation of, 227
wisdom and, 307
gene regulation, 7, 37, 40, 43, 73, 236, 241–245, 250–251
genes. *See also* gene expression; gene regulation; genomes; heritability
agency attributed to, 34
biological theory and, 3, 34, 98–99, 357–358, 375
blueprint formed by, 93, 300
causation and, 51, 357–358, 403
cell membrane and, 56
community formed by, 39
context–dependence and, 100
cytoskeleton and, 53
destiny's relation to, 97
development and, 34, 155–156, 226–227, 283–285
epigenetics' relation to, 100, 233–237
evolutionary role of, 373, 379, 402, 406–407
as first causes of life, 34, 98–99, 375
form of organisms and, 155
genocentric focus on, 111
holism and, 358
identity of, 94
independent life of, 396–397
"instructions" provided by, 3, 53, 99, 104
meaning and, 94
movement and, 51

genes (*continued*)
narrative and, 94
natural selection and, 283–285, 289
over-estimation of, 283–285
particulate conception of, 397
physical interactions and, 51
potential immortality of, 402–404
proteins and, 110–112
as replicating themselves, 409n4
shared across organisms, 94–95
stable variation and, 395–404
as unable to do anything by themselves, 225
unexpected coherence and, 110–112
genetic code, 38, 50, 72, 95–96, 99, 104, 219, 231, 236, 240, 382, 385–386
genetic mutation
creative potential of, 305, 352
development and, 346–347, 351–352
directiveness and, 346–347, 351–352
fitness and, 351–352
holism and, 365
natural selection and, 284
randomness of, 346–347
self–transformation and, 305
wisdom and, 307
genetic recombination, 303, 306
genetic reductionism, 373–374, 389n2
genetic variation, 128, 352, 395, 398, 402–404
genetics. *See* chromatin; chromosomes; DNA; epigenetics; gene expression; gene regulation; genes; genetic code; genetic mutation; genocentrism; genomes; inheritance; population genetics; RNA
genocentrism
agency and, 376
biological theory and, 111, 376–378, 386–387
challenges to, 373–374, 389n1
context–dependence and, 387
controlling factors and, 374
counter–picture to, 406–407
defining evolutionary processes and, 378–379
definition of, 373
digital conception of biology and, 379–381
DNA and, 374–379, 382–386
evo-devo and, 373
genocentrism (*continued*)
gene expression and, 383
gesturing and, 382–386
ideas and, 386–387
materialism and, 387, 388–389
meaning and, 379–380
mechanism and, 374
natural selection and, 374–379
overview of, 373–374, 388–389
particles' distorting effect and, 381–382
persistence of, 373–374
stable variation and, 406–407
view from the bottom or view from the top and, 386–387
genomes. *See also* genes
biological theory and, 96
compartments of, 37
complexity of, 45–46
dynamic nature of, 39, 222
evolutionary theory and, 283
gesturing and, 39
global coherence and, 41–45
as information structures, 56
lawfulness of molecular interactions and, 41–45
methylation and, 236
mitosis and, 38
nucleus and, 38
overview of, 34–35, 45–46
physical interactions and, 42
read-write genomes, 306
self-transformation and, 300, 303–306, 383
sequencing of, 300
shattering of, 109–116
spatial organization of, 39
supercoiling in, 383
unexpected coherence and, 109–116
variation in, 101n1
whole organism and, 305
genomic origami, 42, 162
germline, 303–306, 396
gesturing
cells and, 39
DNA and, 382–386
genocentrism and, 382–386
genomes and, 39
importance of, 39

gesturing (*continued*)
  qualitative biology and, 170, 175
  as substance of biological activity, 63–64
  transcription factors and, 383
goal-directed activity, 16, 19–20, 24, 126
Goethe, Johann Wolfgang von
  capturing a person's inner nature, 8–9
  leaf archetype, 172–173
  morphological studies, 171
  problem of organic form, 171–172
  Romanticism, 427
Goethe's Way of Science (Seamon), 182
Goliath beetle, 394
Gould, Stephen Jay
  agency, 299
  natural selection, 279, 289, 298–299, 375
  punctuated equilibrium, 348
"Great Awakening", 429–430, 432
Great Exhibition (1851), 424
Greek funerary vase, 430
Greek thought, 16, 330, 414–418, 423–424, 429–434

H

H2B (histone), 238, 244
H2Bub1 (histone tail modification), 244
hearts
  causation and, 338
  challenges of, 339–341
  development and, 62, 339–341
  directiveness and, 339
  of horses, 339
  mechanism and, 340–341
  natural selection and, 333n8
  structural divisions of, 62
  wisdom and, 340
heritability, 98, 278, 283, 322, 357–360, 363, 373, 379, 393–395, 401–407. *See also* genes
histones
  core particles of, 36, 42
  definition of, 238
  exchange rate of, 240
  functioning of linker histones, 255n19
  tails of, 242–245
  variants of, 240–241, 248
  visualizations of, 238–239
History in English Words (Barfield), 415
history of ideas, 422–424
Holdrege, Craig
  beavers, 182
  holism of evolutionary theory, 181
  paleontology's lessons, 348
  sloth, 167–171
  whole organism studies, 167–168, 171
holism. *See also* unity
  agency and, 11–12, 78, 83, 149
  biological context as embodied ideas and, 84–89
  biological theory and, 9–10, 78, 81–82, 84–89, 319, 366
  causation and, 9–10, 126–127, 338
  causes and, 9–10
  cells and, 79–83
  Coleridge on, 367–368
  context-dependence and, 9–10, 84–86, 338
  contextual wholes, 87–89
  as counter-movement to analysis, 9
  cytoplasm and, 82
  definition of, 8–10
  development and, 78, 81, 341
  evolutionary theory and, 364–369
  explanatory benefit of, 8–10
  form and, 135–136, 142–143
  genes and, 358
  genetic mutation and, 365
  historical support for, 85
  ideas and, 84–89, 363
  independence and integration of parts and, 364–369
  inheritance and, 360–364
  interiority and, 88
  knowledge of the character of wholes in, 9
  life as demanding reference to, 8–10
  meaning and, 8–10, 81
  mechanism contrasted with, 83
  more than the sum of its parts principle in, 81–83
  narrative and, 8–10, 87–89
  natural selection and, 283–285, 325
  niches and, 79–81
  organ bud transplantation and, 135–137
  organisms and, 149, 358, 369–370
  overview of, 8–10, 78–79
  phenotypic plasticity and, 366
  prediction and, 9

holism (*continued*)
- prioritization and, 79–81
- problem of the whole and, 361–364
- qualitative biology and, 9–10, 194
- relation between parts and wholes in, 9–10
- self–transformation and, 302
- skepticism toward, 9–10, 84
- stable variation and, 396, 399, 406–407
- teleology and, 319, 325
- unexpected coherence and, 111–112

holistic biology, 43, 182

homeostasis, 184, 253n9, 330, 340

horses
- American Pharoah, 339–343, 345, 352–353
- breeding of, 343
- evolution of, 182
- hearts of, 339
- *Hyracotherium* ancestor of, 339–341, 343, 345, 353
- limbs of, 187
- modern form of, 343

How the Leopard Changed Its Spots (Goodwin), 329

"How We Make Sense of the World" (Brady), 465n12

human beings
- biological theory and, 201–202
- chimpanzee DNA overlap with, 95
- epigenetics and, 98
- evolutionary relationship to all other organisms, 4
- face's importance for, 6
- living polarity in, 182–185
- non-human organisms' relation to, 13, 17, 321
- as organisms, 28–29
- qualitative biology and, 182–185
- recent media stories concerning evolution of, 349
- specialization of, 187

Human Genome Project, 46n5, 63, 94

human T cells, 268

Hume, David
- causality, 197n9

hybridization, 321, 342, 349–350, 354, 370n4

hydration shell of myoglobin, 72

Hyracotherium (horse ancestor), 341–343

I

ideas
- biological context as embodied ideas, 84–89
- context-dependence and, 86
- genocentrism and, 386–387
- history of, 422–424
- holism and, 84–89, 363
- interiority and, 363
- interrelationship between, 362–363
- knowledge and, 446
- organic wholes and, 363
- qualities and, 363

immaterial causes, 47n10, 88

"In Defense of 'Nonmolecular' Cell Biology" (Albrecht–Buehler), 262

inanimate world
- appearance and, 448, 450–453
- Cartesian dualism and, 447, 450–454
- cognition and, 454–457
- experience and, 448–450, 452–460
- felt solidity and, 443
- as form of speech, 457–459
- galaxies and, 459–461
- interiority and, 447–450, 460–463
- knowledge of, 443–450, 460–461
- language and, 457–459
- narrative and, 457–459
- overview of, 441–443, 462–463
- participation in creation of, 455–457, 462
- "parts" missing in, 447
- perception and, 445, 447, 449, 451, 454–455, 461
- postmodern understanding of, 447
- quantum physics and, 449–450
- reality and, 450–453
- representation and, 454–455
- sensible qualities and, 443–444
- skepticism and, 448
- unbridgeable difference between living beings and, 2–3, 22, 103, 113, 336–337
- unity of, 447, 452–453

information, 22, 42–43, 50–51, 56–57, 64, 71, 87, 112, 225, 255, 363, 379–381, 399, 418

inheritance
beyond particulate forms of, 388–389
biological theory and, 359–360
central problem of, 358
characters and, 359
context-dependence and, 361
coordination and, 366
development and, 364–366, 379
epigenetics and, 97
genetic research and, 360
heritability and, 359–360, 363
holism and, 360–364
independence and integration of parts and, 364–369
interiority and, 363
meaning and, 363
organizing power and, 360
overview of, 357–359, 369–370
physical interaction and, 363
qualities and, 362
resemblance as a result of, 359
Russell and, 359–360
insects, 27, 69, 167, 170, 192, 301, 347
integral unity of the organism, 13, 16, 369
intelligent design (ID), 18n3, 344, 346
intention. *See also* agency; purposiveness; teleology
agency and, 11–12, 28–29
attributions to non–human organisms of, 13–14
awareness's relation to, 5–6
biological activity and, 7, 358
biological theory and, 201
context-dependence and, 86
definition of, 5, 13–14
development and, 344
form and, 146
individual and collective forms of, 14
interiority and, 6–8
narrative and, 5
organisms and, 8, 13–14, 86
possessing/possessed by distinction in, 5
purposiveness and, 28
types of, 5–6
unexpected coherence and, 111–112
wisdom and, 5
interiority
agency and, 309
aims of current volume regarding, 7–8
appearance and, 158, 192, 324
biological theory's overlooking of, 7–8, 23, 87, 88, 362–363
cells and, 6–7
consciousness and, 15, 416–417, 423–424
definition of, 6
DNA and, 8
evolutionary theory and, 6–8
expression and, 416–417
holism and, 88
ideas and, 363
inanimate world and, 447–450, 460–463
inheritance and, 363
intention and, 6–8
knowledge and, 447–450, 460–461
language and, 416–419
materialism and, 88, 449
matter as inseparable from concept of, 449
meaning and, 6–8, 363
molecules and, 7
narrative and, 6–8
natural selection and, 324–325
organisms and, 23
organization and, 362
outward expression of, 6–8
overview of, 6–8
perception and, 6–7, 15, 23, 417–419
physical interactions and, 7
qualities and, 363
skepticism toward, 7–8
terminology of, 6–7
Interpretation of Development and Heredity, The (Russell), 359
introspective psychology, 424
invisible artist, 161–162

J

jerboa, 189
Jewish thought, 333n6, 423, 430, 432–433
Julie, or the New Heloise (Rousseau), 420
Jumpers (Stoppard), 209

K

Kant, Immanuel
biological wholeness, 132n1
Critique of Judgment, 132n1
epistemology, 465–466n12
Ronald Brady on, 196–197n8–9
knowledge
appearance and, 444, 448, 460–461
biological theory and, 10–11
cells and, 306–307
cognition and, 445–450
conceptual understanding of the world and, 446
as dependent on qualitative appearances, 444
development and, 306–307, 340, 352
effective forms of, 307
empirical science and, 202–203, 212–213, 215
holism and, 9
ideas and, 446
inanimate world and, 443–445, 460–461
interiority and, 447–450, 460–461
meaning and, 446
natural selection and, 280, 293
perception and, 445–450
self-transformation and, 306–307
skepticism and, 448
teleology and, 318
unity and, 447
world lending itself to, 443–445

L

language
abstract words in, 417–418
abuses of, 121–122
agency and, 288–289
biological theory and, 23, 119
causation and, 121–122
consciousness and, 413–419
dualism and, 419, 451–452
evolutionary theory and, 413–419, 424–425
expression and, 416–417
history of, 417
inanimate world and, 457–459
interiority and, 416–419
language (*continued*)
of life, 23, 119
metaphoric language, 125
microworld puzzles and, 261–262
origin of, 458
of scientific description, 7
scientific forms of, 15–16, 417–418
words in, 413–419
Laplace's demon, 83, 90n4
lawfulness, 6, 22, 41–45, 83, 114–115, 146, 194–195, 282, 318, 323, 346, 441
leaf sequence
causation and, 177–181
context–dependence and, 179
dynamic nature of, 178, 180
form and, 177–181
movement and, 177, 196n7
particulars as "coming from" and "passing to" in, 177–179
perception and, 178
qualitative biology and, 172–182
unity and, 179
visualization of, 172–173, 175
Levin, Michael
biological theory's limitations, 140–143
causation, 144
chemistry and cognition conflated, 151n5
control emphasized by, 143
as counter–revolutionary, 143–149
critique of, 147–148
development, 141
form, 140–149
mechanism emphasized by, 143–145, 148–149
novel contributions of, 142
overview of, 140
prediction emphasized by, 143, 148–149
as revolutionary, 140–143
setpoints as causes, 145
tadpole electricity research, 141, 143
teleology in biology, 147
top-down methodological approach, 143–145, 147–149
voltage patterns, 141
Lewontin, Richard
conditions for evolution, 370n5
development, 80–81, 135

Lewontin (*continued*)
DNA as a dead molecule, 35
embryo's flexible coordination, 80–81
need for biologists to take seriously what they already know to be true, 11
photograph of, 365
quasi-independence of traits and variants, 364, 366, 368
life. *See* biological activity; biological theory; evolutionary theory; genes; organisms
lions, 21–22, 116, 144–145, 189, 191
logic circuit, 380

M

mallard ducks, 154
mammals, 122–123, 167–170, 181–182, 186–191, 351, 403
Man and Mammal (Schad), 181
master controllers, 50, 53–54, 120–123, 150, 155, 376
materialism
alienation and, 459–460
biological theory's presumption of, 14–15, 88, 293, 374, 388
Cartesian dualism and, 447, 451
definition of materiality, 14–15
directiveness and, 345
fundamental reality of matter and, 382
genocentrism and, 387, 388–389
inanimate world's relation to, 14
institutionalization of, 298
interiority and, 88, 449
language of dualism and, 451–452
meaning and, 15
physical world distinguished from, 14
physics' presumption of, 30
scientific presumption of, 50, 379
taboo reinforcing, 388
mathematics, 47n10, 116, 202–203, 208, 434
meaning
appearance and, 159, 446
biological activity and, 7, 22, 89, 313, 337
biological theory's taboo against, 15, 20, 87, 328–331
blindsight and, 23
causation and, 88, 124, 125–128, 201
connectedness of, 28
meaning (*continued*)
consciousness and, 28, 412–419, 422–423, 425
cytoskeleton and, 54
definition of, 15
directiveness and, 28
empirical science and, 202
epigenetics and, 96
explanatory dimensions of, 15
form of organisms and, 158
genes and, 94
genocentrism and, 379–380
holism and, 8–10, 81
interiority and, 6–8
knowledge and, 446
materialism and, 15
narrative and, 4–6, 88, 201
natural selection and, 328–331
organisms and, 21–23, 26–30, 87, 346
physical interactions and, 122
proteins and, 66
teleology and, 315, 316–318
unexpected coherence and, 115–116
unity and, 412–416
mechanism
appearance and, 324
biological activity and, 150, 340
biological theory and, 43
bodies and, 62
causation and, 126
chromosomes and, 40–43
DNA and, 62
end-directed nature of, 145–146
explanatory reference to, 10
form and, 145–146
form of organisms and, 158–159
genocentrism and, 374
hearts and, 340–341
holism contrasted with, 83
misunderstandings surrounding, 126
molecules modeled as, 69
movement and, 40
natural selection and, 327–328
organisms modeled as, 18n3, 64, 99, 149–150, 284–285, 293
self-transformation and, 306
teleology and, 327–328

mechanism (*continued*)
unquestioned nature of, 148–149
wisdom and, 307
Mediator complex, 227–228, 253n8–9
meiosis, 303–304
membrane, cell. *See* cell membrane
messaging, 87, 125–127, 387
mice, 79–80, 97, 187, 190, 261, 268, 321
microbiomes, 250, 253, 351–352
microworld puzzles
appearance and, 271
Brownian movement and, 264–265
challenge posed by, 269–270
changes in size and, 263–264
chemical energies and, 266
electrical forces and, 266–267
language used to understand, 261–262
macroworld's relation to, 270–271
movement and, 264–265
non–molecular biology and, 262–264
overview of, 260–261, 271–272
polymerization and, 267–269
sensation and, 274n12
significance of, 271–272
viscosity and, 264
water and, 263–265, 269
mindlessness, 14, 30, 35, 37, 279, 377–379, 407, 412, 422, 445–446, 450–453, 460
mitosis, 38, 43–44, 53, 129
Molecular and Genome Evolution *(Graur),* 377
molecular biology
context–dependence and, 84
"dance" and "choreography" terminology in, 42
digital interpretation of, 379
epigenetics in, 94
form of organisms and, 182
gene expression and, 219–220
high era of, 73
information in, 111
life in, 34
microstate tracking in, 143
non–molecular biology contrasted with, 262–264
promise of, 219–220
RNA splicing discovered in, 104
terminology for, 42
molecules. *See also* atoms; DNA; RNA; proteins
behavior of, 7
biological activity and, 136
biological theory and, 12
biomolecules, 63–64, 71–72
bodies as molecular machines, 62
cell membrane and, 55
context-dependence and, 84
cytoskeleton, 52
definition of, 12
DNA as dead molecule, 35
experience and understanding of, 12
explanatory role of, 34
form and, 137–140, 142
freedom of movement of, 181
gene expression and, 221–225, 228
global coherence and, 41–45
interiority and, 7
lawfulness of interaction of, 41–45
mechanistic conception of, 69
movement and, 63–64, 164n5, 181
organisms constituted by, 22
particulate understanding of, 381
transcription factors, 221–223
visualizations of, 381
water as molecule of life, 72–73
Mona Lisa (Da Vinci), 421
monotheism, 432
"Morphodynamics of Dinosaurs, Other Archosaurs, and Their Trackways, The" (Lockley), 192–193
moths, 157, 167–168, 170
movement
biological activity as based on, 63–64
bodies and, 61–63
Brownian movement, 264–265
causation and, 177, 179–181
cell membrane and, 55–56
cells and, 41, 51–52, 55–56, 62–63, 81
chromosomes and, 38, 40, 42
circulatory system as example of, 61–62
cytoplasm and, 55
cytoskeleton's facilitating of, 51–54
development and, 61–62
difference and sameness in formative movement, 175
DNA and, 42, 63, 385

movement (*continued*)
formative movements, 174–177, 196n7
gene expression and, 38, 41, 232
genes and, 51
mechanism and, 40
microworld puzzles and, 264–265
molecules and, 63–64
narrative and, 5
nucleosomes and, 38, 240–241, 245–248
perception and, 175–176
phase transitions and, 70
polarity and, 183–184
proteins and, 64–66
qualitative biology and, 174–184
randomness and, 164n5
RNA and, 232–233
sensed power and, 179–180
stem cells and, 247
structure's relation to, 62
unity and, 179
mRNA (messenger RNA), 64, 104–107, 112–113, 232, 234, 245, 249
mutations. *See* genetic mutation

N

narrative
agency and, 26–29
biological activity and, 79
biological theory and, 4–6, 10–11
causation and, 88, 201
context-dependence and, 86
development and, 162
episodes as composing, 5
evolutionary theory and, 4–6, 30
gene expression and, 227
genes and, 94
goal-like behavior and, 5
holism and, 8–10, 87–89
imprint of evolution and, 5
inanimate world and, 457–459
intentional and teleological nature of life stories in, 5
interiority and, 6–8
living narratives, 87–89
meaning and, 4–6, 88, 201
movement and, 5
organisms and, 4–10, 26–30, 87
overview of, 4–6

narrative (*continued*)
regulation and, 70
storytelling as the being of organisms, 29–30
natural selection
agency misattributed to, 287–290, 292–294, 298–299, 322–328
algorithmic conception of, 279–281, 287–290, 344, 375
arrival of the fittest problem for, 290–292, 325
artificial selection's relation to, 288, 344
begging the question problem for, 325–326
biological activity and, 326
biological theory's dependence on, 3, 8, 288–289, 293–295, 322, 324
causation and, 325
challenges for, 325–328
conditions for, 277–279, 326, 332, 374–375
conventional view of, 289, 294–295
creativity of, 298–300
development and, 292–293, 295, 350
directiveness and, 281–282
explanatory challenge of, 3, 277–280, 284, 290–292
fitness and, 182, 194, 278, 283, 294, 305, 328, 346–347, 351–354, 364, 374–375
future-orientation and, 281–282
genes and, 283–285, 289
genetic mutation and, 284
genocentrism and, 374–379
guaranteed results of, 280–287
hearts and, 333n8
holism and, 283–285, 325
interiority and, 324–325
isolation and, 350
knowledge and, 280, 293
meaning and, 328–331
mechanisms missing for, 327–328
metaphoric understanding of, 288–289
as omnipotent explanatory force, 3, 280, 287, 289
overview of, 277–280, 294–295
purposiveness and, 322–328
relaxation of, 350
selection *for* and *against* in, 395

natural selection (*continued*)
- self-transformation and, 298–300
- simplicity of, 277–281
- syllogistic core of, 279
- teleology and, 20, 322–331
- topoisomerases and, 48n11
- trivial validity of, 286
- types and, 283–285
- unanswered questions about, 285–287
- unity and, 283–285
- variation and, 281–282

Nature (Emerson), 415
Nature of Selection, The (Sober), 277
new science, 30, 171
New Testament, 414
niches of organisms, 79–81, 160, 350
non-molecular biology, 262–264
non-protein-coding RNAs, 249–250
nucleosomes
- core particles in, 238–239, 242
- definition of, 36
- DNA and protein conjoined in, 237–242
- DNA breathing on surface of, 245–246
- dynamic nature of, 240–241
- functions of, 246–248
- gene expression and, 237–248, 383–384
- movement of, 38, 240–241, 245–248
- proteins and, 237–242
- rhythm and, 245–248
- ribbon diagram of, 242
- role in DNA packing of, 238
- from static to dynamic understanding of, 246–248
- stem cells and, 246–248
- structure of, 237–240
- tails of histones and, 242–245
- transcription factors and, 246
- visualizations of, 237–239

nucleus, the
- biological emphasis on, 50
- choreography of, 42
- control function of, 50
- cytoskeleton and, 53
- definition of, 40
- DNA in, 40
- genomes and, 38
- unexpected coherence and, 112–113

Nut (sky goddess), 415

O

organelles, 50–51, 54–56, 66, 70, 82, 136, 350
organisms. *See also* form of organisms
- activity of, 10–12, 30, 73–74
- adaptation to environment, 23
- agency of, 10–11, 21–23, 26, 30, 147, 149, 304–305, 320–321, 358, 376, 400
- atoms constituting, 22
- attempts to correct or repair errors by, 22–23
- behavior and, 22
- biological theory and, 329
- bodies' relation to, 29–30
- bringing back of, 100–101
- causation and, 125, 146, 223
- coherence as not mysterious for, 116
- complex organisms in biological theory, 17
- context–dependence and, 89–90, 99–100, 361
- death and, 22
- definition of, 21, 26, 28
- development and, 4–5, 22, 24, 344–345
- digital form of, 279
- directiveness and, 346, 353
- drama of life in, 3
- dynamic capacities of, 394
- ends are more constant than the means of attaining them 23–26, 83, 103, 136
- environment expressed by, 167–171
- errors and, 22–23
- essence of, 29
- evolutionary theory and, 3, 308–309
- as focal agents, 6
- form and, 146, 181–195
- function and, 22
- germline management in, 303–305
- holism and, 149, 358, 369–370
- human beings as, 28–29
- information and, 56–57
- intention and, 8, 13–14, 86
- interiority and, 23
- interpretation of environment by, 21–22
- living trajectory of, 146
- material existence of, as narrative, 31
- meaning and, 21–23, 26–30, 87, 346
- mechanistic model of, 18n3, 64, 99, 149–150, 284–285, 293

organizations (*continued*)
molecules constituting, 22
narrative form of, 4–10, 26–30, 87
perception and, 22, 407
as physical interactions, 50
purposiveness and, 19, 21–30, 223
qualitative biology and, 167–171
reflexes and, 22
regulation and, 70
responses and, 21–22
self–motion as distinguishing feature of, 21
self–transformation and, 124, 305–308
stable variation and, 404–407
storytelling as the being of, 29–30
struggle to survive and reproduce of, 20
teleology and, 19–20, 319, 326, 329–330
unity of, 13, 16, 358, 363, 369–370
whole organism studies, 167–168, 171
wisdom of, 3, 26, 305–308, 345
organizing powers, 89, 360, 394, 407
Origin of Species, The (Darwin), 153, 291, 368
original participation, 424–427, 432–433

P

p53 120–121
Panathenaic prize amphora, 430
pan-psychism, 463n1
Parapsychology Laboratory, 24
particles, ultimate, 395–399
PAX gene, 403
peacock feather, 153, 158
perception
of ambiguous images, 212
atoms and, 210, 239
biological theory and, 23
collective illusions and, 213
consciousness and, 15, 422, 425, 432–434
cytoskeleton and, 51–54
deception and, 208–214
development and, 211–212
doubt surrounding, 210
Earth and sun relationship and, 208–209
empirical science and, 202–214
error and, 208–214
inanimate world and, 445, 447, 449, 451, 454–455, 461
interiority and, 6–7, 15, 23, 417–419
perception (*continued*)
interpretation and, 22
knowledge and, 445–450
leaf sequence and, 178
movement and, 175–176
organisms and, 22, 407
qualitative biology and, 175–176
sense-perception, 204, 210–211, 239, 414, 418, 447
thought–aspect of, 213, 445
Persephone, myth of 416
phase separation, 68–69, 75n7
phase transitions, 69–70, 250
physical interactions
agency contrasted with, 11–12
biological theory and, 158
form of organisms and, 155–163
gene expression and, 239
genes and, 51
genomes and, 42
interiority and, 7
meaning and, 122
multi-dimensional reasons for things and, 131
organisms as, 50
regulation unable to be accounted for by, 70
teleology and, 315–316
physics, 8, 30, 114, 130–131, 159, 201, 262, 401, 449–450
form and, 163
See also quantum physics
plasma membrane. *See* cell membrane
pneuma, 414–415, 417
Poetic Diction (Barfield), 413
polarity, 57, 182–185, 191, 368
polymerization, 267–269
population genetics, 376–378
populations as primary subject of evolution, 321–322, 346–353
potter wasps, 1–5
pre-initiation complex (PIC), 226–228, 231
principle of change (versus stability), 400, 405–406, 408–409
proteins
causation and, 126, 222–223
cell membrane and, 375–376
crystalline order contrasted with, 64–66

proteins (*continued*)
  computer simulations of, 72
  cytoplasm and, 67
  cytoskeleton and, 52–53
  DNA and, 224, 237–242
  dynamic nature of, 64–66
  fluid-like form of, 64
  functions of, 64–65
  gene expression and, 222–223, 228–229
  genes and, 110–112
  intrinsically disordered proteins, 64–66, 70
  lock-and-key conception of, 64
  meaning and, 66
  movement and, 64–66
  nucleosomes and, 237–242
  phase transitions and, 70
  potential functional shapes of, 346
  RNA and, 69
  substrate binding of enzymes, 71, 269
  surface-molten form of, 64–65
  tata-binding protein (TBP), 228–229
  unexpected coherence and, 110–112
psychoanalysis, 424
punctuated equilibrium, 347–351
purposiveness. *See also* agency; directiveness; intention; teleology
  adaptability in activities based on, 24–25
  anthropomorphism and, 25
  biological activity and, 16, 24–25, 325–326, 331, 336
  biological theory and, 15, 19–20, 25–26, 201
  blindsight and, 19
  causation and, 26, 223
  cells and, 24–25
  cessation of activities following, 24–25
  consciousness and, 28
  definition of, 15
  development and, 344–345
  as distinguishing feature of animate world, 24–25
  DNA and, 346
  evolutionary theory and, 4, 19–21
  explaining away, 19–20
  intention's role in, 28
  natural selection and, 322–328
  organisms and, 19, 21–30, 223
purposiveness (*continued*)
  persistence in activities based on, 24–25
  teleology and, 316–318, 322–328
  terminology for, 25
  typical features of behavior based on, 24
puzzles of the microworld. *See* microworld puzzles

Q

qualitative biology. *See also* biological theory
  atoms and, 166
  biology of form and, 181–195
  causation and, 177, 179–181
  contemporary biology contrasted with, 180–181
  context–dependence and, 179
  as contradictory, 166–167
  development and, 180–181, 187–188, 191, 193
  difference and sameness in, 175
  evolutionary theory and, 181–195
  form and, 171–181
  fully qualitative dimension of, 9
  gesturing and, 170, 175
  holism and, 9–10, 194
  human beings and, 182–185
  interpretation of threefold organization and, 185–186
  lawfulness and, 194–195
  leaf sequence and, 172–182
  mammalian threefold organization and, 186–190
  manifestation through particulars and, 176–177
  materialist biology contrasted with, 180–181
  movement and, 174–184
  organisms express qualities of the environment, 167–171
  overview of, 166–167, 195–196
  particulars as "coming from" and "passing to", 177–179
  perception and, 175–176
  problem of organic form and, 171–181
  sensed power and, 179–180
  sloths and, 69, 167–171
  starting point for, 182–185
  subtle interweaving and, 190–191

qualitative biology (*continued*)
threefoldness and, 181–195
unity and, 179
whole–organism studies and, 167–171
qualities, sensible, 166, 202, 260, 362–363, 443–444
quantum physics, 260, 449–450, 466n13

R

randomness, 30, 90n5, 164n5, 328, 371n7
Ranunculus acris (buttercup plant), 172–173, 177
regulation. *See* gene expression; gene regulation
Renaissance, 421, 427, 433–434
representation, 454–457
reproduction, 20, 35, 123, 160, 292, 323, 329, 345, 358, 386
ribosomes, 220, 249, 385
RNA. *See also* DNA; transcription; transcription factors
ceRNAs (competing endogenous RNAs), 235, 255n16
cytoplasm and, 53, 67, 104, 220, 249
definition of, 36
development and, 234
dynamics of, 232–233
functional significance impacted by editing of, 254n14
gene expression role of, 53, 219–231, 234, 249–250
movement and, 232–233
mRNA, 64, 104–107, 112–113, 232, 234, 245, 249
non-protein-coding RNAs, 249–250
proteins and, 69
RNA polymerase and, 220, 226, 229–231, 233, 240, 246, 254n13
scaffolding structure provided by, 249–250
splicing of, 104–109, 114, 117n6, 139–140, 181, 233, 237, 327, 346, 387
transcription and, 36, 231
unexpected coherence and RNA splicing, 104–109, 112–114
wisdom and, 307
"Role of Nuclear RNA" (Michieletto and Gilbert), 250
Romanticism, 426–427, 437n15
roundworm (*Caenorhabditis elegans*), 94, 124–128
Russell, E. S.
agency, 147
directiveness, 12–13
Directiveness of Organic Activities by, 13
end as more constant than means of attaining it, 26, 83, 136
inheritance, 359–360
Interpretation of Development and Heredity, 359
on Johannsen's problem, 359–360
narrative connectedness of behavior, 27–28

S

S.B. (organ transplant patient), 204–205, 211
Schad, Wolfgang
aims of, 199n17
biological theory's overlooking of form, 182, 194–195
mammalian threefold organization, 186–190
Man and Mammal, 181
organismal systems, 184–185
patterns of evolution, 191–192
polarity as organizational principle of organisms, 183–184, 189
relative immobility of the head, 185
rigid schematic notion of threefoldness ruled out, 198n12
specialization of organisms, 187
threefold organization, 185–186
Threefoldness in Humans and Mammals, 181
science. *See* biological theory; empirical science; evolutionary theory; materialism; quantum physics
Scientific Revolution, 166, 207, 421, 434, 464n6
selection. *See* natural selection
self-transformation
agency and, 304–305
biological theory and, 305
cell metamorphosis and, 302–303
cells and, 302–303
development and, 302–305

self-transformation (*continued*)
directiveness and, 305
epigenetics and, 303–304
genetic mutation and, 305
genomes and, 300, 303–306, 383
germline management and, 303–305
holism and, 302
insects and, 300–302
knowledge and, 306–307
magical appearance of, 300–302
mechanism and, 306
meiosis and, 303–304
natural selection and, 298–300
organisms and, 124, 305–308
overview of, 298–300, 308–309
power of orchestration and, 300
punctuated equilibrium and, 310n3
unity and, 302
wisdom and, 305–308
sense perception. *See* perception
sensible qualities, 166, 202, 260, 362–363, 443–444
"Shape-Shifting Blobs That Rule Biology" (Dolgin), 68. *See also* phase transitions
signaling, 52, 75n4, 125–126, 339
"Simulating Water and the Molecules of Life" (Gerstein and Levitt), 71–72
skepticism, 404, 448
sloths, 116, 167–171, 191, 195, 321, 362
Sober, Elliott
agency, 292
conditions for evolution, 281
natural selection, 277, 281–283, 292
Nature of Selection, 277
sol-gel transitions, 75n5
solidity
as illusion, 209, 211
quality of, 443, 448–449, 463n2
Speaker's Meaning (Barfield), 464n6
spliceosome, 105–108
stable variation
agency and, 400–401
bodies and, 404
causation of, 400, 403
Dawkins and, 395–407
dynamism and, 404–405
genes and, 395–404
genocentrism and, 406–407
stable variation (*continued*)
holism and, 396, 399, 406–407
"immortal" genetic variation and, 402–404
organisms and, 404–407
organizational powers and, 401
overview of, 393–395, 408–409
principle of change and, 400, 405–406, 408–409
ultimate particles of biology and, 395–399
unity and, 404–405
whole cells and organisms and, 406–407
stem cells, 47n8, 79–80, 227–228, 246–248
Story of Art, The (Gombrich), 429
survival of the fittest, 277–279, 286, 290, 311, 325, 374
System of Transcendental Idealism (Schelling), 465n9
systems biology, 43

T

tadpoles, 135, 141, 148, 181, 300, 305, 347
tata-binding protein (TBP), 228–229, 254n12
teleology. *See also* agency; directiveness; intention; purposiveness
agency and, 313, 314–322
arrival of the fittest problem and, 325
begging the question problem and, 325–326
biological theory threatened by, 15, 19–21, 30, 315
causation and, 312–313, 315–318
cells and, 312, 314–322
coordinating power and, 317
death and, 312
development and, 16, 312–314, 316, 318, 321–322, 347
end-directedness, 7, 13, 20, 30, 87, 201, 227, 272, 283, 305, 315, 337, 342
entelecheia and, 330
evolutionary theory and, 314–322, 331–332, 347
function and, 322–323
holism and, 319, 325
individual cells and evolutionary populations and, 314–322

teleology (*continued*)
knowledge and, 318
meaning and, 315, 316–318
mechanism and, 327–328
narrative and, 5
natural selection and, 20, 322–331
observation and, 316, 324–325
organisms and, 19–20, 319, 326, 329–330
overview of, 312–314, 331–332
physical interactions and, 315–316
purposiveness and, 316–318, 322–328
reluctance to use terms of, 19
skepticism surrounding, 313, 315
telos-realizing activity, 13, 15, 147, 314, 316, 323, 326, 331, 336, 347, 441
things themselves, 455, 462
thought relations, 203–208, 445–450,, 457–459
threefoldness
development of, 198n11
form of organisms and, 181–195
interpretation of, 185–186
mammalian threefold organization, 186–190
polarity and, 190
proposal of, 185–186
qualitative biology and, 181–195
rigid schematic notion of, 198n12
subtle interweaving and, 190–191
Threefoldness in Humans and Mammals (Schad, 181
tigers, 189
"Time for Epigenetics" (*International Journal of Biochemistry & Cell Biology*), 99
topoisomerases, 41, 43–45, 106, 284, 306–307
torsion, 40, 49
Tracking Dinosaurs (Lockley), 192
transcription. *See also* transcription factors
form and, 221–222
initiation of, 225–226
molecular "village" required for, 221–223
non-protein-coding genes and, 232
post–transcriptional regulation following, 232
as rhythmical performance, 231
RNA and, 36, 231
termination of, 232
transcription factors
causation and, 222–223
DNA and, 383
gene expression and, 219–222, 383–385
gesturing and, 383
nucleosomes and, 246
promiscuity of, 222–223
transcription regulation network of Parkinson's disease, 223
transformation, self-. *See* self–transformation
transformative series. *See* leaf sequence
translation, 53, 104, 220, 249, 414–415

U

ultimate particles, 395–399
unconscious, the, 424
unexpected coherence
centralized control and, 111
DNA coordination of splicing activity and, 108–109
epigenetics and, 104
genes and, 110–112
genomes and, 109–116
holism and, 111–112
intention and, 111–112
meaning and, 115–116
as not mysterious, 116
nucleus and, 112–113
overview of, 103–104, 116
as problem or solution, 112–116
proteins and, 110–112
reductionism and, 114
reversibility of RNA splicing and, 107–108
RNA and, 104–109, 112–114
shattering of genomes and, 109–116
spliceosome and, 105–107
translation and, 104
unity. *See also* holism
biological theory and, 319
of inanimate world, 447, 452–453
knowledge and, 447
leaf sequence and, 179
meaning and, 412–416
movement and, 179
natural selection and, 283–285

unity (*continued*)
organisms and, 13, 16, 358, 363, 369
primordial forms of, 412–416
qualitative biology and, 179
recovering the unity of the world, 452–453
self-transformation and, 302
stable variation and, 404–405

V

value as taboo in biological theory, 20
van den Berg, Jan Hendrik
Changing Nature of Man, 420
European sense of nature, 420–421
human individuation, 437n14
Mona Lisa, 421
Romanticism, 427
variation, stable. *See* stable variation
view from the top/bottom, 386–387
vinclozolin (fungicide), 97
viscosity, 264–265

W

water
as biomolecule, 71
bodies composed of, 62, 71–73
cells and, 55, 84, 112, 308
DNA's interaction with, 382
enzymes and, 71
flow mediated by, 71–73
as information–rich, 71
as mediator of flow, 71–73
microworld puzzles and, 263–265, 269
as molecule of life, 72–73
proteins and, 284
Weiss, Paul
agency, 147
cell's preservation of its character, 115
development as transformation, 163
holism, 81–83, 112, 283
life as dynamic process, 29
on mechanistic conception of organisms, 83
on models of molecules, 381
movement of cells , 55
whole as less variant than activities of parts, 136
western civilization, 424–425
What Genes Can't Do (Moss), 34, 389n1
whole organisms. *See* holism; organisms
wildebeests, 21–22
wisdom
biological activity and, 313
biological theory and, 253n5, 306
DNA and, 307–308
effective power of, 308
gene expression and, 307
genetic mutation and, 307
hearts and, 340
intention and, 5
of machines, 18n3
mechanism and, 307
of organisms, 3, 26, 305–308, 345
RNA and, 307
self–transformation and, 305–308
terminology for, 307
wound healing, 138

www.ingramcontent.com/pod-product-compliance
Lightning Source LLC
LaVergne TN
LVHW060015210926
848481LV00013B/3716

9780974490687